AF598588

Topics in Applied Physics
Volume 108

Topics in Applied Physics

Topics in Applied Physics is a well-established series of review books, each of which presents a comprehensive survey of a selected topic within the broad area of applied physics. Edited and written by leading research scientists in the field concerned, each volume contains review contributions covering the various aspects of the topic. Together these provide an overview of the state of the art in the respective field, extending from an introduction to the subject right up to the frontiers of contemporary research.

Topics in Applied Physics is addressed to all scientists at universities and in industry who wish to obtain an overview and to keep abreast of advances in applied physics. The series also provides easy but comprehensive access to the fields for newcomers starting research.

Contributions are specially commissioned. The Managing Editors are open to any suggestions for topics coming from the community of applied physicists no matter what the field and encourage prospective editors to approach them with ideas.

Manuel Cardona Roberto Merlin (Eds.)

Light Scattering in Solids IX

Novel Materials and Techniques

With 215 Figures and 14 Tables

Professor Dr. Manuel Cardona

MPI für Festkörperforschung
Abt. Experimentelle Physik
Heisenbergstr. 1
70569 Stuttgart, Germany
m.cardona@fkf.mpg.de

Professor Roberto Merlin

University of Michigan
Dept. Physics
450 Church Street
Ann Arbor, MI 48109-1040, USA
merlin@umich.edu

Library of Congress Control Number: 82017025

Physics and Astronomy Classification Scheme (PACS):
78.30.-j; 78.47.+p; 78.70.Ck; 63.22.+m; 78.67.-n

ISSN print edition: 0303-4216
ISSN electronic edition: 1437-0859
ISBN-10 3-540-34435-7 Springer Berlin Heidelberg New York
ISBN-13 978-3-540-34435-3 Springer Berlin Heidelberg New York

Springer is a part of Springer Science+Business Media

springer.com

Typesetting: DA-TEX · Gerd Blumenstein · www.da-tex.de
Production: LE-TEX Jelonek, Schmidt & Vöckler GbR, Leipzig
Cover design: WMXDesign GmbH, Heidelberg

Printed on acid-free paper 57/3100/YL 5 4 3 2 1 0

Preface

The series "Light Scattering in Solids" was launched in 1975 with the eighth volume of the Springer Collection "Topics in Applied Physics." That volume dealt mainly with Raman and Brillouin spectroscopy of low-frequency elementary excitations (phonons, electronic excitations including plasmons and, magnetic excitations such as magnons) The advent of the laser around 1960, its commercial availability a few years later, and concomitant developments in light-dispersion and -detection systems have positioned light-scattering spectroscopy among the most powerful techniques for the investigation and characterization of condensed matter, particularly solids. To date, a wide range of material systems, including bulk semiconductor (crystalline and amorphous), semiconductor nanostructures (e.g., superlattices), fullerites, nanotubes, high-T_c superconductors and, very recently, more conventional superconductors (boron-doped diamond, magnesium diboride), have been successfully investigated using light-scattering techniques.

The first volume of the series was meant to be a "one-shot affair". However, due to the scarcity of literature in the field it soon ran out of print and a new edition was called for. It appeared in 1983 under the title *Light Scattering in Solids I*, thus signaling the beginning of a series. Volume I (LSS I) contains eight articles covering the fundamentals of several applications of light-scattering spectroscopy, both spontaneous and stimulated, to the investigation and characterization of solids. Both editions together have been cited a total of 1050 times in "source journals" (see Chap. 1) as of June 25, 2006. *Light Scattering in Solids II* appeared in 1982. It contains three articles dealing with resonant scattering, multichannel detection and nonlinear scattering techniques. These topics reflect advances in experimental equipment that followed the introduction of lasers in the field. LSS II has been cited about 1000 times. Series volumes III, IV, V, and VI cover mostly investigations of novel materials, such as intercalated graphite, superionic conductors, doped semiconductors, molecules on rough metallic surfaces (SERS), superlattices and high-T_c semiconductors LSS VII covers crystal-field d and f electron excitations and magnetic excitations. LSS VIII covers fullerenes, surfaces, and a novel technique: time-domain spectroscopy using coherent phonons.

In the earlier volumes (I–VIII), it was repeatedly emphasized that laser-Raman spectroscopy only reveals, in crystals, excitations with wavevector $\boldsymbol{q} = 0$. It cannot beat, in this respect, inelastic neutron scattering (INS) that

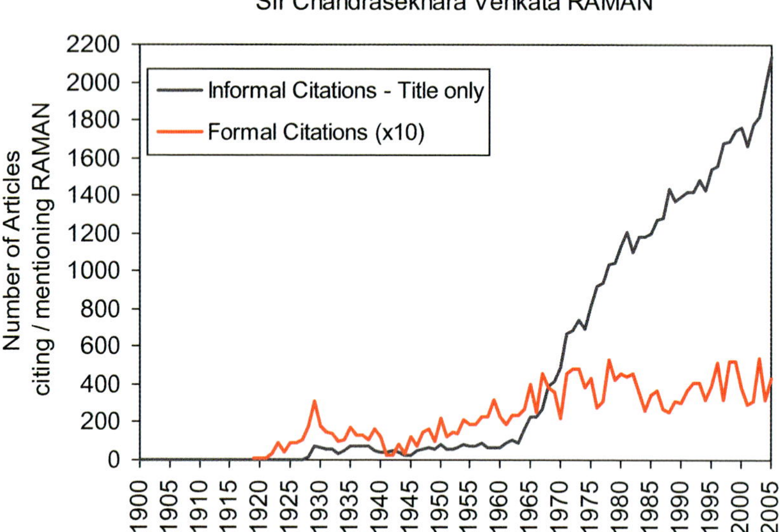

Fig. 1. Courtesy of Werner Marx

can access the whole Brillouin zone (BZ) of usual crystals. However, superlattices, with a supercell and a mini-BZ, can give access to wavevectors over the full mini-BZ (see LSS V). Standard crystals, with primitive-cell dimensions of the order of 5 Å, have more recently been investigated by means of inelastic X-ray scattering (IXS), using highly monochromatized synchrotron radiation. Chapters 5 and 6 of the present volume (LSS IX) discuss recent results involving, respectively, spontaneous and time-domain coherent IXS. Chapter 2 summarizes work on scattering enhanced by means of electromagnetic and acoustic resonances in superlattices. Chapter 3 discusses carbon nanotubes. Scattering by acoustic phonons in quantum dots is presented in Chap. 4.

Since the appearance of LSS VIII, bibliometric techniques and, in particular citation analysis, have become available as a research tool to most light-scattering researchers. The most common platform is the Web of Science (WoS), based on the citations index. These techniques can be used for literature searches and for impact studies of publications and authors. While details are given in Chap. 1, some results have already been mentioned above. We would like to conclude this preface with an analysis of the citations received by Sir C.V. Raman and by L.N. Brillouin. The citation history of Raman's publications is shown in Fig. 1. It displays a clear peak in the year 1928 (the Raman effect was discovered early that year). It then drops to an average of ≈ 10/year, to take off again at the end of World War II. From 1965 till now it oscillates about 40/year, a surprising fact in view of the already

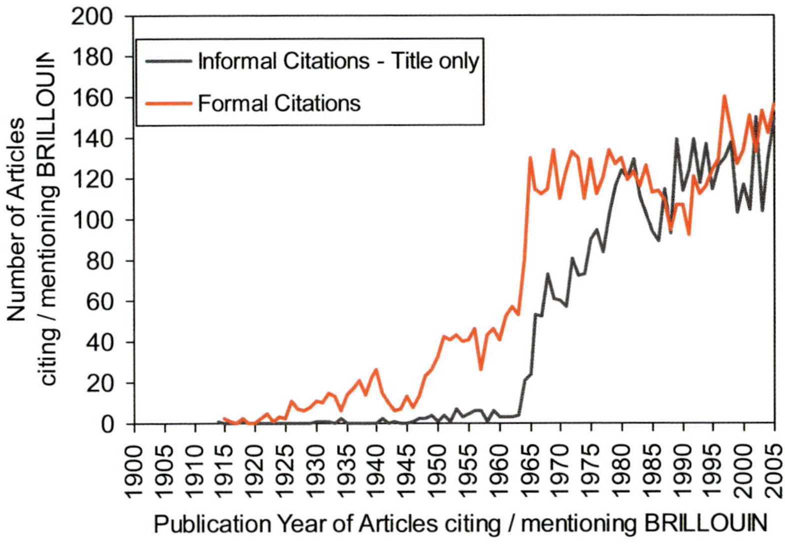

Fig. 2. Courtesy of Werner Marx

mentioned increase in Raman spectroscopy studies after the advent of the laser. The reason is easy to surmise. When a technique, theory or material becomes a household word associated with a particular person (in our case either Raman or Brillouin) future scientists replace mentioning the original (formal) reference and simply give the name of the item together with that of its author, i.e., Raman spectroscopy, Brillouin scattering, etc. The WoS enables the study of such "informal" references by entering the name of the author in question as a topic in the General Search Mode. The history of the informal citations of Raman so obtained is also shown in Fig. 1. These informal citations, limited to the titles of source articles, (if abstract and keywords were included we would get about three times as many informal citations) overtake the formal ones after 1928, the year of Raman's discovery. The total number of formal citations in Fig. 1 is 2270, that of informal ones 52 700 ($\approx$ 150 000 if abstracts and keywords mentioning Raman are included).

For comparison we display in Fig. 2 a similar citation history for Leon N. Brillouin. In this case the number of informal citations per year (a total of 4350, titles only) catches up with the formal ones around 1980. If abstracts and keywords were included, the informal citations would catch up with the formal ones around 1968, about the time of the development of Sandercock's multiple-pass spectrometers.

Figures 1 and 2 illustrate the tremendous growth of Raman and Brillouin spectroscopy that has taken place since 1965, putting these scientists in the category of the few authors cited mainly informally. With this volume we hope to contribute to that growth.

Stuttgart and Ann Arbor
July 2006

Manuel Cardona
Roberto Merlin

Contents

Light Scattering in Solids IX
Manuel Cardona and Roberto Merlin 1
1 General Considerations and Contents of Previous Volumes 1
2 The International Conferences on Raman Spectroscopy (ICORS) 3
3 Bibliometric Considerations Concerning the Raman Effect 4
4 The Nobel Prize 10
5 Contents of the Present Volume 12
References 13
Index 14

Raman Scattering in Resonant Cavities
Alejandro Fainstein and Bernard Jusserand 17
1 Motivation 18
2 Historical Background 20
3 Fundamentals of Optical Resonant Raman Scattering 22
3.1 Basics of Optical Microcavities 22
3.1.1 Distributed Bragg Reflectors 23
3.1.2 Fabry–Perot Resonators 27
3.1.3 Semiconductor Microcavities 29
3.1.4 Coupled Microcavities 32
3.2 Basics of Raman Amplification by Optical Confinement 33
3.3 Samples and Their Characterization 36
3.4 Angle Tuning in One-Mode Cavities 39
3.5 Two-Mode Cavities 40
3.6 Models of Optical Resonant Amplification 43
3.7 Raman Scattering with Confined Photons 48
4 Experimental Results on Raman Amplification by Optical Confinement 49
4.1 Spectral Selectivity and Amplification 49
4.2 Angular Selectivity 50
4.3 Cavity Performance 52
5 Research on Phonons of Nanostructures Using Optical Microcavities 57
5.1 Optical Phonons of Distributed Bragg Reflectors 58

5.2 Standing Optical Phonons in Finite SLs 64
5.3 Acoustic Phonons of SLs: Finite-Size Effects 68
5.4 Acoustic Nanocavities 74
5.4.1 Phonon-Cavity Design 75
5.4.2 Raman Amplification with Light–Sound Double Resonators 76
5.4.3 Raman Spectra of Acoustic Cavities 78
5.4.4 Raman Scattering with Confined Photons 80
5.4.5 Finite-Size Effects 81
5.4.6 Phonon-Mode Parity................................ 81
5.4.7 Cavity vs. Mirror Modes 82
6 Strongly Coupled Microcavities: Polariton-Mediated Resonant Raman Scattering 83
6.1 Cavity Polaritons in Strongly Coupled Microcavities 86
6.2 Theory of Polariton-Mediated Raman Scattering in Planar Cavities 88
6.3 Experimental Evidence of Polariton Mediation of Light Scattering in Cavities 91
6.3.1 Resonant Raman Scans 91
6.3.2 Double-Resonant Angle-Tuning Experiments 93
6.3.3 Outgoing Resonant Raman Experiments 96
6.3.4 Multibranch Polaritons: Finite-Lifetime Effects 98
7 Conclusions.. 102
References 104
Index.. 109

Raman Scattering in Carbon Nanotubes

Christian Thomsen and Stephanie Reich 115
1 Introduction 115
2 Structure and Growth 117
3 Lattice Dynamics 126
3.1 Symmetry 126
3.2 Radial-Breathing Mode 130
3.2.1 Chirality Dependence 132
3.2.2 Double-Walled Nanotubes 134
3.3 Tangential Modes 136
4 Electronic Properties 144
4.1 Zone Folding 144
4.2 Electronic Band Structure 148
5 Optical Properties 150
5.1 Band-to-Band Transitions 151
5.2 Kataura Plot: Nanotube Families and Branches............. 155
5.3 Excitons 158
6 Raman Effect in Carbon Nanotubes 162
6.1 Raman Spectra of Single-Walled Nanotubes 162

6.2 Single Resonances 164
6.3 Double-Resonant Raman Scattering 169
6.4 Anisotropic Polarizability 174
7 Assignment 177
8 Metallic and Semiconducting Nanotubes 187
8.1 Phonon–Plasmon Coupled Modes 188
8.2 Dispersion of Nanotube Phonons 192
8.3 Electrochemical Doping of Carbon Nanotubes 193
9 Electron–Phonon Interaction 196
9.1 Electron–Phonon Coupling of the Radial-Breathing Mode 196
9.2 Interference Effects 202
9.3 Kohn Anomaly and Peierls Distortion 204
10 *D* Mode, High-Energy Modes, and Overtones 207
10.1 *D* Mode and Defects 208
10.2 Picotubes 211
10.3 Two-Phonon Modes 212
10.4 Phonon Dispersion 215
11 Conclusions 215
References 217
Index 232

Resonant Raman Scattering by Acoustic Phonons in Quantum Dots
Adnen Mlayah and Jesse Groenen 237
1 Introduction 237
2 Introduction to Wavefunction Localization and Electron–Acoustic Phonon Interaction 239
2.1 Low-Frequency Resonant Raman Scattering in Quantum Wells 240
2.2 Diffraction and Interference 242
2.3 Scattering Efficiency 245
2.4 Intermediate-State Selection 247
3 Raman Interferences in Self-Assembled Quantum Dots 248
3.1 Self-Assembly and -Ordering of Quantum Dots 249
3.2 Spatial Localization 250
3.3 Spatial Correlation 255
3.3.1 Layer Stacking 255
3.3.2 Vertical Ordering 262
3.3.3 In-plane Ordering 267
3.3.4 Impurity Distributions 270
3.4 Surface Effects 273
3.4.1 Acoustic Mirror 273
3.4.2 Acoustic Cavity 277
4 Semiconductor and Metal Particles in the Strong Acoustic Confinement Regime 282

4.1 Eigenfrequencies and Eigenmodes ... 283
4.1.1 Lamb's Model ... 283
4.1.2 Anisotropy in Lamb's Model ... 284
4.1.3 Limitations of Lamb's Model and Microscopic Approaches ... 284
4.1.4 Matrix Effects ... 288
4.2 Optical Studies of Embedded Particles ... 293
4.3 Interactions between Confined Phonons and Electronic Excitations ... 297
4.3.1 Exciton–Phonon Interaction in Semiconductor Particles ... 298
4.3.2 Plasmon–Phonon Interaction in Metal Particles ... 298
5 Conclusions ... 306
References ... 307
Index ... 314

Inelastic X-Ray Scattering from Phonons
Michael Krisch and Francesco Sette ... 317
1 Introduction ... 317
2 Scattering Kinematics and Inelastic X-ray Cross Section ... 319
3 Instrumental Principles ... 324
4 Phonons in Single Crystals ... 327
4.1 Large-Bandgap Materials ... 328
4.1.1 SiC ... 328
4.1.2 GaN and AlN ... 329
4.2 Optical Phonons in Diamond ... 331
4.3 Optical Phonon Dispersion in Graphite ... 333
4.4 Quantum Systems ... 334
4.5 Correlated Electron Systems ... 336
4.5.1 Superconductors ... 336
4.5.2 Charge-Density Wave Systems ... 341
4.5.3 Actinides ... 342
5 Phonons in Polycrystals ... 344
5.1 Ices ... 345
5.2 Ice Clathrates ... 346
6 Phonon Studies at High Pressure ... 349
6.1 Zone-Boundary Modes in Cesium ... 351
6.2 Acoustic Branch Softening in SmS ... 353
6.3 Sound Velocities in hcp Iron ... 354
6.4 Elasticity of cobalt to 39 GPa ... 357
6.5 Lattice Dynamics of Molybdenum ... 359
7 Further Applications ... 360
7.1 Determination of the Phonon Density-of-States ... 360
7.2 Phonons in Surface-Sensitive Geometry ... 362
8 Concluding Remarks ... 363

References 364
Index 369

Ultrafast X-Ray Scattering in Solids
David A. Reis and Aaron M. Lindenberg 371
1 Introduction 371
2 Ultrafast X-Ray Sources 372
3 X-Ray Diffraction from Coherent Phonons 375
3.1 Coherent Acoustic Phonons 378
3.2 Coherent Optical Phonons 393
4 X-Ray Studies of Ultrafast Phase Transitions 398
4.1 Nonthermal Melting 398
4.2 Solid-Solid Phase Transitions 408
5 Concluding Remarks 409
References 411
Index 422

Index 423

Light Scattering in Solids IX

Manuel Cardona[1] and Roberto Merlin[2]

[1] Max-Planck-Institut für Festkörperforschung, Heisenbergstr. 1,
70569 Stuttgart, Germany
m.cardona@fkf.mpg.de
[2] Department of Physics, University of Michigan, 450 Church Street,
Ann Arbor, MI 48109-1040 USA
merlin@umich.edu

Dios (he dado en pensar) pone un empeño
en toda esa inasible arquitectura
que edifica la luz con la tersura
del cristal y la sombra con el sueño.

God (I've come to think) has left his mark
in all this unfathomable architecture
built up by light through the perfection
of the crystal and by darkness through dreams.

Jorge Luis Borges
Los Espejos, in *El Hacedor*
(Emecé, Buenos Aires, 1960)

Abstract. We briefly review the contents of this as well as previous volumes of the series Light Scattering in Solids, and present a chronological account of the International Conference on Raman spectroscopy (ICORS).

A bibliometric study of early publications on the Raman effect is also presented together with a succint historical outline of the discovery of the effect and information recently made available about the Nobel Prize awarded to Sir Chandrasekhar Venkata Raman in 1930.

1 General Considerations and Contents of Previous Volumes

This is the ninth volume of a series devoted to light scattering in solids, with special emphasis on semiconductors and superconductors. Volume I of the series was first published in 1975, followed by a second edition in 1982 [1].

The editor of Vol. I had in mind writing an introduction to light scattering in solids at a time when highly sophisticated equipment had become commercially available at affordable prices and the spreading of light-scattering spectroscopy (usually called Raman spectroscopy, although the term combination spectroscopy is common in the Russian literature. We must also include here Brillouin spectroscopy) for basic studies as well as for the characterization

M. Cardona, R. Merlin (Eds.): Light Scattering in Solid IX,
Topics Appl. Physics **108**, 1–16 (2007)

of solids could already be predicted. The main fact that contributed to converting light scattering from a highly specialized tool, available only in a few laboratories around the world, into a widely available spectroscopic technique was the replacement of incoherent light sources (sunlight in the first of Raman's publications on the matter [2], followed by gas-discharge lamps already in his second publication [3] and, later, the so-called Toronto arc [4]) to coherent laser sources in the 1960s [5]. The editor of Vol. I, overwhelmed by the growth of the field and the wide acceptance of the 1975 volume (it soon went out of print), relabeled its new edition (1982) as *Light Scattering in Solids I* and thus converted it, with the help of Springer Verlag, into the first of a series of which the present volume is the ninth. Gernot Güntherodt joined Manuel Cardona as series editor for Vols. II–VIII.

A second volume was published in 1982. It deals with what was then current advances in the field, including electronic resonance phenomena (tunable cw lasers had become commercially available), multichannel detection and developments in the field of nonlinear light-scattering spectroscopy (coherent Raman and hyper-Raman techniques). Volumes I and II have been translated into Russian by Mir (Moscow) in 1979 and 1983, respectively.

A listing of the contents of Vols. I to V is given at the end of Vol. VI. We shall avoid here the repetition of this list and make only a few remarks about the contents of Vols. VII and VIII, especially in view of the fact that the Introduction Chapter of Vol. VII also includes a detailed description of the contents of Vols. I to VI.

Volume VII includes a chapter about crystal-field excitations in materials containing transition and rare earth metals. Such excitations take place either from d-like to d-like states or from f-like to f-like states, respectively, and have therefore even parity in isolated atoms or in crystal atoms placed at centers of inversion. Consequently, they are rather weak in infrared spectroscopy and relatively strong (allowed) in Raman spectroscopy. Thus, Raman spectroscopy has become a powerful tool for their investigation, in particular in high-T_c superconductors and related oxides. The last chapter in Vol. VII is devoted to Brillouin scattering in layered magnetic structures. This research represents a continuation of work presented in Vols. III and V that incorporates advances in instrumentation and materials preparation.

Volume VIII discusses vibrational spectroscopy of fullerenes and is thus an excellent introduction to the work on nanotubes presented in Chap. 3 of the present volume. It also discusses scattering by coherent phonons measured in the time domain using femtosecond laser pulses. This work has been extended in the present volume to recent advances achieved with ultrafast X-ray pulses (Chap. 6). Volume VIII also discusses Raman scattering by surface phonons. Advances in instrumentation (multichannel detection, extreme resonant conditions) had made possible the investigation of single atomic layers, and thus to detect, by Raman spectroscopy, vibrations associated with semiconductor surfaces.

2 The International Conferences on Raman Spectroscopy (ICORS)

Our Light Scattering in Solids series is probably the most comprehensive series available covering Raman and Brillouin spectroscopy of solids. The reader may also find it interesting to consult the proceedings of the International Conferences on Raman Spectroscopy (ICORS) now being held biennially. It will be held in August of this year (2006) in Yokohama, Japan. The first conferences were devoted mainly to molecular spectroscopy and also to instrumentation. New subjects were added as the realm of applications of light scattering broadened. Among them we mention biology, medical applications and, since the 1984 X-ICORS, condensed matter physics and chemistry. The ICORS proceedings constitute an excellent historical record of the development of the field; some of the volumes, however, are difficult to find. Hence, we decided to list in what follows all the 29 ICORS, their venues, and where the proceedings were published.

The first ICORS was held in Ottawa (Canada) in 1969 and the second in Oxford (England) the following year, 1970 (the only exception to the biennial frequency). Unfortunately, for these first two conferences no proceedings were published and, thus, the program and the booklet with the abstracts are rather difficult to find.

The third ICORS (III-ICORS) was held in Reims (France) in 1972. It was the first one for which proceedings were published [6]. The next conference (IV- ICORS, 1974) was held in Bowdoin College, Brunswick, Maine (USA), an idyllic New England setting. It was the last ICORS for which no proceedings were published.

The fifth ICORS (1976) was held in another idyllic setting, the Black Forest, in the town of Freiburg im Breisgau (Germany). The proceedings were published by Schulz Verlag [7]. It was followed by VI-ICORS (1978), held in Bangalore, India. Raman had made Bangalore the center of his activities after being appointed director of the Indian Institute of Science in 1933. Having passed away in 1970, he was no longer present to appreciate the colossal development of his initial discovery made 50 years earlier, in 1928. The venue was chosen to commemorate this anniversary. The proceedings were published by Heyden [8].

The VII-ICORS returned to Ottawa (Carleton University) in 1980 and the proceedings were published by North-Holland (Amsterdam).

The VIII-ICORS was held in Bordeaux (France). The proceedings were published by Wiley [9], a pattern that has continued almost uninterrupted (except for the IX-ICORS, see below) till the present day.

The IX-ICORS was held in Tokyo in 1984. The proceedings were published by the Chemical Society of Japan in the same year. They list as editors the Organizing Committee of the IX-ICPS.

The X-ICPS was held at the University of Oregon (Eugene, Oregon, USA) in 1986. From now on, by contractual agreement, Wiley took over the publication of the proceedings. It was stipulated that the corresponding volume, containing two-page extended abstracts, would be available at the Conference [10].

The XI-ICORS (1988) moved back to Europe, being held in London [11].

The XII-ICORS (1990) was held in Columbia, the capital of South Carolina in the so-called Deep South of the United States. It required some stamina to resist the heat and the humidity in the open spaces in the middle of the summer. Fortunately, the excellent conference facilities were well air conditioned. Proceedings were again by Wiley [12].

The XIII-ICORS (1992) was held at the University of Würzburg, the capital of Franconia in Germany, and a jewel of baroque architecture. It is also at the center of the region producing some of the best German wines, a fact that the participants had the opportunity to enjoy. The University has had for quite some time several prominent groups working on Raman spectroscopy. For the proceedings, see [13].

The XIV-ICORS (1994) was held in Hong Kong, at the then newly founded and rather impressive Hong Kong University of Science and Technology (HKUST). Proceedings were again by Wiley [14].

The XV-ICORS (1996) went back to the United States (Pittsburgh). (See proceedings [15].)

The XVI-ICORS (1998) moved, for the first time, to the southern hemisphere. It was held in Cape Town, South Africa, being hosted by the University of Cape Town, an institution undergoing at the time profound social changes. The proceedings were edited by a South African Raman expert, *A. M. Heyns*, from the University of Pretoria [16].

The next conference, the XVII-ICORS, was held in Beijing in the year 2000, to coincide with the 80th birthday of Kun Huang, one of the internationally best-known Chinese condensed-matter physicists (his book the "Dynamical Theory of Crystal Lattices (University Press, Oxford, 1954)", written with M. Born, has been cited over 6000 times). (See proceedings [17].)

Back to Europe, the XVIII-ICORS was held in Budapest (Hungary) in 2002. (For proceedings, see [18].)

Finally, the XIX-ICORS moved back to down under. It was held in Gold Coast, Queensland, Australia in 2004 [19].

At the time of writing this introduction, we are looking forward to the XX-ICORS, to be held in Yokohama (Japan) on August 20–25 of 2006.

3 Bibliometric Considerations Concerning the Raman Effect

In this section we present a few remarks on the impact of inelastic light-scattering publications as measured by the number of citations that appeared in the literature. The most complete citation data banks are based on the *Citations Index*, provided by ISI-Thomson (Philadelphia) in particular in the form of the Web of Science (WoS). These data banks have been extended in the year 2005 to cover publications in so-called *source journals*[1] all the way back to the year 1900. They can also reveal citations of nonsource publications provided the citations have appeared in source journals.

A discovery as important as the Raman effect is seldom made in a vacuum. It nurtures on a number of prior discoveries, some of which will be analyzed next.

The basis of light scatering is already found in Einstein's Nobel Prize winning quantum hypothesis of light, which he published in the *annus mirabilis* 1905 [20]. It has been cited up to January 2006 (the date of the searches in this section) 410 times, a number that seems rather low, especially for a Nobel Prize winning publication. The reason for the low number of citations is twofold.

1. In those early days researchers cited very little. For instance, in [20] *Einstein* only cited two colleagues, whereas in the special relativity article he cited none.
2. The second reason is more subtle. When a physical theory or effect, or its author, becomes very important, it becomes a household word. A formal citation to its original literature site is no longer given. One simply refers to it informally (e.g., Einstein's equation, the Raman effect, Brillouin scattering). We shall discuss this point again in connection with Raman's discovery.

The main thrust of [20] concerned the explanation of the photoelectric effect on the basis of the quantization of the electromagnetic energy content of light.[2] Once this is done, Einstein discusses the Stokes shifts observed in photoluminescence (fluorescence) as due to some loss of energy before the excitation recombines to re-emit a photon. This photon will then contain less energy than the incident one, i.e., it will be redshifted. Einstein even goes beyond that and realizes that the material on which the excited light impinges may already be excited, e.g., because its temperature may not be zero. In this case the emitted photon will increase its energy with respect to the incident one and *anti-Stokes emission* will result. In this manner, Einstein had qualitatively predicted inelastic light scattering.

The next important step in the field of inelastic light scattering was taken by *L. Brillouin* [22, 23]. In a short note [22] he theoretically predicted the inelastic light scattering by acoustic waves (phonons) that now bears his name. Brillouin was then drafted into the French Army on account of the war. Upon returning after five years he completed his doctoral thesis, and successfully submitted it to the University of Paris. He then wrote a longer

[1] These are about 6000 journals selected by ISI-Thomson, from about 100 000, as being relevant to the progress of science.

[2] The term photon did not appear in the literature till the mid-1920s. It seems to have been coined by *Lewis* [21].

article with the details of his calculation [23]. Reference [22] has been cited 34 times, whereas [23] has been cited about 360 times. Among the reasons mentioned above for the low number of citations, we must add the fact that these articles were written in French. In the early days this did not much matter but nowadays it is easier to mention "Brillouin scattering" than to go and search for the appropriate reference in French: Brillouin scattering has become a household word.

The search software available with the WoS can also be used to find out how many times a given concept or "household word" is mentioned in the titles of source journal articles (after 1990 one can also include in the search abstract and keywords). Concerning "Brillouin scattering" we find by this procedure that it has been cited in such an *informal* way 4200 times. Querying Google for "Brillouin scattering" one finds 576 000 entries!!

At this point we must mention that in the Russian literature Brillouin scattering is referred to as *Mandelstam–Brillouin scattering* although the first formal publication by *Mandelstam* is found in 1926 [24], in an article that has been cited 52 times[3]. The term Mandelstam–Brillouin appears as informal citation in the WoS 107 times and in Google 73 times. The overwhelming majority of the authors using this term are Russian.

We do not know the basis for the priority given to Mandelstam by Russian authors. In a recent article, *Ginzburg* and *Fabelinskii* [25] mention that:

> "the splitting of the Rayleigh scattering had been predicted by Mandelstam in 1918 ..."

Without giving any reference – neither formal not informal – to support this claim (see also [26]). We must therefore conclude that priority should be given to Brillouin's 1914 article [22]. Inside information or possible internal reports do not suffice to set priorities.

The first observation of Brillouin scattering was, however, performed in Leningrad (now St. Petersburg) by *E. Gross* using a low-pressure mercury arc as a light source [27,28]. The two Nature articles in [27] have been cited a total of 192 times. The reader of [26] will notice another controversy involving priorities, this time between two Russian groups, Gross (Leningrad, now St. Petersburg) and the Mandelstam–Landsberg team (Moscow). *Fabelinskii* [26] suggests that the former did not play completely fair with the Moscow colleagues with whom he was in close contact. He rushed into publication as a single author while there were still some dubious points about the experimental results. The Moscow authors were extremely careful and refused to join him, thus losing any claims to joint priority with Gross. A similar pattern will become evident when we discuss the priority question concerning Raman scattering.

---

[3] Including several variations of the transliteration of Mandeltam's name, such as Mandelshtam.

For the history of the early days of Brillouin scattering, see the reviews by two early pioneers from both sides of the Atlantic, *Fabelinski* [26] and *Rank* [29].

After this brief review of the history of Brillouin scattering, we move on to Raman scattering. An Austrian physicist by the name of *Adolf Smekal* is often credited with having worked out first the theory of inelastic scattering by optical phonons in 1923 [30]. Not unexpectedly, in the Austro-German world an (unsuccessful) attempt was made to refer to the effect as the Smekal–Raman effect. Reference [30] has been cited 103 times, whereas the term Smekal–Raman has appeared 9 times as "informal reference" having been used mostly by German and Austrian authors. The question of priorities concerning Smekal and Raman seems to be clear, Smekal being responsible for the theoretical development, worked out full five years prior to Raman's experimental discovery. Smekal's work was not mentioned in the original work of *Raman* [2, 3, 31][4] who was probably not aware of it. The term Smekal–Raman is found as a *topic* 9 times in the WoS, the first time, surprisingly, appears in the Physical Review [32]. The authors worked in the United States, at Berkeley and the University of St. Louis. The article appeared in 1929, before the award of the Nobel Prize to Raman. Of these 9 *informal* citations, 5 were by K. W. F. Kohlrausch, an Austrian physicist. Smekal succeeded Kohlrausch in 1953 as president of the Austrian Physical Society. The most recent mention of the Smekal–Raman term in a source journal took place in 1999 [33]. Using Google one finds the term Smekal–Raman 194 times.

There is, however considerable acrimony and confusion concerning the discovery by Raman and a seemingly independent discovery by Landsberg and Mandelstam. The first report of the observations by *Raman* is found in [2]. This article was submitted to Nature on Feb. 16, 1928 and appeared on March 31 of that year. At face value, the date of submittal takes precedence over the earliest date of *observation*, February 21, 1928 reported by *Landsberg* and *Mandelstam* [25,26]. According to [25], Landsberg and Mandelstam did not submit their manuscript for rather dramatic reasons: a relative of Mandelstam had been arrested and sentenced to be shot. Mandelstam had to devote most of his time to try to commute that sentence at the time he should have been writing the manuscript. He lost editorial priority but was successful in saving his relative's life.

The advocates of the Russian authors also argue that the first of Raman's publications ([2], coauthored with *K. S. Krishnan*) is rather vague, a point we have to agree with. Not enough details of the experiment are given that would allow its reproduction without additional information. It is mentioned that the Sun was used as a light source, but neither quantitative data (nor figures) are presented. The details of the two glass filters used are not given

[4] This article was based on a lecture delivered on March 16, 1928 at the South India Science Association in Bangalore. He submitted the article on March 22 immediately on his return to Calcutta. See [31].

nor is it mentioned which liquids were measured. It is mentioned in passing that "spectroscopic confirmation is also available" without any details about its nature. However, this is the article that is usually mentioned to buttress the discovery. It has been cited 192 times.

In the second publication of *Raman* on the effect ([3], single author) more details are given. It is mentioned that the Sun, as a light source, was replaced by a mercury arc with sharp spectral lines that allowed the observation of discrete Raman lines. But again no figures or numbers are given. Much more detailed information is found in [31], a rather long article submitted in March 22, 1928, where the date of Raman's first observation is given as Feb. 28, 1928. According to Raman "publicity was given the following day". Thus, if the date given in [25] for the observation of line spectra by the Russian team stands (February 21, 1928), their observation would precede that of Raman by only one week. However, for whatever reason, the Russian workers delayed submission of their manuscript ([24]) till May 6 and thus lost formal priority. The original Russian manuscript has been cited 75 times. In spite of the clear formal priority of Raman, Russian colleagues refer to Raman scattering as *combination scattering*, a rather clumsy term that probably refers to the presence of combined rotational–vibrational lines in the molecular spectra.

The comprehensive article by *Raman* [31] displays one figure with the inelastic scattering spectrum observed by him for benzene. The quality of figures available in old publications is rather poor, especially concerning reproductions of photographic raw data. We have therefore contacted the Raman institute in Bangalore and have obtained a scanning of a rather good photographic plate available to them. We reproduce it in Fig. 1. A similar figure can be found in [34].

Fortunately, an original photographic plate exposed on February 23–24 for 15 h – by Landsberg and Mandelstam – has been reproduced in [35, 36] (Fig. 2). A microdensitometer trace of the photographed spectrum of quartz taken with the incident 2536 Å line of mercury, made by A. A. Sychev, is also shown in [31]. We have reproduced it in our Fig. 2. The upper trace represents the spectrum of the incident light. The lower trace has two peaks: the stronger one is the Rayleigh scattering, the weaker one the Raman satellite shifted by $\sim 470\,\mathrm{cm}^{-1}$. The Russian authors checked that the frequency shift was the same for 3 different incident mercury lines.

The reader may be surprised by the small number of citations received by these pioneering papers. As already mentioned, the most cited of them is that by *Raman* and *Krishnan* [2] (162 citations) whereas its Russian counterpart [37] has been cited only 80 times. This is mainly due to Raman scattering having become a household word. Querying the WoS for "Raman scattering" as a topic (as contained in article titles and, since 1990, also abstracts and keywords) we find an overwhelming number of 25 900 "informal citations". Raman alone is mentioned 97 000 times.

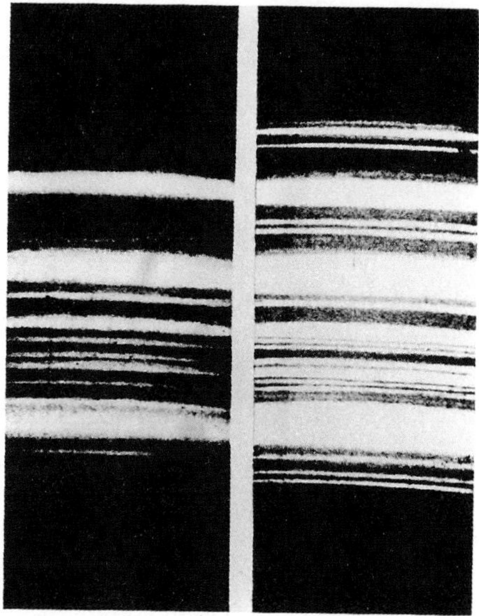

Fig. 1. Spectrum of a mercury arc lamp (*left*) and spectrum emitted by the illuminated benzene (*right*). The additional *lines* not present in the spectrum on the left are inelastic (Raman) scattered lines. The two *lines in the extreme bottom* are probably anti-Stokes lines. We are thankful to G. Madhavar and K. Krishnama Raju, of the Raman Research Institute for providing a print and a scanning of the original plate taken by *Raman* [31]

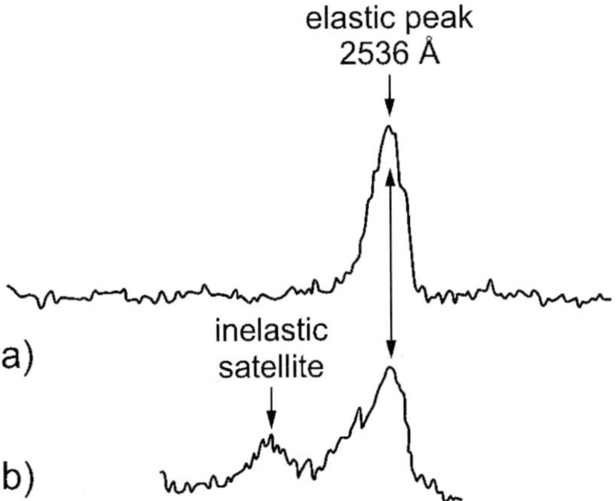

Fig. 2. (**a**) Photographic plate trace showing a line of a mercury arc (2536 Å) used to obtain the scattering spectrum of a quartz sample (**b**). The satellite to the *left* in (**b**) represents inelastic (combination according to the Russian nomenclature) scattering. Taken during February 23–24, 1938 by *Landsberg* and *Mandelstam* [35, 36]

4 The Nobel Prize

Information about nominations for the prize, as well as the deliberations of the various committees, are kept highly secret, at least for fifty years. Extensive information about nominators and nominees, covering the 1901–1950 period is now available [38]. The "Raman" case has required especial attention from the part of the Nobel foundation in view of the question of priorities (involving next to Raman, Smekal, Landsberg and Mandelstam, and possibly even Brillouin). Smekal and Brillouin were actually never nominated. Hence, according to the rules of the Nobel Foundation, they did not qualify for the prize. This, however, should not affect the question of priorities, for which they were not fully recognized, especially Smekal.[5] The official comments of the Foundation are reproduced in [38].

We first mention that Raman was nominated for the 1930 prize by 10 of the world's most famous physicists (E. Bloch, N. Bohr, L. de Broglie, M. de Broglie, J. B. Perrin, R. Pfeiffer, E. Rutherford, J. Stark, C. T. R. Wilson, and a Russian colleague, O. Khvol'son). He had already been nominated twice for the 1929 prize (by Fabry and Bohr). Landsberg and Mandesltam were nominated in 1929 by two Russians, N. Papaleski and O. Khvol'son (the latter had also nominated Raman). Mandelstam was also nominated by Papaleksi who explicitly excluded Landsberg. Khvol'son had proposed that half of the prize should go to Raman, the other half to Landsberg and Mandelstam. The reader can nowadays follow the acrimony concerning the exclusion of Mandelstam and Landsberg by the Nobel Foundation as a result of the release of records (see [25] and [34]). The authors of [25] and [34] criticize the weak evidence presented in the two Nature papers by *Raman* ([2] and [3]). The authors of [25] go as far as to say

> "... more detailed analysis ... shows that in the visual observations conducted in Ref. 30, the discovery of such radiation is scarcely possible. Raman and Krishnan assumed that there is some frequency-reduced radiation and that this is what they had seen."

As already mentioned, the second publication [3] is more credible, while [32] contains considerable detail, including the date of the first photographic observation of a line spectrum by Raman (our Fig. 1): "the lines of the spectrum of the new radiation were first observed on February 28, 1928", a week later than that claimed for the observation of Landsberg and Mandelstam (our Fig. 1 is a scanning of the photographic plates obtained for benzene).

Immediately after submitting [3] Raman made the discovery public (his own words). According to *Ginzburg* and *Fabelinskii* (Ref. [25]), he obtained 2000 reprints of his manuscript on March 22 "and sent them to all eminent physicists, including those working on light scattering in France, Germany,

---

[5] The reader may want to look at the biography of Alfred Smekal using Google, in particular his connections with the Third Reich.

Russia, Canada and the United States, as well as to scientific institutions all over the world, thereby consolidating his priority" (see [33] where the source of this statement is cited). According to [33], one of Raman students narrates that he (Raman) approached Rutherford, Siegbahn and Bohr concerning support for the Nobel award. He was very concerned about the question of priority after the work of Landsberg and Maldestam was published, but considered the matter settled in his favor after the German physicist *Pringsheim* [39] reproduced Raman's experiments and coined the term "Raman effect" [40].

The critics of the Nobel decision recognize the right of the foundation (not necessarily its wisdom) in setting the rules to establish priority on the basis of submittal dates and dates of publication. On the basis of these rules, Raman certainly wins. However, the critics are particularly chagrined (see [25]) by the following statement from the foundation: "*Mandelstam and Landsberg cannot argue to have obtained their results independently.*" The authors of [25] interpret this statement as a frontal accusation of plagiarism against their colleagues and countrymen, Mandelstam and Landsberg, whom they consider to be highly competent and top notch, extremely honest human beings. We must admit that the statement of the foundation is tactless, to say the least.

The authors of [25] proceed to analyze the reasons for the failure of the Foundation to include their Russian colleagues among the 1930 winners. The poor performance of Russian scientists as international prize winners has been scrutinized, during the past few years, by *Vitaly L. Ginzburg*, a Nobel laureate himself (2003) [41]. The authors of [25] emphasize the importance of international contacts and nominations that were rather difficult (and probably dangerous) to entertain within the communist system. Moreover, Russian scientists did not seem to be very keen on lobbying for their deserving colleagues, let alone nominating them when entitled. Again, this may be a dangerous enterprise if the nominee is not to the taste of present or future rulers. Moreover, the physics prize has never been awarded jointly to more than three scientists, up to 1956 (Bardeen, Brattain and Shockley for the transistor) to more than two. Hence, in 1930 the Committee either had to change the rules (or the practice) or choose one among the three nominees for inelastic light scattering. Even if they had changed the rules, the question would have remained concerning Brillouin, Smekal, and Krishnan who could have been excluded because they had not been nominated but, would that have been fair? In any case, the question remains as to why Brillouin and Smekal had not been nominated at least by their countrymen (some of whom later on tried to change the name of the effect to include Smekal's). After all, the problem of lack of readiness to nominate colleagues may not only be a Russian problem. Is it a European one? And, why did Raman not include Krishnan in his lobbying although he had coauthored his first publication on the subject? Notice that, at the end of [31] *Raman* says "I owe much to the valuable cooperation in this research of Mr. K. S. Krishnan ...".

As a prize winner Raman was entitled to nominate candidates for the Prize. He nominated O. Stern in 1934. Stern was awarded the Nobel Prize in 1944 after he had received about 80 nominations. Raman also nominated Fermi in 1938 and Lawrence in 1939. Both were awarded the Prize the year of their nomination; none of his nominees failed. This proves that Raman kept well in touch with the top developments in physics. Leon Brillouin was asked at least once to submit nominations by the Nobel Foundation. He nominated Pauli and Goudsmit in 1935 (Pauli received the Prize in 1945). Brillouin's father, Marcel, seems to have been asked at least three times. He submitted a total of 19 nominations in 1910, 1916 and 1922. Only two of them were successful: Einstein's and Perrin's. With the exception of Einstein, all of Marcel Brillouin's nominees were French. Till 1950 there were 51 nominations in physics from Russia and the USSR. However, all the 10 USSR-Russia laureates received the prize after 1950.

5 Contents of the Present Volume

The present volume reviews recent developments concerning mainly semiconductor nanostructures and inelastic X-ray scattering. Raman studies of resonant cavities is the subject of Chap. 2, by *A. Fainstein* and *B. Jusserand*, which also includes a substantial discussion of resonant enhancement and acoustic cavities. In a broad sense, this Chapter is an extension of the Chapter by C. Weisbuch and R. G. Ulbricht in Vol. III of this series, where light scattering by bulk exciton polaritons is discussed. Spontaneous emission and, more generally, the quantum behavior of the radiation field in confined geometries is a topic of much current interest. The properties of the coupled exciton–photon system in semiconductor structures can resemble those of an atom in an optical cavity, and this analogy leads to the possibility of exploring such an interaction in a regime of parameters that is normally not accessible to atomic experiments.

Chapter 3, by *C. Thomsen* and *S. Reich*, centers on one-dimensional carbon nanotubes. Previously, Raman scattering by graphite intercalation compounds and fullerenes were covered, respectively, in Vol. III of this series by M. S. Dresselhaus and G. Dresselhaus and in Vol. VIII by J. Menendez and J. B. Page. This Chapter provides an extensive discussion of the electronic and optical properties of semiconducting and metallic nanotubes, and it reviews the crucial role Raman scattering plays in their identification.

Raman scattering by acoustic phonons in quantum dots is discussed in Chap. 4 by *A. Mlayah* and *J. Groenen*. Here, the focus is the interplay between the atomic-like structure of the electronic spectrum and the classical behavior of the acoustic modes. This review includes both, spontaneous Raman and time-resolved optical experiments.

Inelastic X-ray scattering by phonons is discussed in Chap. 5, by *M. Krisch* and *F. Sette*. Unlike conventional (visible and UV) Raman methods, which probe primarily excitations at the center of the Brillouin zone, inelastic X-ray scattering allows one to explore the whole phonon dispersion. The remarkable capabilities of this technique manifest themselves in the case studies presented by the authors, particularly those on semiconductors such as SiC and correlated electron systems. This Chapter provides also an overview of the technical requirements, its implementation in third-generation synchrotron radiation sources, and a detailed comparison between X-ray and inelastic neutron scattering.

Chapter 6, by *D. Reis* and *A. Lindenberg*, discusses very recent developments in time-domain X-ray methods. Following a review of ultrafast X-ray sources, this Chapter provides an extensive discussion of recent work on stimulated X-ray scattering by acoustic modes and coherent optical phonons generated by subpicosecond optical pulses. The Chapter closes with an overview of studies at higher excitation intensities involving phase transformations and, in particular, the so-called nonthermal melting transition.

References

[1] M. Cardona (Ed.): *Light Scattering in Solids I: Introductory Concepts*, vol. 8, 2nd ed., Topics in Appl. Phys. (Springer, Berlin, Heidelberg, New York 1982)
[2] C. V. Raman, K. S. Krishnan: Nature **121**, 501 (1928)
[3] C. V. Raman: Nature **121**, 619 (1928)
[4] B. P. Stoicheff: Can. J. Phys. **32**, 330 (1954)
[5] S. P. S. Porto, D. L. Wood: J. Opt. Soc. Am. **52**, 251 (1962)
[6] J. P. Mathieu (Ed.): *Advances in Raman Spectroscopy* (Heyden and Son Ltd., London 1973)
[7] (H. F. Schulz Verlag, Freiburg i. Br. 1976)
[8] (Heyden and Son Ltd., London 1976)
[9] J. Lascomb, P. V. Huong (Eds.): *Proceedings of the 8th International Conference on Raman Spectroscopy* (Wiley 1982)
[10] W. L. Peticolas, B. Hudson (Eds.): *Proceedings of the 10th International Conference on Raman Spectroscopy* (Wiley 1986)
[11] R. J. H. Clark, D. A. Long (Eds.): *Proceedings of the 11th International Conference on Raman Spectroscopy* (Wiley 1988)
[12] J. R. Durig, J. F. Sullivan (Eds.): *Proceedings of the 12th International Conference on Raman Spectroscopy* (Wiley 1990)
[13] W. Kiefer, M. Cardona, G. Schaak, W. W. Schneider, H. W. Schrötter (Eds.): *Proceedings of the 13th International Conference on Raman Spectroscopy* (Wiley 1992)
[14] N. Yu, X. Li (Eds.): *Proceedings of the 14th International Conference on Raman Spectroscopy* (Wiley 1994)
[15] S. Asher, P. Stein (Eds.): *Proceedings of the 15th International Conference on Raman Spectroscopy* (Wiley 1996)
[16] A. M. Heyns (Ed.): *Proceedings of the 16th International Conference on Raman Spectroscopy* (Wiley 1998)

[17] W. Kiefer, B. F. Yu (Eds.): *Proceedings of the 17th International Conference on Raman Spectroscopy* (Wiley 2000)
[18] M. Janos, G. Jalsavszky (Eds.): *Proceedings of the 18th International Conference on Raman Spectroscopy* (Wiley 2002)
[19] P. M. Fredericks, R. L. Frost (Eds.): *Proceedings of the 19th International Conference on Raman Spectroscopy* (Wiley 2004)
[20] A. Einstein: Ann. d. Physik **17**, 132 (1905)
[21] G. N. Lewis: Nature **118**, 874 (1926)
[22] L. Brillouin: Compt. Rend. (Paris) **158**, 1331 (1914)
[23] L. Brillouin: Ann. Phys. (Paris) **17**, 88 (1922)
[24] L. I. Mandelstam: J. Russ. Chem. Soc. **58**, 381 (1926)
[25] V. L. Ginzburg, I. L. Fabelinskii: Herald of the Russian Academy of Sciences **73**, 152 (2003)
[26] I. L. Fabelinskii: Usp. Fiz. Nauk. **170**, 93 (2000)
[27] E. Gross: Nature **126**, 201 (1930) ibid. **126**, 400 (1930)
[28] E. Gross: Z. Phys. **63**, 685 (1930)
[29] D. H. Rank: J. Acoust. Soc. Am. **49**, 937 (1971)
[30] A. Smekal: Naturwiss. **11**, 873 (1923)
[31] C. V. Raman: Indian J. Phys. **2**, 387 (1928)
[32] B. Podolsky, V. Rojansky: Phys. Rev. **34**, 1367 (1929)
[33] S. Arrivo: Chem. Eng. News **77**, 7 (1999)
[34] R. Sing, F. Riess: Notes and Records of the Royal Society **55**, 267 (2002)
[35] I. L. Fabelinskii: Ups. Fiz. Nauk. **126**, 124 (1978)
[36] I. L. Fabelinskii: Sov. Phys. Uspekhi **21**, 780 (1978)
[37] G. S. Landsberg, L. I. Mandelstam: Naturwiss. **16**, 557 (1928)
[38] E. Crawford: *The Nobel Population, 1901–1950 A census of the nominators and the nominees for the prizes of physics and chemistry* (Universal Academy Press, Tokyo 2002)
[39] P. Pringsheim: Naturwiss. **16**, 597 (1928)
[40] R. Singh, F. Riess: Current Sci. **74**, 1112 (1998)
[41] V. L. Ginzburg: *On Science, Myself and Others* (Taylor and Francis 2004)

Index

Bangalore, 8
benzene, 8
Brillouin
  spectroscopy, 1

Combination Scattering, 8

Einstein, 5, 12

Ginzburg, 6
Gross, 6

ICORS, 1, 3

Krishnan, 11

Mandelstam–Brillouin scattering, 6
Mandelstam–Landsberg, 6

Nobel Prize, 10

quartz, 8

Raman
  effect, 11
Rayleigh scattering, 6, 8

Smekal, 7
synchrotron radiation, 13

Third Reich, 10

Toronto arc, 2

Web of Science, 5

Raman Scattering in Resonant Cavities

Alejandro Fainstein[1] and Bernard Jusserand[2]

[1] Instituto Balseiro and Centro Atómico Bariloche, Av. E. Bustillo 9500, 8400 S. C. de Bariloche Río Negro, Argentina
afains@cab.cnea.gov.ar

[2] Institut des Nanosciences de Paris, UMR CNRS 7588, Université Pierre et Marie Curie, Campus Boucicaut, 140 Rue de Lourmel, 75015 Paris, France
bernard.jusserand@insp.jussieu.fr

Abstract. The modification of the optical properties of matter due to photon confinement in optical microcavities has been an active field of research in the last ten years. This review addresses the problem of Raman scattering in these optically confining structures. Two completely different regimes exist and will be discussed here. Firstly, a situation in which the action of the microcavity is basically to enhance, and to spatially and spectrally confine, the photon field, but otherwise the light–matter interaction process remains unaltered. This is the case when the laser and scattered photon energies are well below those of the excitonic transitions in the structure, or when the coupling between the latter and the cavity mode can be treated in a "weak coupling" approximation. This regime will be labeled here as "optical resonant Raman scattering". And secondly, a "strong-coupling" regime in which exciton and cavity-photon modes cannot be treated separately, leading to coupled excitations, so-called cavity polaritons. In this second case, when the laser or scattered photons are tuned to the excitonic energies, and thus an electronic resonant Raman-scattering process is achieved, the Raman-scattering process has to be described in a fundamentally different way. This regime will be labeled as "cavity-polariton-mediated Raman scattering".

The Chapter will begin, after a brief historical introduction, with a review of the fundamental properties of optical microcavities, and of the different strategies implemented to achieve double optical resonant Raman scattering in planar microcavities. A section will be then devoted to experimental results that highlight the different characteristics and potentialities of Raman scattering under optical confinement, including, in some detail, an analysis of the performance of these structures for Raman amplification. A subsequent section will present a series of research efforts devoted to the study of nanostructure phonon physics that rely on microcavities for Raman enhancement. In particular, emphasis will be given to recent investigations on acoustic cavities that parallel their optical counterparts but that, instead, confine hypersound in the GHz–THz range. The Chapter will then turn to a quite different topic, that of cavity-polariton-mediated scattering. The theory developed in the 1970s for bulk materials will be briefly introduced, and its modifications to account for photon confinement in planar structures will be addressed. Finally, a series of experiments that demonstrate the involvement of polaritons in the inelastic scattering of light in strongly coupled cavities will be reviewed. The Chapter will end with some conclusions and prospects for future developments on this subject.

M. Cardona, R. Merlin (Eds.): Light Scattering in Solid IX,
Topics Appl. Physics **108**, 17–114 (2007)

1 Motivation

Raman scattering is a powerful spectroscopic technique that has found applications in such diverse fields as basic and applied physics, chemistry, and biology, medicine, sensing, and forensic sciences. It can be used to study a broad spectrum of molecular and condensed-matter excitations, including molecular vibrations, phonons, plasmons, inter- and intraband electronic excitations, crystal-field transitions, magnons and other spin excitations. One of its major weaknesses, however, is related to the usually small cross section of inelastic scattering processes. This limitation has been partially overcome by the access to lasers in a broad spectral range, and the use of very efficient spectrometers and detectors. However, it remains a source of concern when small scattering volumes are to be studied (as, e.g., in the domain of nanoscience or single-molecule sensing), or when other more intense emission sources are present in the same spectral range.

Basically, two techniques have been developed to deal with this problem, namely, "resonant Raman scattering (RRS)" [1–5], and "surface-enhanced Raman scattering (SERS)" [6–8]. RRS is related to the amplification of the Raman-scattering amplitude when the laser or scattered-field energies are tuned to electronic interband transitions of the molecule or solid. Raman amplifications using RRS can go typically up to $\sim 10^5$. From a macroscopic point of view, this amplification is related to the resonant critical points of the dielectric function or optical response of the system, which are modulated by the excitation under study to give the inelastically scattered wave. In a microscopic quantum-mechanical picture, on the other hand, these resonances appear as zeros of denominators of scattering probabilities calculated from third-order perturbation theory (second order in light–matter and first order in electron-excitation interactions).[1] Besides being used to amplify Raman cross sections, RRS is also a powerful instrument to study interband optical transitions, and electron-excitation interactions. No technique is, however, universal and RRS is no exception. Firstly, only excitations that are related to (that is, that modulate) the resonant electronic optical transition are selectively amplified. Secondly, since these electronic transitions need to be dipolar allowed to be strongly resonant, in many cases they are also accompanied by strong optical emission that competes with, and can obscure, the Raman-scattered peak. And thirdly, laser lines tuned to the material electronic transitions are required, something that is not always readily accessible.

SERS, on the other hand, exploits the huge enhancement of the electromagnetic field in the near-field of specifically designed metallic (typically Au and Ag) nanostructures, when their plasmon excitations are resonantly excited. Different strategies have been pursued to tailor these nanostructures,

[1] These applies to *phonon* Raman scattering, which will be the main subject of this Chapter. Inelastic scattering by *electronic* excitations, on the other hand, follows from second-order perturbation theory [1].

including chemically roughened surfaces, nanoparticle colloids, e-beam lithography, and self-assembled nanoparticle and metallic cavity structures. The main mechanism by which SERS accomplishes its task is the amplification of the field sensed by the molecule under study (the so-called "electromagnetic effect"), but the role played by the electronic coupling between plasmons and molecular electronic excitations (the so-called "chemical effect") is also relevant [9]. The nanostructures have typical sizes in the range 2–50 nm, and the lasers required for resonant excitation fall, e.g., in the green for Ag and in the red for Au. Until quite recently it was believed that the SERS amplification also attained values of the order of 10^5. However, it has been realized quite recently that in many cases most of the scattered signal comes from very few selected "hot-spots" for which the enhancement can go as high as 10^{15} [10,11]! This has opened the way to single-molecule Raman spectroscopy and ultrasensitive molecular identification, an area of research of enormous potential and present activity. The most important issues that attract considerable work these days are related to the understanding of hot-spots, the fabrication of SERS substrates with controllable properties, and the differential response of different molecules. Once these difficulties are overcome, probably the most important limitation of SERS will be the fact that the system under study has to be put in close contact with a metallic specifically prepared substrate. This can be a difficult task for the study of some molecules, but it is usually not applicable for the study of solid-state matter.

In this Chapter we will address an alternative approach to Raman amplification. Namely, Raman enhancement by optical confinement in resonant cavities. Or in short, "*optical* resonant Raman scattering". We will describe monolithic planar structures based in distributed Bragg reflectors (DBRs) and constituted by semiconductor materials grown using molecular beam epitaxy (MBE) techniques. The concepts, however, can and have been extended to other kinds of mirrors and materials. After giving a brief historical survey in Sect. 2, we will introduce in Sect. 3 the fundamentals of optical cavities and we will discuss the main ingredients and strategies for Raman efficiency amplification in these structures. The most important characteristics of Raman amplification by light confinement will then be illustrated with experiments in Sect. 4. The Chapter will then continue in Sect. 5 with a series of examples of applications of optical microcavities for the study of phonons in nanostructures. A special emphasis will be given to nanocavities that are conceptually similar to their optical counterparts, but that confine instead acoustic waves in the GHz–THz range. Section 6 will then be devoted to Raman scattering in "strongly coupled" microcavities, structures in which the cavity modes are not optical but a mixture of photon and exciton states (the so-called "cavity polaritons"). Besides the added enhancement related to the exciton resonance in the Raman process, these structures are very interesting from a fundamental point of view for the understanding of polariton-mediated Raman scattering in solids. At the end some conclusions and prospects for future work will be given.

2 Historical Background

It has been realized for some time that fundamental atomic properties such as light emission lifetime and Lamb-shift are not intrinsic characteristics of an atom but of the coupled atom-vacuum-field system [12–15]. In fact, the radiation characteristics of an emitter can be altered by modifying the vacuum-field fluctuations in its vicinity, e.g., by placing it near a surface [16, 17], within a microcavity [14, 18–21], or in a photonic bandgap material [22]. These modifications of the electric fields can be exploited in both first order in light–matter interaction processes: absorption *and* emission. Concerning the *absorption* of an emitter, a physical way to picture the effect of a nearby reflecting surface is to realize that the reflected light self-interferes, changing the excitation field pattern (spatial distribution and amplitude). Similarly, the field *emitted* by the dipole (direct *plus* reflected) at the dipole position drives it determining its decay rate [6, 7, 23–27]. An equivalent way to present the problem is by realizing that the interfaces modify the density of optical modes at the dipole position and hence, through Fermi's golden rule, also its emission and absorption characteristics [17]. In particular, the decay of an excited atom into a definite photon mode can be strongly enhanced/inhibited by placing it in maxima/minima of the photon field spatial distribution [28]. Similarly, the spontaneous and stimulated emission can be increased by an enhancement of the light field at the *excitation* energy using, e.g., a cavity [29]. The consequent modifications of the spectroscopic properties (luminescence and absorption) of dipoles close to a metallic surface have been at length discussed in the past and studies have been performed as a function of dipole–interface distance [23–27] and surface morphology and excitations [6, 7]. Recently, these topics have acquired renewed interest in the context of near-field optical microscopy with metallic tips.

Absorption and emission are complementary first order in light–matter interaction process.[2] Raman scattering, on the other hand, involves two light–matter interactions: laser excitation *and* emission of the (inelastically) scattered photon. Thus, extending the above ideas to higher-order processes a strong enhancement/inhibition of the Raman cross section of an active material embedded in a cavity may be envisaged through the modification of one or two of the intervening light–matter vertices. A first proposal to use the positive interference of laser and inelastically scattered light to enhance the Raman spectra of a thin film dates to some 25 years ago [30, 31]. In this work *Nemanich* and coworkers [30, 31] deposited the thin film on a transparent dielectric film that itself rested on a metallic reflector. The thickness

[2] Here, as in the discussion of Raman scattering, we are referring to the order required to evaluate amplitudes. Absorption and emission are first order in amplitude, second order in cross section. Similarly, electronic Raman scattering is second order in amplitude, fourth order in cross section. And last, for phonon, plasmon and other boson scattering, Raman is third order in amplitude and sixth order in cross section.

of the thin film and dielectric spacer were chosen so that the reflectance of the trilayer be zero, and thus most of the light would be absorbed at the thin film. Furthermore, the interference condition for the scattered radiation was such that inphase addition of the forward and backscattered components occurred. With such a scheme, Raman amplifications of up to 20 were demonstrated for a thin film of polycrystalline tellurium [30,31]. The method, termed "interference-enhanced Raman scattering" (IERS), can be basically understood as based on a low-Q half-cavity, limited below by a single bottom metallic mirror and above by air. Cavities with giant Q-factors, on the other hand, have been demonstrated in the context of Raman scattering in liquid micrometer-size droplets [32–35]. By total internal reflection at the liquid air interface, the microdroplets lead to "morphology-dependent resonances" provided by whispering-gallery modes with Q-factors larger than 10^8. Through these resonances stimulated Raman scattering with high-order Stokes peaks both with pulsed [32, 33] and continuous-wave [34, 35] excitation have been demonstrated. The use of purposely designed planar cavities in the search for quantum electrodynamical effects in light scattering has been pioneered by the group of *De Martini* [29, 36]. Field-confinement effects at the Stokes energy, including scattering enhancement, inhibition, and directionality have been observed for liquid benzene in a piezoelectrically controlled microcavity [36–38].

Parallel to the above developments, research on semiconductor microcavities and photonic bandgap structures found great impulse after the initial proposal by *Yablonovitch* [22] that light emission could be controlled in solid-state devices by appropriate engineering of the optical modes. The last 15 years have seen enormous developments in the design and construction of photonic bandgap materials at optical wavelengths, and on the study of fundamental and applied concepts that involve planar and nanostructured microcavities [15, 18–22]. The highest-quality microcavities are grown using semiconductors by different kinds of epitaxial techniques. Typically, mirrors and sample are grown monolithically, and reflecting metallic surfaces are replaced by lossless distributed Bragg reflectors (DBRs). By using these DBRs, optical microcavities with Q-factors in the 10^3–10^4 range have been constructed displaying increased emission (and thus reduced laser thresholds) [15,18,19] and novel optical phenomena [15, 20, 21]. These include the observation of cavity polaritons in strongly coupled cavities, parametric amplification involving polaritons, possibly "boser" emission [39,40], single-photon quantum dot emission, etc. [15, 20, 21]. Concerning Raman-scattering, different strategies have been conceived to use such planar microcavities for single [41] and double [42,43] optical resonant Raman-scattering enhancement. Raman efficiency amplifications in excess of 10^5 have been demonstrated [44, 45]. Besides this major trend, Raman scattering in cavities has also been reported on structures made of organic molecules using mixed dielectric–metal mirrors [46], and with electrochemically grown porous silicon [47]. Raman scattering in microcavities has been used to study nanostructure phonon [42, 43, 48–50] and electronic excitations [51].

As commented above, high-quality semiconductor cavities display the so-called "strong-coupling" regime, characterized by coupled exciton–photon normal modes. These mixed optical excitations have been termed "cavity-polaritons" and have led to strong and fruitful research activity [15, 20, 21]. In a Raman-scattering process when the laser or scattered energies are tuned to excitonic energies in addition to the cavity optical resonances, a completely different regime that involves the participation of polaritons is observed. The fundamental problem of polariton-mediated inelastic light scattering in bulk materials has been investigated for a long time [52–58]. In fact, proof of the involvement of polaritons in first-order Raman scattering by optical phonons has been sought for several decades [52–54, 56, 57, 59, 60]. However, only in planar microcavities has the real need to take into account polaritons as an intermediate step of inelastic scattering found definitive demonstration [61–64]. Besides being an interesting issue in its own right, polariton-mediated Raman scattering has been shown to provide valuable information on cavity–polariton physics, e.g., to disclose the intervening dephasing mechanisms [62–64].

3 Fundamentals of Optical Resonant Raman Scattering

After the brief summary on the main historical developments relevant to optical resonant Raman scattering presented in the previous section, we will address here the fundamentals of resonant Raman scattering in optically confined structures. We will first describe the DBRs and cavity design and properties. The basic equations required to understand RRS in microcavities and the concept of single and double optical resonances will then be addressed, followed by a discussion of the actual strategies used for double-resonant optical enhancement in planar structures. Namely, the angle tuning of the photon dispersion in single-mode structures, and the two-mode double resonators for near-backscattering resonant geometries. A model will then be presented to describe the amplification of Raman scattering in cavities that treats separately and on different grounds the processes of excitation and emission. The consequences of using optical standing waves in the kinematics of the Raman process will then be briefly addressed.

3.1 Basics of Optical Microcavities

While the quantization of electromagnetic waves in microwave cavities has been known for a long time, it is only recently that it has been transposed in optical wavelength-scale structures. The field of atomic physics has been a pioneer in such effects based on very high finesse confocal vacuum microcavities (for a review see, for instance, [65]). We are interested here in a more recent device, the semiconductor planar microcavity. This device is based on the fabrication by molecular beam epitaxy of well-controlled, atomically flat semiconductor layers at the nanometer scale. The basic ingredient to obtain high-finesse planar microcavities is the high reflectivity distributed Bragg reflector (DBR), the optical properties of which we will first describe before considering planar microcavities.

3.1.1 Distributed Bragg Reflectors

The distributed Bragg reflector is a well-known optical device [66, 67]. Contrary to metallic mirrors, DBRs only contain nonabsorbing dielectric materials and their performance fully relies on the optimization of light interferences inside the reflector layers. The full structure is not absorbing, the reflectivity being "distributed" across several interfaces over micrometers in the structure, and all the nonreflected intensity is finally transmitted on the other side of the mirror. Also contrasting with metals, DBRs usually reflect light only inside some specific wavelength ranges but very high reflectivities at almost arbitrary wavelengths can be achieved by a proper choice of the building materials and layer sequence. Moreover, this wavelength depends on the angle of incidence and thus DBRs can be viewed as dispersive mirrors. These DBRs have been more recently included in new structures devoted to the confinement of light at the wavelength scale, giving rise to the famous optical microcavities [68]. We present in this section some basic properties of optical DBRs and microcavities that play an essential role in the Raman optical resonances described in Sects. 4 and 5 of this Chapter.

Moreover, based on the analogy between the propagation equation for light and sound in solids, it has been previously suggested that the well-known concept of optical DBRs could be easily transposed to design acoustic DBRs [69]. We demonstrated recently [50] that acoustic microcavities are also realistic, powerful devices to manipulate acoustic waves at the acoustic wavelength scale, i.e., at the nanometer range for THz ultrasound. The second aim of this subsection is thus to recall some well-known properties of optical DBRs and microcavities in a form that emphasize analogies with the acoustic superlattices and microcavities [4].

The most common DBR is a periodic multilayer quarter-wavelength stack made of the periodic alternance of two dielectric compounds with contrasting refractive index and working at normal incidence. It is well known that the largest stop band around a given wavelength λ is then obtained when the width of each building layer equals $\lambda/4$. The main parameter governing the performance of a DBR is the refractive index contrast between the individual layer refractive indexes n_1 and n_2 ($n_1 > n_2$). The width of the stop band and the corresponding reflectivity approximately equal:

$$\frac{\Delta\lambda}{\lambda} = 2\frac{n_1 - n_2}{\pi\sqrt{n_1 n_2}}, \tag{1}$$

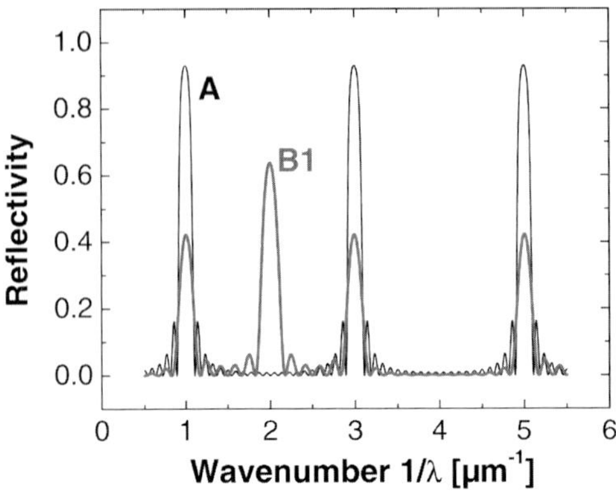

Fig. 1. Reflectivity of Bragg reflectors centered at 1 μm with $(\lambda/4, \lambda/4)$ (A) and $(3\lambda/4, \lambda/4)$ (B1) stacks, respectively

and:

$$R = 1 - 4\frac{n_{\text{ext}}}{n_{\text{s}}}\left(\frac{n_2}{n_1}\right)^{2N}, \tag{2}$$

where n_{ext} and n_{s} are the refractive indices at the top and the bottom of the DBR, respectively. The number of periods N is assumed to be large enough such that $\left(\frac{n_2}{n_1}\right)^N \ll 1$. For GaAs/AlAs DBRs, the typical values for n_1 and n_2 are 3.6 and 3.0 in the near infrared ($\lambda \sim 0.9\,\mu\text{m}$) and the relative stop bandwidth amounts to 0.116. Assuming a DBR grown epitaxially on a GaAs substrate, typical reflectivities are 0.97 for $N = 10$ and 0.999 for $N = 20$. The reflectivity of a DBR with 10.5 $(\lambda/4, \lambda/4)$ periods sandwiched between semi-infinite GaAs layers on both sides is shown in Fig. 1. In this figure and other illustrations of Sect. 3.1, we choose because of practical interest a reference wavelength of 1 μm. However, since the structure is also defined in terms of λ, the plots are basically in dimensionless units and thus the same picture applies for any choice of the reference wavelength.

The quarter-wavelength stack actually is one particular example of DBR. An illuminating method to describe the properties of a given DBR at any wavelength is to solve the propagation equation of light in an infinite superlattice. The corresponding light dispersion $\omega(q)$ satisfies the following equation:

$$\cos(qd) = \cos(q_1 d_1)\cos(q_2 d_2) - \frac{1}{2}\left(\frac{n_1}{n_2} + \frac{n_2}{n_2}\right)\sin(q_1 d_1)\sin(q_2 d_2)\,, \tag{3}$$

in which $d = d_1 + d_2$ is the period of the superlattice, and $q_1 = \omega n_1/c$ and $q_2 \omega n_2/c$ are the local wavevectors at frequency ω in the constituting layers 1 and 2, respectively. Note that (3) gives directly q as a function of ω,

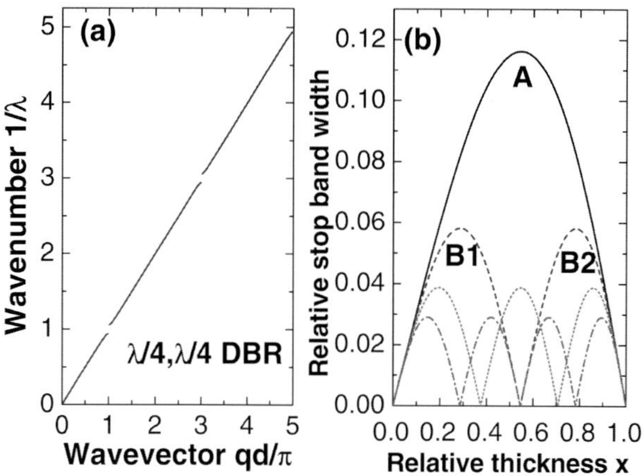

Fig. 2. (**a**) Photon dispersion in a $(\lambda/4, \lambda/4)$ DBR showing the gaps at $1/\lambda = 1$ and 3, as shown in curve A of Fig. 1. (**b**) Variation of the stop bandwidths, normalized to center energies, as a function of the relative thickness $x = d_2/(d_1 + d_2)$ of the constituting layers. The *labels* A, B1 and B2 correspond to $(\lambda/4, \lambda/4)$, $(3\lambda/4, \lambda/4)$ and $(\lambda/4, 3\lambda/4)$ stacks, respectively. The *solid*, *dashed*, *dotted* and *dash-dotted curves* correspond to $p = 1, 2, 3$ and 4, respectively

not ω vs. q. For acoustic phonons one gets a similar equation, given by Equation (3.5) of [4]. The dispersion curve corresponding to (3) is shown in Fig. 2a in the extended Brillouin zone scheme for a specific choice of the layer thicknesses. Bandgaps appear at any integer multiple p of the lowest gap:

$$\omega_p = \frac{p\pi c}{n_1 d_1 + n_2 d_2}, \tag{4}$$

or

$$\lambda_p = \frac{2(n_1 d_1 + n_2 d_2)}{p}. \tag{5}$$

These gaps are the stop bands obtained in the above optical description. Assuming a small amplitude of refractive index modulation, an approximate expression for the bandgaps can be derived analytically:

$$\frac{\Delta\omega_p}{\omega_p} = 2\frac{n_2 - n_1}{p\pi\sqrt{n_1 n_2}} \sin\left(p\pi \frac{(1-x)n_1}{(1-x)n_1 + x n_2}\right), \tag{6}$$

where $x = d_2/(d_1 + d_2)$. The variation of $\Delta\omega_p/\omega_p$ on x is shown in Fig. 2b for $p = 1$–4.

For $p = 1$ (solid curve in Fig. 2b), we recover the well-known result that the widest stop band (A) is obtained when $(1-x)n_1 = x n_2$, i.e., for quarter-wavelength layers. For $p = 2$ (dashed curve), there are two optimum values for x (B1 and B2), corresponding to $(\lambda/4, 3\lambda/4)$ and $(3\lambda/4, \lambda/4)$ stacks, respectively. Note that for quarter-wavelength layers the gaps with even index (shown for $p = 2$ and $p = 4$ in Fig. 2b) vanish exactly. This can also be verified from curve A in Fig. 1, where only one of every two reflectivity stop bands appear. The DBR reflectivity in the case B1, also shown in Fig. 1, displays the $p = 1$–3 gaps but is zero for $p = 4$ (as expected from the $p = 4$ dot-dashed curve in Fig. 2b).

In a real DBR structure with a finite number of periods, the residual transmission across the device can be understood as a tunneling effect across the corresponding bandgap of the infinite structure and the magnitude of the transmission T (or the reflectivity $R = 1 - T$) directly reflects the magnitude of the evanescent wavevector within the gap. At the center of the $p = 1$ gap, and assuming an optimized stack, this evanescent wavevector is maximum and can be obtained from the following relation:

$$\cos(qd) = -\frac{1}{2}\left(\frac{n_1}{n_2} + \frac{n_2}{n_1}\right) < -1\,. \tag{7}$$

The amplitude of the optical field at the Bragg energy thus decays in the Bragg reflector over a finite number of periods N_d equal to:

$$N_d = 1/\ln(\frac{n_1}{n_2})\,, \tag{8}$$

i.e., ~ 5.5 for the GaAs/AlAs structure. The reflectivity for a finite structure then reads:

$$R = 1 - \exp(-2N/N_d) = 1 - \left(\frac{n_2}{n_1}\right)^{2N}\,. \tag{9}$$

Another very important property of Bragg mirrors is the angular dependence of the reflectivity, which leads to two different phenomena: firstly, the wavelength at the stop-band center varies as a function of the angle of incidence, and secondly the reflectivity at a given wavelength can only be large within a finite acceptance angle. These two properties reflect the fact that the quantity $\omega \cos\theta$ that satisfies the condition for Bragg reflection remains a constant in the material when the external angle θ_0 is varied. This gives the following angular dispersion for the Bragg modes:

$$\omega_p(\theta_0) = \frac{\omega_p(0)}{\cos(\theta_0 n_0/n_{\mathrm{eff}})}\,, \tag{10}$$

where n_0 is the outside refractive index, and n_{eff} is an effective index for the multilayer. For the specific case of a $(\lambda/4, \lambda/4)$ stacking we note that $n_{\mathrm{eff}} = \sqrt{n_1 n_2}$.

Taking into account the energy width of the stop band, this expression leads to the following acceptance angle at the Bragg energy ω_p:

$$\cos(\theta_0 n_0/\sqrt{n_1 n_2}) > 1 - \frac{n_1 - n_2}{p\pi\sqrt{n_1 n_2}}\,. \tag{11}$$

With GaAs/AlAs parameters, this corresponds to an angular acceptance of 20° for $p = 1$ and $n_0 = \sqrt{n_1 n_2}$. On the other hand, using a reasonable angle of incidence of 50° in vacuum allows an increase in the energy of the resonance by 3.6 %, i.e., 50 meV at 1.4 eV.

3.1.2 Fabry–Perot Resonators

A planar microcavity consists of a thin layer of semiconductor inserted between two DBRs, leading to a monolithic single-mode, or few-modes, high-finesse Fabry–Perot (FP) resonator. This is a particular case of the more general problem of an impurity mode in a one-dimensional periodic system [70]. If we insert a layer with a different thickness and/or a different refractive index inside a DBR, local modes are expected to appear in the stop bands. Considering again the case of GaAs/AlAs DBRs with a GaAs layer of different thickness, it can be easily shown that such local modes move across the stop bands when the defect thickness is varied [70].

A very particular situation appears when the defect thickness equals $m\lambda/2$, with m an integer number. This is the only case where the standing wave associated to the local mode satisfies the boundary conditions for a Fabry–Perot resonance for any interface in the structure, leading to a maximum enhancement of the light field confined in the defect layer. This new situation is described as a microcavity, even though there is no fundamental difference from a standard Fabry–Perot resonator, except in the use of optimized Bragg reflectors instead of normal mirrors. Structures with thicknesses slightly off this optimum situation actually do not display qualitative changes in their optical properties but only some decrease of their performance.

Before describing the specific properties of semiconductor microcavities, let us first recall some basic results for standard planar Fabry–Perot structures. Confocal FP resonators with concave mirrors also exist, which provide a three-dimensional confinement of light. As we are not aware of corresponding constructs in the context of photons or phonons in semiconductors, we only consider here a FP with one-dimensional confinement properties made of a thin layer of thickness d and refractive index n surrounded by planar mirrors of high reflectivity R. R is taken independent of the wavelength λ and the angle of incidence θ. For simplicity we also assume the same reflectivity on each side of the structure. Optical modes are confined in the cavity at wavelengths $\lambda_p = 2nd\cos\theta/p$. The reflectivity pattern of a FP cavity, shown in Fig. 3 is then given by the Airy function A [71]:

$$R(\lambda,\theta) = 1 - A(\alpha) = 1 - \frac{1}{1+\mu\sin(2\alpha)} = \frac{\mu\sin(2\alpha)}{1+\mu\sin(2\alpha)}\,, \tag{12}$$

with $\mu = 4R/(1-R)^2$ and $\alpha(\lambda,\theta) = 2\pi nd\cos\theta/\lambda$.

This reflectivity pattern displays narrow dips at the confined energies. A fundamental number to describe a cavity is the finesse F that relates the separation between successive modes and the width of the dips:

$$F = \frac{\pi\sqrt{R}}{1-R} . \tag{13}$$

This quantity describes the performance of the FP as a passive optical device, i.e., when one is only interested in optical fields outside the cavity. It gives the wavelength resolution $\lambda/\Delta\lambda$ when the FP is used as a high-resolution spectrometer. It also gives the angular acceptance of a mode at a given wavelength.

A second important property of a Fabry–Perot resonator is the strong enhancement of the optical field inside the cavity and the strong modification of the light emission by a dipole located inside the cavity [72]. The field inside the cavity can be obtained in the same way as that used for the derivation of the Airy function for the reflectivity. The maximum *intensity* enhancement is given by:

$$Q = \frac{4\sqrt{R}}{1-R} , \tag{14}$$

which defines the "quality factor" of the structure. The average enhancement inside the whole cavity then equals $Q/2$. For the typical reflectivities considered previously (0.97 and 0.999), the corresponding enhancement of the optical intensities at the antinodes in the FP cavity then reach 133 and 4000, respectively. One should also note that at antiresonance, when the wavelength is between two FP modes, the intensity inside the cavity becomes, on the contrary, very strongly depressed. It reduces to $4\sqrt{R}(1-R)/(1-R)^2$, that is, 0.03 and 0.001 for the previous cases.

The same modification applies to the radiation pattern of an emitter located inside the FP cavity. If we consider the simple case of an emitter located at the central plane of the cavity, the modification of the intensity is given by [67]:

$$I/I_0 = \frac{4\sqrt{R}}{1-R} A(\alpha) . \tag{15}$$

At the maximum of the Airy function, the emitted intensity is enhanced by the same factor Q as the incident radiation is amplified inside the cavity. This is an illustration of the reversibility principle. The angular radiation pattern is strongly affected by the cavity, with narrow emission lobes around well-defined emission angles defined by the Airy function.

Another interesting parameter to introduce is the finite lifetime of the photon inside the cavity due to the finite transmission of the mirrors. This lifetime can be viewed as the average number of oscillations of the photon in the cavity before it escapes, times the period of such oscillations:

$$\tau_c = \frac{nd}{c(1-R)} \,. \tag{16}$$

3.1.3 Semiconductor Microcavities

All the above-described FP properties qualitatively apply to semiconductors with distributed Bragg reflectors with only moderate modifications. This is particularly true for the enhancement of the incident field intensity and of the radiation pattern in the cavity that are amplified by the same factor Q. The main differences concern the specific reflection properties of the Bragg mirrors: 1. the reflectivity is only high in some wavelength bands, the stop bands, at a given angle and conversely in some angular sectors at a given wavelength, 2. the reflectivity is distributed and the optical field penetrates into the mirrors over several periods, and 3. different refractive indices in the different constitutive layers have to be considered.

The third point leads to several technical complications because the different optical modes experience different refractive index patterns due to their finite spatial extension. Because the quantity $n\cos\theta$ does not display the same variation from layer to layer, a microcavity optimized at a given angle, e.g., at normal incidence, is no longer a perfect quarter-wavelength microcavity at other angles. However, these complications can be neglected to a good approximation when the refractive indices of the different materials in the device are not too different. One then uses a single effective refractive index n_{eff} for the cavity modes, which is usually defined from the angular variation of the energies deduced from exact calculations.

The effect of using Bragg reflectors is illustrated in Fig. 3 in which we compare the reflectivity from a microcavity with that of a Fabry–Perot resonator with the same cavity thickness d and the same mirror power reflectivity R. We used a refractive index $n_1 = 3.6$ for the cavity (GaAs), and $n_1 = 3.6$ and $n_2 = 3.0$ (AlAs) for the mirrors. We also chose for simplicity all the thicknesses so as to achieve the lowest-energy cavity mode at $1.0\,\mu$m. The effect of the limited stop bands of Bragg reflectors is very clear in the figure: the reflectivity is only high around some of the FP modes, the modes with even index being absent because the corresponding Fourier components of the modulation are vanishing in $(\lambda/4, \lambda/4)$ stacks (see Fig. 2). A more striking effect appears close to the cavity dips. The linewidth of the modes is significantly smaller in microcavities as compared to FP resonators, with a value of finesse of the order of 250 instead of 43, though the power reflectivity is exactly the same. The origin of this difference is more subtle and reflects the fact that the Bragg mirrors are distributed mirrors. As a consequence, the phase of the reflection at the cavity boundaries displays significant variations around the cavity-mode energy [71] where it vanishes for a well-designed

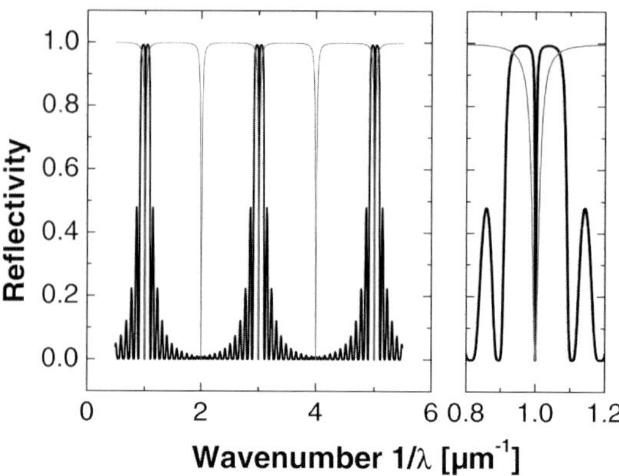

Fig. 3. Reflectivity of a $(\lambda/4, \lambda/4)$ Bragg microcavity centered at 1 μm (*thick solid curve*), compared to a Fabry–Perot with standard mirrors with the same reflectivity (*thinner curve*). The *right panel* is a detail of the left one around 1 μm^{-1}

structure with coincident cavity and DBR energies. This phase varies linearly around this energy with a slope proportional to the penetration depth in the mirror L_{DBR}. The width of the microcavity mode in wave numbers $(1/\lambda)$ is reduced accordingly and can be written as:

$$\delta = \frac{(1-R)}{2\pi nc(d+L_{\mathrm{DBR}})} . \tag{17}$$

When compared to the mode separation, an effective finesse is obtained that equals the FP finesse times the ratio $(d+L_{\mathrm{DBR}})/d$. The latter is of the order of 5 in the structure being considered. We recall that, in a FP resonator, the finesse (13) and the intensity enhancement (14) only differ by a numerical factor and display the same dependence with the physical parameters. This is no longer true in a DBR cavity, due to the penetration of the field in the distributed mirrors [71].

We show in Fig. 4 the spatial variation of the optical intensity of the lowest-energy mode in the same cavity. This mode extends over a finite width well outside the cavity and the enhancement at the maximum amounts to 50, corresponding to the quality factor of the structure. This property has deep implications in the context of "strong coupling", where one cannot neglect the modification of the internal emitter properties due to the cavity, as we have assumed in the previous subsection. Such alteration takes place when the optical-mode density becomes sufficiently large at the emitter location. In a DBR microcavity, the effective size of the mode L_{eff} cannot be reduced arbitrarily, giving rise to some limitations in the strong coupling performance of such devices. In this subsection, we mainly focus on the enhancement of the Raman-scattering efficiency due to the modification of the optical properties and we are not concerned with strong coupling, which will be

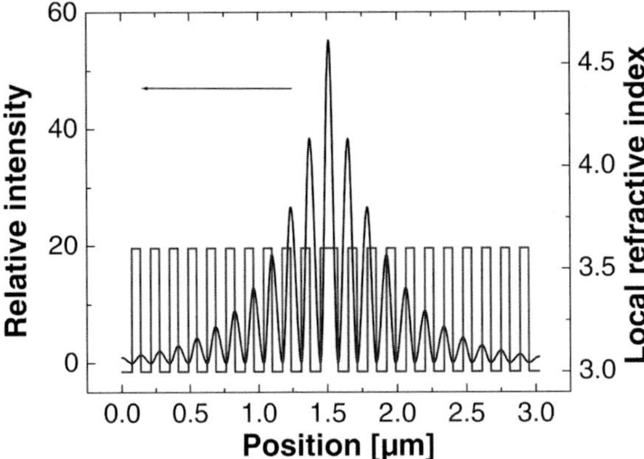

Fig. 4. Variations along the stack axis of the intensity of the optical field confined in a microcavity. The corresponding variations of the refractive index are shown with *solid thin lines*

treated in Sect. 6. The main impact of the distributed reflectivity in this context is to reduce the ultimate enhancement at the center of the cavity and the spatial selectivity of this enhancement. The volume of the sample in which the enhancement is significant not only contains the material at the center of the cavity but includes several periods of the mirrors as well. Moreover, the spatial optical field pattern displays an interesting specific feature depending on whether the cavity refractive index is higher or lower than that in the first layer of each DBR. Assuming again the simple case of a device made of only two materials, GaAs and AlAs, the displacement pattern displays either minima or maxima at the cavity edge. In the usual case, where a large optical field is requested at the center of the cavity, $\lambda/2$ cavities satisfy well this requirement with small-index cavities (e.g., AlAs), but λ cavities have to be considered if the spacer material (e.g., GaAs) is the larger-index material [73].

Let us finally mention that DBR microcavities display strong limitations when multimode cavities are needed. Bragg reflectors indeed are excellent mirrors in some very limited energy ranges and it is usually impossible to realize a good cavity at two different arbitrary wavelengths. Some alternative strategies have been developed. The first idea was to take advantage of the stop-band edge minima, at which the optical field displays some modifications as well [74]. It has been suggested that such edge minima could be used, beside the cavity modes, to enhance the optical signal in some devices as, e.g., for second-harmonic generation. We verified that this idea is not effective in the context of optically enhanced Raman scattering and that these side minima can even lead to experimental errors when determining the optimum resonance condition. The second idea is to use mirrors with more complex multilayer stacks instead of two quarter-wavelength layers and to design them

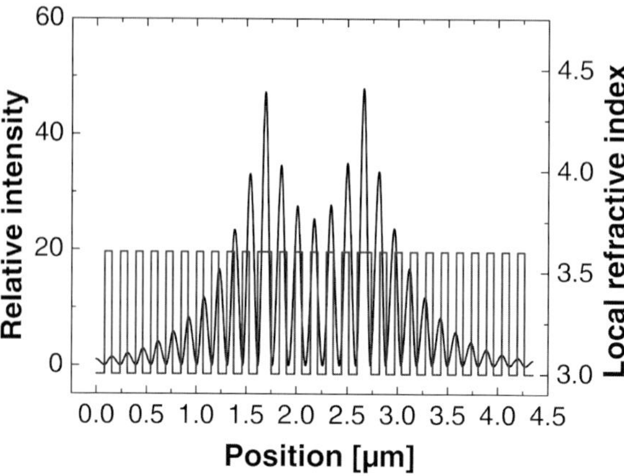

Fig. 5. Same as Fig. 4 with a double-cavity structure

to display high reflectance in broader bands or at two completely different arbitrary wavelengths [75]. The third solution is to couple several cavities through intermediate low-reflection mirrors [76]. We will now describe the properties of these latter coupled microcavity devices.

3.1.4 Coupled Microcavities

The confined optical mode in a microcavity can be considered as the building block for multiple microcavities as confined electron levels in quantum wells can be used to build double quantum wells or superlattices. The only difference lies in the fact that photon barriers are obtained from Bragg mirrors, while tunnel barriers from a compound with a larger bandgap are used in the context of electrons. When two identical photon cavities become coupled through an intermediate mirror with a finite reflectivity $R_{\rm int}$, two coupled modes appear with either a symmetric or an antisymmetric optical pattern. The spatial distribution of the optical intensity in a double cavity based on the parameters of the microcavity shown in Fig. 4 and a central mirror with $N_{\rm int} = 5$ GaAs/AlAs layers is shown in Fig. 5.

The separation between these two modes (in wave numbers) is given by:

$$\Delta = \frac{\sqrt{1 - R_{\rm int}}}{2\pi nc(d + L_{\rm DBR})}, \tag{18}$$

which depends on the transmission through the central mirror and the effective width of the confined mode in the single cavity $d + L_{\rm DBR}$. The reflectivity pattern around the first cavity doublet is shown in Fig. 6 for $N_{\rm int} = 5$ as well as variations of this quantity with the number of periods in the central mirror.

The extension of such ideas to multiple coupled cavities has been recently introduced to build one-dimensional photonic crystals, and coupled

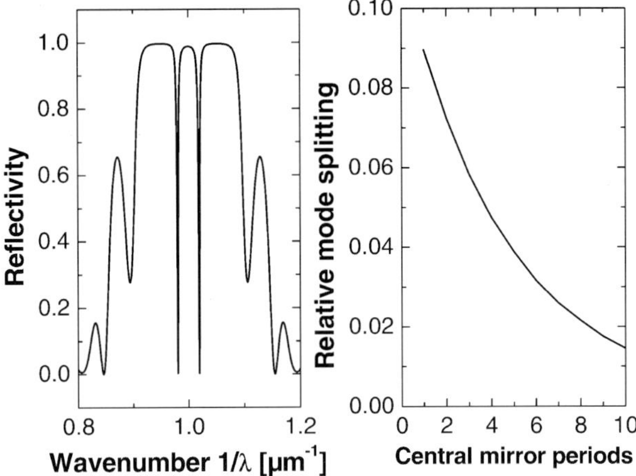

Fig. 6. *Left* Reflectivity of a double cavity around the doublet mode. *Right*: variation of the doublet splitting as a function of the central mirror design

cavities with shifted confined energies have been suggested to mimic electron Bloch oscillation with photons [77–79]. Similar ideas in the context of acoustic phonons have been discussed in [80]. To summarize, optical microcavities are a very powerful and versatile tool to manipulate optical fields in semiconductors. We will now focus on the methods to take advantage of these properties in the context of Raman scattering.

3.2 Basics of Raman Amplification by Optical Confinement

Raman scattering is one of the dominant processes for the interaction of light with matter. Contrary to absorption or emission where one single photon is created or annihilated in exchange of a quantum of vibration or of any other dipole-allowed excitation in the sample, Raman scattering involves two photons with two different energies. During the Raman process, the incident photon with energy E_{i} is annihilated and a new photon with a different energy E_{s} is created. Depending on whether E_{s} is smaller or larger than E_{i}, this corresponds to the creation (Stokes process) or annihilation (anti-Stokes) of a quantum of energy in the sample. As Stokes and anti-Stokes processes are basically similar, we will continue the discussion assuming a Stokes process. The latter does not require that the system be in an excited state before the inelastic scattering, and is also usually more intense. The fundamentals of light scattering have been described at length in previous volumes of this series [1–3]. Let us recall here that Raman scattering is a second-order process in the light–matter interaction and is governed by squared matrix elements in the form of:

$$\sigma \propto \left| \sum_b \frac{\langle f\,|H_{\mathrm{EL}}|\,b\rangle\langle b\,|H_{\mathrm{EL}}|\,i\rangle}{E_b - E_{\mathrm{i}} - \hbar\omega_{\mathrm{i}}} \right|^2 , \tag{19}$$

where $|i\rangle$, $|b\rangle$, and $|f\rangle$ are initial, intermediate and final states of the scattering process, respectively, with energies E_{i}, E_b, and E_{f}, and the sum is performed over all possible intermediate states. These are many-body states of the material system, including for instance both electronic and vibrational degrees of freedom. As in any second-order process, the intermediate state is a virtual state and the energy is only conserved between the initial and final state. Wavevector, on the other hand, is conserved at any step as it is an intrinsic property of the matrix elements in a system with translational invariance (i.e., a crystal). H_{EL} is the Hamiltonian of the light–matter interaction that is treated as a perturbation within the dipolar approximation. Note that the above expression relies on the existence of a weak coupling between light and matter, an approximation that is no longer valid in some specific situations in which the concept of a polariton must be introduced. These are strongly coupled excitations described as a mixed entity involving both light and matter. We will discuss in Sect. 6 the impact of polaritonic strong coupling in the context of Raman scattering. In this introductory part we restrict ourselves to the perturbative approximation.

A second step in the treatment of (19) is to separate the electronic and vibrational coordinates in the above expression. This is done by using the adiabatic approximation and the Raman-scattering process is then obtained from a new perturbative expression as:

$$\sigma \propto \left| \sum_{a,b} \frac{\langle f\,|H_{\mathrm{EL}}|\,b\rangle\langle b, n+1\,|H_{\mathrm{EP}}|\,a,n\rangle\langle a\,|H_{\mathrm{EL}}|\,i\rangle}{(E_a - E_{\mathrm{i}} - \hbar\omega_{\mathrm{i}})\,(E_{\mathrm{f}} - E_b - \hbar\omega_{\mathrm{s}})} \right|^2 . \tag{20}$$

Here H_{EP} is the electron–phonon interaction Hamiltonian. $|a,n\rangle$ and $|b,n+1\rangle$ are the weakly coupled virtual intermediate electron–phonon states in which a and b describe the electronic part and n and $n+1$ represent the number of phonons. We have not included the phonon numbers in the external (light–matter) matrix elements as they are not affected, in this approximation, by these steps of the scattering process in a crystal. The denominator now contains two energy factors involving the incident and scattered photon energies, respectively. This general expression is the basic tool for discussing the Raman-scattering process. Based on (20), Raman scattering is usually described as a process that involves the destruction of an incident photon and creation of an exciton ($\langle a\,|H_{\mathrm{EL}}|\,i\rangle$), the subsequent scattering of this exciton to another state with emission of a phonon ($\langle b, n+1\,|H_{\mathrm{EP}}|\,a,n\rangle$), and finally the recombination of the latter with emission of the (Stokes-shifted) photon

($\langle f\,|H_{\rm EL}|\,b\rangle$) [3]. One has to keep in mind that all the steps involved in the above sum are virtual processes and that neither the incident nor the scattered photons have to coincide with a real transition in the material. They can be, and often are, below the bandgap. Moreover, there is no need, and usually no possibility, for the phonon in the intermediate step to fulfill both energy and wavevector conservation required for the transition of the exciton from state $|a,n\rangle$ to $|b,n+1\rangle$. Finally, in a virtual process, time ordering of the intermediate events is arbitrary as causality does not apply to the intermediate states. The process schematically described actually illustrates an exceptional situation that can lead to a double resonance when both the incident and scattered photons coincide with *real* electronic transitions [81, 82]. In this case, the two denominators in the Raman-scattering tensor exactly vanish when finite broadening of the states are not included and this single term completely dominates in the sum over intermediate states. This example is an extreme situation of a more general one in which one of the photons coincides with a dipole-active real transition. This situation is described as "resonant Raman scattering" (RRS) and has been, and continues to be, an extensively active field of research [3].

We are concerned in this Chapter with a completely different mechanism of resonance: that is, optical resonant Raman scattering. Such resonance only modifies the external matrix elements of the Raman tensor, involving the light–matter interaction (e.g., $\langle f\,|H_{\rm EL}|\,b\rangle$ in (20)). In the dipolar approximation, these matrix elements are proportional either to the incident or scattered light-field amplitude. Thus, the mechanisms of spatial amplification of light in microcavities can be used to enhance the Raman-scattering amplitude through the modification of these matrix elements. In the adiabatic approximation, this modification is completely independent of the electron–phonon interaction included in the central matrix element, i.e., standard (electronic) resonant Raman scattering and optical resonant Raman scattering can be treated separately. Moreover, the dipole matrix elements can be assumed to remain unaffected by the confinement of light and only reflect the local amplitude of the light-field as in the absence of confinement. Based on all these properties, the scattered amplitude can be locally described as the product of a microscopic factor, independent of the field amplitude, but reflecting the Raman mechanisms involved, and of two other factors reflecting the amplitude of the incident and scattered optical field, respectively. We have seen in Sect. 3.1 that the microcavity confinement modifies both the amplitude of an optical field incident on the device and the emission pattern of an emitter located inside the cavity. We have also seen that the two effects are governed by the same figure of merit, because of the reversibility principle. In fact, the same amplification Q-factor (given for a FP resonator by (14)) applies for the intensity of the incident field when measured at the center of the cavity, and for the emission intensity rate, when detected at the optimum collection direction. Thus, as follows from (20), the maximum scattering efficiency amplification in a perfect double optical resonance process (DOR, i.e., a process in which *both* incident *and* scattered photons are tuned to an optical cavity mode) equals Q^2. Or, equivalently, it is proportional to the fourth power of the light-field cavity enhancement. Consequently, if an experimental configuration can be realized where the amplification of the incident field is combined with the amplification of the scattered emission, very large enhancement can be expected for the external Raman efficiency.

The amplification power of a microcavity can be tested by the simultaneous measurement of the efficiency in the absence and in the presence of cavity enhancement. Such tests involve the comparison of separate experiments and samples, and thus uncertainties are unavoidable. They have, however, been performed and will be described in Sect. 4.3. On the other hand, what one can easily analyze, both experimentally and theoretically, is how the enhancement is modified as a function of the detuning of the microcavity close to the double optical resonance condition. Such detuning can be done either by changing the photon energy in a tunable laser or by displacing the spot position in a tapered sample. However, we recall that in a Raman-scattering process photons at two *different* energies are involved: the incident energy $E_{\rm i}$ and the scattered energy $E_{\rm s}$ are separated by the energy of an internal excitation of the sample. Thus, in order to perform double optical resonance experiments a configuration is required in which two confined modes with a tunable separation exist in the same microcavity. This has been done in two different ways, namely by angle tuning in a single-mode cavity, and by specifically designing double cavities. These two strategies will be described, respectively, in Sects. 3.4 and 3.5 below.

3.3 Samples and Their Characterization

The microcavity samples used for the different experiments reviewed in this Chapter will be described in the corresponding sections. However, we would like at this introductory stage to give some general remarks that will be useful for the discussion to follow.

The microcavities we will be talking about are grown using semiconductor materials with standard epitaxial techniques such as, e.g., molecular beam epitaxy (MBE). Typically these consist of a spacer, with the "active" material embedded in it, and surrounded by DBRs made also of semiconductor materials that are MBE compatible (i.e., lattice matched) with the active part. Probably the best-quality samples, in terms of growth, have been made of III–V semiconductors, involving GaAs, AlAs, $Ga_{1-x}Al_xAs$ alloys, and also $Ga_{1-x}In_xAs$. However, also very good samples are available from II–VI materials, typically CdTe and its solid solutions with Mg and Mn, $Cd_{1-x}Mn_xTe$ and $Cd_{1-x}Mg_xTe$. The growth of the latter are usually not as well under control as their III–V counterparts, and sometimes the cavities display localized exciton states that can complicate the interpretation of the experiments. Notwithstanding this limitation, II–VI materials have relatively larger oscillator strengths (something relevant to the "strong-coupling" regime) and also display larger optical-phonon Raman efficiencies. In general, the choice of the materials making the DBRs and spacer has to be made such that they are transparent in the energy range where the active part needs to be studied. For example, in samples in which the spacer is GaAs, and where the "active" part consists of $Ga_{0.85}In_{0.15}As$ quantum wells, the DBRs are made of GaAs/AlAs multilayers or, if the spacer as a whole is to be investigated, e.g., with $Ga_{0.8}Al_{0.2}As$/AlAs. Another obvious but important criterion for the choice of material, when phonon Raman scattering is to be performed, is to select the DBRs materials so that their vibrations do not overlap with those of the active part.

It is also usually the case that the "active" material embedded in the spacer possesses dipole-active optical transitions in the energy range where the cavity mode falls. These correspond, e.g., to the lower energy gap related exciton states of a quantum well (QW), or a multiple quantum well (MQW), or the spacer as a whole. The domain of "strong coupling" describes the situation in which exciton and cavity modes are degenerate, or close to degeneracy. In this case, due to continuous absorption and re-emission before tunneling out of the cavity, exciton and photon cannot be distinguished leading to coupled excitations called "cavity-polaritons". On the other hand, most of the Raman experiments we will describe here were performed with the cavity modes well below any excitonic transition in the spacer. In this latter case, the cavity mode is considered to be "purely optical" (i.e., with no exciton component).

For the purpose of evidencing the coupling between cavity-photons and excitons, or to adjust the energy of the cavity modes with the available lasers for resonant excitation in a Raman experiment, a mechanism of tunability of one or various of the involved energies is required. One way to tune the energy of the cavity modes with respect to the exciton states is by varying the temperature. The exciton energy usually increases as the temperature is lowered. Concomitant with this increase, the index of refraction at a given energy decreases because the resonant gaps move farther away. Since the cavity-resonance condition is given by $d = m\lambda/2n$, with d the spacer thickness, n the spacer refractive index, and m an integer number, it turns out that the resonant wavelength decreases as the temperature is lowered. Consequently, the cavity-mode energy increases as the temperature is reduced, i.e., it shifts in the same direction as the exciton energy does. However, cavity-mode and exciton energies do not vary in the same way, and consequently some tunability is possible by controlling the temperature. In any case, this is not the most convenient way to control the energies involved, since relevant physical processes depend on temperature, and thus the ideal situation is to keep this as a free parameter. Another possibility is to change the angle, exploiting the different dispersion of the exciton and cavity-mode energies with the in plane wavevector $k_{\parallel}$. However, it is in many cases convenient to study the inplane dispersion of the polaritons, or of the excitations probed by Raman scattering, and thus an alternative way to set the energy of the cavity mode, independently of $k_{\|}$, is also desirable. The alternative, most extended and convenient way to change the energies involved is to grow the sample with a small taper, so that the layer thicknesses change with position in the wafer. This is quite easily done by exploiting the typical nonuniformity of the material deposition in an MBE chamber. In the most general case, *all* layers are tapered. Thus, by moving the laser spot on the sample surface the DBRs stop band and the cavity-mode energy move rigidly following the cavity thickness as $\lambda = 2nd/m$ (i.e., its energy *increases* with decreasing d). On the other hand, the exciton state of the QWs only depends slightly on d because of the related minor change in confinement energy. Typically, within a 2-inch wafer, cavity modes set at around 1.5 eV can be tuned as much as 100 meV by displacing the laser spot on the wafer surface without significant reduction of quality of the samples due to inhomogeneous broadening. In this case, the laser energy, temperature, and $k_{\|}$, remain as free parameters that can be set as required for the specific experiment.

In order to perform an experiment, a "cartography" of the cavity and exciton states on the wafer is required so as to be able to tune the laser with respect to the cavity mode and to define clearly what is the situation concerning the exciton energies involved. This is done in one of two ways, reflectivity or photoluminescence measurements. By photoluminescence experiments one detects the light that is emitted from the microcavity, and reaches the exterior through the cavity modes or through optical stop-band side oscillations. This emission is originated on the exciton states or on residual below-gap emission. For this experiment the excitation is typically done above the stop-band energy, but below the gaps of the materials making the DBRs. In Raman experiments performed in the transparency region of the spacer materials, the laser used for Raman excitation is not very efficient in generating photoluminescence. In these cases it is sometimes convenient to clearly define the energy of the cavity mode for resonant excitation, to excite at the same spot with a second above-gap laser. For example, in experiments performed using a Ti-sapphire laser on a GaAs cavity with $Ga_{0.8}Al_{0.2}As/AlAs$ DBRs, a HeNe laser can be used to evidence the spectral position of the cavity mode. Photoluminescence is indeed quite convenient as a characterization tool, because basically it comes with no cost together with the Raman spectra. However, it has two limitations that can be sometimes of concern. On the one hand, only the lowest-energy states can be probed by photoluminescence. On the other hand, the detected amplitude of these states is strongly influenced by the carrier population that is not necessarily thermalized and thus complicates the interpretation of the experiments. In these cases, reflectivity measurements that can be performed, e.g., with a conventional light source and a spectrometer, a Fourier transform infrared spectrometer or with a tunable laser, are the preferred choice.

3.4 Angle Tuning in One-Mode Cavities

As discussed above, high-finesse cavities, as required for maximum field enhancement, are obtained by the use of almost total reflectivity mirrors. Although this implies no limitation for luminescence studies, in which the excitation is made at energies above the cavity stop band, it poses a problem for Raman experiments where photons differing by the energy of the excitation under study must get into, or leave, the microcavity. The experiments reported by *Cairo* and coworkers using a piezoelectrically controlled cavity [36], commented on in Sect. 2, relied on a cavity-single-resonance condition: while the bottom mirror of the cavity was reflecting at both excitation and Stokes energies, the top mirror *only* confined the Stokes radiation. The mentioned stimulated Raman scattering studies on droplets [32–35], on the other hand, made use of relatively large cavity diameters (10–30 μm) and thus the high mode densities enabled the tuning of the laser frequency (or the droplet size) so that cavity double resonance was possible.

Typical planar semiconductor microcavities as the one depicted in Fig. 3, have only one cavity vertical mode within the stop band. Figure 3 displays the cavity reflectivity while the corresponding electric field distribution at normal incidence tuned to the cavity mode for that specific cavity is shown in Fig. 4. The aim is to exploit these spectrally dependent modifications of the photon-field amplitude and spatial distribution to produce enhanced Raman signals. In Raman-scattering experiments performed on the face normal to the growth z-axis of such a structure, both the incoming and outgoing photons (separated by the energy of the excitation under study, e.g., 36 meV for a GaAs optical phonon) must resonate with optical modes of the cavity in order to get into, or leave, the structure. For a planar semiconductor microcavity with a single vertical mode within each stop band, such as that depicted in Fig. 3, this can be accomplished by taking advantage of the continuum of inplane optical modes [42]. In fact, if ω_0 is the frequency of the cavity mode along z, and $k_\|$ the photon inplane wavevector, the mode dispersion is given by

$$\omega(k_\|) = \sqrt{\omega_0^2 + (ck_\|/n_{\text{eff}})^2}\,. \tag{21}$$

This is the k-space equivalent of the angular dependence of the DBR energies ω_p given by (10). An optical double resonance can hence be achieved by tuning the angle (with respect to z) between incoming and scattered rays [42]. For example, if light is collected along z, the required incidence angle θ_0 (measured in air) for Stokes scattering of an excitation of frequency ω_{ph} will be given approximately by

$$\theta_0 \approx n_{\text{eff}} \arccos\left(\frac{\omega_0}{\omega_0 + \omega_{\text{ph}}}\right)\,. \tag{22}$$

Here, n_{eff} is an effective refractive index for the cavity.

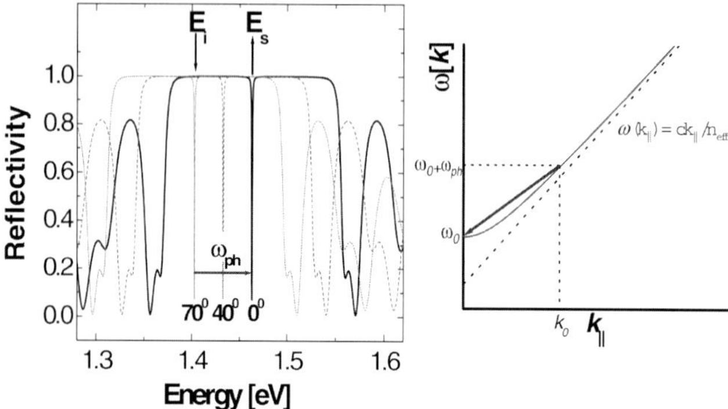

Fig. 7. Scheme of the double optical resonant angle tuning geometry in single optical-mode cavities. *Left*: Optical reflectivity for different angles with respect to the heterostructure's normal z. E_i and E_s indicate the incoming and scattered fields, respectively. *Right*: scheme of the cavity-mode inplane dispersion and of the relevant energies and wavevectors. ω_0, ω_ph and k_0 label the 0° cavity-mode energy, the phonon energy, and the laser inplane wavevector, respectively

These concepts are illustrated in Fig. 7. In the example shown light is incident at an angle defined by the inplane photon dispersion and the energy of the studied excitation (ω_ph), and the scattered radiation is collected along z. The left panel shows how the optical reflectivity depends on the angle with respect to z, and how this can be exploited to tune the incoming (E_i) and scattered (E_s) fields with optically confined modes. The right panel, on the other hand, represents the same idea but in k-space. The cavity-mode inplane dispersion given by (21) is shown, together with the energies and wavevectors relevant for a double optical resonant Raman process. Note that both Stokes and anti-Stokes (i.e., emission or absorption of an excitation by the incoming photon, respectively) configurations are feasible, depending on whether the excitation or the scattered light angle is the largest. The example shown corresponds to a Stokes process. For anti-Stokes scattering the roles of E_i and E_s have simply to be interchanged.

3.5 Two-Mode Cavities

We have shown in the previous section that cavity-enhanced Raman scattering can be obtained through double optical resonances in standard microcavities designed with only one confined optical mode [42]. In such situations, the tuning of both laser and Stokes (nondegenerate) photons can be accomplished by tuning the incidence and scattered angles with respect to the growth axis (z). This scattering geometry is appropriate for the study of small-energy excitations, but becomes cumbersome above $\sim 35\,\mathrm{meV}$ (e.g., for optical phonons) where incidence angles typically larger than $\sim 50^\circ$ have to be used. On the other hand, note that in principle *both* the excitation and

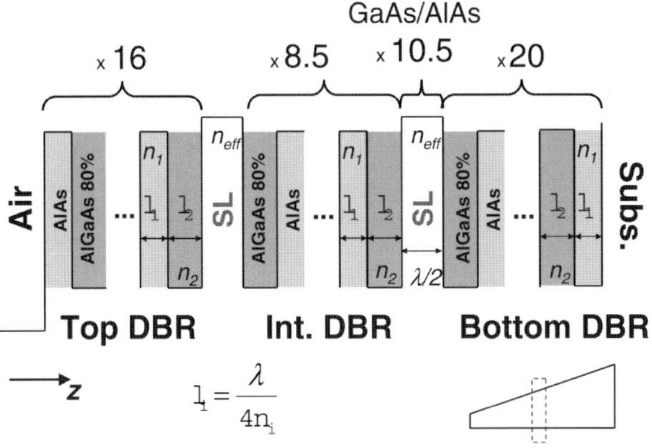

Fig. 8. Scheme of the double cavity consisting of two $\lambda/2$ spacers made by 10.5-period GaAs(8.5 nm)/AlAs(4.4 nm) MQWs, with 20 DBR $Ga_{0.8}Al_{0.2}As/AlAs$ pairs on the *bottom*, 16 on *top*, and 8.5 in *between*

collection directions may be varied according to (21). This opens up the possibility to measure the inplane dispersion of the excitations by changing both angles together. However, for the study of optical phonons or other high-energy excitations, in the single-mode cavity scheme discussed up to now the large angle required between incoming and scattered fields implies a strong limit for the range of inplane wavevectors that can be probed. We are thus going to describe here an alternative strategy for DOR Raman-scattering enhancement designed for almost exact backscattering geometries.

The idea relies on a double-cavity structure, as described conceptually in Sect. 3.1, specifically designed for optical-phonon DOR Raman-scattering experiments in a backscattering geometry. Double microcavities were originally proposed for dual-wavelength laser emission [83]. A coupled double cavity consists of two *identical* $\lambda/2$ spacers enclosed by top and bottom DBRs, and separated by a lower-reflectivity mirror. In Fig. 8 a scheme of a double cavity designed for the study of optical phonons in finite-size superlattices is presented. It has two $\lambda/2$ spacers made by 10.5-period GaAs(8.5 nm)/AlAs(4.4 nm) MQWs, with 20 DBR $Ga_{0.8}Al_{0.2}As/AlAs$ pairs on the bottom, 16 on top, and 8.5 in between. The choice of the alloy in the DBRs is defined, on the one hand, so that the DBRs are transparent at the MQWs exciton energy, and on the other hand because the longitudinal optical (LO) GaAs-like phonons of $Ga_{0.8}Al_{0.2}As$ are spectrally well separated from the MQWs confined GaAs-like vibrations. Like the simpler structure described in the previous section, these double cavities can be purposely grown with a slight taper so as to tune the cavity modes by displacing the spot position on the wafer.

In Fig. 9 the measured cavity modes of the structure depicted in Fig. 8 are shown. In the inset we display a typical reflectivity spectrum clearly show-

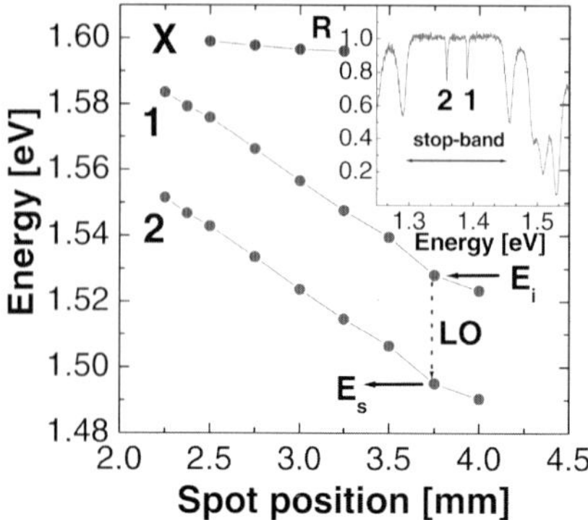

Fig. 9. Luminescence measurements at 77 K. X, 1, and 2 label the MQWs exciton, and the upper and lower cavity modes, respectively. The DOR Raman scattering by an LO phonon is also schematically shown. *Inset*: Room-temperature reflectivity displaying the two confined optical modes within the DBRs stop band

ing the two cavity modes. We also present in the main panel the variation of the energy of the cavity modes when displacing the laser spot position in the surface of the tapered structure. These positions have been deduced from luminescence measurements as explained above. Note how the two initially degenerate optical modes of the identical cavities are split by the mutual coupling. The splitting across the structure is almost constant, around 32–33 meV. The average energy, on the other hand, can be tuned over a large range. The square modulus of the field distribution (which is the same for the two modes if the two cavities are identical) displays maxima in the two cavity spacers, as illustrated in Fig. 10a. The structure constitutes a "photonic molecule" in the same sense as used for atoms. In fact, the two split cavity modes are completely equivalent to the bonding and antibonding states of, e.g., a hydrogen molecule. As for a H_2 molecule, the two modes differ in their parity (symmetric and antisymmetric) with respect to the center of the structure [84]. This is shown in Fig. 10b, where the physical reason for the difference in energy between the two cavity modes can also be understood: E_1 and E_2 distribute their amplitudes differently in the two materials. Thus, the mode with amplitude larger in the material with higher refractive index will be blueshifted with respect to that concentrated in the lower-index regions.

The splitting of the bonding and antibonding states of H_2 can be varied by changing the overlap of the electronic clouds corresponding to the individual atoms. Equivalently, the splitting between photonic states can be tuned by changing the cavity-mode field penetration in the neighboring cavity. This is accomplished by reducing or augmenting the reflectivity of the intermediate mirror, as shown in Fig. 6 (right). In such a way, the mode splitting can be tuned to the energy of the excitation under study. For the structure described here the splitting was set to be slightly smaller than the optical-phonon energy of the spacer MQWs. As schematized in Fig. 9, the Stokes-scattering

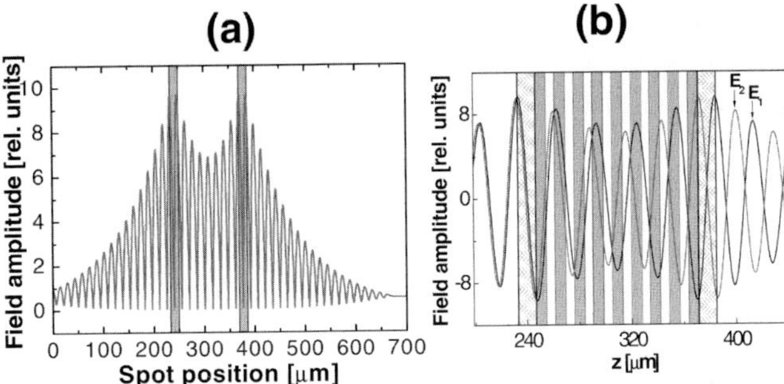

Fig. 10. Field distribution in the doublex-cavity structure. In (**a**) the square modulus of the fields (which is identical for the two confined modes) is shown. In (**b**) a detail of field amplitude at the double cavity and intermediate DBR region is shown. Note the different parity of the modes with respect to the center (mirror plane) of the structure, and the different distribution of amplitudes in the two DBR materials

DOR is attained by exciting and collecting the scattered light at the upper and lower cavity modes, respectively, using an almost backscattering geometry where incident and scattered photons are collinear [85]. A small-angle tuning can be used in addition to selectively amplify different regions of the spectra, in the same way as explained for the single-mode cavities. Moreover, an anti-Stokes experiment can also be performed by simply reversing the role of E_{i} and E_{s} in the scheme. Note that it is important that the two cavities be identical for achieving a DOR. Only in such a case are the two optical eigenmodes distributed equivalently in the two cavities, giving the necessary overlap between the laser and Stokes fields.

3.6 Models of Optical Resonant Amplification

In the previous sections we have described the main characteristics of optical resonant Raman scattering in microcavities, and we have presented two alternative strategies to attain such a condition in planar structures. In this section we will provide a more quantitative discussion of the Raman efficiency introducing a model to describe the modifications due to optical confinement in a planar microcavity. The model considers the two photon vertices of the Raman process as independent, that is, takes into account the presence of the cavity through its added effect on the excitation and emission steps. As discussed in Sect. 3.1, the enhancement of excitation and emission in a cavity are given by the same factor Q. Thus, it is in principle possible to evaluate the amplification of a Raman process because of coupling through a cavity mode in a cavity simply by evaluating one of these quantities. Because of simplicity, this is typically done by calculating the enhancement of the excitation-laser field at the position in the cavity where the scatterer resides [44]. However, in this section we will present a description that considers the excitation and emission separately. This picture provides a more complete perspective on optical properties of DBR cavities that is important to fully grasp the implications of using interference mirrors for light confinement. For this purpose, we will discuss the spectroscopical properties of dipoles in DBR cavities, i.e., their emission pattern and excitation-field spatial distribution. They are shown to strongly depend on photon frequency leading either to enhancement (for resonance conditions) or inhibition. This discussion is the equivalent, for a layered semiconductor system, of the analysis performed in Sect. 3.1 for Fabry–Perot resonators. A transfer-matrix formalism is used to evaluate the *exciting* field at the cavity spacer. Since this is a standard procedure, we refer the readers to the reference list for further details [66, 67]. The *emission* step of the scattering process, on the other hand, is modeled in terms of dipoles modified by the presence of the cavity. As we will briefly review in what follows, this is done by evaluating the electric field (direct plus reflected by the Bragg mirror) at the site of the emitting dipole [24–27, 73]. Standard methods of classical antenna theory based on a TE and a TM plane and evanescent-mode decomposition of the dipole emission are used to obtain these fields [86].

The basis for this theory was developed very early by *Sommerfeld* and others [87–89] in order to describe the radiation from an antenna in the presence of a partially conducting earth. It was later extended to treat the problem of spontaneous emission lifetime modifications by *Chance* and coworkers, including the problem of a molecule emitting near a partially reflecting surface and between two parallel mirrors [25–27]. A more recent analysis of single-mode and planar cavities, and the problem of spontaneous emission lifetime and strong coupling in cavities was given by *Abram* and *Oudar* [90]. The basic ideas can be understood starting from the classical equation of motion of a dipole (assumed to be a harmonically bound charge) [25–27]:

$$\ddot{\mu} + \Omega^2 \mu = (e^2/m) E_{\mathrm{R}} - \gamma_0 \dot{\mu} \,, \tag{23}$$

where Ω is the dipole oscillation frequency in the absence of damping, m and e are the effective mass and charge of the dipole, respectively, E_{R} is the electric field produced by the dipole and reflected back to the dipole, and γ_0 is the damping constant (inverse lifetime) for the dipole in an infinite medium. E_{R} oscillates at the same frequency as μ. Using the ansatz $\mu = \mu_0 \mathrm{e}^{\mathrm{i}\omega t}$, it is straightforward to see that

$$\omega = -\mathrm{i}\gamma_0/2 + \Omega(1 - \gamma_0^2/4\Omega^2 - e^2 E_{\mathrm{R}}/\mu_0 m \Omega^2)^{1/2} \,. \tag{24}$$

Assuming that the interference-induced corrections to the dipole oscillation are small, it follows from (24) that both the decay rate γ_0 and the real part of E_{R} contribute to a shift of the emission frequency, while the imaginary part of E_{R} modifies the decay rate in the presence of self-interference:

$$\gamma = \gamma_0 + (e^2/\mu_0 m \Omega) Im(E_{\mathrm{R}}) \,. \tag{25}$$

In the process of spontaneous emission, the electric field that is reflected back to the dipole (E_R) drives it and modifies its decay rate. Thus, the modifications of the emission introduced by the cavity walls can be taken into account through a relatively simple calculation of the spatial distribution of the field of a dipole in stratified media, a problem that is treated in many textbooks on electromagnetic theory [86]. This is done by writing the solutions to the wave equation in each medium in terms of superposition of TE and TM wave components (the boundary conditions at each interface lead to the well-known Fresnel reflection coefficients), and using the so-called Sommerfeld identity that expresses the spherical waves generated by the dipole in terms of cylindrical coordinates appropriate for the geometry of the layered system [86]. It is important in this procedure to consider both plane and evanescent waves. For the latter the wavevector is imaginary, defining a penetration depth. Thus, in homogeneous media these do not contribute to the far field. However, if close to the source there is another medium with a refractive index larger than that where the dipole resides, the evanescent waves that reach the boundary can later propagate freely as plane waves. These contributions are essential to treat the so-called leaky waves in stratified dielectrics that we will show below characterize the emission of dipoles in DBR structures at intermediate angles [45, 73].

Theoretical results using this dipole-emission model are illustrated with the example of a low-finesse asymmetric microcavity consisting of a $3\lambda/2$ GaAs/AlAs MQW grown on a *single* AlGaAs/AlAs DBR. The structure consists of a 36-period GaAs/AlAs (70 Å/30 Å) MQW deposited on 28 periods of quarter-wavelength $Al_{0.33}Ga_{0.67}As$/AlAs (617 Å/707 Å) layers acting as the DBR [41]. The sample is limited above by air, and below by GaAs. The total thickness of the MQW is $3\lambda/2$. We begin by describing the emission characteristics of dipoles located at different positions of the spacer MQWs, and emitting at an energy corresponding to the cavity mode along z (i.e., with wavelength λ). Figure 11a shows the calculated emission intensity as a function of azimuthal angle θ (with respect to the growth axis, z) for dipoles oriented parallel to the layers and located at a distance $d_0 = \lambda/2$ (top trace) and $d_0 = \lambda/4$ (bottom curve) from the MQW–DBR interface. The total emission integrated over the MQW thickness is also shown (thicker solid curve). All curves are normalized to the dipole emission in an infinite medium of equal refractive index. Note that in calculating the average emission the emitted intensities have been summed and not the complex fields, that is, incoherent radiation is assumed. This assumption is equivalent to considering that the oscillator phases are uncorrelated, and models a system in which intermediate phonon vibrations add an additional source of phase incoherence, even if the excitation of all dipoles at different layers is done coherently. Three regions can be distinguished in Fig. 11a: Along z and up to $\theta \approx 16°$, the radiation leaves the sample towards air. Between 16° and $\approx 65°$, total internal reflection inhibits the latter, but energy leaks to the DBR and GaAs substrate.

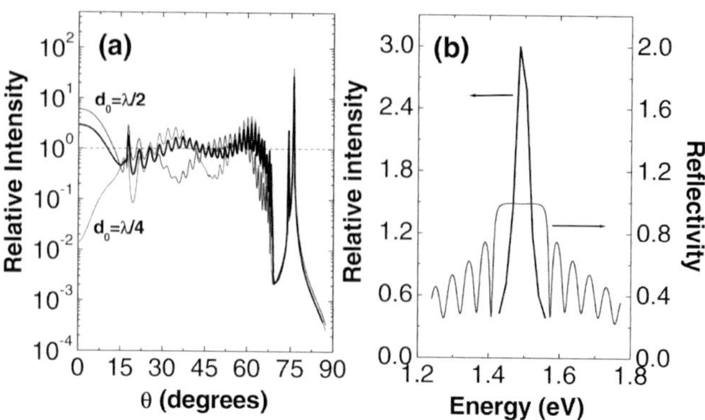

Fig. 11. (**a**) Calculated radiation pattern as a function of azimuthal angle for horizontal dipoles located at $d_0 = \lambda/2$ (*top curve*) and $d_0 = \lambda/4$ (*bottom*) from a DBR. The *thicker solid curve* corresponds to the emission integrated through the $3\lambda/2$ thickness of the MQW. (**b**) Normal incidence DBR reflectivity (*dashed*) and MQW-integrated z-axis emission intensity as a function of photon energy. The *curves* are given relative to the emission of a similar dipole in an infinite medium of equal refractive index

Finally, above $\approx 65°$ radiation can no longer leave the sample except through guided modes (peaks at $\approx 76°$).

We are interested here mainly in the emission along and close to z ($\theta \sim 0°$), where light is collected by the optics. According to Fig. 11a, this emission in the half-cavity can be enhanced up to a factor ≈ 6 or inhibited down to ≈ 0.01 depending on whether the dipole is located at a distance $d_0 = \lambda/2$ (antinode) or $d_0 = \lambda/4$ (node) from the DBR, respectively. Since there is no phase change on reflection at the MQW/DBR and MQW/air interfaces, these distances correspond to constructive or destructive interference, respectively, of the direct and reflected emitted light at the dipole location. We note that the factor of 6 enhancement is due to the presence of *both* MQW/DBR and MQW/ air interfaces. In fact, a dipole at a distance $d_0 = \lambda/2$ of the same DBR mirror but located on an infinite half-space only enhances its z-axis emission by a factor 2. The z-axis emission integrated over the whole MQW, on the other hand, is enhanced by a factor of ≈ 3. However, these values are critically dependent on frequency, as we show in Fig. 11b. There, the integrated z-axis emission is presented for several photon wavelengths around λ. The DBRs reflectivity, evaluated by the standard method of transfer matrices [66, 67] is also shown for comparison. The emission enhancement drops rapidly away from λ, leading to relative inhibition out of resonance.

The cavity-induced self-interference of the laser excitation is based on the same physical principle as the above-discussed emission and thus displays a qualitatively similar frequency behavior. As we have discussed above, in a Raman-scattering experiment the excitation is performed at an energy *differ-*

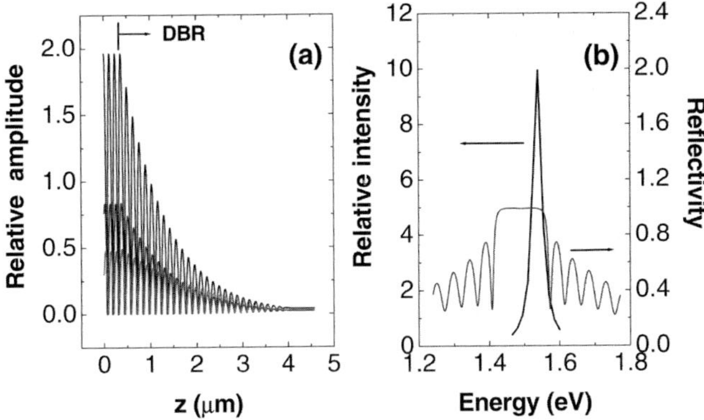

Fig. 12. (**a**) Calculated optical-field amplitude at the MQW and DBR for several photon wavelengths around λ for a plane wave of amplitude 1 incident from air at $\theta_0 = 52°$. (**b**) Normal-incidence DBR reflectivity and MQW-averaged 52° incidence excitation intensity as a function of photon energy. The *curves* are given relative to the normal incidence field intensity in a MQW without DBR

ent from that of the emission (higher or lower depending on whether Stokes or anti-Stokes processes are studied [1–3]). Thus, Fig. 11b indicates that for normal incidence (exact backscattering) the excitation would be inhibited. As we have argued in Sect. 3.4, this problem can be circumvented by tuning the incidence angle [42]. Figure 12a presents the optical electric-field amplitude within the MQW and DBR for a wave incident at an angle $\theta_0 = 52°$. These curves, correspond to three different wavelengths close to λ. Standing-wave patterns with constructive and destructive interferences, depending on the photon energy, are formed. They decay exponentially within the DBR. Contrary to a MQW without DBR, where the transmission factor of the surface reduces the excitation by ≈ 2, the insertion of a DBR leads to an *enhancement* by a factor 2. In Fig. 12b we show the *intensity* of the incident field averaged in the MQW region as a function of photon frequency, relative to the field intensity in a MQW without DBR (i.e., grown as is standard on GaAs, and limited above by air). A relative excitation enhancement of up to a factor 10 may be attained for the described low-Q cavity, which drops as the photon frequency is varied. Note that the maximum of the curve is no longer centered at the normal incidence stop band, also shown on the figure. It is actually centered at the 52° one. In this example ($\theta_0 = 52°$), the shift between the excitation and emission enhancement curves (35 meV) corresponds to the frequency of a GaAs optical phonon.

A phonon Raman-scattering process may be separated into three steps, consisting of the (virtual) creation of an exciton, followed by the emission of a vibration by that exciton, and finally by the recombination of the (also virtual) final state exciton. The two described mechanisms of excitation and emission enhancement thus contribute to such a process if the incidence angle is appropriately tuned. This leads, for the analyzed half-cavity and considering Figs. 11b and 12b, to a double-resonance excitation–emission enhancement of ≈ 30 (with respect to a MQW without DBR). Moreover, an even larger enhancement exists if there is spatial coherence between excitation and emission, as is the case in a Raman process. In this case, calculated by multiplying the excitation and emission enhancements *before* averaging through the MQW, the overall amplification is 50 % larger, i.e., it could attain a factor of ≈ 50 for such a low-Q cavity. In Sect. 4.3 these concepts will be used to describe the experimental amplification capabilities of both low- and high-Q cavities, and enhancements in excess of 10^5 will be demonstrated.

In ending this section we note that if the emission in all directions needs to be calculated (e.g., for the evaluation of emission lifetimes) [73, 90], or if the *total* Raman cross section is required, a model such as the one presented above that accounts for the cavity-modified emission in all directions is needed. On the other hand, if only the amplification on the scattered light–matter interaction vertex along one direction defined by the cavity mode is evaluated (e.g., z) then, as argued in the beginning of this section following the discussion on Sect. 3.1, it is sufficient to calculate the enhancement of the electric field with respect to the nonconfined situation for this particular direction. In such a case, the Raman amplification for a double optical resonant experiment is essentially given by the fourth power of the field enhancement in the structure [44, 49, 51]. This simpler approach will be used in Sect. 5.4 to describe the Raman amplification in photon–acoustic-phonon double resonators.

3.7 Raman Scattering with Confined Photons

Thus far we have discussed various features of Raman scattering in microcavities related to the involvement of optically confined states. Namely, the spectral discretization of photon states, the mode spatial confinement and amplification, and the angular patterns of the mode, as evidenced both in excitation and emission. Another important aspect of Raman scattering under optical confinement is related to the fact that the incident and scattered photons are no longer plane waves (or cannot be reasonably well described as plane waves), as in typical Raman-scattering experiments. In fact, the Raman efficiency in a cavity can be written as [43]:

$$\sigma \propto \left| \int \left(\mathrm{e}^{\mathrm{i}k_{\mathrm{i}}z} + \mathrm{e}^{-\mathrm{i}k_{\mathrm{i}}z} \right)^* \phi_{\mathrm{ph}} \left(\mathrm{e}^{\mathrm{i}k_{\mathrm{s}}z} + \mathrm{e}^{-\mathrm{i}k_{\mathrm{s}}z} \right) \right|^2 , \tag{26}$$

where the first and third terms in the r.h.s. describe the incident and scattered fields, respectively, and ϕ_{ph} is the displacement pattern associated with a specific phonon. Note that the photon fields are themselves optical standing waves set up by the incident and reflected components. By construction, for the cavity optical modes $k_{\mathrm{i}} = k_{\mathrm{s}} = \pi/D$ [42]. Thus, (26) can be separated into two scattering components $\sigma_{\mathrm{b}} \propto \left|2\int \cos(2\pi z/D)\phi_{\mathrm{ph}}^{\mathrm{b}}\right|^2$ and $\sigma_{\mathrm{f}} \propto \left|2\int \phi_{\mathrm{ph}}^{\mathrm{f}}\right|^2$, respectively. Here, b and f stand for backscattering and forward scattering, respectively. That is, besides the backscattering component (transferred wavevector $k_z = 2k_{\mathrm{i}} = 2\pi/D$) expected for the considered scattering geometry, the fact that photons in a cavity are standing waves leads to the observation, in addition, to a forward-scattering contribution (transferred wavevector $k_z = 0$). In a photon picture this can be simply understood considering that both laser and scattered photons are reflected back and forth in a cavity due to reflection at the mirrors, and finally escape by tunneling. In such a picture, light collected for both positive and negative z is equivalent. We note that forward-scattering experiments in semiconductor heterostructures are seldom performed because they imply the complete removal of the substrate [91, 92]. As will become clear in Sect. 5, this feature of microcavities will find great application in the study of phonons in semiconductor nanostructures.

4 Experimental Results on Raman Amplification by Optical Confinement

We review in this section a series of experimental results on Raman scattering in semiconductor microcavities that evidence the consequences of optical confinement discussed in the previous section. We begin with the first reported example of double optical resonances in a cavity that highlights the amplification potential of the structure and the power for spectrally selecting specific phonon features. We then illustrate the angular dependence of excitation and emission in high-Q cavities and we end the section by describing a series of single- and double-resonance experiments conceived to test quantitatively the amplification power of planar semiconductor resonators.

4.1 Spectral Selectivity and Amplification

In order to illustrate the concept of optical amplification of Raman scattering in cavities we start with Raman spectra obtained in a microcavity consisting of a half-wave AlAs spacer enclosed by GaAs/AlAs quarter-wave layers (13.5 pairs above, 20 below) acting as distributed Bragg reflectors (DBRs), grown by molecular beam epitaxy on a GaAs wafer [42, 73]. Two 12 nm wide strained $\mathrm{In_{0.14}Ga_{0.86}As}$ quantum wells are located at the center of the AlAs cavity, each enclosed by 1 nm wide layers of GaAs. A finesse of 300 was estimated from reflectivity curves for this structure [73]. A slight taper exists in all layers thus enabling the tuning of the cavity mode by displacing radially the spot under examination. For comparison, a second sample similar to the first but with its growth stopped just after the AlAs cavity (i.e., without the top DBR) was also studied [45].

Figure 13 shows the Raman spectra of both the cavity (top panel) and test samples (bottom panel), taken for the crossed $z'(x,y)\bar{z}$ configuration at 1.37 eV (below the QW gap ≈ 1.385 eV). $x(y)$ indicates incident (scattered) light polarized along the crystal axis (100)[(010)], and $\bar{z}$ a scattered photon propagating along z. z' indicates light incident, for this case, at 54° (external) with respect to z. For the 20.62 mm spectrum (highlighted with a thicker curve), θ_0 and the spot under observation are such that both Stokes and excitation photons are simultaneously resonant to the cavity mode and differ in energy by the longitudinal optical phonon of GaAs (LO ≈ 295 cm^{-1}). Compare this spectrum with that taken from the test (no top DBR) sample, shown in Fig. 13 (bottom). Note in particular the difference in amplitude: enormous scattering enhancement with respect to the test sample is obtained by cavity optical confinement. We postpone a quantitative analysis of this cavity enhancement to Sect. 4.3, remaining qualitative here with the purpose of highlighting the main characteristics of optically confined Raman scattering. One may selectively enhance different parts of the spectrum by collecting data from different spot positions (i.e., the cavity mode resonant with *different* Stokes energies) but leaving the angle fixed. By doing this we slightly detune the laser photons, but taking advantage of the width of the geometrical resonance almost double resonance is preserved. Note in Fig. 13 how all the secondary structures are selectively amplified by displacing the cavity mode through the Stokes region, from ≈ 304 cm^{-1} in the lower spectrum to ≈ 278 cm^{-1} in the upper one. In fact, at least eight Raman peaks, some with doublets, can be clearly identified within the LO (≈ 295 cm^{-1}) and TO (transverse optic phonon, ≈ 272 cm^{-1}) range. It is possible to relate these modes to specific regions in the heterostructure, i.e., the surfaces limiting the DBRs, the DBRs themselves, or the central QWs. They reflect an interesting problem of interface-like DBR modes discretized by finite-size effects that has been treated in detail in [42] and [41] and will be the subject of Sect. 5.1.

4.2 Angular Selectivity

According to (10) and (21), the Raman enhancement in a planar microcavity should depend critically on the incidence angle, θ_0. This is in fact the case, as is shown in Fig. 14 where we present the scattering intensity as a function of θ_0 for a fixed laser energy and with the cavity mode along z in resonance with the LO-phonon Stokes frequency (a) or with a lower-energy interface phonon mode (b) [42]. Note that in (a) maximum intensity is obtained for θ_0 set to 54°, corresponding to a DOR at the LO-phonon energy, while in (b) the maximum is at $\theta_0 \sim 52°$, tuned for DOR at the lower-energy interface mode. These two panels demonstrate that for highlighting specific peaks, one has simply to tune the cavity mode with the desired Stokes energy and vary the angle for double resonance. In fact, by doing this, peaks that appear secondary on Fig. 14a can be made even larger than the LO peak (as in Fig. 14b). Note that more than two orders of magnitude variations of the

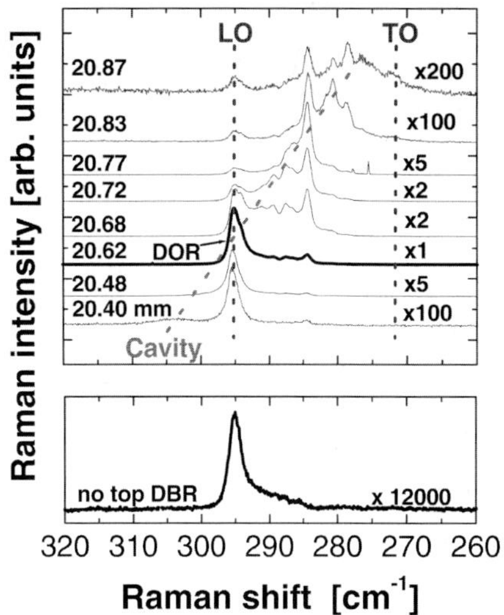

Fig. 13. Raman spectra taken with $z'(x, y)\overline{z}$ configuration for both the cavity (*top*) and a test QW sample, i.e., without top DBR (*bottom panel*). z' corresponds to the laser incident at 54° with respect to z (external to the sample). The laser energy was 1.37 eV, below the QW's exciton. In the *top panel* Raman-scattering spectra collected for different positions in the wafer are shown, that is, with selective enhancement of different parts of the Stokes region. The incident angle was held fixed, for exact double resonance slightly below the LO phonon

Raman peak intensity are observed by rotating the incident angle less than four degrees, thus going from double to single (outgoing) resonance. The angular dependence half-width at half maximum (HWHM) is of the order of 1°, which is consistent with the finesse of the measured reflectivity optical mode in the same structure (0.3 nm) [73].

Another important aspect of light scattering in microcavities, also directly related to the confined photon angular dispersion given by (10) and (21), is the angular directionality of the emitted light [41]. This can be seen in Fig. 15, where the Raman intensity as a function of emission angle is shown. With the sample at the focus (80 mm) of the $f/2$ aperture collecting lens used in the experiments, the scattered light propagates parallel as a 40 mm cross section beam up to the spectrometer, where it is once more focused to the entrance slit. Thus, within the small angular interval spanned by the collecting lens, the angular distribution of the scattered light can be determined by radially collimating the beam. The cavity mode was tuned (along z) to the lower-energy peak (i.e., larger Raman shift). It is evident that the emission is highly concentrated along a direction determined by the outgoing resonance condition, along z for the lower-energy mode, and at a finite angle for the higher-energy peak (i.e., lower Raman shift). We note that, for a low-Q half-cavity and within the collection aperture used in the experiments, the

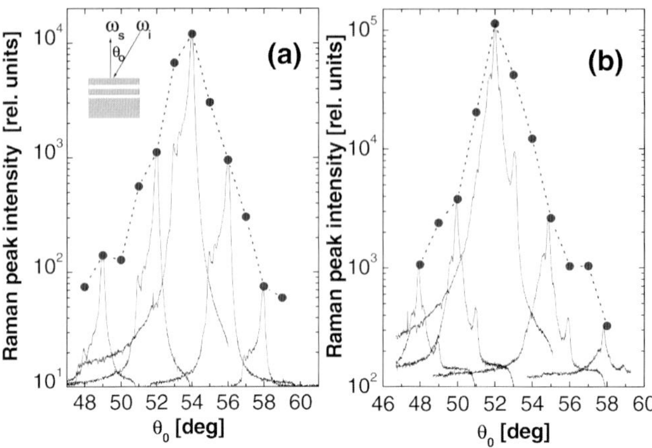

Fig. 14. Dependence of the Raman peak intensity on θ_0 (measured external to the sample), and normalized to the test sample (*full dots*). In (**a**) the cavity is tuned along z to the LO Stokes energy, while for (**b**) the mode was set to one interface mode at lower Raman shifts. The *dotted lines* are a guide to the eye joining the experimental points. Raman spectra are shown at a few different angles for illustration, shifted so that the most intense peak (LO or interface phonon) coincides with the respective angle. The *inset* shows the scattering configuration

emission pattern was observed to be isotropic. The calculated angular dependence of the emitted radiation, presented in Sect. 3.6 can be used to evaluate the emission patterns of dipoles in a cavity. As an example, this is done for the Stokes energies of the Raman peaks labeled with solid and empty circles in Fig. 15, and displayed by solid curves in the right panel [41].

4.3 Cavity Performance

In order to test the theoretical model presented in Sect. 3.6 and the actual performance of real cavities for Raman amplification, single and double optical resonance experiments in both high-Q and low-Q type cavities will be discussed [41, 45]. The relative scattering enhancement for the four possible configurations is determined and compared with calculations of the respective cavity-induced Raman-efficiency modifications. Two cavities are studied: one complete (high-Q), the other without top DBR but otherwise identical (low-Q), both grown on GaAs substrates. The structures consist of a 28-pair $AlAs/Al_{0.33}Ga_{0.67}As$ bottom DBR followed by the active spacer layer made of a $3\lambda/2$ total thickness (36 period) 70 Å/30 Å GaAs/AlAs MQW, terminated, in the case of the complete cavity, with a 25-period $AlAs/Al_{0.33}Ga_{0.67}As$ top DBR. The complete and half-cavities differ in their Q-factors, the latter having a finesse about 50 times smaller than the former. The MQWs optical and folded-acoustic phonons provide specific and well-localized probes. In addition, the signals are much larger than for single-QW structures, enabling a

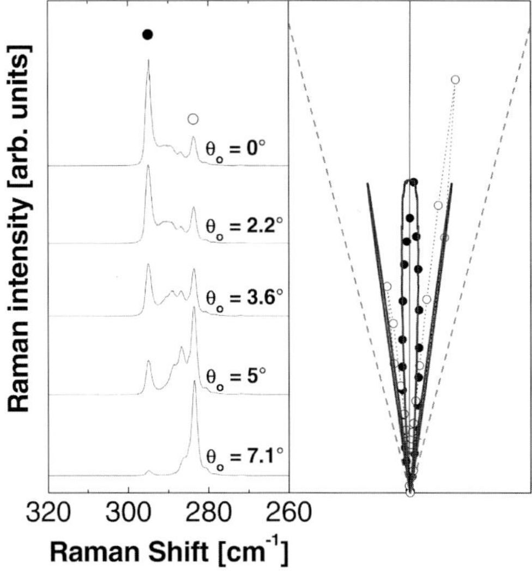

Fig. 15. *Left*: Cavity optical phonon Raman spectra as a function of collection angle. *Right*: Angular pattern of the emission intensity corresponding to the peaks labeled in (**a**), respectively, with *solid* and *open circles*. *Full curves* are calculated (see text). The *dotted curve* is a guide to the eye joining the experimental points. The *dashed straight lines* denote the collection aperture limits

clear determination of the cavity effects also for low-Q structures. The Al concentration on the alloy making the DBRs is chosen to be the smallest possible for growth-quality reasons, compatible with two requirements: transparency at the MQWs exciton energy and, at the same time, an average refractive index smaller than that of the MQWs so that light propagating along the MQW layers is waveguided. The latter condition is meant to enable single optical resonance experiments, as will be explained below.

The excitation (at 1.53 eV and focused to a circular spot of $\approx 50\,\mu$m diameter with standard $f = 10$ cm optics) was performed at 77 K, well below the MQWs fundamental gap (1.6 eV), that is, under optical but not excitonic resonance conditions [3]. Different Stokes signals *specific* to the MQWs are monitored: the GaAs-like optical vibrations, longitudinal (LO $\approx 295\,\mathrm{cm}^{-1}$) and transverse (TO $\approx 272\,\mathrm{cm}^{-1}$), and the first pair of folded acoustic phonons (FA $\approx 15\,\mathrm{cm}^{-1}$). For the single optical resonance experiments a right-angle geometry was used: the excitation is done parallel to the cavity axis (z), close to a freshly cleaved edge to reduce absorption of the scattered (waveguided) light that is collected *along* the MQW layers from the side of the sample. Thus, only the incoming photon is resonant with a vertical cavity mode. On the other hand, for the double-resonance geometry the angle-tuning configuration described in Sect. 3.4 was used. The scattered light was collected along z, while the incident angle θ_0 was tuned for resonance depending on the phonon under study ($\theta_0 \sim 50°$ for the optic phonons, while $\theta_0 \sim 20°$ for the folded acoustic vibrations).

Figure 16 (bottom) presents the full-cavity *single-resonance* Raman intensity due to LO, TO and FA phonons, as a function of ω_0, the frequency of the cavity mode along z (varied by shifting the spot position in the tapered cavity). For comparison, the reflectivity (monitored at the same time as the Raman signals) is also presented (top). Various features of the data should be noted: first, all three phonon resonance scans coincide and, as evidenced by the reflectivity curve, their maxima correspond to ω_0 being tuned to the laser energy (we recall that the excitation here takes place along z). The lack of dependence of the Raman resonance scans on the energy of the Stokes photon (which is different for the three vibrations), reflects the fact that only the incoming photon resonates. Second, largely amplified signals are seen in resonance and extinction (below the noise level) on detuning ω_0. Third, the FWHM of the curves on Fig. 16 ($\approx 0.7\,\text{meV}$) reflect the cavity finesse (≈ 2000). The half-cavity single-resonance data (not shown in the figure), on the other hand, is qualitatively similar, with a FWHM $\approx 32\,\text{meV}$ (i.e., a finesse ≈ 45), and a maximum signal about 20 times smaller. Note the clear asymmetry of both the Raman resonance scan and the reflectivity, which display a larger tail towards lower cavity-mode energies. This asymmetry arises from the fact that the laser light is focused, and thus it impinges on the cavity with a distribution of angles around $\theta_0 = 0$. Consequently, when the cavity mode along z is set to a value *smaller* than the laser energy, because of the mode angular dispersion given by (10) and (21), part of the excitation light can still enter the cavity through an angle different from zero.

We now turn to the double-resonance results. In Fig. 17 we present the LO and FA phonon Raman-scattering intensity, obtained under optical double-resonance conditions, as a function of ω_0. θ_0 equals $\sim 52°$ and $\sim 20°$ for the LO and FA resonance scans, respectively. Data for both the complete (circles) and half (squares) cavities are shown. The maxima of the curves occur in this case when ω_0 coincides with the Stokes energy of each vibration, as expected for light collected along z. Note the observed dynamical range larger than 10^5 for the complete cavity, leading to huge signals under exact double resonance and extinction (as for the single resonance above) for detunings of about 3–4 FWHM. On the contrary, for the low-Q half-structures the dynamical range falls to values lower than 10^2, and no complete suppression is observed. The overlapping of the low-Q cavity LO and FA curves implies that signals are observed even if the cavity is completely detuned. This fact was used in [41] to evidence the added effect of the incoming and outgoing resonances in the scattering process.

A quantitative comparative analysis of the Raman intensities for the four different configurations, single and double optical resonances, in both high- and low-Q cavities, is possible since the active media leading to the Raman signals, i.e., the MQWs, are the same. Only the photon density of the incoming and/or outgoing channels at the MQWs location is varied in a controlled

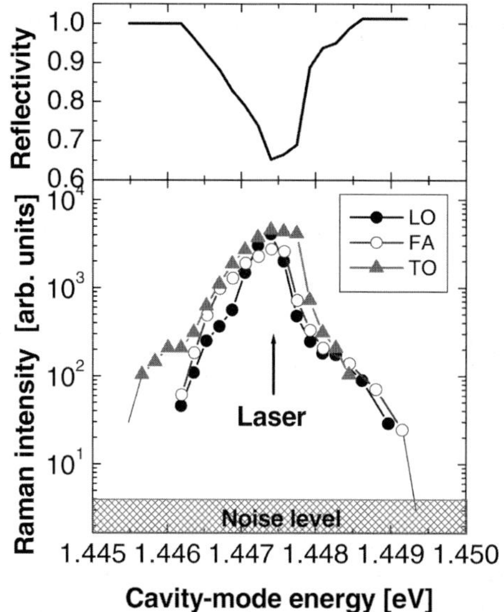

Fig. 16. *Bottom*: Raman intensity *single*-resonance scans obtained by monitoring different phonon (LO, TO and FA) signals as a function of the cavity-mode energy. *Top*: Reflectivity curve showing that the resonance maxima correspond to the laser energy being tuned to ω_0

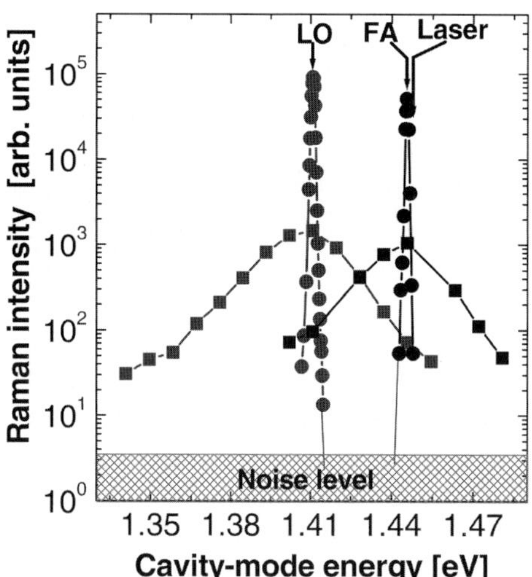

Fig. 17. Raman intensity LO and FA phonon *double*-resonance scans on both the high-Q (*circles*) and low-Q (*squares*) cavities. Note that the scans for the LO and FA phonons are obtained with different incident laser angles, tuned for the respective double resonance (52° and 20°, respectively)

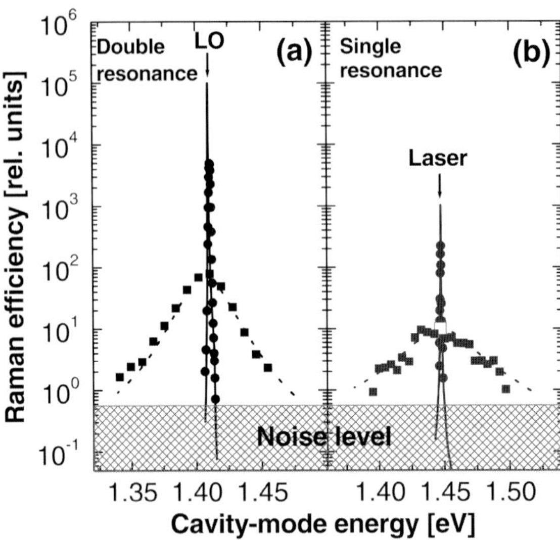

Fig. 18. Experimental (*solid symbols*) and theoretical (*curves*) LO-phonon double (**a**) and single (**b**) resonance scans, given with respect to the Raman efficiency calculated for a MQW without any optical confinement. The only adjusting parameter for the four experimental configurations is the maximum of the half-cavity single resonance scan, labeled with a larger *red solid square* in (**b**)

way [45]. Hence, the Raman efficiency of three of the structures can be calculated, parameter-free, in terms of the fourth, of which the scattering intensity at a given ω_0 is adjusted to fit the experiment. In this process it is assumed that, for the single-resonance configurations, no enhancement exists on the outgoing channel. In Fig. 18 the calculated Raman-scattering optical resonance scans for the four reported configurations are presented and compared with the LO-phonon experimental data. The scattering efficiencies are given with respect to that of a similar MQW directly grown on a GaAs substrate. Unfortunately, this is just a theoretical value because no detectable signal could be observed from LO phonons in a conventional optically unconfined MQW under the same experimental conditions (below-gap excitation). There is a single adjusting parameter for all experimental curves, which was chosen to be the maximum of the half-cavity single-resonance scan (labeled with a larger solid red square in Fig. 18b). There is an overall excellent agreement between theory and experiment, particularly in the spectral position and width of all resonance scans, and also on the magnitude of the half-cavity double-resonant enhancement. However, the quantitative agreement is not perfect for the performance of the high-Q cavity. We will address this point next.

To obtain the curves in Fig. 18 an angular average of the emission within the aperture defined by the collecting lens was performed (see Fig. 15) [45].

In addition, a spatial average of the scattering efficiency through the MQWs $3\lambda/2$ thickness was done. Theoretically, scattering enhancements larger than 10^6 (with respect to the optically unconfined situation) can be attained in such a cavity for a spatially localized scattering object (such as, e.g., a QW, a layer of quantum wires, dots, or a single interface). However, both the spatial extension of the active medium and the finite collection aperture contribute to reduce the relative amplifying power of the high-Q cavity. In fact, nodes and antinodes of both excitation and emission light fields exist for different positions within the spacer. On the other hand, the effect of the angle integration is clear from Fig. 15: due to the highly directional emission, further increasing the collection aperture in a high-Q cavity does not lead to larger signals. Even allowing for the uncertainty of Raman-intensity measurements (especially when different samples and scattering configurations were compared), the results suggest that a limiting factor not considered in the calculations is acting on the full-cavity efficiency. This conclusion is further supported by the good quantitative agreement of theory and experiment observed for the half-cavity. The smaller full-cavity single- and double-resonance efficiency as compared with theory (a factor of 5 and 20, respectively) can be attributed to the acceptance angle on the excitation channel given by (11). This angular selectivity is related to the finesse by the relation $\frac{\Delta\lambda}{\lambda_o} = \sin\theta_0 \frac{\Delta\theta}{\theta_0}$. This implies that although increasing the finesse further confines the optical field within the cavity, and thus enhances the theoretical Raman efficiency, an intrinsic limitation arises due to the correspondingly reduced angular acceptance ($\Delta\theta$) of the structures. This implies that an optimized cavity design must be used if large energy densities need to be applied, as, e.g., with a microscope objective.

5 Research on Phonons of Nanostructures Using Optical Microcavities

Having discussed in the previous section the characteristics and potentialities of Raman scattering in microcavities, we will review in this section research performed using optical confinement. Besides being interesting examples of nanostructure phonon physics, they serve to illustrate the different features that are characteristic of optically confined Raman spectroscopy. The section begins with the description of the interface phonons of the layered media constituting the DBRs and eventually embedded quantum wells [42, 48]. The use of microcavities is highlighted here by the selective enhancement through angle tuning in a single-mode cavity of finite-size discretized vibrations characteristic of different sections of the structure [42, 48]. We then proceed to describe standing optical vibrations in MQWs made by just a few periods and limited by a material where the optic phonons cannot propagate [43]. In this case a double cavity is used that, besides providing a 10^5 Raman-efficiency enhancement [85], enables through the use of almost exact backscattering geometries the study of the inplane dispersion of the vibrations [43]. While in many cases optical phonons cannot propagate from one semiconductor to the other, leading to confined optical modes and to the standing optical vibrations described here, the acoustic dispersion relations overlap and these waves can propagate from one material to the other. However, even in this case the existence of a new periodicity introduced by a MQW leads to the different characteristic behavior evidenced by the folding of the phonon dispersion and the opening of Brillouin zone-center and zone-edge minigaps, similar to those in Fig. 2a [4]. Microcavities have enabled, in this case, the study of MQWs of only a few periods, evidencing interesting finite-size effects that need to be fully taken into account both in the photoelastic description of the Raman process as in the model of the intervening acoustic modes [93]. Acoustic vibrations are described by wave equations similar to those applicable to photons when the index of refraction is replaced by the acoustic impedance $Z = \rho v$ (here ρ is the material density and v its sound velocity). Thus, many of the concepts used in the optics of confined structures can be translated to acoustics in order to design cavities that confine sound. The section ends with a review of the properties of these acoustic cavities, and their study using double resonators that confine both light and sound [44,50].

5.1 Optical Phonons of Distributed Bragg Reflectors

We begin by describing the optical vibrations of the layered structure consisting of the microcavity and, eventually, some QWs embedded at its core [42, 48]. The spectra observed in the optical-phonon-frequency region, as illustrated in Fig. 13, consists of a complex series of peaks related to the interface-like optical-phonon modes of the finite-size layered media. In this section we will briefly review the main characteristics of these spectra. The observed interface modes are always present in a semiconductor cavity, independently of the specific active medium embedded at its core, and thus their detailed understanding is a key issue for using Raman spectroscopy to study weakly scattering objects in cavity geometries. In addition, a quantitative analysis of the spectra can serve to monitor the field distribution within these photonic structures [42, 48].

The collective excitations of layered polar dielectric materials, both electronic and vibrational, have been extensively discussed in the past [4,94–107]. In particular, simple dielectric continuum models [94], modified dielectric continuum models that take into account mechanical boundary conditions besides the electrostatic ones [95,96], and microscopic calculations [97,98] have been reported for the description of the optical vibrations. References [4,94–98] describe periodic infinite structures. In addition, single slabs [99, 100], *finite* (but otherwise periodic) structures [101–103] and light scattering by plasmons in the latter [104–107], have been also discussed within the simpler dielectric continuum models. Finite-size effects basically manifest themselves in the appearance of "surface modes" at the limiting surfaces of the sample, and in a discretization of the wavevector along the heterostructure symmetry axis. Also, a relaxation of the wavevector conservation along the growth axis is expected in a scattering process. In fact, such effects have been observed in the case of plasmons by *Pinczuk* et al. [107]. On the other hand, experiments in single layers have been reported that demonstrated the existence of "surface vibrations" [99, 100].

The evaluation of the optical phonon modes in a microcavity is comparatively more complicated in that it is a nonperiodic massive layered structure and thus no simple analytic expressions for the phonon-dispersion relations are obtainable. The optical-phonon modes have been calculated extending the usual dielectric continuum model [101] by the use of a matrix method closely related to that used to describe the propagation of light in layered media [66, 67]. This method, especially suitable for heterostructures with an arbitrary sequence of layers, is described in [41]. We note here that this simplified model does not take into account the mechanical boundary conditions at the interfaces, which are known to be important to correctly describe the coupling of "interface" and confined phonon modes in thin QWs [95, 96]. In fact, strictly speaking the distinction between "confined" and "interface" modes does not apply when more elaborate models are used. However, the layers making the cavity DBRs are usually relatively thick and hence phonon-confinement effects are not expected to be important. The interface modes described by the simple dielectric continuum model correspond to the first-order confined vibrations that, due to their larger macroscopic electric field, display a marked anisotropy as a function of propagation angle (for a discussion on confined and the so-called interface phonons see the clarifying work of *Shields* et al. [108] and references therein).

Lattice motions create an electric field that may be expressed (in the absence of retardation) in terms of a scalar potential ϕ. For a layered system, with z its axis and isotropic in the xy-plane, this potential should be of the form $\phi(\boldsymbol{r}) = \phi(z)\exp(\mathrm{i}q_{\|}.r_{\|})$ and satisfy [101]:

$$\epsilon(\omega)\left[\frac{\partial^2\phi(z)}{\partial z} - q_{\|}^2\right] = 0\,, \tag{27}$$

where $\epsilon(\omega)$ is the dielectric function of the medium (different for each layer), and $q_{\|}$ is an inplane wavevector. As usual, the dielectric function of each material is taken as $\epsilon_j = \epsilon_{\infty j}\left(\omega_{\mathrm{L}j}^2 - \omega^2\right) / \left(\omega_{\mathrm{T}j}^2 - \omega^2\right)$, with $\omega_{\mathrm{L}j}$ and $\omega_{\mathrm{T}j}$ the LO (longitudinal optical) and TO (transverse optical) frequencies, respectively, and $\epsilon_{\infty j}$ the high-frequency dielectric constant of material j. For the frequencies that correspond to zeros of $\epsilon(\omega)$, the modes are bulk-like and confined to each layer [101]. More interesting here are those modes for which $\epsilon(\omega) \neq 0$, which are extended modes specific of the layered system. These vibrations are conventionally termed "interface modes" [4].

We thus consider the problem of finding solutions to (27), in an arbitrary layered material, subject to the boundary conditions that require continuity of ϕ and of the normal component of the field displacement at the interfaces. In addition, the localization of ϕ in space is required, i.e., the potential should be zero at $\pm\infty$. The electrostatic potential within layer n, defined by $z_n \leq z \leq z_{n+1}$, takes the general form:

$$\phi(z) = a_n \exp q_{\|}(z - z_n) + b_n \exp -q_{\|}(z - z_n) . \tag{28}$$

The coefficients a_n and b_n in layer n, expressed as a two-component vector, can be related with those in layer $n+1$ by multiplication of the latter by the following "transfer" matrix derived by application of the boundary conditions to (28):

$$T_{n,n+1} = \frac{1}{2} \begin{pmatrix} \left(1 + \frac{\epsilon_{n+1}}{\epsilon_n}\right) \exp^{-q_{\|} d_n} & \left(1 - \frac{\epsilon_{n+1}}{\epsilon_n}\right) \exp^{-q_{\|} d_n} \\ \left(1 - \frac{\epsilon_{n+1}}{\epsilon_n}\right) \exp^{q_{\|} d_n} & \left(1 + \frac{\epsilon_{n+1}}{\epsilon_n}\right) \exp^{q_{\|} d_n} \end{pmatrix} , \tag{29}$$

where ϵ_n is the dielectric function of layer n and d_n its thickness.

The coefficients of the first (0, e.g., *air*) and last (N, e.g., *substrate*) layers can hence be related by the closed relation

$$\begin{pmatrix} a_0 \\ b_0 \end{pmatrix} = T \begin{pmatrix} a_N \\ b_N \end{pmatrix} , \tag{30}$$

where T is simply the matrix product of the $N-1$ transfer matrices (29) defined by the structure. By further noting that, due to the localization of the potential, a_N and b_0 must be zero, the implicit equation:

$$T_{22} = 0 \tag{31}$$

follows. Equation (31) is solvable numerically for the eigenfrequencies of the interface-like modes of the structure. Once the eigenfrequencies are obtained, the potential at each layer can be calculated in terms of the normalization constant b_N by using (28) with the respective coefficients obtained by successive application of the transfer matrix (29). Finally, the normalization coefficient may be determined from energy considerations once the potential has been evaluated.

The study of these interface phonons of layered media will be illustrated with the example of the AlAs cavity described in Sect. 4.1 [42, 73]. It has a half-wave AlAs spacer with two 12 nm wide strained $In_{0.14}Ga_{0.86}As$ quantum wells at the center of the AlAs cavity, which is enclosed by GaAs/AlAs DBRs. This structure is schematized in the inset of Fig. 19. For comparison, a second "test AlAs" sample identical to the AlAs cavity but with its growth stopped just after the AlAs spacer (i.e., without the top DBR) is also discussed. The spectra of these cavities with selective enhancement of different peaks was shown in Fig. 13.

In Fig. 19 we show two selected Raman spectra of the cavity sample, together with that of the test (half-cavity) sample. These were obtained for the

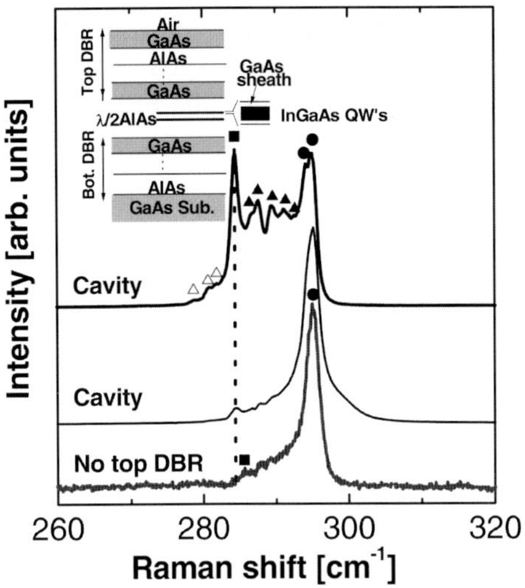

Fig. 19. GaAs-like optical-phonon Raman spectra taken with $z'(x,y)\overline{z}$ configuration for both the AlAs-cavity and test (without top DBR) samples. The laser energy was 1.35 eV, incident at 54° with respect to z (external to the sample). Stokes photons are collected along z. θ_0 was tuned for cavity optical double resonance slightly below the LO peak $\approx 295\,\mathrm{cm}^{-1}$. The two AlAs-cavity spectra (*top curves*) correspond to the selective enhancement of different Stokes regions. The same symbols used to label the peaks are shown as experimental points in Fig. 20. The amplitudes of the spectra are normalized. A sketch of the cavity sample is shown in the *inset*

crossed $z'(x,y)\overline{z}$ configuration with an excitation energy of 1.35 eV (well below the QWs gap ≈ 1.385 eV). z' describes the propagation direction of light incident not normal to the cavity but, for this case, at 54° (external) with respect to z. The spectra in Fig. 19 were taken with the incident angle tuned for optical double-resonant Stokes scattering by phonons of frequency slightly below that of the longitudinal optical phonon of GaAs (LO $\approx 295\,\mathrm{cm}^{-1}$). For the two full-cavity spectra different regions of the spectra were selectively amplified. The latter enables a precise determination of the peak frequencies. The test-sample spectrum was taken under exactly the same experimental conditions but, due to the reduced scattering efficiency [42], much longer integration times were required (600 s, compared with 1 s for the cavity spectra).

The cavity spectra shown in Fig. 19 display a complex series of peaks, whose relative amplitudes can be varied by different enhancement conditions. These appear between the LO and TO ($\sim 272\,\mathrm{cm}^{-1}$) frequencies of bulk GaAs. Two bands can be identified according to their intensity, one very strong and delimited above by the LO phonon and below by a second peak at $\sim 284\,\mathrm{cm}^{-1}$ (see the solid circles and squares, respectively, on the upper spectrum of Fig. 19), the other much weaker between the latter and the TO frequency (peaks labeled with open triangles). Note that the most intense peaks occur at the LO phonon and at $\sim 284\,\text{cm}^{-1}$, several smaller ones (labeled with full triangles) appearing between them. The test sample spectrum, on the other hand, also displays a main peak located at the bulk-GaAs longitudinal optical-phonon frequency with, in addition, some smaller features at lower energies ranging from $\sim 286\,\text{cm}^{-1}$ to the LO energy. Note, however, that the Raman shifts of these secondary peaks are *not* exactly the same for the two samples. Most notably, the lower-frequency peak in the test sample spectrum (labeled with a solid square) appears at an energy clearly *above* that of the respective peak in the cavity structure.

The calculated phonon dispersion in the GaAs optical phonon range and as a function of the inplane wavevector $q_{\|}$ (small dots) for the full-cavity structure is displayed in Fig. 20. One mode is observed for each interface, contributing to two almost continuous bands plus other clearly separated branches [101]. It is possible to relate these modes to specific regions in the heterostructure, i.e., the surfaces limiting the DBRs, the DBRs themselves, or the central QWs, by evaluating their associated electrostatic potential [48]. This assignment (shown for identification purposes in Fig. 20) is essentially correct for large inplane wavevectors. In a Raman-scattering experiment with infrared light and for the geometry being discussed, $q_{\|} \leq 6 \times 10^{-4}\,\text{cm}^{-1}$. For this relatively small value, the vibrations are not completely localized but tend to be extended throughout the structure. The so-called "finite-size" effects are apparent in the phonon dispersion in Fig. 20. Besides the existence of separate modes related to surface states, each discrete curve in the DBR bands corresponds to a different quantized wavevector along z: if D and d are the total size of the structure and the DBRs period, respectively, for the upper DBR band q_z ranges from the smallest value (π/D) at the bottom to the largest possible one (π/d) at the top. Note, for comparison, that in an infinite layered structure the interface modes form two *continuous* bands corresponding to $0 < q_z < \pi/d$ [4]. These bands, characteristic of layered media, reflect the strong anisotropy introduced by the boundary conditions on the Coulomb polarization fields.

The position of the experimental peaks are displayed in Fig. 20 on top of the dispersion curves with the same labels to facilitate their identification. There is essentially no fitting parameter: the calculated dispersion curves depend only on the layer's thickness (nominal values were used) and the TO- and LO-phonon frequencies [4]. Note that the transferred inplane wavevector, shown with a vertical dashed line in Fig. 20, is uniquely determined by the scattering configuration (i.e., the incidence and collection angles).

The main experimental peaks can be assigned to scattering by the QWs, DBR/AIR surface and *upper* DBR-band modes. The observation of intense peaks originating mostly from the *upper* DBR-band and QW modes can be simply understood from symmetry considerations. The modes in the upper(lower) half of Fig. 20 are mainly polarized along $z(q_{\|})$. Thus, the observation of intense scattering only between the bulk LO frequency and $\approx 285\,\text{cm}^{-1}$

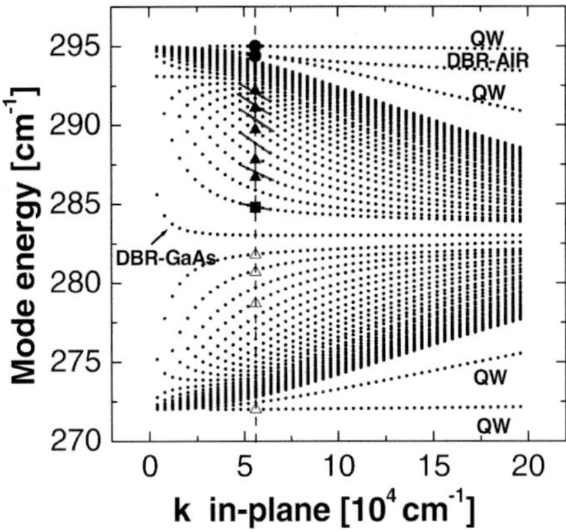

Fig. 20. Interface–phonon dispersion (*small dots*) as a function of inplane wavevector $q_{\|}$ calculated for the AlAs $\lambda/2$ cavity. The structure is limited above by air and below by the GaAs substrate. The *vertical dashed line* denotes the experimental transferred $q_{\|}$. *Symbols* correspond to the experimental points. The *short solid lines* indicate the modes that are allowed to scatter due to parity

follows from the Raman selection rules for deformation-potential-mediated scattering in crystals belonging to the D_{2d} point group that, for $z'(x,y)\overline{z}$ configuration, correspond to z-polarized $B_2(z)$ symmetry phonons [3]. It is well established, in fact, that deformation-potential-scattering dominates in off resonance light-scattering experiments [109]. The observation of the weaker peaks labeled with open triangles in Figs. 19 and 20 can be understood from the fact that for $q_{\|} \neq 0$ a small z-component exists also for the displacements corresponding to modes in the lower DBR band [42, 48].

The experimental peaks within the DBR-band (full triangles) are reasonably well accounted for by theory (to within $1\,\mathrm{cm}^{-1}$) if one of every two modes scatter in the AlAs cavity, a requirement not present for the test sample. This is a consequence of a parity selection rule that derives from the optical field and phonon-mode symmetry. Since the light-field distribution is approximately *even* with respect to the center of the microcavity, parity forbids deformation-potential-mediated scattering by modes with *odd* displacement distributions (along z). In fact, consecutive values of q_z correspond to modes of alternating parity [41]. A similar analysis enables the assignment of the doublet of the cavity LO-like peak (solid circles on the upper spectrum in Fig. 19) to the lower QW vibration and to the DBR–air-surface mode [48].

Besides the existence of a DBR–air-surface mode, and the discretization of the DBR spectra due to q_z quantization, finite-size effects provide a straightforward explanation for the relative shift of the lowest mode in the upper DBR band of the two samples (labeled by a solid square in Fig. 19). These modes correspond to the smallest quantized q_z, which we recall is given by π/D, with D the total thickness of the cavity. For the full cavity D is larger, hence the minimum q_z smaller, and thus the corresponding mode frequency is lower [41]. As discussed in [48], this shift is precisely accounted for by the dielectric continuum model discussed here. Note that these modes, which propagate almost normal to z (i.e., inplane), are usually not observable in a backscattering configuration. However, in the full-cavity spectra the latter is as intense as the signal at the bulk-like LO frequency. As will become clear next, this is due to the fact that in a microcavity, photons are multiply reflected due to the DBRs, leading to forward scattering (see Sect. 3).

Although the z wavevector component is not strictly conserved because of the lack of full periodicity, *partial* conservation [107] may be expected. The transferred q_z, under this latter assumption, depends on whether the process of excitations is through forward or backward scattering. For the former, $\boldsymbol{q}$ is mainly perpendicular to z (i.e., $q_z \approx 0$), while for the latter it is mostly along z [3]. The Raman signal can thus be separated into two contributions coming from forward and backward scattering, respectively. Both couple to the QWs modes, which are independent of q_z, leading to scattering close to the bulk LO frequency ($\approx 295\,\mathrm{cm}^{-1}$). Concerning the DBRs, forward-scattering couples to the lowest branch of the upper band ($\approx 285\,\mathrm{cm}^{-1}$, full square in Fig. 20), which corresponds to a discretized $q_z \approx 0$. The backscattering contribution, on the other hand, may only be derived after some guess regarding the magnitude of q_z is made. If the scattering proceeds for GaAs-like vibrations mainly at the GaAs layers, then $q_z \approx 2\pi n_{\mathrm{GaAs}}/\lambda$, where n_{GaAs} is the GaAs refractive index and λ the laser wavelength in vacuum. This should occur for deformation-potential-mediated scattering if the displacements are confined to the GaAs layers. In this case, coupling should be more intense for those modes with discretized q_z closest to $q_z \approx 2\pi n_{\mathrm{GaAs}}/\lambda$ [107]. To decide which modes these are, one simply has to calculate the interface-phonon dispersion for an *infinite* DBR corresponding to that q_z. This curve crosses the experimental $q_{\|}$ at $\approx 289\,\mathrm{cm}^{-1}$, where the most intense secondary peaks are indeed observed [41].

5.2 Standing Optical Phonons in Finite SLs

In the previous subsection, we relied on a single value of the inplane wavevector to identify the phonon modes of the structure and to discuss the effect of the inplane wavevector in the Raman spectra. We will describe here a phonon Raman study that profits from the use of a double-cavity structure as a means to attain large scattering intensities, to access forward-scattered modes, and to follow the inplane dispersion of the excitations [43]. The cavity is used to study the phonon spectra of a finite GaAs/AlAs superlattice. "Standing optical vibrations" are demonstrated that involve the GaAs confined phonons with a standing-wave envelope determined by the superlattice thickness.

The double cavity under consideration (modes shown in Fig. 9) has two identical $\lambda/2$ spacers made by 10.5-period GaAs (8.5 nm)/AlAs (4.4 nm)

MQWs, with 20 $Ga_{0.8}Al_{0.2}As$/AlAs DBR pairs at the bottom, 16 at the top, and 8.5 in between. Note that, in contrast to the structure described in the previous section, the mirrors are made of materials different from those making the MQWs to spectrally resolve their contribution to the Raman signal. In fact, the longitudinal optical GaAs-like phonons of $Ga_{0.8}Al_{0.2}As$ are spectrally well separated from the MQWs confined GaAs-like vibrations [110]. On the other hand, in GaAs/AlAs-based materials the optical phonons form narrow bands that do not overlap and thus the vibrations propagate in one of the constituents and not in the other (where they become evanescent waves). Consequently, the optical modes in GaAs/AlAs MQWs are confined to individual layers, this confinement being characterized by the integer mode order m [4, 109]. In addition, and contrary to standard MQWs grown on GaAs, in the structure under study the MQWs are limited by an AlGaAs alloy where the GaAs-like optical modes cannot propagate. It will be this confinement to the full thickness of the finite MQWs that results in standing optical vibrations [43].

The reported DOR Raman-scattering experiments were performed well within the transparency region ($\sim 80\,\mathrm{meV}$ below the MQWs exciton energy X) [43]. The amplification of specific features in the spectral region of the GaAs-like optical phonons (34–36 meV) was accomplished by slightly adjusting the incidence angle ($\theta_0 \sim 2$–$5°$) [42]. Once the DOR is attained, the inplane dispersion of the excitations can be followed by turning the sample with respect to the incidence (and collection) angle, and slightly displacing the laser spot on the sample for retuning. Due to the backscattering geometry and the huge DOR enhancement (which for this cavity is $\sim 10^5$, independent of angle) the spectra can be followed without any difficulty up to $\theta_0 \sim 80°$.

Spectra for varying sample angle (that is, the transferred inplane wavevector $q_{\|}$) are displayed in Fig. 21. For the smallest $q_{\|}$ (topmost spectrum) basically one split Raman peak is observed at the bulk LO energy of GaAs ($\sim 295\,\mathrm{cm}^{-1}$), plus a broad and asymmetric feature at $\sim 286.5\,\mathrm{cm}^{-1}$. As $q_{\|}$ increases, the large GaAs-like mode splitting develops into two separate peaks. One intense component remains basically unaltered, while the second, also intense, displays a strongly dispersive behavior. As a function of increasing $q_{\|}$ the dispersive component decreases in energy and generates a complex anticrossing behavior with a series of smaller peaks appearing at lower energies. As follows from their energy, these weaker peaks are related to the odd-parity m-order GaAs-like confined modes [4, 109]. The transfer of intensity between the peaks involved is a clear signature of mode coupling. The energies of the peaks are shown in Fig. 22.

Dielectric continuum model calculations of the interface modes, as discussed in Sect. 5.1 [4, 94, 99, 101], were used to understand the measured giant dispersion and anticrossings [43]. We recall that these calculations do not take into account the mechanical boundary conditions, and thus cannot predict the existence of the split m-order confined modes [95, 96]. They basically include the macroscopic electric polarization effects on the $m = 1$ vibration.

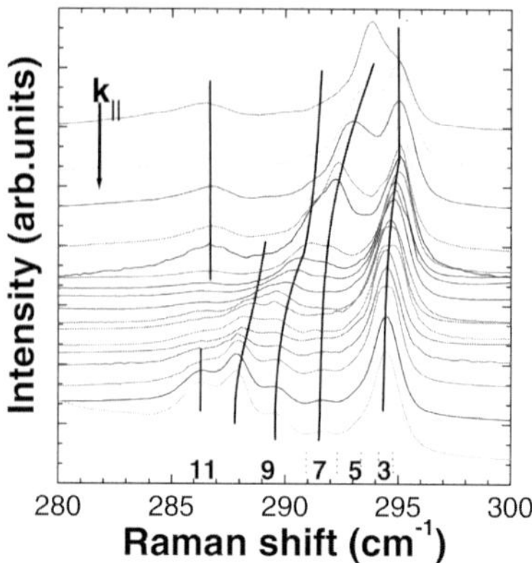

Fig. 21. Raman phonon spectra taken under DOR conditions for varying $q_{\|}$. The *solid lines* are guides to the eye following the shift of the Raman peaks. The numbers 3 to 11 indicate the energy of the m confined phonons

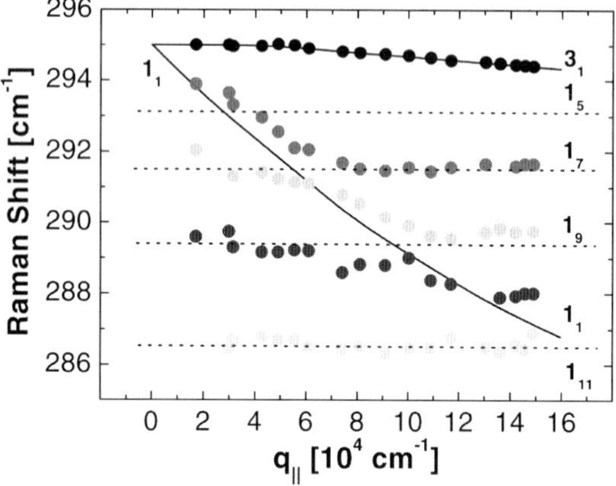

Fig. 22. Position of Raman peaks of Fig. 21 vs. $q_{\|}$. *Dashed lines* indicate the calculated energy of the m confined phonons. *Solid lines* correspond to the dielectric continuum model. N_m labels the order of standing wave N, confined mode m [43]

The calculated double-cavity $q_{\|}$-dispersion is displayed in Fig. 23. This figure also shows the displacement along z for some selected vibrations. For presentation clarity, the patterns shown were obtained for a simplified double cavity with the same number of QWs but reduced DBR layers. The DBRs GaAs-like vibrations of the alloy constitute the dense discretized modes occupying the spectral region between 272 and 287 cm^{-1}. In the spectral region corresponding to GaAs-like LO MQWs phonons, on the other hand, several modes are observed that have been labeled with the integer number $N = 1, 2, 3$, and as air/DBR and sub/DBR. As follows from the associated displacement, the latter correspond to vibrations localized at the air/DBR and DBR/substrate interfaces. The cavity-mode field in these regions is small, and thus no sizable scattering from such modes is to be expected [43]. On the other hand, N labels the order of "standing optical phonons", vibrations characterized by

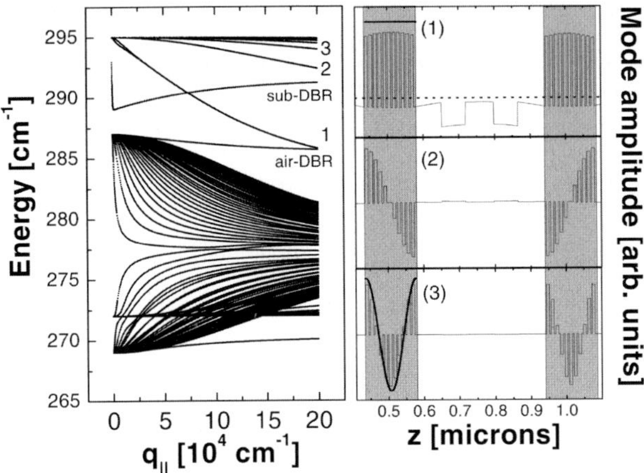

Fig. 23. *Left*: Double-cavity optical phonon dispersion calculated with a dielectric continuum model. The numbers correspond to the labels in the *right panel*. *Right*: Standing optical phonon $N = 1, 2$ and 3 displacement pattern in the central region of the double cavity. The *gray zones* correspond to the MQWs spacers. The *white region* in between represents a 5-layer DBR. The *thick solid line* in 1(3) indicates the wavevector-conserving (modulating) photon-field distributions

a standing-wave envelope of displacements that are spatially limited to the GaAs regions of the MQWs. The first-order $N = 1$ mode is the one that carries the largest macroscopic polarization and thus displays the strongest $q_{\|}$ dispersion.

Within the framework given by the dielectric model, most of the experimental results in Figs. 21 and 22 can be understood. In fact, the $q_{\|}$ dispersion of the $N = 1$ and $N = 3$ standing optical modes is well accounted for by the model (see the solid curves in Fig. 22). In addition, though not contemplated by the dielectric continuum model, the anticrossings with the odd-parity confined m modes are expected from symmetry grounds and in correspondence with the results on confined optical vibrations in standard MQWs [95, 96, 108, 111]. The expected spectral position of the m-order confined vibrations are indicated in Figs. 21 and 22. Note that the $q_{\|}$ where the anticrossings occur is nicely reproduced by the calculations. The broad asymmetric feature below $\sim 287\,\mathrm{cm}^{-1}$, on the other hand, is clearly due to the GaAs-like DBR alloy vibrations described in Sect. 5.1. However, an important issue remains to be addressed, namely why are the $N = 1$ and $N = 3$ modes (and *only* these standing-wave modes) observed with such strong intensity. The explanation for this observation relies again on the fact that both, forward and backscattering components, are observed in Raman experiments in cavities. It follows from (26) that the phonon-displacement pattern of the $N = 3$ mode overlaps exactly with the field distributions associated with the wavevector-modulated (backscattering) component $(\cos(2\pi z/D))$.

The $N = 1$ dispersive mode, on the other hand, presents an almost uniform displacement that corresponds with the *constant* wavevector-conserving (forward) scattering-field distribution (see Fig. 23). From these same arguments it follows that the scattering by all other "standing waves" should not be observed [43].

5.3 Acoustic Phonons of SLs: Finite-Size Effects

We have shown above that optical phonon confinement in (GaAs/AlAs)-based materials leads to interesting phenomena that include the phonon quantization, the existence of a phonon standing-wave envelope determined by the MQWs full thickness, and strong anisotropies that arise from the layering-induced modification of the long-range electrostatic fields. On the contrary, due to the spectral overlap and small elastic impedance mismatch, the acoustic-phonon branches in an ideal superlattice (SL) made of a periodic sequence of semiconductor layers can propagate throughout the structure [4]. The resulting spectra can be described by backfolding the phonon dispersion of an average bulk solid and opening of small minigaps at the zone center and reduced new Brillouin-zone edge [4]. However, in a Raman process performed in an infinite ideal SL in the transparency region crystal momentum is conserved. For a backscattering geometry along the growth direction this implies that the wavevector transferred is $q_z = k_\mathrm{L} + k_\mathrm{S}$, with $k_\mathrm{L}(k_\mathrm{S})$ the laser (scattered) wave number. Thus, in a Raman-scattering experiment involving longitudinal acoustic (LA) phonons, characteristic doublets are observed that reflect the folding of the phonon dispersion [4].

The above picture changes in *finite* SLs made by only a few periods, for which Brillouin-zone schemes and crystal-momentum conservation do not hold [93, 112, 113]. As originally demonstrated by *Lockwood* and coworkers for Si/Ge SLs, finite-size effects result in a broadening of the folded phonon peaks and in the appearance of side oscillations [112]. These features are conceptually similar to a diffraction pattern, where the peak widths depend on the number of illuminated grooves and the satellites correspond to the side maxima. More recently, *Giehler* and coworkers [113] have reported a study of "mirror-like" SLs that, besides these effects, demonstrate that dips in the spectra can appear due to interferences in the Raman-scattered light. We present next an investigation of finite-size effects for Raman scattering by acoustic phonons in GaAs/AlAs SLs that relies on an optical microcavity to enhance the Raman efficiency and to access the $q_z = 0$ forward-scattering contribution [93]. The results are analyzed using a photoelastic model for the Raman efficiency that takes into account the superlattice's finite size, the cavity-confined optical field, and the acoustical phonon displacements including the elastic modulation. Issues concerning the SLs limited number of periods and the implication of photon confinement in the Raman-scattering process are addressed. The studied structure consists of a 10.5-period (21 layers) GaAs/AlAs SLs of 85 Å/44 Å nominal width that acts as the spacer of a microcavity enclosed by $Ga_{0.8}Al_{0.2}As/AlAs$ DBR mirrors. Since the acoustic-folded phonon excitations are in the range 10–30 cm^{-1} (i.e., $\sim$ 1–4 meV), the angular tuning of the double optical resonance using a single-mode cavity [41, 42, 45] is straightforward. Both Stokes and anti-Stokes acoustic phonon spectra have been collected using microcavity geometries. For the Stokes (anti-Stokes) spectra the collection (excitation) was set along z, while the excitation (collection) angle was adjusted around $\sim 15°$ for double optical resonant enhancement.

Figure 24 presents typical Stokes and anti-Stokes spectra obtained at 80 K with 850 nm excitation using the cavity mode for Raman amplification. The excitation was performed below the gaps of both the SL and the GaAs substrate to avoid residual luminescence. The cavity-mode linewidth, determined mainly by the collection solid angle and the laser-spot diameter, was around 10–12 cm^{-1}. This implies that no peak is selectively amplified, and thus that the relative intensity of the different spectral features is intrinsic [42, 51]. Note that for the optical-phonon spectra described in Sect. 4 and for the study of two-dimensional electron gases in [51] the opposite limit was used. For the latter, the mode width was much smaller than the spectral region covered by the excitations under study, so that by appropriate angle tuning specific parts could be selectively enhanced. Coming back to Fig. 24, note that *three* main peaks (and not two as in standard backscattering folded acoustic-phonon experiments in SLs) are observed at $\sim 10.9\,cm^{-1}$, $\sim 12.4\,cm^{-1}$, and $\sim 14.5\,cm^{-1}$, respectively. These have been labeled with thick down-pointing arrows in Fig. 24. Besides the main peaks, weaker equally spaced oscillations are well resolved, mainly towards low energies (indicated with up arrows in Fig. 24). The peaks in the spectra are relatively broad. The linewidths are intrinsic, as verified by increasing the spectral resolution with no noticeable change in the spectral shape [93].

The acoustic-phonon Raman spectrum of the finite-size SL has been modeled using a photoelastic description of the Raman efficiency [93]. Within this model the efficiency for scattering by longitudinal acoustic phonons is given by [4]:

$$I(\omega_{q_z}) \propto \left| \int \mathrm{d}z E_{\mathrm{L}} E_{\mathrm{S}}^{*} p(z) \frac{\partial \Phi(z)}{\partial z} \right|^2 , \tag{32}$$

where $E_{\mathrm{L}}(E_{\mathrm{S}})$ is the laser(scattered) field, $p(z)$ is the spatially varying photoelastic constant, and Φ describes the (normalized) phonon displacement [4, 113, 114]. Note that $p(z)$ is, above gaps or excitons, a complex number. In the case we are discussing, where the laser frequency was tuned below the fundamental gap of the SLs, $p(z)$ was taken as a real quantity. Equation (32) basically reflects the coherent sum of the light scattered from the different layers due to a modulation of the dielectric function by the acoustic phonons through a photoelastic mechanism. For an infinite periodic SL the wavevector q_z is a good quantum number. Thus, for a standard Raman-scattering geometry, with laser and scattered fields given by

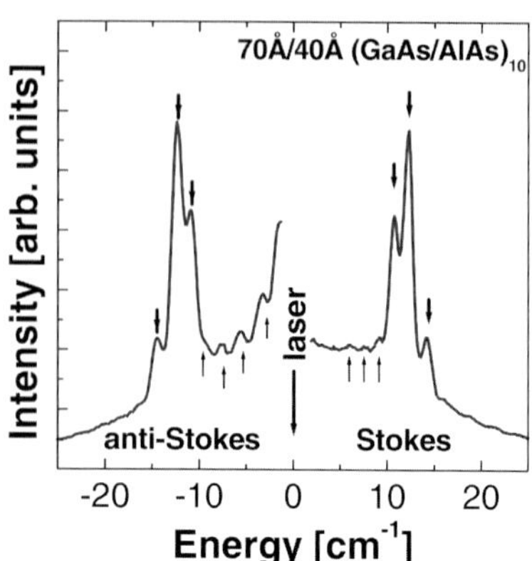

Fig. 24. Stokes and anti-Stokes spectra in the energy region corresponding to the first folded acoustic phonons, obtained with 850 nm excitation using a cavity mode for Raman amplification. The anti-Stokes spectrum has been collected with the amplification slightly tuned towards smaller Raman shifts thus amplifying preferentially the low-energy oscillations

plane waves ($E_\mathrm{L}E_\mathrm{S}^* = \mathrm{e}^{\mathrm{i}(k_\mathrm{L}-k_\mathrm{S})z}$), $I(\omega_{q_z})$ leads to the usual phonon doublets with transferred wave number given by the conservation law $q_z = k_\mathrm{L} - k_\mathrm{S}$. For the case we are discussing, however, q_z is only partially conserved due to the finite size of the structure [93, 112, 113]. In addition, as in (26) E_L and E_S correspond to the cavity-confined photons $E_\mathrm{L} = E_\mathrm{S} = \mathrm{e}^{\mathrm{i}k_z z} + \mathrm{e}^{-\mathrm{i}k_z z}$ where, by construction, $k_z = \pi/D$ with D the spacer width [42]. Consequently, $E_\mathrm{L}E_\mathrm{S}^* \propto 2 + 2\cos(2\pi z/D)$. Thus, both a wavevector-conserving (the term 2) and a wavevector-modulating contribution ($2\cos(2\pi z/D)$) are coherently added to the Raman efficiency $I(\omega_{q_z})$.

For the evaluation of the acoustic phonons in the layered structure a standard elastic continuum description of the sound waves as originally proposed by *Rytov* [115] is used [4]. This model can be implemented using a matrix method, appropriate for nonperiodic elastic multilayers as described next [116]. A solution in the form $u_m(z) = a_m\mathrm{e}^{\mathrm{i}q_m z} + b_m\mathrm{e}^{-imq_m z}$ is proposed for each layer m. Here q_m is the local wavevector given by $q_m = \omega/v_m$. The sound velocity is $v_m = \sqrt{C_m/\rho_m}$, with C_m and ρ_m the elastic constant and density of layer m, respectively. Imposing as boundary conditions at each interface the continuity of the displacement and strain,

$$u_m(d_m) = u_{m+1}(0)\,, \tag{33}$$

$$C_m \frac{\mathrm{d}u_m(d_m)}{\mathrm{d}z} = C_{m+1}\frac{\mathrm{d}u_m(0)}{\mathrm{d}z}\,, \tag{34}$$

with d_m the thickness of layer m, a recursive relation for the coefficients a_m and b_m can be obtained:

$$\begin{pmatrix} a_{m+1} \\ b_{m+1} \end{pmatrix} = T_m \begin{pmatrix} a_m \\ b_m \end{pmatrix} . \tag{35}$$

In (35), the transfer matrix T_m is given by,

$$T_m = \frac{1}{2} \begin{pmatrix} (1+Z_m)\mathrm{e}^{\mathrm{i}q_m d_m} & (1-Z_m)\mathrm{e}^{-\mathrm{i}q_m d_m} \\ (1-Z_m)\mathrm{e}^{\mathrm{i}q_m d_m} & (1+Z_m)\mathrm{e}^{-\mathrm{i}q_m d_m} \end{pmatrix} , \tag{36}$$

where $Z_m = \frac{\rho_m v_m}{\rho_{m+1} v_{m+1}}$ represents the acoustic impedance mismatch between two successive layers.

To derive the displacement pattern as a function of the phonon energy used as input for the evaluation of the Raman efficiency, (32) one has to define, for each particular problem, what are the appropriate boundary conditions. For example, a semiconductor-air surface can be modeled as a free boundary, that is, $\mathrm{d}u(0)/\,\mathrm{d}z = 0$. From this condition it follows that $a_0 = b_0 = u_0$. With this initial value fixed, all other coefficients can be expressed as a function of u_0 using (35). u_0, in turn, can be obtained through the energy normalization [116]:

$$\int \rho(z)\omega^2 |u(z)|^2 \,\mathrm{d}z \propto \hbar\omega n(\omega)\,\text{Bose Einstein population factor} . \tag{37}$$

With the above model one can also obtain the transmission and reflection of acoustic waves in the structure. For this purpose the appropriate boundary conditions at the first and last (substrate) layers are, respectively, $a_0 = 1$ and $b_{\text{subs}} = 0$. (a_{subs},b_{subs}) can be expressed in terms of (a_0,b_0) through successive iterations of (35):

$$\begin{pmatrix} a_{\text{subs}} \\ b_{\text{subs}} \end{pmatrix} = T \begin{pmatrix} a_0 \\ b_0 \end{pmatrix} , \qquad T = T_N T_{N-1} \ldots T_0 . \tag{38}$$

From these equations it follows that:

$$R = |r|^2 = \left| \frac{b_0}{a_0} \right|^2 = \left| \frac{T_{21}}{T_{22}} \right|^2 , \tag{39}$$

and

$$T = |t|^2 = \left| \frac{a_{\text{subs}}}{a_0} \right|^2 = \left| \frac{1}{T_{22}} \right|^2 . \tag{40}$$

The described matrix method is equivalent to that used for the calculation of the cavity-photon field also required in (32) [66, 67]. For the latter, the transfer matrix is basically the same as (36), with Z replaced by the refractive-index mismatch $Z = n_m/n_{m+1}$. The finite size of the SL leads to

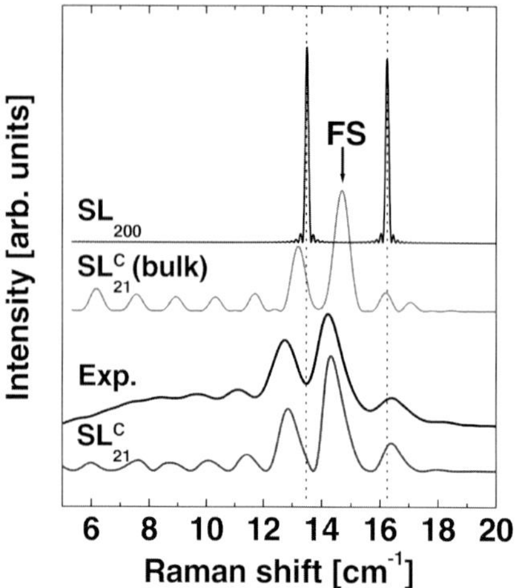

Fig. 25. From *top* to *bottom*: 1. a calculated spectrum corresponding to an "infinite" SL made by 100 periods and a backscattering standard configuration (SL_{200}), 2. spectrum of a single SL in a microcavity calculated considering the linear average media acoustic-phonon dispersion ($\mathrm{SL}_{21}^{\mathrm{C}}$(bulk)), 3. a typical experimental curve (Exp.), and 4. the full calculation for a microcavity embedded SL with phonon modes derived including the acoustic modulation ($\mathrm{SL}_{21}^{\mathrm{C}}$). In all cases $\lambda = 850\,\mathrm{nm}$. FS labels the forward-scattering contribution

modifications both on the phonons of the structure and on the way that radiation couples to these modes and leads to the scattered light. In order to discern these effects, the Raman efficiency using a simplified average medium description of the phonon spectra will also be presented. Within this latter description the phonon-displacement pattern is taken to be $\Phi(z) = \cos(q_z z)$, with the dispersion relation given by $\omega_q = |q_z|\, v_{\mathrm{ac}}$ with v_{ac} the average sound velocity [4, 113].

We now proceed to compare the experimental Raman spectra of the finite SL structure embedded in a cavity with the photoelastic model calculations. For this purpose, Fig. 25 shows from top to bottom: 1. a calculated spectrum corresponding to an "infinite" SL made by 100 periods and a standard backscattering configuration. 2. A SL in a microcavity calculated considering the (bulk-like) linear average media acoustic-phonon dispersion. 3. A typical experimental curve and, 4. the full calculation for a microcavity embedded SL with phonon modes derived including the acoustic modulation.

The "infinite" SL curve (SL_{200}) defines the standard spectrum with two narrow doublet peaks separated by $\sim 2.8\,\mathrm{cm}^{-1}$. For an infinite SL and $q_z = 0$ the two phonon modes corresponding to the center-zone doublets differ in their parity with respect to the center of the layers [4]. This leads, following (32), to one of these modes being Raman forbidden. Reminiscent of this $q_z = 0$ selection rule, for the finite but small wavevector corresponding to the backscattering spectra the (−1) component at $\sim 13.5\,\mathrm{cm}^{-1}$ in Fig. 25 is observed to be slightly stronger [4]. Note that weak oscillations are discernible close to the major peaks, much like in a diffraction grating pattern. These oscillations are, however, usually below the detection limit of Raman spectrometers. The cavity spectrum calculated with average "bulk"-like phonons ($\mathrm{SL}_{21}^{\mathrm{C}}$(bulk)), on the other hand, helps to identify those features that derive solely from the interferences in the Raman cross section due to the SLs finite size, and from the cavity-photon confinement (effects due to the real phonon structure only appear when the acoustic modulation is fully taken into account and will be discussed below). Firstly, the peaks are broader as expected from a diffraction-like pattern with fewer source layers. Concomitant with this, oscillations appear at the low-energy side, as in the experimental spectra. In addition, the backscattering maxima are shifted with respect to the infinite SL-doublet position as in the experimental data, although a quantitative agreement for the peak positions is not obtained. This lineshift is a consequence of the finite-size-related oscillatory modulation of the doublet-like broad peaks [93, 113]. Finally, a peak is observed between the backscattering doublet. This peak can be identified with a forward-scattering contribution arising from the standing-wave character of the cavity-mode field and corresponding to a transferred wavevector $q_z = 0$. Note that there is a single peak centered with respect to the doublet because the "bulk-like" phonon dispersion does not include the minigaps induced by in the acoustic modulation. In fact, such a linear dispersion is strictly valid only away from the reduced Brillouin-zone edges. This contrasts with the experimental curve, where clearly an asymmetric noncentered peak is observed. In any case, the simplified model with the bulk-like phonons accounts for many of the experimental results, including the presence of three main peaks, the broadening of the peaks, the increase of the doublet separation, and the existence of low-energy oscillations. We note that these low-energy oscillations add coherently for the backwards- and forward-scattering contributions, thus explaining their strong intensity in the experimental spectra.

The asymmetry and shift of the central forward-scattering contribution is an important point that requires further discussion. In fact, this feature can be fully accounted for with a proper description of the phonon spectra (that is, including the acoustical modulation). Such a calculation is displayed as the lowest spectrum in Fig. 25 (curve $\mathrm{SL}_{21}^{\mathrm{C}}$). As commented above for an infinite superlattice and due to a parity-selection rule the $q_z = 0$ spectra display only one component of the center-zone doublet. For $q_z \neq 0$, on the other hand, this selection rule is partially relaxed and the two components of the doublet are observed, their splitting and relative intensity varying with the transferred wavevector [4]. Consequently, in a *finite*-size SL, partial nonconservation of q_z leads to the observation of the "forbidden" component of the doublet in the forward-scattering ($q_z = 0$) contribution. This explains the observed asymmetry of the main peak at $\sim 14.5\,\mathrm{cm}^{-1}$. For the backscattering part, on the other hand, wavevector nonconservation basically leads only to a line broadening. Note in Fig. 25 that for the full calculation, including the acoustic modulation, the agreement with the experiment is remarkable without any adjustable parameter. This includes the features described above for the simpler average media model, but also almost every other detail of the experimental results: the existence of three main peaks, the asymmetry of the forward-scattering contribution, the appearance of oscillations and their period, the spectral position and relative intensity of all peaks (including the weak peak at $\sim 18\,\mathrm{cm}^{-1}$), and the linewidths. The agreement with the photoelastic calculations is so good, and the understanding of the phenomena is such, that complex phonon devices can be designed and their properties predicted with great confidence [75].

The observation of forward scattering due to the cavity confinement deserves a brief comment. In fact, although it is customary to assume for Raman scattering in bulk materials that the transferred wavevector is almost zero in comparison to the bulk Brillouin-zone edge ($4\pi/\lambda \ll \pi/a$, with λ the laser wavelength and a the lattice parameter), it is non-negligible in terms of the reduced SL Brillouin zone. Consequently, the observed Raman doublets typically reflect the bulk-dispersion folding but are not influenced by the opening of the minigaps. In order to probe the minigaps two strategies have been pursued in the literature. On the one hand, the zone edge can be sensed by a proper design of the SL structures [114, 117]. On the other hand, forward-scattering geometries have been used [91]. The cavity geometry discussed here provides direct access to forward-scattering ($q_z = 0$) excitations. This implies that zone-center minigap modes are accessible to Raman scattering in cavity geometries. This will prove, in the next section, to be essential for the observation of confined acoustic vibrations in phonon cavities [93].

5.4 Acoustic Nanocavities

Resonant cavities are the basis of numerous physical processes and devices. In the domain of optics, this includes, e.g., lasers, amplifiers, engineering of photon spectral and spatial densities, and control of light–matter interaction [15]. All these fundamental concepts can be extended to the domain of phonon physics. In fact, phonon amplification [118] and "lasing" [119] have been investigated for many years and, recently, phonon engineering has been proposed as a means to control phonon lifetimes [120] and the electron–phonon interaction [121]. It is thus of interest to develop cavities for acoustic waves, and resonators where both light and sound are confined and their interaction enhanced. We will discuss in this section the fundamentals of acoustic cavities and present, in some detail, experimental results on cavity-confined acoustic phonons obtained using cavity-confined Raman scattering. The section provides a comprehensive illustration of many of the different features of Raman scattering in cavities discussed in previous sections, namely, double optical amplification, a simplified model that accounts for the observed 10^5 efficiency enhancement, the observation of $q_z = 0$ modes based on the standing-wave character of the participating confined photons, and finite-size effects.

5.4.1 Phonon-Cavity Design

Phonon cavities are meant to provide space for wave propagation within reflective walls, supporting a standing wave resonant with a particular wavelength. For such a device to be useful, the mirrors must let sound tunnel in and out of the device. As discussed in Sect. 3.1, such mirrors, based on the destructive interference of waves, can be made using multilayers. As shown in Figs. 1 and 3, a planar periodic stack of two materials with different refractive indices and optical thickness $\lambda/4$ reflects photons propagating normal to the layers within a stop band around λ [15]. Similarly, a periodic stack of materials with contrasting acoustic impedances reflects sound [69, 122, 123]. A *phonon cavity* can be thus constructed by enclosing between two superlattices (SLs) a spacer of thickness $d_c = m\lambda_c/2$, where λ_c is the acoustical phonon wavelength at the center of the phonon minigap [50, 124]. The first $k = 0$ folded phonon minigap in a SL is maximum for a layer thickness (d) to velocity (v) ratio given by $d_1/v_1 = d_2/3v_2$ (see Fig. 2 (right), and Equation (3.14) and Fig. (3.8) in [4]). The relevant parameter in the design of this mirror is the acoustic impedance mismatch, defined as $Z = v_1\rho_1/v_2\rho_2$. v_j and ρ_j are, respectively, the sound speed and density of material j. For the acoustic cavity mode to be centered at the phonon minigap, d_c must satisfy $d_c = 2v_c d_1/v_1 = 2v_c d_2/3v_2$ ($m = 1$) [44, 50]. In Fig. 26 a scheme of a phonon-cavity structure is presented on shaded gray tones, based in semiconductor GaAs and AlAs materials. The phonon DBRs consist of 11-period 74 Å/38 Å GaAs/AlAs SLs, and enclose a 50 Å $Al_{0.8}Ga_{0.2}As$ spacer.

The phonon mirror reflectivity at the center of the minigap is given, for a SL made of N periods, by $R = 1 - 4Z^{2N} + O(Z^{4N})$, with the acoustic impedance mismatch defined as $Z = v_1\rho_1/v_2\rho_2 < 1$. v_j and ρ_j are, respectively, the sound speed and density of material j. Note that this is the acoustic equivalent of (2). According to formulae used for optical cavities, the finesse for a planar resonator is given by $F = (\pi/(1 - \sqrt{R_1 R_2})(L_{\text{eff}}/d_c)$ [15]. Here, $L_{\text{eff}} \sim d_c + 4\lambda/2$ is the effective size of the cavity mode, larger than d_c because of penetration into the mirrors. Assuming that the cavity is enclosed by two identical mirrors, then $F \sim (\pi/4Z^{2N})(L_{\text{eff}}/d_c)$, which gives the finesse as a function of Z and the number of mirror periods. By the same token, extending equivalent equations applied to optical cavities, it turns out that the phonon lifetime within the cavity is given by $\tau \sim \frac{L_{\text{eff}}}{v}\frac{1}{\ln R}$. Note that L_{eff}/v is the transit time of the phonon through an "effective" cavity of width L_{eff}. Thus, $N_\tau = 1/\ln R$ gives the number of times that the phonon is reflected back and forth before leaving the cavity by tunneling through the mirrors. We note that, while only a modest variation of the acoustic impedance mismatch can be obtained for materials common in semiconductor technology (e.g., $Z = 0.73$ and 0.84 for Si/Ge and GaAs/AlAs, respectively), the impact

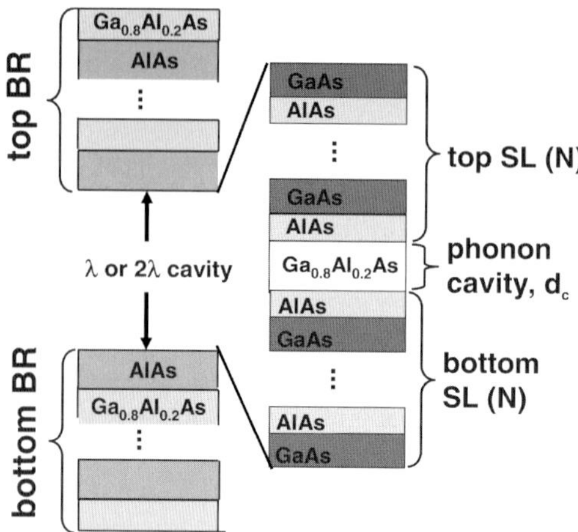

Fig. 26. Scheme of a phonon cavity embedded within an optical cavity. The phonon "mirrors" consist of N-period 74 Å/38 Å GaAs/AlAs SLs, and enclose a $Ga_{0.8}Al_{0.2}As$ spacer of thickness d_c. The phonon-cavity structures, on the other hand, constitute the spacers of an optical microcavity enclosed by $Ga_{0.8}Al_{0.2}As$/AlAs optical DBRs

on the cavity performance is enormous. In fact, using 15-period phonon mirrors N_τ can range from less than 50 for GaAs/AlAs to over 3000 for Si/Ge. In addition, other materials such as, e.g., piezoelectric and ferroelectric oxides of the perovskite family of $SrTiO_3$ and $BaTiO_3$ allow for much larger acoustic impedance mismatches (in the range of 0.65), together with multifunctional acoustic, electronic and optical properties.

5.4.2 Raman Amplification with Light–Sound Double Resonators

The standing-wave phonons described above can be generated and studied using standing-wave photons through Raman-scattering processes by placing the acoustic cavity inside an optical cavity (see Fig. 26) [44, 50, 124]. For this purpose and as shown in Fig. 26, the phonon cavity fits precisely as the spacer of a λ optical cavity defined by bottom(top) optical Bragg reflectors consisting of 20(16) $Ga_{0.8}Al_{0.2}As$/AlAs pairs. As we have argued in previous sections, such a geometry provides access to "forward-scattering" (FS, $q_z = 0$) Raman processes and strongly enhances the interaction between light and phonons. In the experiment an incident photon is coupled to the optical cavity, a phonon is generated in the resonant acoustic-cavity state, and a Raman photon is scattered out through an optical-cavity mode [44, 50, 124]. As we will show here, the forward-scattering component is essential to couple the light with the confined phonon modes of a planar acoustic cavity [50].

To illustrate the amplification of the light–sound interaction in these devices [44], we show in Fig. 27 the dependence of the double optical resonance

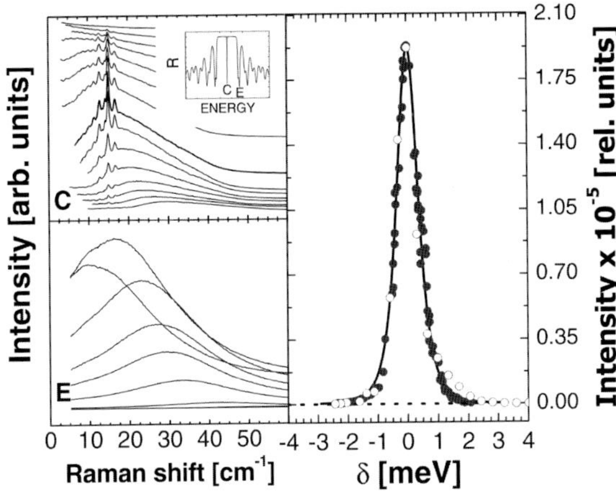

Fig. 27. *Left panel*: Raman spectra obtained while detuning the laser energy away from DOR for the cavity (C) and edge (E) modes. C and E are indicated in the optical reflectivity shown in the inset. *Right panel*: Raman amplification scan as a function of δ obtained for the laser (*full circles*) and angle-detuning (*open circles*) experiments. The *full* (*dotted*) *curve* corresponds to the photoelastic model calculations for the cavity (edge) modes, normalized to a similar phonon cavity without an optical cavity

(DOR) on detuning δ. To model these data a simplified model is used that, by assuming a situation of double optical resonance and collection of the scattered light only along the cavity mode, evaluates the cavity amplification simply as the fourth power of the cavity electric-field enhancement (see Sect. 3.6 for more details). If the optical field in the cavity is expressed as $E = E_0 f(\delta)$, where δ represents the laser detuning, then the Raman intensity (which is proportional to the product of the squared incoming and scattered fields) depends on δ as $f^4(\delta)$ [42]. Figure 27 (left) displays Raman spectra obtained while detuning the laser energy out of DOR for the cavity mode, or through a reflectivity minimum on one of the optical stop-band edges (see the inset in Fig. 27). Note that no Raman peaks can be distinguished from luminescence and noise when coupling through the stop-band edge (E), while clear spectra can be extracted by exploiting the cavity enhancement (C). The Raman amplification scan obtained from these spectra is displayed with full circles as a function of δ in Fig. 27 (right). The cavity can also be detuned by varying the incidence angle with the laser energy fixed. In this case only the incoming vertex is detuned, and consequently the DOR evolves into a single outgoing resonance. Hence, this second experiment depends on the detuning as $f^2(\delta)$. Note that an angle detuning can be related to an energy detuning by the cavity-mode dispersion $\omega_c(\theta_0) = \omega_c(0)/\sqrt{1-(\sin(\theta_0)/n_{\text{eff}})^2}$. Here, ω_c is the cavity-mode energy, θ the azimuthal angle, and n_{eff} an effective index of refraction [42]. Such experimental results are represented in Fig. 27 (right)

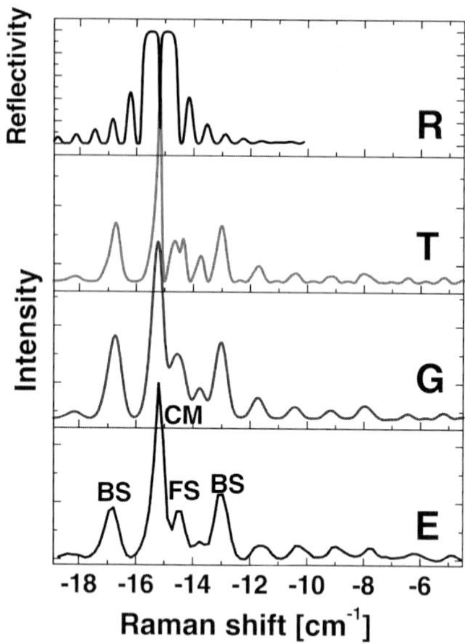

Fig. 28. Calculated reflectivity (R) and calculated (T) and measured (E) Raman anti-Stokes spectra of an acoustic cavity displaying one phonon cavity mode within the first zone-center folded phonon minigap. A calculated Raman spectra convoluted with a Gaussian of FWHM $= 0.3\,\mathrm{cm}^{-1}$ that accounts for the experimental resolution is also shown (G). The *labels* of the peaks are explained in the text

by open circles. These have been converted by expressing angle variations as energy detunings, and squaring the measured intensity so that the intensity depends on δ also as $f^4(\delta)$.

The experimental results can be compared with theoretical curves obtained using the photoelastic model for the Raman efficiency as given by (32) [4]. Such calculations, performed for both the optical cavity and edge modes are shown in Fig. 27 with full and dotted curves, respectively. The curves have been normalized with respect to a similar phonon cavity grown *without* an optical cavity. The experimental results, in turn, have been multiplied by a constant to fit the maximum intensity of the cavity-mode amplification curve. The agreement between experiment and theory is excellent. The light–sound interaction can be enhanced over five orders of magnitude in the described photon–phonon cavities [44]. Moreover, the predicted Raman intensity for coupling through the edge mode is $\sim 10^{-5}$ times smaller than that resonant with the cavity mode, thus explaining its nonobservation in the experiment.

5.4.3 Raman Spectra of Acoustic Cavities

Figure 28 presents the calculated acoustic reflectivity and the calculated and measured (80 K) anti-Stokes Raman spectra for a phonon cavity made with

11-period 74 Å/38 Å GaAs/AlAs phonon mirrors and a $Ga_{0.8}Al_{0.2}As$ spacer of thickness $d_c = 49.6$ Å ($\lambda/2$). The nominal finesse of this acoustic cavity is $F \sim 140$. The calculations are based on the photoelastic model for the Raman efficiency (32) [4]. In addition to the calculations shown for the nominal structures, Fig. 28 also displays curves convoluted with a Gaussian of full width at half-maximum FWHM $= 2\sigma = 0.3\,\mathrm{cm}^{-1}$ that accounts for the experimental resolution of the used triple spectrometer operated in subtractive mode at 830 nm. We note that at $\sim$ 830 nm, that is, below the exciton energy of the phonon mirror quantum well (QW), folded phonon Raman spectra in standard semiconductor SLs are difficult to observe because of their low intensity. On the other hand, because of the $\sim 10^5$ double optical resonant amplification almost noiseless spectra can be obtained with a few seconds of CCD integration using optical microcavities as done here.

The experimental spectra in Fig. 28, which can be almost perfectly reproduced by the calculations, display a series of peaks that are labeled as CM, BS and FS. CM corresponds to the confined cavity mode, and is coincident with the dip in the reflectivity stop band. BS and FS, on the other hand, stand for the "backscattering" and "forward-scattering" contributions. The FS component coincides with the phonon stop-band edge, and corresponds to one of the two $q = 0$ minigap states. The other folded phonon at $q = 0$ is Raman forbidden because of parity [4]. These assignments will be further discussed below [50].

Depending on d_c, a different number of confined modes can be subtended by the cavity within a specific stop band. Phonon cavities with a spacer of thickness $m\lambda/2$ display, with increasing m, two additional cavity modes besides the mode at the stop-band center. These enter symmetrically into the high-reflectivity region leading to three-mode structures [75]. A two-mode cavity, on the other hand, can be constructed by tuning a certain mode separation by selection of the appropriate m, and then slightly adjusting the spacer thickness to shift rigidly the confined modes. Calculated acoustic reflectivity and measured and calculated Stokes Raman spectra for two- and three-mode cavities grown in such a way are shown in Fig. 29. These are made, respectively, with 19- and 16-period 74 Å/38 Å GaAs/AlAs phonon mirrors, and $Al_{0.2}Ga_{0.8}As$ spacers of thickness $d_c = 794$ Å ($\sim 16\lambda/2$) and $d_c = 1489$ Å ($\sim 30\lambda/2$) [44, 75]. The number of phonon-mirror periods is chosen so as to make the structures match precisely 2λ spacers of optical cavities. For this sample the Raman spectra were collected using the highest resolution of a triple spectrometer operated in additive mode. At $\sim$ 830 nm, spectra can be collected down to a few wave numbers and with typical resolution better than $0.15\,\mathrm{cm}^{-1}$. In addition to the calculations shown for the nominal structures, Fig. 29 also displays curves convoluted with a Gaussian of full width at half-maximum FWHM $= 2\sigma = 0.15\,\mathrm{cm}^{-1}$, in order to simulate the experimental resolution. The main features of the experimental spectra are well reproduced by the calculations without any fitting parameter. We note that many of the details of the reported two- and three-mode cavity spectra

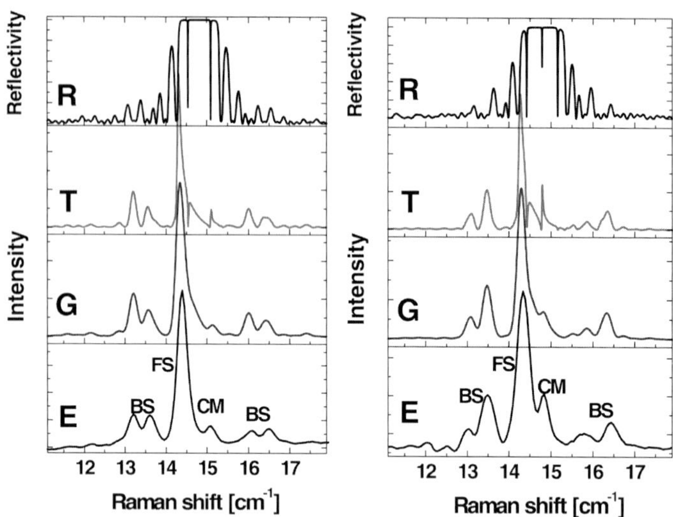

Fig. 29. Calculated reflectivity (R) and calculated (T) and measured (E) high-resolution Raman Stokes spectra of acoustic cavities displaying two (*left*) and three (*right*) phonon cavity modes within the first zone-center folded phonon minigap. Calculated Raman spectra convoluted with a Gaussian of FWHM $= 0.15\,\mathrm{cm}^{-1}$ that accounts for the experimental resolution are also shown (G). The *labels* of the peaks are explained in the text

cannot be discerned using a standard subtractive-mode Raman configuration [44] but do require the higher resolution used to collect the spectra in Fig. 29.

5.4.4 Raman Scattering with Confined Photons

As we have argued before, in a cavity both forward (FS) and backscattering (BS) components contribute to the scattered light due to the standing-wave character of the confined optical mode [41,42]. This is demonstrated in Fig. 30 where we compare the one-mode cavity measured Stokes spectrum with calculated curves. The latter correspond to the FS (transferred $q_z = 0$), BS ($q_z = 2k_\mathrm{L}$, with k_L the laser wave number) and total Raman intensities. From these spectra it is straightforward to identify the cavity mode (CM), and the separate FS and BS contributions. The BS contribution is related to the standard folded-phonon spectra observed in Raman scattering in SLs [4], while the FS component corresponds to one of the two $q = 0$ minigap states that is Raman allowed [4]. Its position coincides with the phonon stop-band edge. Note, on the other hand, that the CM is *only* observed through the FS component of the total scattering.

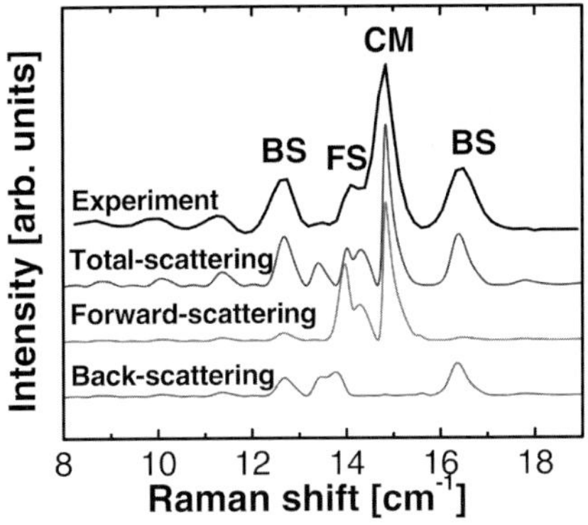

Fig. 30. Measured one-mode phonon cavity spectra compared to the total, backscattering (BS) and forward-scattering (FS) calculated spectra

5.4.5 Finite-Size Effects

Besides the FS and BS related peaks, several other spectral features appear in Figs. 28 and 29. Namely, the low-energy oscillations (particularly clear in Fig. 28), and the splitting of the BS and FS peaks (evident in the BS component of the two-mode cavity spectra in Fig. 29). The origin of these features can be traced down to finite-size effects that originate in the existence of the cavity spacer on the otherwise periodic structure, and in the reduced number of periods that make the phonon mirrors [93]. This is demonstrated in Fig. 31, where calculated spectra corresponding to a phonon mirror with a very large (300) number of periods is compared with the spectrum of the one-mode phonon cavity with ten-period mirrors. As for a diffraction pattern with a few illuminated groves, the peaks are broadened and side oscillations become evident. In addition, the existence of the cavity spacer leads to interferences between the light scattered from the two mirrors that explains the appearance of dips in the spectra and, consequently, of apparent peak splittings. These interferences in so-called "mirror superlattices" have been discussed by *Giehler* and coworkers using a simplified model for the sound dispersion [113]. A complete model that takes into account the acoustic modulation and includes strain-zero boundary conditions at semiconductor/air interfaces has been addressed by *Pascual Winter* et al. [125].

5.4.6 Phonon-Mode Parity

Note that in the calculated spectra in Fig. 29 the confined phonons display alternate peaks and dips. This reflects the fact that the modes alternate between allowed and forbidden, respectively. This can be understood by analyzing the displacement distribution $u(z)$ associated with each of these modes, and shown for the three-mode cavity in Fig. 32. Note that while the first and third modes are even with respect to the center of the cavity, the second

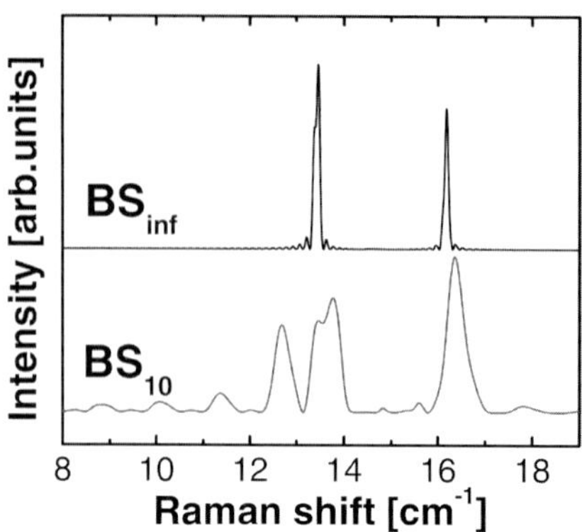

Fig. 31. Calculated backscattering Raman spectra for a 300-period phonon mirror (BS_{inf}) compared with that of the one-mode cavity with 10-period mirrors (BS_{10})

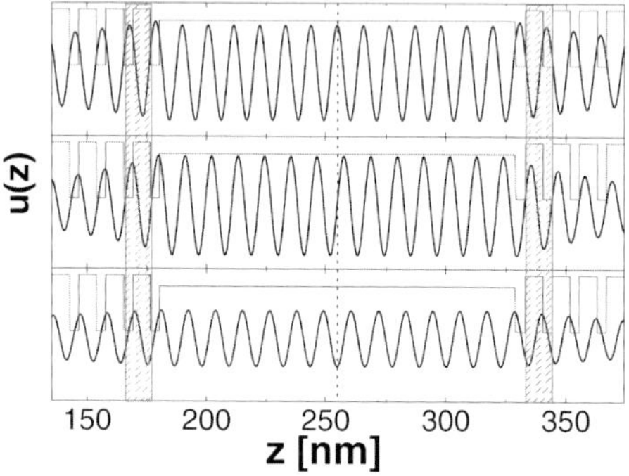

Fig. 32. Displacement distribution $u(z)$ for the three confined modes of the three-mode phonon cavity. The *thin lines* identify the layers through their photoelastic constants. The *vertical dashed line* indicates the center of the phonon cavity. The *shadowed regions* highlight a mirror period

(central) mode is odd. On the other hand, for the photoelastic mechanism it is the strain (du/dz) that is relevant, which has the parity reversed with respect to $u(z)$. Noting that the Raman cross section is also proportional to the optical-mode intensity that is even with respect to the center of the cavity (see (32)) [4, 50], it is thus straightforward to conclude that only modes with odd $u(z)$ (i.e., even strain) are Raman allowed. Hence, subsequent standing phonon modes are allowed or forbidden according to their parity.

5.4.7 Cavity vs. Mirror Modes

While in the single-mode cavity (Fig. 28) the main feature corresponds to the confined acoustic mode, for the two- and three-mode cavities the spectra

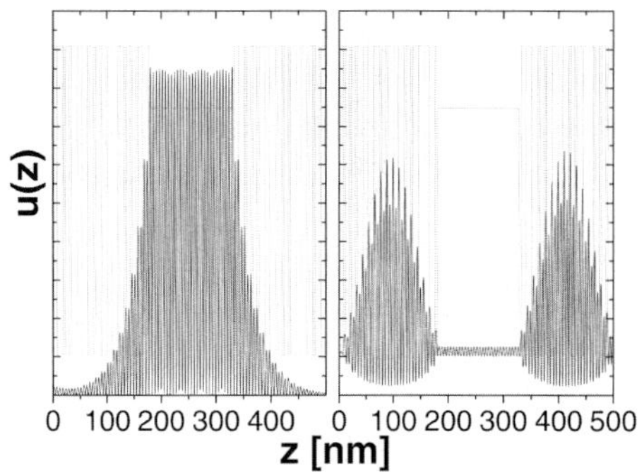

Fig. 33. Calculated displacement distribution $u(z)$ for the cavity (*left*) and forward-scattered modes (*right*) of the three-mode phonon cavity. The *thin lines* identify the layers on the basis of their photoelastic constants

are dominated by the FS contribution. The two and three-mode cavities have larger phonon mirrors in order to fit in broader optical cavities (2λ vs. λ for the one-mode phonon cavity). On the other hand, the cavity and FS modes have essentially different character. In fact, while the acoustic cavity mode is confined within the cavity spacer and decays exponentially into the mirrors, the phonon mode that leads to the FS peak is concentrated in the SLs (see the phonon displacements depicted in Fig. 33). Since the SLs making the mirrors are considerably larger in the two- and three-mode cavities, as compared to the one-mode resonator, in the former the corresponding FS contributions are enhanced with respect to the confined mode.

6 Strongly Coupled Microcavities: Polariton-Mediated Resonant Raman Scattering

The Raman scattering of light by optical phonons proceeds through the intermediate step of electronic excitations [1–3]. The standard way to describe this process in semiconductor bulk and confined structures (quantum wells, wires and dots) is as a three-step sequence that involves the creation of a virtual interband exciton, the scattering of the latter by the lattice vibration, and finally the recombination accompanied by the emission of the scattered photon [1–3]. Strictly speaking, however, in highly pure *bulk* semiconductors light does not propagate within the solid as a pure photon but as a combined state involving the oscillating absorption and emission of the photon by an exciton of the same wavevector. The latter, a combined electromagnetic and matter wave, is called a polariton; it is especially important for photon energies close to the exciton absorption [126]. From the polariton point of view inelastic scattering is described as the transformation of a photon at the surface of the solid into a polariton of the same energy, the propagation and scattering of the latter within the solid, and the final propagation and conversion of the scattered polariton into an external photon again at the material boundary [52–57]. Basically, the main difference between the two descriptions is that, in the latter, the radiation–matter interaction is taken into account exactly. The Raman cross section, which follows from third-order perturbation theory in an excitonic description, thus becomes a first-order process in the polariton framework. One should keep in mind, however, that the photons must be converted into polaritons at the surface.

Polariton waves, which basically reflect retardation effects in electric dipole active excitations, are characterized by a coupled-mode energy dispersion that qualitatively differs from that of the pure photon and exciton. This modified dispersion has important consequences in the optical properties of pure bulk semiconductors, which are reflected, e.g., in the light transit time [127], the reflectance and transmission spectra [58, 128], and the luminescence lineshape [59]. Resonant Brillouin scattering, that is light scattering by acoustic phonons, has proved to be a particularly appropriate technique to derive the polariton dispersion [58, 129, 130]. This follows from the linear dispersion of acoustic excitations, which leads to a laser-energy dependence of the Brillouin-peak shifts. This energy dependence can be used to map out the polariton branches [129, 130]. Raman scattering, on the other hand, is essentially different in that optical phonons are mostly dispersionless. Consequently, the involvement of exciton-polaritons in the light-scattering process is more subtly hidden in the resonant Raman peak intensity. It turns out that, in bulk solids, the resonant Raman scans are well described within exciton models that include discrete and continuum states, without the need to rely on polariton effects [52–54, 60].

The situation is radically different in semiconductor microcavities with embedded QWs [15, 20]. In these structures both the QWs excitons and the cavity-photon z-component of the wavevector are quantized. The wavevector component parallel to the surface ($k_{\|}$) is conserved in the photon transmission into the cavity, implying that any external photon couples to one (and only one) polariton mode of the same $k_{\|}$ and of the same energy. One immediate qualitative difference with bulk materials is that the optical spectra are characterized by a large polariton gap. This gap ranges from ~ 3–$25\,\mathrm{meV}$ in semiconductor microcavities, but can go up to $\sim 160\,\mathrm{meV}$ in organic structures [46]. In addition "additional boundary conditions" (ABCs), essential in bulk [131], are not required.

A few papers on resonant Raman scattering in semiconductor microcavities have been published to date [61–64, 132–134]. In [61] large differences in the angle dependence of the resonant Raman intensity between the polaritonic and pure-photonic [42] regimes provided evidence of the involvement of polaritons (as opposed to uncoupled excitons) in the resonant-scattering process. *Tribe* et al. [62], on the other hand, reported in- and outgoing single-resonant Raman scattering from cavity polaritons. Interestingly, a much weaker resonant enhancement was observed for the upper polariton state as compared to the lower state, indicative of the importance of dephasing in the Raman-scattering process. *Stevenson* and coworkers [133] reported the observation of uncoupled exciton-mediated scattering in microcavities in the strong-coupling regime, a result consistent with the angle dependence of polariton-mediated scattering reported in [61]. Raman scattering in samples with multibranch polaritons was also recently reported. These investigations strongly indicated that a theory that includes dephasing in a rigorous way is required [64]. A phenomenological way to include dephasing in the so-called "factorization" model originally developed for bulk semiconductors [52–57] was very recently discussed by *Bruchhausen* and coworkers [134]. These investigations were carried out both in III–V and II–VI semiconductor structures. II–VI cavities are interesting mainly for two reasons. On the one hand, the Rabi gaps and hence the polaritonic effects are much stronger than in GaAs/AlAs-based structures. On the other hand, the Raman efficiency for scattering by optical phonons is larger than in III–V materials, allowing clearer Raman studies resonant with strongly luminescent polaritonic states. A drawback, on the other hand, is that heterostructures made of II–VI materials are less under control, leading sometimes to localized exciton states in the optical transitions spectra that make difficult the separation between exciton- and polariton-mediated Raman-scattering processes.

Raman scattering in semiconductor microcavities is interesting essentially for two reasons. First, in contrast to bulk materials, a polariton description of resonant Raman scattering is unavoidable in microcavities. This provides a rich field in which to investigate the fundamentals of inelastic light scattering in this regime. In addition, in microcavities the exciton–photon coupling and both the exciton and photon strengths can be controlled by external parameters or by sample design. Moreover, cavities both in the strong-coupling and *very* strong coupling regimes ($\Omega < E_\mathrm{b}$ and $\Omega \sim E_\mathrm{b}$, respectively) can be investigated (Ω is the mode splitting and E_b the exciton binding energy). In the latter, and due to the large Rabi splitting, the upper polariton branch overlaps the exciton continuum at relatively small positive detunings. This has critical consequences in the polariton wavefunction, its dephasing and damping [135–137], and consequently should also show up in the Raman efficiency. And secondly, if this research is performed, and adequate theories are developed, Raman scattering could provide invaluable information on the polariton dynamics (interaction with external photons and with phonons) and lifetimes.

In this section we will briefly review some of the investigations that have been reported in this rather unexplored area. Since many excellent reviews have been published on cavity polaritons [15, 20, 21], we will only describe at the beginning those facts that are relevant for the description of cavity polariton-mediated Raman scattering. The section continues with a critical discussion of the available theory of polariton-mediated scattering adapted to planar microcavities. Three different experiments are then described, namely Raman-energy scans, angle-tuning double-resonant experiments, and outgoing resonant detuning investigations of the Raman efficiency. These experimental results are analyzed with the factorization model that provides a qualitative interpretation of the data. One shortcoming of the model is the difficulty of including dephasing and damping in a rigorous way. Experimental evidence of the importance of dephasing, and a phenomenological approach to include this feature in the theory, are presented.

6.1 Cavity Polaritons in Strongly Coupled Microcavities

In previous sections we have described Raman-scattering experiments on microcavities that were devoid of excitonic states in the spectral range probed by the laser and scattered waves. When the cavity mode is resonant or close to an exciton state, and the exciton is not at a node of the confined optical field, a significant interaction occurs between the two, leading to qualitatively different physics. Excitons and photons couple strongly, leading to mixed polariton modes characterized by coherent Rabi oscillations between the exciton and cavity-photon states [15]. For large detunings the modes retain their original (uncoupled) character, while a strong mixing and an energy splitting develops close to degeneracy. Radiatively stable polaritons exist in bulk semiconductors, whereas in quantum wells (QWs) the lack of momentum conservation along the growth direction (z) leads to an intrinsic radiative lifetime. Due to photon confinement the strong coupling regime is recovered in QW-embedded planar microcavities where the coupling strength between two-dimensional excitons and cavity photons exceeds both the cavity-mode and exciton-damping rates [15]. This coupling is determined by the exciton oscillator strength, the amplitude of the cavity field at the QW position, and the number of QWs present in the structure. The consequent modification of the quasiparticle energy dispersion and wavefunction have dramatic consequences on the optical properties, the most conspicuous being the "Rabi" splitting of the otherwise degenerate discrete (for a given $k_{\|}$) bare cavity-mode and exciton energies. This splitting is reflected in an anticrossing (instead of a crossing, which would apply to uncoupled or overdamped oscillators) of the exciton and cavity modes. The latter can be nicely evidenced in planar cavities by tuning the energy of the modes and measuring the luminescence emitted by the QW or its reflectivity [15, 20, 21]. The energy of the cavity mode can be either tuned by varying the spot position in tapered samples, or by measuring as a function of angle and exploiting the cavity mode inplane $k_{\|}$ dispersion. The exciton energy, on the other hand, can be varied by applying electric or magnetic fields [20].

In Fig. 34 we illustrate the mode anticrossing in a strongly coupled microcavity determined through z-axis luminescence experiments performed at 1.7 K. The MBE-grown λ-cavity consists of a $Cd_{0.8}Mn_{0.2}Te$ spacer with three CdTe 88 Å QWs located at the middle and separated by 80 Å. This central region is enclosed by $Cd_{0.4}Mg_{0.6}Te/Cd_{0.75}Mn_{0.25}Te$ Bragg reflectors, 15 periods at the top and 17 periods at the bottom. A thickness gradient in all

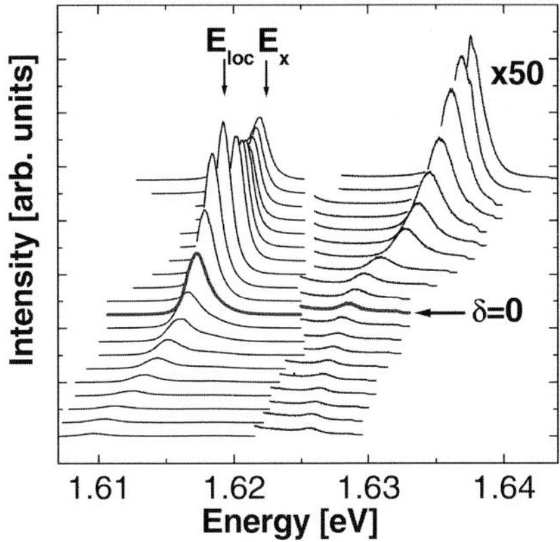

Fig. 34. Microcavity luminescence spectra at 1.7 K for different spot positions. For $\delta = 0$ a Rabi splitting $\Omega = 11$ meV is obtained. The lower branch has maximum intensity at the localized excitons energy (peak not seen due to the chosen scale). Note the much weaker upper peak intensity (multiplied by 50 for clarity) which signals strong thermalization

layers enables a tuning of the cavity mode by displacing the spot under examination. In the figure, spectra for several spot positions are displayed shifted vertically for clarity. The cavity-mode and the free-exciton state (E_X in the figure) should have crossed at zero detuning ($\delta = 0$). Instead, due to the strong coupling the two branches anticross. In fact, for $\delta = 0$ the modes are split by $\Omega = 11$ meV. The spectra corresponding to this situation are highlighted with a thicker curve in the figure. Note that this Rabi splitting is much larger than the cavity and exciton linewidths, a prerequisite for the observation of strong coupling. For this structure, and as we have observed quite generally in II–VI cavities, localized excitons (indicated in Fig. 34 by E_{loc}) appear below the free-exciton energy. These states can be identified in the figure by an increase of the luminescence emitted through the lower polariton mode when both states overlap. On the other hand, they are spectrally clearly separated from the upper polariton branch. Localized states do not Rabi-split through coupling with the cavity mode and thus remain at the uncoupled energy independently of δ.

A very simple model to describe the exciton and cavity modes in a planar cavity is as coupled oscillators, with coupling matrix element $\hbar\Omega/2$. Such a simple model, which captures most of the essential physics, can be solved analytically, and provides useful expressions for the polariton wavefunctions as a function of detuning. The relevant parameters, such as the Rabi splitting, and the cavity and exciton linewidths, are treated phenomenologically and used as fitting parameters or obtained from complementary experiments or theoretical models. The simpler situation, in which a cavity mode couples with a single QW-exciton state, is described by the following 2×2 matrix Hamiltonian:

$$H = \begin{pmatrix} \epsilon_{\rm X} & \hbar\Omega/2 \\ \hbar\Omega/2 & \epsilon_{\rm P} \end{pmatrix} . \tag{41}$$

Here, $\epsilon_{\rm X}$ and $\epsilon_{\rm P}$ are the exciton and cavity photon energies, respectively. The Hamiltonian is straightforwardly diagonalized to give the eigenvalues:

$$\epsilon_{\rm UP,LP} = \frac{\epsilon_{\rm X} + \epsilon_{\rm P}}{2} \pm \sqrt{\left(\frac{\epsilon_{\rm X} - \epsilon_{\rm P}}{2}\right)^2 + \left(\frac{\hbar\Omega}{2}\right)^2} . \tag{42}$$

In this equation UP and LP stand for upper and lower polariton states, respectively. The coupled oscillator model can be used to calculate microcavity inplane dispersions by introducing the $k_{\|}$ dependence on the cavity and exciton energies, or the mode energies as a function of spot position by including the dependence on thickness gradient of the uncoupled mode energies $\epsilon_{\rm X}$ and $\epsilon_{\rm P}$. In addition, it can readily be used to derive the polariton wavefunctions $|\boldsymbol{K}_{\rm UP,LP}\rangle = a_{\rm X}^{\rm UP,LP}|\boldsymbol{X}\rangle + a_{\rm P}^{\rm UP,LP}|\boldsymbol{P}\rangle$, where $a_{\rm X}^{\rm UP,LP}$ ($a_{\rm P}^{\rm UP,LP}$) stands for the exciton (cavity photon) weight of the polariton states. It will be convenient in order to describe the polariton-mediated Raman efficiency to define the exciton and photon strength of the polariton state as $S_{\rm X}^{\rm UP,LP} = |a_{\rm X}^{\rm UP,LP}|^2$ and $S_{\rm P}^{\rm UP,LP} = |a_{\rm P}^{\rm UP,LP}|^2$, respectively. These can be derived from a fit of the polariton energies as is illustrated in Fig. 35 for a $\lambda/2$ cavity with three 72 Å CdTe QWs located at the antinode of the photon field in the $\rm Cd_{0.4}Mg_{0.6}Te$ spacer of a planar cavity. The latter is enclosed by $\rm Cd_{0.4}Mg_{0.6}Te/Cd_{0.75}Mn_{0.25}Te$ Bragg mirrors, 21 pairs on top, and 15.5 on the bottom. A thickness gradient enables tuning of the cavity mode by displacing the laser spot. The corresponding mode energies measured at 77 K are shown in Fig. 35a, together with the fit done using the matrix Hamiltonian of (41). The Rabi splitting for this sample is $\Omega \sim 13$ meV. The mode character of the two branches can be varied from almost pure photonic to pure excitonic by changing the detuning, as is shown through the corresponding strengths for the upper polariton branch in Fig. 35b.

6.2 Theory of Polariton-Mediated Raman Scattering in Planar Cavities

We briefly introduce in this section the so-called "factorization model" of polariton-mediated first-order Raman scattering, discussing its extension for photon-confined structures. Within the factorization models of polariton-mediated RRS, originally developed for *bulk* semiconductors [52–54, 56, 57], the Raman efficiency is calculated by *first-order* perturbation theory using Fermi's golden rule as:

$$\sigma_{\rm pol} \propto T_{\rm i}\tau_{\rm i} \left|\langle \boldsymbol{K}_{\rm i} \left| H_{e-{\rm ph}} \right| \boldsymbol{K}_{\rm s}\rangle\right|^2 T_{\rm s}\rho_{\rm s}\Big|_{E_{\rm i}=E_{\rm s}+\hbar\omega_{\rm ph}} . \tag{43}$$

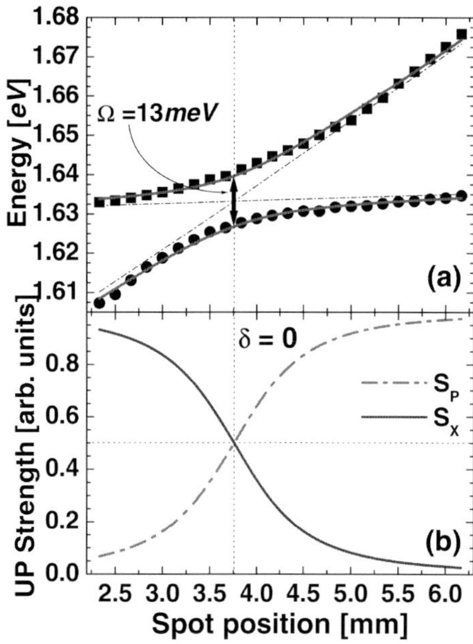

Fig. 35. Polariton mode energies at 77 K. The *solid curve* is a fit with the standard two coupled-mode model. The Rabi splitting for this sample is $\Omega \sim 13\,\mathrm{meV}$. (*b*) Upper mode excitonic (*solid*) and photonic (*dashed curve*) character, derived from the two coupled-mode fit

Equation (43) describes a sequential process that involves (i) the transmission of an incoming external photon and conversion into a polariton at the sample surface (represented by T_i), followed by the phonon-induced scattering of this polariton state ($|\boldsymbol{K}_\mathrm{i}\rangle$) to another ($|\boldsymbol{K}_\mathrm{s}\rangle$) inside the crystal, and terminated by the transmission of the latter to the exterior as a photon where it is detected (T_s). ρ_s is the density of final polariton states, and τ_i is the lifetime of the initial polariton state before it dephases or leaves the sample. These quantities are evaluated considering that energy is conserved in the scattering process, that is, $E_\mathrm{i} = E_\mathrm{s} + \hbar\omega_\mathrm{ph}$, where ω_ph is the energy of the scattered optical phonon. Although being conceptually simple, this model is limited in that the polariton modes $|\boldsymbol{K}\rangle$ describe eigenstates of the system [52–54, 56, 57]. When polariton dephasing and damping are important, as in the microcavity resonant Raman-scattering investigations described here, modifications to the model are required. Reports on resonant Brillouin scattering by acoustic phonons have included dephasing within the W. Brenig, R. Zeyher, and J. L. Birman polariton model by adding an imaginary part to the eigenenergy of the excitons [138]. An alternative approach within the factorization model will be described below.

For the definition of the transmission factors in *bulk* samples, and because of the existence of more than one polariton branch for energies above the exciton energy, additional boundary conditions (ABCs) have to be invoked. This is a complicated procedure, strongly dependent on the model used to describe the excitons and on the surface conditions. In a planar microcavity this problem is not present: only one accessible polariton state exists for a given photon energy and incident photon wavevector $k_{\parallel}$. Thus, the transmission factors T are directly proportional to the photon strength (S_{P}) of the corresponding polariton. On the other hand, polaritons interact with phonons only through their exciton component. Thus, the squared matrix element $|\langle \boldsymbol{K}_{\mathrm{i}} | H_{e-\mathrm{ph}} | \boldsymbol{K}_{\mathrm{s}} \rangle|^2$ will be proportional to the exciton strength (S_{X}) of the participating, polaritons.

The factor ρ_{s} describes, within Fermi's golden rule, the number of accessible final states that satisfy the energy-conservation condition, $E_{\mathrm{s}} = E_{\mathrm{i}} - \hbar\omega_{\mathrm{ph}}$. In a planar microcavity this has to be replaced by the number of final polariton states with $k_{\parallel}$ values *subtended by the collection cone*, and satisfying the energy-conservation condition. This means that, even for a dispersion that is flat along $k_{\parallel}$, the number of final states is not infinite but is limited by the collection optics. The density of states is given by $\mathrm{d}n/\mathrm{d}k = k/2\pi$, and so over the width $\Delta k_{\parallel}$ around $k = 0$ there are $(\Delta k_{\parallel})^2/4\pi$ available states/cm^2. Typically the collection cone can be so small that the polariton energy spread within this $\Delta k_{\parallel}$ is smaller than the homogeneous broadening of the polariton. Thus, ρ_{s} can be replaced by $\rho(k_{\parallel}, \Delta k_{\parallel})_{\mathrm{s}}$. In experiments where the detuning is controlled by displacing the spot on the surface of a tapered sample and light is collected along z, this is a *detuning-independent* constant that is only a function of $\Delta k_{\parallel}$. Otherwise, if the detuning is varied by changing the collection angle and thus $k_{\parallel}$, the dependence on angle of the number of $k_{\parallel}$ states spanned by the collection cone has to be taken into account.

In the literature devoted to polariton-mediated scattering in bulk materials, both the transit and dephasing times of the incident polariton are included ad hoc in τ_{i}. On the other hand, σ_{pol} should also depend on the scattered-polariton lifetime τ_{s}. Noninclusion of polariton-lifetime effects implies that the density of states ρ_{s} in (43) is a delta function of energy centered at the final polariton energy [63,64]. However, experimental results have clearly shown that polariton-lifetime effects are not negligible [62, 64]. Cavity polaritons have a lifetime that is indeed detuning dependent [139, 140]. A phenomenological way to account for this effect within the factorization model [134] is to include in Fermi's golden rule the *spectral density* $P(E_{\mathrm{s}})$, i.e., an additional factor that accounts for the distribution of probability, of the final polariton states (see, for example, [141]). The spectral density can be taken as a Lorentzian probability function with linewidth Γ_{s}, centered at the final polariton energy. In the exact outgoing resonance case, due to the energy-conservation condition $E_{\mathrm{i}} = E_{\mathrm{s}} + \hbar\omega_{\mathrm{ph}}$ this distribution has to be specified at its maximum. Thus $P(E_{\mathrm{s}}) = \Gamma_{\mathrm{s}}^{-1}$. We note that a more rigorous derivation of the proposed way to include dephasing through the spectral density function can be achieved using a Green's function formalism (see [142]).

Summarizing, in this simplified description of polariton-mediated first-order Raman scattering in microcavities, the Raman efficiency can be obtained from the expression:

$$\sigma_{\mathrm{pol}} \propto T_i \tau_{\mathrm{i}} |\langle \boldsymbol{K}_{\mathrm{s}} | H_{e-\mathrm{ph}} | \boldsymbol{K}_i \rangle|^2 T_{\mathrm{s}} \rho^{\mathrm{s}}(k_{\|}, \Delta k_{\|}) P(E_{\mathrm{s}}) \big|_{E_{\mathrm{i}}=E_{\mathrm{s}}+\hbar\omega_{\mathrm{ph}}} . \tag{44}$$

For exact outgoing resonance, light collected along z, and expressing the transmission factors and the exciton–phonon matrix element in terms of the photon and exciton strengths of the incoming and scattered polaritons, (44) reads:

$$\sigma_{\mathrm{pol}} \propto S_{\mathrm{P}}^{\mathrm{i}} \tau_{\mathrm{i}} S_{\mathrm{X}}^{\mathrm{i}} S_{\mathrm{X}}^{\mathrm{s}} S_{\mathrm{P}}^{\mathrm{s}} \Gamma_{\mathrm{s}}^{-1} . \tag{45}$$

As we will see in the following sections, this simple expression retains most of the relevant physics providing clear insight into the involved processes.

6.3 Experimental Evidence of Polariton Mediation of Light Scattering in Cavities

6.3.1 Resonant Raman Scans

A scheme of a cavity polariton-mediated scattering process can be visualized in Fig. 36. This figure displays the z-axis cavity-mode dispersion as a function of spot position, derived from luminescence measurements at 77 K, of a GaAs-spacer λ-cavity with two 12 nm wide $In_{0.14}Ga_{0.86}As$ quantum wells (QWs) at its center, and enclosed by GaAs/AlAs DBRs, 13.5 pairs above, 20 below. In addition, the energy dependence of the cavity modes expected for a finite $k_{\|}$ (i.e., finite θ_0) is schematized by a dashed curve. Note the neat anticrossing of the exciton and cavity modes, with a Rabi splitting $\Omega \approx 4.5$ meV [61]. A Raman-scattering double optical resonance (DOR), as described in the previous sections, is effected by tuning θ_0 so that both incoming and scattered photons become resonant with cavity modes and their frequency difference equals that of an excitation (e.g., phonon) of the system [42]. DOR experiments are performed with E_{L} far below the exciton energy (see the thinner arrow in Fig. 36), so that the observed enhancement derives exclusively from cavity-photon confinement. Cavity-polariton-mediated scattering (CPMS) occurs, instead, when the laser energy is tuned to the polariton region, as exemplified with a thicker arrow in Fig. 36.

Several types of experiments involving CPMS can be performed, and will be described in what follows. In the first and simplest experiment, the laser energy and spot position are varied (keeping θ_0 constant) so that a DOR is scanned through CPMS [61, 143]. In such an experiment, the excitation goes from pure cavity-photon-like to maximum photon–exciton mixing, while the scattered photon is always in an almost pure-photon regime. Note, however, that if θ_0 is kept fixed the laser energy detunes from the polariton modes when passing through the Rabi gap. Such a Raman-intensity scan for a $\lambda/2$

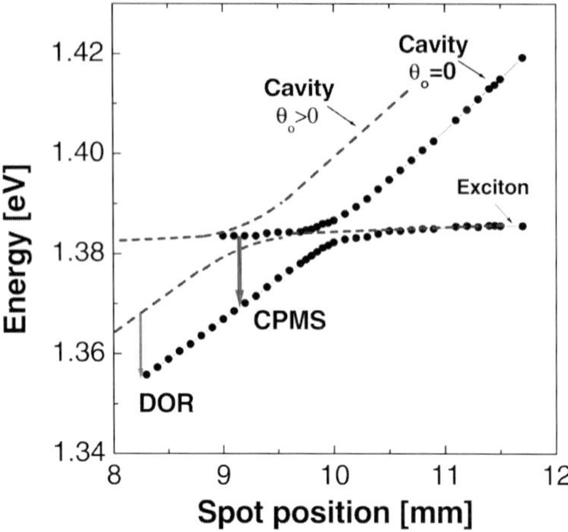

Fig. 36. The *solid circles* indicate experimental z-axis cavity-mode dispersion as a function of spot position derived from luminescence measurements. The *dashed curves* represent the situation for $\theta_0 > 0$, i.e., $k_{||} > 0$. DOR and CPMS processes are indicated. The *arrows end* (tip) indicate the incoming (scattered) photon. The length of the *arrow*, in turn, corresponds to the energy of the scattered excitation

cavity fully filled with a GaAs/AlAs MQW is shown with solid squares in Fig. 37 [143]. The figure also displays the cavity-polariton modes (solid circles), the uncoupled exciton energy (triangles), and the laser energy (empty circles). The energies are given with respect to the exciton energy at zero detuning. The latter were derived from the z-axis emission assuming that excitons do not disperse significantly at the $k_{||}$ corresponding to the used DOR geometry ($\theta_0 \sim 52°$). Note the relatively large mode splitting $\Omega = 11.3\,\text{meV}$ (at $T = 2\,\text{K}$), which has its origin in the fact that many QWs fill up the cavity spacer. With increasing energy, starting from the pure DOR situation, it is seen that the Raman intensity strongly augments as the exciton energy is approached, then decreases when the laser energy detunes from the lower polariton mode, leading again to a second (weaker) maximum when the laser energy approaches the upper polariton branch. Based on (44) and (45) the interpretation of these results is straightforward. As the polariton regime is approached, the Raman efficiency is enhanced following the increase of the exciton strength of the intervening incoming polariton mode. This simply means that the polariton needs to be excitonic to interact with phonons. As the laser energy becomes detuned, light cannot get into the cavity and thus the Raman efficiency decreases, leading to a minimum for zero detuning between the laser and exciton energy. This dependence on laser energy, which basically establishes the cavity as a filter, is represented by the spectral density $P(E)$. More interestingly, the two Raman-efficiency maxima have completely different intensities. Note that the experiment is symmetric on both the upper and lower polariton modes, and thus the transmission factors T

and the exciton strengths should follow the same dependence with laser energy for both modes. Thus, the natural explanation for the observed strongly differing intensity relies on the more rapid dephasing of polaritons in the upper branch compared with those in the lower branch, reflected by $\tau_{\rm i}$ in (44) and (45). The dephasing time is known to be shorter for the upper branch since these polaritons can scatter to lower polariton and exciton states of a different $k_{\|}$, a channel that cannot occur for the lower branch [20, 62].

The first report of such a difference between the UP and the LP in Raman-scattering experiments was reported by *Tribe* and coworkers [62] through outgoing resonant experiments performed with laser light incident at $\theta_0 = 0°$, and scattered light also collected along $\theta_0 = 0°$. The latter were resonant with the polariton modes, while the laser entered the sample through mirror losses. For such an outgoing resonant scattering geometry the difference in intensity for the two branches is represented in (45) by the differing $\Gamma_{\rm s}^{-1}$. We note that in outgoing CPMS experiments the separation between the scattered Raman photons and luminescence has been shown to be extremely difficult, the experiments of Tribe et al. being the only ones reported to date. It is possible that these became possible because of the inbuilt electric field present in the studied p-i-n diode structure. Plausibly, luminescence is quenched more strongly than Raman scattering, allowing the resonant Raman experiments to be performed. On the other hand, incoming resonant experiments similar to those displayed in Fig. 36 but performed in III–V microcavities with few QWs, and hence much smaller Rabi splittings, were also reported in [61] and [62]. In these experiments, however, only a single Raman intensity peak close to the energy of the free excitons was observed. It was suggested that this result might be due to Raman scattering proceeding also through weakly coupled localized excitons [61] or because of screening of the exciton–photon interaction at the higher incident power conditions employed in [62] for the ingoing resonant experiments.

6.3.2 Double-Resonant Angle-Tuning Experiments

A different aspect of the mediation of cavity-polariton modes as intermediate steps in the scattering process can be evidenced from the dependence of the resonant Raman intensity on angle of incidence [61, 143]. Such an experiment, described in Fig. 38, is more conveniently explained in k-space. For this purpose, panel (b) in the figure shows the inplane dispersion of the cavity polaritons for a detuning such that the cavity mode at $k_{\|} = 0$ lies one LO-phonon energy below the uncoupled exciton ($E_{\rm x}$). The proposed experiment is a *double-resonance* experiment in which, at *fixed* laser energy ($E_{\rm L} = E_{\rm x}$), the incidence angle (θ_0) is varied up to the angle of anticrossing corresponding to that detuning. In this experiment the *outgoing* channel does not change, being always resonant with the cavity mode at $\theta_0 = 0°$. With $E_{\rm L}$ tuned to the exciton energy ($E_{\rm x}$), when $k_{\|}$ is varied by changing θ_0 starting from zero (represented by dashed arrows in Fig. 38b), the incoming channel evolves

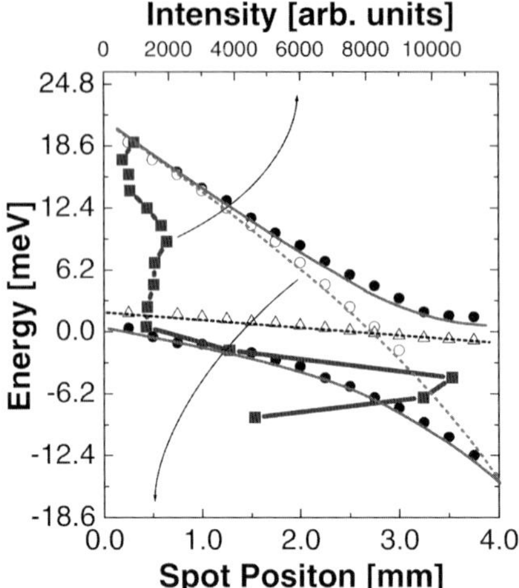

Fig. 37. Raman-intensity scan for a $\lambda/2$ cavity fully filled with a GaAs/AlAs MQW and $\Omega = 11.3\,\text{meV}$ (*solid squares*). The cavity-polariton modes (*solid circles*), the uncoupled exciton energy (*triangles*), and the laser energy (*empty circles*) are also displayed. The *solid*, *dashed* and *dotted curves* are guides to the eye joining the experimental points. The laser and mode energies are given with respect to the exciton energy

continuously from almost pure exciton-like character to maximum exciton–photon mixing (CPMS). However, since the laser energy is not adjusted in the angle variation, close to the anticrossing angle the double-resonance condition is not strictly satisfied. For comparison purposes, Fig. 38a also shows the same experiment under pure-photonic DOR conditions, that is, for $E_L \ll E_x$. For this situation, detuning from DOR the Raman efficiency should decrease according to the finesse of the cavity, as demonstrated in Fig. 14.

Figure 39 displays the Raman intensity as a function of θ_0, recorded with $E_L = E_x$ (solid symbols) for the same large Rabi gap MQW microcavity used for the experiments on Fig. 37 [143]. For comparison, an angular dependence obtained with $E_L \ll E_x$, i.e., in the pure optical DOR regime, is also shown (open circles). The latter decreases steeply on detuning θ_0, and gets below the detection noise level within $\sim 20°$ of the DOR condition. In contrast, for E_L at the exciton resonance the curves decrease much more smoothly and no complete suppression is observed even for $\theta_0 = 0°$. If cavity and exciton were *not* coupled, as is the case when $E_L \ll E_x$, the Raman intensity should fall steeply with θ_0, reflecting the cavity finesse through its filtering properties [42]. The Raman signal, proportional to the light intensity at the QWs, provides a straightforward explanation for the steep decrease of intensity of the $E_L \ll E_x$ data. However, it is clear that a different regime exists for $E_L \approx E_x$. As we will argue next, the cavity-polariton description

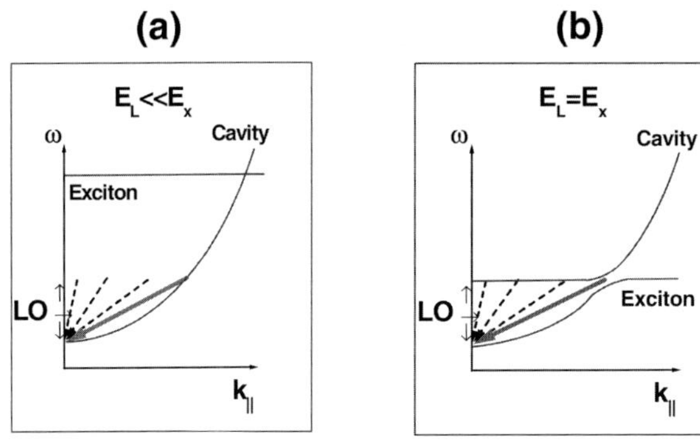

Fig. 38. Schemes of RRS geometries. The *arrow end* (tip) indicates the incoming (scattered) photon. The angle dependence of the incoming channel in a double-resonance experiment is followed for (**a**) the purely photonic ($E_{\mathrm{L}} \ll E_{\mathrm{x}}$) and (**b**) the polariton ($E_{\mathrm{L}} = E_{\mathrm{x}}$) regimes. In (a) the Raman efficiency when detuning out of DOR (*thick arrow*), should follow the finesse of the cavity. In the polariton regime (**b**) "cavity-polariton-mediated scattering" corresponds to the maximum exciton–photon mixing (indicated with a *thick arrow*), while the angular detuning goes from maximum exciton–photon coupling to almost pure exciton character (*thinner dashed arrows*)

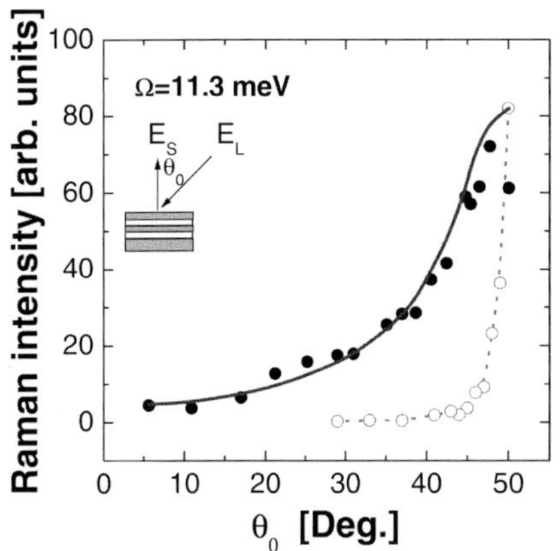

Fig. 39. Angular dependence of the resonant Raman intensity for a MQW cavity. *Solid symbols* correspond to $E_{\mathrm{L}} = E_{\mathrm{x}}$. *Open symbols* were obtained for $E_{\mathrm{L}} \ll E_{\mathrm{x}}$ (i.e., pure optical DOR). The *solid curve* corresponds to the CPMS model, adjusted to fit the small-angle data. The *dashed curve* is a guide to the eye joining the experimental points

of RRS provides a natural explanation for such a different resonant angular dependence.

Concerning (45), note that the outgoing channel in the angle-dependence experiment remains fixed, and thus T_s (i.e., S_P^s), S_X^s, and Γ_s^{-1} are constant. In going from the CPMS condition to the pure-exciton behavior (that is, decreasing θ_0), S_X^i grows from 1/2 to 1. This dependence only provides a small variation of the Raman-scattering efficiency that, in fact, has the opposite sign to that observed in Fig. 39. On the other hand, $T_i \propto S_P^i$ falls with decreasing angle from 1/2 to 0, giving the strongest angular dependence. Finally, the uppper cavity-polariton lifetime in similar structures has been shown to increase only by a factor ≈ 2 in going from zero to maximum detuning [139]. Neglecting this latter small contribution that requires an independent measurement to be precisely determined, the resonant Raman-scattering intensity can be simply approximated as $\sigma_{\mathrm{pol}}(\theta_0) \propto S_X^i S_P^i$. This behavior, with the photon and exciton strengths of the incoming polariton deduced from a fit of the polariton dispersion, is shown by the solid curve in Fig. 39, with the constant prefactor in σ_{pol} fitted to the Raman intensity at small angles. The RRS intensity decay with decreasing angle (i.e., $k_{\|}$) is mainly determined by the concomitant reduction of S_P^i. The model nicely describes the data except for very close to the anticrossing. Here, due to the fixed laser energy, the Raman intensity falls below the model due to the incoming detuning produced by the opening of the Rabi gap.

Similar experiments were reported for a cavity with a much smaller Rabi gap in [61]. In this case, in the region of small detuning the theoretical CPMS curve falls *below* the experimental data. This discrepancy can be understood by arguing that CPMS is accompanied by a weak-coupling contribution that falls steeply like the finesse. For cavities with small Rabi gap, because of mode overlap the transmission at zero detuning is significant. Thus, in such cases, Raman processes mediated by weakly coupled localized excitons may contribute a large part of the total resonant Raman-scattering signal. This could explain for these samples the observation of a single peak (instead of two) in the ingoing CPMS scans as described in the previous section [61]. In that sense, in small Rabi gap structures, the angle experiment helps to *extract* the cavity-polariton-mediated component from the total Raman cross section.

6.3.3 Outgoing Resonant Raman Experiments

The role of cavity polaritons as intermediate states in the scattering process can be fully evidenced if the above-discussed resonant enhancement is studied as a function of exciton cavity–photon detuning. That is, if the intensity of the scattered Raman signal tuned to a polariton mode is followed continuously as this mode varies from pure photon, to largely mixed, and finally to pure exciton. Such experiments, exploiting the intense Raman signals and low photoluminescence emission at the upper polariton branch, have been reported for II–VI microcavities using backscattering geometries at normal incidence. In these experiments the scattered photons are tuned resonantly

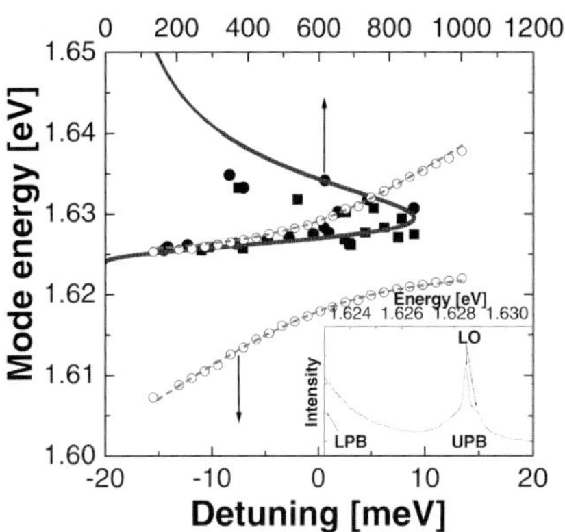

Fig. 40. QW CdTe and spacer $Cd_{0.8}Mn_{0.2}Te$ LO-phonon Raman intensity at outgoing resonance with the upper polariton branch, as a function of the energy of the latter. The polariton-mode energies taken from Fig. 34 are also shown. The *thin solid lines* are fits to the coupled oscillator matrix Hamiltonian, from which the exciton and photon strengths are derived. The *solid curve* corresponds to the CPMS model described in the text. The *inset* shows the two Raman peaks at resonance with the upper polariton branch (UPB)

to the UP, while the laser excitation enters the structure through mirror losses [63, 64, 132, 134].

Data for the II–VI cavity characterized by the photoluminescence data of Fig. 34 are shown in Fig. 40. Results for the longitudinal optical phonons at 171 cm^{-1} (CdTe QW mode) and at 166 cm^{-1} (CdTe-like vibrations of the spacer $Cd_{0.8}Mn_{0.2}Te$ layer) are shown by solid circles and squares, respectively, together with the polariton branches derived from the spectra of Fig. 34. Both Raman lines display a resonance behavior (the intensities have been normalized). However, the maxima of these resonance scans *is not* observed at the exciton energy, as for standard exciton-mediated Raman scattering [3], but at the energy where both polariton branches are closer, i.e., where exciton–cavity–phonon mixture is maximum. Moreover, the Raman intensity has a *minimum* and becomes unobservable when the polariton mode approaches the exciton energy. The same happens when the mode becomes pure photon-like.

The above results agree with the theory of cavity-polariton-mediated Raman scattering as expressed in (45). For the single outgoing resonance geometry being discussed τ_i can be taken as a constant, independent of detuning, determined by the intervening exciton lifetime. This follows from the fact that the incoming laser enters the cavity through the mirror residual transmission, and its energy falls within the exciton continuum. For the same reason, T_i can be taken as a constant proportional to the transmission of the Bragg mirror. The incoming vertex of the Raman process is not resonant and is thus not expected to depend strongly on the laser energy. Thus, one can assume that the detuning dependence of the exciton–phonon matrix element in (44) follows from that of the exciton strength of the *scattered* polariton. Consequently, for this particular experiment the Raman efficiency reads $\sigma_{\mathrm{pol}} \propto S^{\mathrm{s}}_{\mathrm{X}} S^{\mathrm{s}}_{\mathrm{P}} \Gamma_{\mathrm{s}}^{-1}$. This means that the process is optimized when the detuning is such that the scattered polariton couples best with the exterior continuum photons (i.e., when it is more photon-like) while, at the same time, it couples best with the optical phonons (that is, when it is more exciton-like). Thus, a compromise is obtained that predicts a maximum efficiency for zero detuning, decreasing towards the pure exciton and pure photonic limits. In nonoptically confined semiconductors the maximum efficiency occurs at the exciton energy, while in microcavities, polaritons with large exciton character cannot escape efficiently from the cavity. In Fig. 40 the solid curve represents an evaluation of $\sigma_{\mathrm{pol}} \propto S^{\mathrm{i}}_{\mathrm{X}} S^{\mathrm{i}}_{\mathrm{P}}$, adjusted to fit the maximum intensity of the resonance scan. The general behavior is well reproduced, particularly the maximum intensity for zero detuning ($\delta = 0$) and the decrease towards the pure exciton and photon limits. Note, however, that a clear departure from the model is observed for positive detunings. Based on (45) this can be explained by a decrease of the polariton lifetime (Γ_{s}^{-1}) due to the opening of additional relaxation channels when the upper polariton branch approaches the exciton continuum. These, and other finite-lifetime, effects on the Raman efficiency have been quite clearly evidenced through Raman scans in microcavities displaying multibranch anticrossings. Because of the richness of these results we will describe them in what follows in some detail.

6.3.4 Multibranch Polaritons: Finite-Lifetime Effects

Let us consider an outgoing resonant Raman experiment as described in the previous section but involving a strongly coupled microcavity displaying anticrossings between the cavity-photon mode and the first and second hydrogenic exciton states ($1s$ and $2s$).[3] In order to describe this system the Hamiltonian of (41) has to be extended to a 3×3 matrix including the three coupled modes. In doing this, the scattered polariton wavefunction becomes $|K_{\mathrm{s}}\rangle = a_{\mathrm{P}}\, |1_{\mathrm{P}}, 0_{\mathrm{X}_{1s}}, 0_{\mathrm{X}_{2s}}\rangle + a_{\mathrm{X}_{1s}}\, |0_{\mathrm{P}}, 1_{\mathrm{X}_{1s}}, 0_{\mathrm{X}_{2s}}\rangle + a_{\mathrm{X}_{2s}}\, |0_{\mathrm{P}}, 0_{\mathrm{X}_{1s}}, 1_{\mathrm{X}_{2s}}\rangle$, where, e.g., $|1_{\mathrm{P}}, 0_{\mathrm{X}_{1s}}, 0_{\mathrm{X}_{2s}}\rangle$ indicates the state involving one cavity photon and no excitons, and the coefficients a_l give the (detuning dependent) photon or exciton weight of the polariton state. For such an eigenvector in an outgoing resonant Raman process, the squared exciton–phonon matrix element in (44) is proportional to $|a^s_{\mathrm{X}_{1s}} + \alpha a^s_{\mathrm{X}_{2s}}|^2 \propto S^s_{1s} + \alpha^2\ S^s_{2s} + \alpha\ \beta\ S^s_{1s,2s}$ [64]. Here $S^s_n = |a^s_{\mathrm{X}_n}|^2$ are the exciton strengths, and $S^s_{1s,2s} = a^{s\,*}_{\mathrm{X}_{1s}} a^s_{\mathrm{X}_{2s}} + a^s_{\mathrm{X}_{1s}} a^{s\,*}_{\mathrm{X}_{2s}}$ is an interference term. $\alpha = \langle X^s_{2s}|H_{\mathrm{e-ph}}|\boldsymbol{K}_{\mathrm{i}}\rangle / \langle X^s_{1s}|H_{\mathrm{e-ph}}|\boldsymbol{K}_{\mathrm{i}}\rangle$, takes into

[3] Three-mode anticrossings involving the cavity-confined photons and the X_{1s} and X_{2s} states have been observed in II–VI cavities using reflectivity measurements. See, e.g., [144].

account the difference of the scattering-matrix elements between the common exciton initial state and the different exciton components of the final polariton state. The coherence factor β has been introduced to describe partial coherence between the polarization associated with the two intervening exciton levels [64]. It basically defines whether the contributions from X_{1s} and X_{2s} must be added before or after squaring in the Raman efficiency. A similar description has been discussed by *Menéndez* and *Cardona* [145], to account for the interference between allowed and forbidden Raman scattering by LO phonons in bulk GaAs.

It can also be shown that the line broadening as a function of detuning is, to a good approximation, $\Gamma_{\mathrm{s}} = S^s_{\mathrm{P}}\gamma_{\mathrm{P}} + S^s_{1s}\gamma_{1s} + S^s_{2s}\gamma_{2s}$ [146]. Here, γ_n correspond to the cavity-photon or exciton homogeneous linewidths. From (44), the polariton-mediated outgoing RRS efficiency for a three-polariton branch cavity can then be expressed as:

$$\sigma_{\mathrm{pol}} \propto \Gamma_{\mathrm{s}}^{-1} S^s_{\mathrm{P}} (S^s_{\mathrm{X}_{1s}} + \alpha^2 S^s_{\mathrm{X}_{2s}} + \alpha\beta S^s_{\mathrm{X}_{1s,2s}}) . \tag{46}$$

In Fig. 41a the polariton-mode energies as a function of spot position are shown for a $\lambda/2$ II–VI cavity grown with a thickness gradient and displaying a three coupled mode behavior at low temperatures [64]. The cavity has three 72 Å CdTe QWs located at the antinode of the photon field in the $\mathrm{Cd_{0.4}Mg_{0.6}Te}$ spacer. The cavity spacer is enclosed by $\mathrm{Cd_{0.4}Mg_{0.6}Te/Cd_{0.75}Mn_{0.25}Te}$ Bragg mirrors, 21 pairs on top, and 15.5 at the bottom. Both the experimental data (full circles) and a fit using a three coupled mode model, with Rabi splittings $\Omega_{1s} = 13.4\,\mathrm{meV}$ and $\Omega_{2s} = 4.5\,\mathrm{meV}$ (solid curves) are shown. The upper polariton mode was not observable at the low temperature used (4.5 K). The exciton and photon strengths for the medium polariton (MP) branch derived from the fit in (a) are displayed in Fig. 41b. The dashed lines represent the exciton strength (S_{X1} and S_{X2}), the dotted-dashed curve the interference term ($S_{\mathrm{X1,X2}}$), while the thin solid line corresponds to the photon strength (S_{P}). The thick solid curve, on the other hand, is the photon strength multiplied by the factor Γ_{s}^{-1} that accounts for the lifetime contribution (for clarity of the figure Γ_{s}^{-1} has been normalized to unity). For this figure, the cavity-photon and exciton linewidths have been taken, respectively, as $\gamma_{\mathrm{P}} = 0.4\,\mathrm{meV}$, $\gamma_{\mathrm{X1}} = 1\,\mathrm{meV}$, and $\gamma_{\mathrm{X2}} = 1\,\mathrm{meV}$. MP resonant Raman profiles, calculated using (46) and the parameters displayed in panel (b), are shown in Fig. 41c with (full curve) and without (dashed curve) damping factor Γ_{s}^{-1}. For these outgoing resonant medium polariton Raman profiles the interaction of each exciton state with the involved LO phonons was taken to be equal ($\alpha = 1$), while their contributions to the Raman-scattered wave were assumed to contribute incoherently ($\beta = 0$). The dependence of the Raman profiles on β has been described in [64].

For interpreting the theoretical results of Fig. 41 let us first consider the situation without damping. The energies for which S_∂ intersects the excitonic strengths $S_{\mathrm{X}_{1s}}$ and $S_{\mathrm{X}_{2s}}$ correspond to an almost optimized scattering efficiency and thus to maxima (two in this case) in the resonant Raman scan.

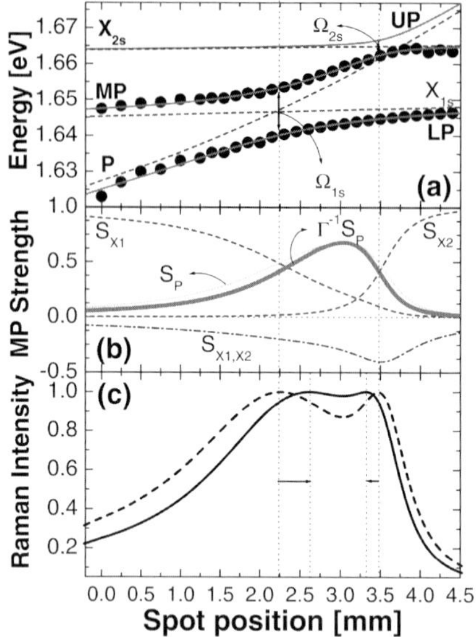

Fig. 41. (**a**) Polariton-mode energies at $T = 4.5\,\mathrm{K}$ as a function of spot position on a tapered II–VI microcavity. Experimental data (*full circles*) and a fit using a three coupled oscillator model (*solid curves*) are shown. (**b**) Medium polariton (MP) strengths derived from the fit in (**a**). The *dashed lines* represent the exciton strength (S_{X1} and S_{X2}), while the interference term ($S_{\mathrm{X1,X2}}$) is indicated with a *dash-dotted curve*. The *thin solid line* corresponds to the photon strength (S_{P}). The *thick solid curve* is the latter multiplied by Γ_{s}^{-1} (normalized to unity for illustration purposes). (**c**) MP-resonant Raman profile calculated using the parameters derived in (**b**), with (*full curve*) and without (*dashed curve*) damping factor Γ_{s}^{-1}. The linewidths have been taken as $\gamma_{\mathrm{P}} = 0.4\,\mathrm{meV}$, $\gamma_{\mathrm{X1}} = 1\,\mathrm{meV}$, $\gamma_{\mathrm{X2}} = 1\,\mathrm{meV}$, and $\alpha = 1$, $\beta = 0$

Since in each crossing the exciton strength of the third (not involved) state is relatively small, the energies of the maxima basically coincide with the intersections. However, note that if X_{1s} and X_{2s} were closer in energy, or their corresponding Rabi splittings were larger, the resonant Raman scan maxima would not necessarily be given by the condition $S_{\mathrm{p}} = S_{\mathrm{X}_{1s}}$ and $S_{\mathrm{p}} = S_{\mathrm{X}_{2s}}$. As for the simpler two-level case analyzed before, well away from these crossings and towards the pure excitonic limits of the MP (either X_{1s} or X_{2s}), the Raman efficiency drops to zero. Although in these regimes polaritons can interact well with phonons, their participation in a Raman-scattering process is hampered by the inefficiency to couple with the outside continuum photons. As follows from Fig. 41c, when damping is considered the two maxima shift, leading to a reduced peak separation. Note that in the case considered, $\gamma_{\mathrm{P}} < \gamma_{\mathrm{X1}}, \gamma_{\mathrm{X2}}$. Thus, the medium polariton lifetime ($\tau = \hbar/\gamma$) is mainly determined by exciton dephasing (τ_{X}) and not by the cavity-photon lifetime (τ_{P}). In this case it is to be expected that the maximum Raman efficiency will be obtained not for $\delta = 0$ (i.e., for $S_{\mathrm{p}} = S_{\mathrm{X}}$) but for a polariton state

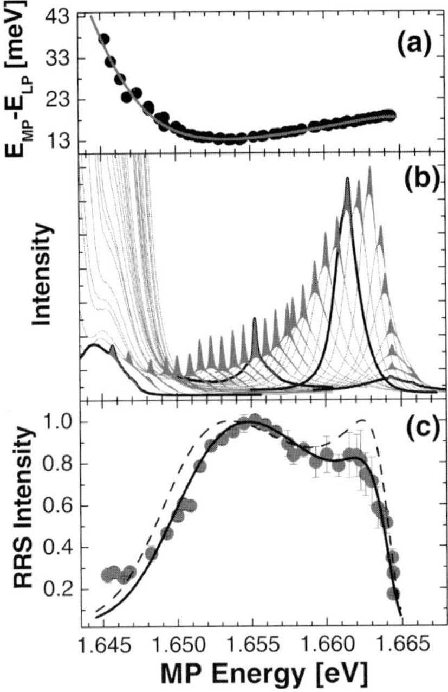

Fig. 42. (**a**) Energy difference between the medium (MP) and low (LP) polariton modes (*full circles*) of the cavity described in Fig. 41. The *solid curve* is the energy difference calculated from a fit to the polariton dispersion using a three coupled oscillator model. (**b**) Raman spectra at outgoing resonance with the MP for various values of detuning. The Raman peak for each detuning is highlighted on top of the MP luminescence. (**c**) Normalized resonant Raman intensity as a function of the MP mode energy derived from the spectra in (**b**). The *solid (dashed) curve* is a calculation with the CPMS model including (neglecting) lifetime effects

with a detuning such that $S_{\mathrm{p}} > S_{\mathrm{X}}$ (predominant photonic character) since in this case the corresponding polariton lifetime will be larger. Hence, the effect of damping is to *reduce* the separation of the Raman intensity peaks that appear as the detuning is varied. If, on the other hand, $\gamma_{\mathrm{P}} \gg \gamma_{\mathrm{X1}}, \gamma_{\mathrm{X2}}$ the shift occurs towards the opposite direction.

Figure 42b presents the Raman spectra of the cavity described in Fig. 41 at outgoing resonance with the MP, for various values of detuning. The Raman peak for each detuning is highlighted on top of the MP luminescence. The energy difference between the medium (MP) and low (LP) polariton modes derived from the data shown in Fig. 41a is displayed as a function of the MP energy with full circles in panel (a). The solid curve is the difference calculated from the fit using the three-oscillator model. The normalized resonant Raman intensity as a function of the MP mode energy derived from the spectra in (b) is shown in panel (c). The solid curve in panel (c) is a calculation with the CPMS model including lifetime effects and using as parameters the photonic and excitonic strengths derived from the fit with the three-oscillator model, and $\gamma_P = 0.4\,\text{meV}$, $\gamma_{X1} = 2\,\text{meV}$, $\gamma_{X2} = 2\,\text{meV}$, and $\alpha = 1$, $\beta = 0.4$. This fit gave the best description of the experimental data. The dashed curve in Fig. 42c is the resonant Raman profile calculated by neglecting polariton damping and taking $\alpha = 1$ and $\beta = 0$. Two important features should be highlighted from these data. First, a double-peaked Raman intensity scan is observed, with efficiency tending to zero at the pure excitonic limits. Again, this is a clear result that contrasts with what would be expected for QWs without a cavity, for which Raman maxima occur precisely *at* the exciton energies. This result finds a natural explanation within the CPMS model. Secondly, note from the experimental data in panels (c) and (a) that the lower-energy Raman intensity peak is clearly blueshifted from the minimum of the LP–MP mode separation (i.e., cavity-X_{1s} zero detuning). This is a clear signature of a lifetime effect. In fact, although the CPMS model without damping provides a qualitative description of the data, it cannot account for the Raman scan peak positions. The latter can be well accounted for by theory when the contribution of Γ_s^{-1} is included in the model.

We note that the choice of damping factors γ_P and γ_X is not arbitrary. On the one hand, they can be obtained from independent measurements. In fact, the values chosen are compatible with those published in the literature [139, 147, 148]. On the other hand, Γ_s^{-1} is reflected on the detuning dependence of the spectral density P_E in (44), a parameter that can be readily accessed by performing Raman scans as a function of laser (or, equivalently, scattered) energy for fixed values of detuning. In fact, the values $\gamma_P = 0.4\,\text{meV}$, $\gamma_{X1} = 2\,\text{meV}$, $\gamma_{X2} = 2\,\text{meV}$ are compatible with the measured MP spectral densities as reflected on the outgoing resonance Raman scan as a function of laser energy tuned to the medium polariton (MP) mode shown for two different detunings ((a) MP at 1.653 eV and (b) MP at 1.660 eV) in Fig. 43. In this figure the Raman intensities, shown as a function of the Stokes energy, have been normalized to one. The solid curves are fits assuming a Lorentzian probability function with linewidth Γ_s, clearly providing a consistent description of all the data.

7 Conclusions

In this review Chapter we addressed the potentialities of optical microcavities for the study of weakly scattering objects, and the richness of the physics that microcavities can generate, involving either nanostructure excitations or cavity polaritons. In concluding, we enumerate what we believe are the most interesting and challenging open problems related to Raman scattering in microcavities.

As shown throughout this Chapter, optical confinement using planar microcavities can lead to Raman amplification in the range of 10^4–10^6. This

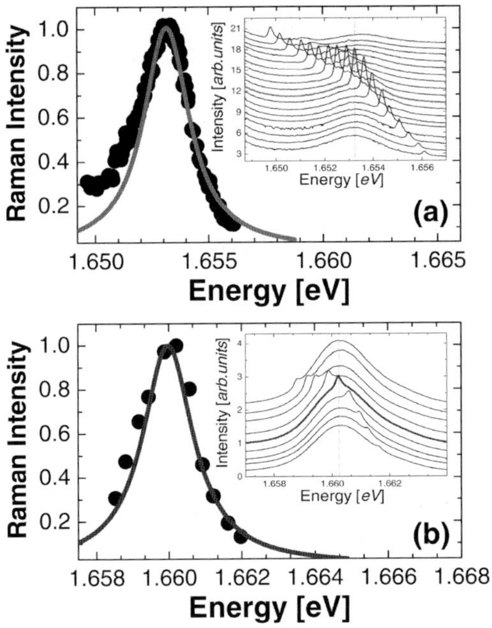

Fig. 43. Outgoing resonance Raman scan as a function of laser energy tuned to the medium polariton (MP) mode, for two different detunings: (**a**) MP at 1.653 eV and (**b**) MP at 1.660 eV. The *insets* show the CdTe QWs first-order Raman LO-phonon peak ($\sim 172\,\mathrm{cm}^{-1}$) on top of the MP luminescence. Note the resonant enhancement of the Raman peak as the scattered light is tuned with the polariton mode. The *solid curves* are fits using the factorization model and taking $\gamma_{\mathrm{P}} = 0.4\,\mathrm{meV}$, $\gamma_{\mathrm{X1}} = 2\,\mathrm{meV}$, and $\gamma_{\mathrm{X2}} = 2\,\mathrm{meV}$

optical resonant amplification can be accompanied, in addition, by exciton-related enhancements of about 10^3, leading to achievable Raman efficiency increases of the order of 10^9. Such enhancements open the way to a series of challenging possibilities. First, we believe that an interesting open problem is that of the Raman spectroscopy (of phonons or electronic excitations) of *single* low-dimensional nanostructures such as, e.g., single monolayers, wires, and even dots. The Raman investigation of a single semiconductor dot is, we believe, a stimulating project that could benefit from the potentialities of cavity enhancement. Such investigations would avoid the inhomogeneous broadening inherent in having a large number of "samples", increasing enormously the potentialities of Raman scattering as a microscopic tool able to access the different excitations of these newly developed nanostructures. Secondly, stimulated Raman scattering could also benefit from optical confinement to reduce the scattering volumes by as much as the Raman efficiency is enhanced, e.g., from 6 to 9 orders of magnitude. It remains to be demonstrated whether planar semiconductor microcavities can lead to Raman gain under pumped or continuous-wave excitation. Thirdly, Raman stimulation implies also a stimulated emission of coherent phonons. It is an interesting issue whether acoustic and optical cavities can contribute essential ingredients for the development of phonon "lasers". Acoustic cavities based on distributed Bragg reflectors as described here could provide the required feedback mechanism, while optical microcavities could be exploited to efficiently seed the stimulated-emission process.

All the experiments reviewed here, and to the best of our knowledge all the reported investigations of Raman amplification in condensed-matter cavities, rely on planar one-dimensional confining structures. However, and in the same way as micrometer-size liquid droplets have led to enormous Raman gain, it is possible to conceive higher-dimensional optical confining structures that could be exploited for Raman amplification. These include pillars that, based on 1D cavities, achieve 3D optical confinement by lateral nanostructuring. Also whispering gallery modes in high-Q disk-like cavities can be a possible choice in the search of Raman signals from single, or a few, semiconductor dots. It would also be interesting to search for Raman amplification using high-Q modes related to defects and high density of states extended states in 2- and 3-dimensional photonic bandgap devices.

Last but not least, the reported experiments dealing with cavity-polariton-mediated scattering are extremely few, and the available theories quite crude. We believe that this remains an interesting and quite unexplored territory for research on a very basic and fundamental light–matter interaction process.

Acknowledgements

Many colleagues and students have contributed through the years to the reported research and have collaborated through stimulating and enlightening discussions. All of them are deeply acknowledged, including Richard Planel, Veronique Thierry-Mieg, Jacqueline Bloch, Bernard Sermage, Izo Abram, Regis André, Thomas Freixanet, Florent Perez, Mariano Trigo, Axel Bruchhausen, Paul Lacharmoise, Florencia Pascual Winter, Daniel Lanzillotti Kimura, and Guillermo Rozas. The work presented here is the result of a collaboration supported by SECyT-ECOS(Sud) from Argentina and France. A. F. would also like to thank the John Simon Guggenheim Memorial Foundation for its generous support.

References

[1] A. Pinczuk, E. Burstein: in M. Cardona (Ed.): *Light Scattering in Solids*, vol. 1 (Spinger, Berlin 1975) p. 23

[2] L. Falicov, P. Martin: in M. Cardona (Ed.): *Light Scattering in Solids*, vol. 1 (Springer, Berlin, Heidelberg 1975) p. 79

[3] M. Cardona: in M. Cardona, G. Güntherodt (Eds.): *Light Scattering in Solids*, vol. 2 (Springer, Berlin, Heidelberg 1982) p. 19

[4] B. Jusserand, M. Cardona: in M. Cardona, G. Güntherodt (Eds.): *Light Scattering in Solids*, vol. 5 (Springer, Berlin, Heidelberg 1989) p. 49

[5] N. Esser, W. Richter: in M. Cardona, G. Güntherodt (Eds.): *Light Scattering in Solids*, vol. 8 (Springer, Berlin, Heidelberg 2000)

[6] H. Metiu: Prog. Surf. Sci. **17**, 153–320 (1982)

[7] A. Otto: in M. Cardona, G. Güntherodt (Eds.): *Light Scattering in Solids*, vol. 4 (Springer, Berlin, Heidelberg 1984)

[8] K. Kneipp, H. Kneipp, I. Itzkan, R. R. Dasari, M. S. Feld: J. Phys.: Condens. Matter **14**, 597 (2002)

[9] R. Zeyher, K. Arya: in M. Cardona, G. Güntherodt (Eds.): *Light Scattering in Solids*, vol. 4 (Springer, Berlin, Heidelberg 1984)

[10] K. Kneipp, Y. Wang, H. Kneipp, L. T. Perelman, I. Itzkan, R. R. Dasari, M. S. Feld: Phys. Rev. Lett. **78**, 1667 (1997)

[11] P. Etchegoin, R. C. Maher, L. F. Cohen, R. J. C. Brown, H. Hartigan, M. J. T. Milton, J. C. Gallop: Chem. Phys. Lett. **375**, 84 (2003)

[12] E. M. Purcell: Phys. Rev. **69**, 681 (1946)

[13] D. Kleppner: Phys. Rev. Lett. **47**, 233 (1981)

[14] R. J. Thompson, G. Rempe, H. J. Kimble: Phys. Rev. Lett. **68**, 1132 (1992)

[15] E. Burstein, C. Weisbuch (Eds.): *Confined Electrons and Photons: New Physics and Applications* (Plenum, New York 1995)

[16] R. E. Kunz, W. Lukosz: Phys. Rev. B **21**, 4814 (1980)

[17] E. Snoeks, A. Lagendijk, A. Polman: Phys. Rev. Lett. **74**, 2459 (1995)

[18] Y. Yamamoto, R. E. Slusher: Phys. Today **46**, 66 (1993)

[19] R. E. Slusher, C. Weisbuch: Solid State Commun. **92**, 149 (1994)

[20] M. S. Skolnick, T. A. Fisher, D. M. Whittaker: Semicond. Sci. Technol. **13**, 645 (1998)

[21] Semicond. Sci. Technol. **18** special number, edited by J. Baumberg and L. Viña, dedicated to the physics of microcavities

[22] E. Yablonovitch: Phys. Rev. Lett. **58**, 2059 (1987)

[23] K. H. Drexhage: J. Lumin. **12**, 693 (1970)

[24] K. H. Tews: J. Lumin. **9**, 223 (1974)

[25] R. R. Chance, A. Prock, R. Silbey: J. Chem. Phys. **60**, 2744 (1974)

[26] R. R. Chance, A. Prock, R. Silbey: J. Chem. Phys. **62**, 771 (1975)

[27] R. R. Chance, A. Prock, R. Silbey: in I. Prigogine, S. A. Rice (Eds.): *Advances in Chemical Physics* (Wiley, New York 1978) pp. 1–65

[28] Z. L. Zhang, M. Nishioka, C. Weisbuch, Y. Arakawa: Appl. Phys. Lett. **64**, 1068 (1994)

[29] F. De Martini, P. Mataloni, L. Crescentini: Opt. Lett. **17**, 1370 (1992)

[30] G. A. N. Connell, R. J. Nemanich, C. C. Tsai: Appl. Phys. Lett. **36**, 31 (1980)

[31] R. J. Nemanich, C. C. Tsai, M. J. Thompson, T. W. Sigmon: J. Vac. Sci. Technol. **19**, 685 (1981)

[32] J. B. Snow, Shi-Xiong Qian, R. K. Chang: Opt. Lett. **10**, 37 (1985)

[33] Shi-Xiong Qian, R. K. Chang: Phys. Rev. Lett. **56**, 926 (1986)

[34] H.-B. Lin, J. D. Eversole, A. J. Campillo: Opt. Lett. **17**, 828 (1992)

[35] H.-B. Lin, A. J. Campillo: Phys. Rev. Lett. **73**, 2440 (1994)

[36] F. Cairo, F. De Martini, D. Murra: Phys. Rev. Lett. **70**, 1413 (1993)

[37] F. De Martini, M. Marrocco, C. Pastina, F. Viti: Phys. Rev. A **53**, 471 (1996)

[38] M. Marrocco: Phys. Rev. A **54**, 3626 (1996)
[39] P. Senellart, J. Bloch: Phys. Rev. Lett. **82**, 1235 (1999)
[40] F. Boeuf, R. André, R. Romestain, L. Si Dang, E. Péronne, J. F. Lampin, D. Hulin, A. Alexandrou: Phys. Rev. B **62**, R2279 (2000)
[41] A. Fainstein, B. Jusserand, V. Thierry-Mieg: Phys. Rev. B **53**, 13287 (1996)
[42] A. Fainstein, B. Jusserand, V. Thierry-Mieg: Phys. Rev. Lett. **75**, 3764 (1995)
[43] A. Fainstein, M. Trigo, D. Oliva, B. Jusserand, T. Freixanet, V. Thierry-Mieg: Phys. Rev. Lett. **86**, 3411 (2001)
[44] P. Lacharmoise, A. Fainstein, B. Jusserand, V. Thierry-Mieg: Appl. Phys. Lett. **84**, 3274 (2004)
[45] A. Fainstein, B. Jusserand: Phys. Rev. B **57**, 2402 (1998)
[46] A. I. Tartakovskii, M. Emam-Ismail, D. G. Lidzey, M. S. Skolnick, D. D. C. Bradley, S. Walker, V. M. Agranovich: Phys. Rev. B **63**, 121302 (2001)
[47] L. A. Kuzik, V. A. Yakovlev, G. Mattei: Appl. Phys. Lett. **75**, 1830 (1999)
[48] A. Fainstein, B. Jusserand: Phys. Rev. B **54**, 11505 (1996)
[49] A. Mlayah, O. Marco, J. R. Huntzinger, A. Zwick, R. Carles, V. Bardinal, C. Fontaine, A. Muñoz-Yagüe: J. Phys.: Condens. Matter **10**, 9535 (1998)
[50] M. Trigo, A. Bruchhausen, A. Fainstein, B. Jusserand, V. Thierry-Mieg: Phys. Rev. Lett. **89**, 227402 (2002)
[51] T. Kipp, L. Rolf, C. Schüller, D. Endler, C. Heyn, D. Heitmann: Phys. Rev. B **63**, 195304 (2001)
[52] B. Bendow, J. L. Birman: Phys. Rev. B **1**, 1678 (1970)
[53] B. Bendow: Phys. Rev. B **2**, 5051 (1970)
[54] B. Bendow: Polariton theory of resonance raman scattering in solids, in *Springer Tracts in Modern Physics*, vol. 82 (Springer, Berlin, Heidelberg 1978)
[55] R. Zeyher, C.-S.Ting, J. L. Birman: Phys. Rev. B **10**, 1725 (1974)
[56] M. Matsushita, J. Wicksted, H. Z. Cummins: Phys. Rev. B **29**, 3362 (1984)
[57] M. Matsushita, M. Nakayama: Phys. Rev. B **30**, 2074 (1984)
[58] C. Weisbuch, R. G. Ulbrich: in M. Cardona, G. Güntherodt (Eds.): *Light Scattering in Solids*, vol. 3 (Springer, Berlin, Heidelberg 1989)
[59] F. Ashary, P. Yu: Solid State Commun. **47**, 241 (1983)
[60] A. Nakamura, C. Weisbuch: Solid State Commun. **32**, 301 (1979)
[61] A. Fainstein, B. Jusserand, V. Thierry-Mieg: Phys. Rev. Lett. **78**, 1576–1579 (1997)
[62] W. R. Tribe, M. S. Skolnick, T. A. Fisher, D. J. Mowbray, D. Baxter, J. S. Roberts: Phys. Rev. B **56**, 12429 (1997)
[63] A. Fainstein, B. Jusserand, R. André: Phys. Rev. B **57**, 9439 (1998)
[64] A. Bruchhausen, A. Fainstein, B. Jusserand, R. André: Phys. Rev. B **68**, 205326 (2003)
[65] J. M. Raimond, S. Haroche: in E. Burstein, C. Weisbuch (Eds.): *Confined Electrons and Photons* (Plenum, New York 1995) p. 383
[66] A. Yariv, P. Yeh: *Optical Waves in Crystals* (Wiley, New York 1984)
[67] M. Born, E. Wolf: *Principles of Optics* (Plenum, New York 1980)
[68] C. Weisbuch, M. Nishioka, A. Ishikawa, Y. Arakawa: Phys. Rev. Lett. **69**, 3314 (1992)
[69] V. Narayanamurti, H. L. Störmer, M. A. Chin, A. C. Gossard, W. Wiegmann: Phys. Rev. Lett. **43**, 2012 (1979)

[70] R. P. Stanley, R. Houdré, U. Oesterle, M. Ilegems, C. Weisbuch: Phys. Rev. A **48**, 2246 (1993)
[71] G. Panzarini, L. C. Andreani, A. Armitage, D. Baxter, M. S. Skolnick, V. N. Astratov, J. S. Roberts, A. V. Kavokin, M. R. Vladimirova, M. A. Kaliteevski: Phys. Rev. B **59**, 5082 (1999)
[72] A. Kastler: Appl. Opt. **1**, 17 (1962)
[73] I. Abram, S. Iung, R. Kuszelewicz, G. Le Roux, C. Licoppe, J. L. Oudar, E. V. K. Rao, J. I. Bloch, R. Planel, V. Thierry-Mieg: Appl. Phys. Lett. **65**, 2516 (1994)
[74] Y. Dumeige, I. Sagnes, P. Monnier, P. Vidakovic, I. Abram, C. Mériadec, A. Levenson: Phys. Rev. Lett. **89**, 043901 (2002)
[75] A. Fainstein, N. D. Lanzillotti Kimura, B. Jusserand: Proceedings of the SPIE, in (2005) p. 254
[76] R. P. Stanley, R. Houdré, U. Oesterle, M. Ilegems, C. Weisbuch: Appl. Phys. Lett. **65**, 2093 (1994)
[77] G. Malpuech, A. Kavokin: Semicond. Sci. Technol. **16**, 1 (2001)
[78] R. Sapienza, P. Costantino, D. Wiersma, M. Ghulinyan, C. J. Oton, L. Pavesi: Phys. Rev. Lett. **91**, 263902 (2003)
[79] V. Agarwal, J. A. del Río, G. Malpuech, M. Zamfirescu, A. Kavokin, D. Coquillat, D. Scalbert, M. Vladimirova, B. Gil: Phys. Rev. Lett. **92**, 097401 (2004)
[80] N. D. Lanzillotti Kimura, A. Fainstein, B. Jusserand: Phys. Rev. B **71**, 041305 (2005)
[81] A. Alexandrou, M. Cardona, K. Ploog: Phys. Rev. B **38**, 2196 (1988)
[82] A. Alexandrou, C. Trallero-Giner, G. Kanellis, M. Cardona: Phys. Rev. B **40**, 1013 (1989)
[83] P. Pellandini, R. P. Stanley, R. Houdré, U. Oesterle, M. Ilegems: Appl. Phys. Lett. **71**, 864 (1997)
[84] A. Armitage, M. S. Skolnick, V. N. Astratov, D. M. Whittaker, G. Panzarini, L. C. Andreani, T. A. Fisher, J. S. Roberts, A. V. Kavokin, M. A. Kaliteevski, M. R. Vladimirova: Phys. Rev. B **57**, 14877 (1998)
[85] B. Jusserand, T. Freixanet, A. Fainstein: Physica E **7**, 646 (2000)
[86] J. A. Kong: in *Electromagnetic Wave Theory* (Wiley-Interscience, New York 1986) p. 309
[87] A. Sommerfeld: Ann. Phys. (Leipzig) **28**, 665 (1909)
[88] H. Weyl: Ann. Phys. (Leipzig) **60**, 481 (1919)
[89] B. Vander Pol: Physica (Utrecht) **2**, 843 (1935)
[90] I. Abram, J. L. Oudar: Phys. Rev. A **51**, 4116 (1995)
[91] B. Jusserand, F. Mollot, D. Paquet: Surf. Sci. **228**, 151 (1990)
[92] B. Jusserand, D. Paquet, F. Mollot, F. Alexandre, G. Le Roux: Phys. Rev. B **35**, 2808 (1987)
[93] M. Trigo, A. Fainstein, B. Jusserand, V. Thierry-Mieg: Phys. Rev. B **66**, 125311 (2002)
[94] K. Huang, B. Zhu: Phys. Rev. B **38**, 13377 (1988)
[95] R. Pérez-Alvarez, F. G. Moliner, V. R. Velasco, C. Trallero-Giner: J. Phys.: Condens. Matter **5**, 5389 (1993)
[96] M. P. Chamberlain, M. Cardona, B. K. Ridley: Phys. Rev. B **48**, 14356 (1993)
[97] E. Richter, D. Strauch: Solid State Commun. **64**, 867 (1987)
[98] S. Baroni, P. Giannozi, E. Molinari: Phys. Rev. B **41**, 3870 (1990)

[99] D. J. Evans, S. Ushioda, J. D. McMullen: Phys. Rev. Lett. **31**, 369 (1973)
[100] M. Nakayama, M. Ishida, N. Sano: Phys. Rev. B **38**, 6348 (1988)
[101] R. E. Camley, D. L. Mills: Phys. Rev. B **29**, 1695 (1984)
[102] B. L. Johnson, J. T. Weiler, R. E. Camley: Phys. Rev. B **32**, 6544 (1985)
[103] P. Hawrylak, G. Eliasson, J. J. Quinn: Phys. Rev. B **34**, 5368 (1986)
[104] J. K. Jain, P. B. Allen: Phys. Rev. Lett. **54**, 947 (1985)
[105] J. K. Jain, P. B. Allen: Phys. Rev. Lett. **54**, 2437 (1985)
[106] G. Eliasson, P. Hawrylak, J. J. Quinn: Phys. Rev. B **35**, 5569 (1987)
[107] A. Pinczuk, M. G. Lamont, A. C. Gossard: Phys. Rev. Lett. **56**, 2092 (1986)
[108] A. J. Shields, M. Cardona, K. Eberl: Phys. Rev. Lett. **72**, 412 (1994)
[109] A. K. Sood, J. Menendez, M. Cardona, K. Ploog: Phys. Rev. Lett. **54**, 2111 (1985)
[110] S. Adachi: J. Appl. Phys. **58**, R4 (1985)
[111] M. Zunke, R. Schorer, G. Abstreiter, W. Klein, G. Weimann, M. P. Chamberlain: Solid State Commun. **93**, 847 (1995)
[112] P. X. Zhang, D. J. Lockwood, J. M. Baribeau: Can. J. Phys. **70**, 843 (1992)
[113] M. Giehler, T. Ruf, M. Cardona, K. Ploog: Phys. Rev. B **55**, 7124 (1997)
[114] J. He, J. Sapriel, J. Chavignon, R. Azoulay, L. Dugrand, F. Mollot, B. Djafari-Rouhani, R. Vacher: J. Phys. (Paris) **48-C5**, 573 (1987)
[115] S. M. Rytov: Akust. Zh. **2**, 71 (1956)
[116] J. He, B. Djafari-Rouhani, J. Sapriel: Phys. Rev. B **37**, 4089 (1988)
[117] H. Brugger, H. Reiner, G. Abstraiter, H. Jorke, H. J. Herzog, E. Kasper: Superlattices and Microstructures **2**, 451 (1986)
[118] P. A. Fokker, J. I. Dijkhuis, H. W. de Wijn: Phys. Rev. B **55**, 2925 (1997)
[119] I. Camps, S. S. Makler, H. M. Pastawski, L. E. F. Foa Torres: Phys. Rev. B **64**, 125311 (2001)
[120] M. Canonico, C. Poweleit, J. Menéndez, A. Debernardi, S. R. Johnson, Y. H. Zhang: Phys. Rev. Lett. **88**, 215502 (2002)
[121] J. Chen, J. B. Khurgin, R. Merlin: Appl. Phys. Lett. **80**, 2901 (2002)
[122] P. V. Santos, L. Ley, J. Mebert, O. Koblinger: Phys. Rev. B **36**, 4858 (1987)
[123] N. M. Stanton, R. N. Kini, A. J. Kent, M. Henini, D. Lehmann: Phys. Rev. B **68**, 113302 (2003)
[124] J. M. Worlock, M. L. Roukes: Nature **421**, 802 (2003)
[125] M. F. Pascual Winter, A. Fainstein, M. Trigo, T. Eckhause, R. Merlin, C. Chen: Phys. Rev. B **71**, 085305 (2005)
[126] J. J. Hopfield: Phys. Rev. **182**, 945 (1969)
[127] M. Nawrocki, R. Planel, Benoît à la Guillaume: Phys. Rev. Lett. **36**, 1343 (1976)
[128] V. A. Kisilev, B. S. Razbirin, I. N. Uraltsev: phys. stat. sol. (b) **72**, 161 (1975)
[129] R. G. Ulbrich, C. Weisbuch: Phys. Rev. Lett. **38**, 865 (1977)
[130] J. Wicksted, M. Matsushita, H. Z. Cummins, T. Shigenari, X. Z. Lu: Phys. Rev. B **29**, 3350 (1984)
[131] R. Zeyher, J. L. Birman, W. Brenig: Phys. Rev. B **6**, 4613 (1972)
[132] A. Bruchhausen, M. Trigo, A. Fainstein, B. Jusserand, R. André: in *Proceedings of the XXVI-International Conference on the Physics of Semiconductors* (Edinburgh (Scotland, UK) 2002)
[133] R. M. Stevenson, V. N. Astratov, M. S. Skolnick, J. S. Roberts, G. Hill: Phys. Rev. B **67**, 081301 (2003)

[134] A. Bruchhausen, A. Fainstein, B. Jusserand, R. André: in *Proceedings of the XXVII-International Conference on the Physics of Semiconductors* (Flagstaff (USA) 2004)
[135] J. Bloch, T. Freixanet, J. Y. Marzin, V. Thierry-Mieg, R. Planel: Appl. Phys. Lett. **73**, 1694 (1998)
[136] J. B. Khurgin: Solid State Commun. **117**, 307 (2001)
[137] H. Deng, G. Weihs, C. Santori, J. Bloch, Y. Yamamoto: Science **298**, 199 (2002)
[138] J. C. Merle, R. Sooryakumar, M. Cardona: Phys. Rev. B **30**, 3261 (1984)
[139] B. Sermage, S. Long, I. Abram, J. Y. Marzin, J. Bloch, R. Planel, V. Thierry-Mieg: Phys. Rev. B **53**, 16516 (1996)
[140] P. Borri, J. R. Jensen, W. Langbein, J. M. Hvam: Phys. Rev. B **61**, R13377 (2000)
[141] J. J. Sakurai: in *Modern Quantum Mechanics* (Benjamin/Cummings, Menlo Park, California 1985) pp. 325–335
[142] M. L. Goldberger, K. M. Watson: *Collision Theory* (Wiley, NY 1964)
[143] A. Fainstein, B. Jusserand, V. Thierry-Mieg, R. André: phys. stat. sol. (b) **215**, 403 (1999)
[144] F. Boeuf: Ph.D. thesis, Université Joseph Fourier-Grenoble 1, CNRS, Boîte Postale 87, F-38402, St. Martin d'Heres Cedex, France
[145] J. Menéndez, M. Cardona: Phys. Rev. Lett. **51**, 1297 (1983)
[146] J. Wainstain, C. Delalande, D. Gendt, M. Voos, J. Bloch, V. Thierry-Mieg, R. Planel: Phys. Rev. B **58**, 7269 (98)
[147] R. P. Stanley, R. Houdré, C. Weisbuch, U. Oesterle, M. Ilegems: Phys. Rev. B **53**, 10995 (1996)
[148] M. Müller, J. Bleuse, R. André: Phys. Rev. B **62**, 16886 (2000)

Index

absorption, 20
acoustic
  cavities
    raman spectra, 78
  impedance
    mismatch, 75
  phonons
    finite-size effects, 68
additional boundary conditions (ABCs), 84, 90
adiabatic approximation, 35
Airy function, 28
AlAs, 29, 31, 36, 60
AlAs/$Al_{0.33}Ga_{0.67}As$, 52
AlGaAs/AlAs, 45
amplification
  light–sound interaction, 76
  optical resonant, 43
angle tuning, 36, 37
  one-mode cavities, 39
angular selectivity, 50
antenna effect, 44
anti-Stokes process, 33
anticrossings, 67
  multibranch, 98

backscattering, 79
$BaTiO_3$, 76
benzene, 21
Bloch oscillation
  electron, 33
  photons, 33
boundary conditions
  electrostatic, 58
  mechanical, 58

Bragg
  microcavity
    reflectivity, 30
  mirror, 29, 88
    angular dependence of the reflectivity, 26–27
  reflector, 26
    distibuted (DBR), 19, 21, 23
Brillouin
  zone
    center, 58
    edge, 58
    extended scheme, 25
    schemes, 68
broadening
  homogeneous, 90
  inhomogeneous, 38

cavity
  acoustic, 17
  cartography, 38
  double, 36
  double structure, 32
  GaAs, 38
  high-Q, 52, 57
  high-finesse, 39
  II–VI, 98
  low-Q, 52
  multimode, 31
  multiple coupled, 32
  piezoelectrically controlled, 39
  polariton-mediated scattering, 91
  polaritons, 22
  Q-factors, 21
  Raman spectra, 53
  resonance condition, 37
  two-mode, 40
$Cd_{1-x}Mg_xTe$, 36
$Cd_{1-x}Mn_xTe$, 36
CdTe, 36
  QW, 97, 99
  QW mode, 97
collection cone, 90
colloids, 19
confinement
  photon, 17
conservation
  energy, 35
  wavevector, 35
coulomb polarization fields, 62
coupled microcavities, 32
coupled oscillators, 87
coupling
  between cavity-photons and excitons, 37
  strong, 17
  weak, 17
crystal-field transitions, 18
crystal-momentum conservation, 68

damping
  polariton, 102
DBR, 23
  acoustic, 23
  GaAs/AlAs, 60
  penetration depth, 30
  reflectivity, 26
  structure, 26
  transmission, 26
dephasing time, 93
dielectric
  continuum model, 58, 59
  function
    resonant critical point, 18
  model, 67
dipole
  damping constant, 44
dispersion relations
  phonon, 59
distributed Bragg reflectors
  optical phonons, 58
double optical resonance (DOR), 65, 76
double-resonance, 35
double-resonant
  angle-tuning, 93
double-resonators
  light–sound, 76

electronic excitations
  inter- and intraband, 18
electronic transitions
  dipolar allowed, 18
emission
  boser, 21
  dipole, 20
  single-photon, 21
  spontaneous, 20, 45
  stimulated, 20

energy conservation
in the scattering process, 89
excitations
plasmon, 18
exciton
of the QWs, 38

Fabry–Perot
confocal resonator, 27
performance, 28
Q, 28
resonators, 44
standing wave, 27
far field, 45
Fermi's golden rule, 20
final states
accessible, 90
finesse, 49, 75
effective, 30
finite-size effects, 81
folded
acoustic phonons, 52
forward-scattered modes, 64
forward-scattering, 76, 79
frequencies
LO, 59
TO, 59
Fresnel reflection coefficients, 45

$Ga_{0.8}Al_{0.2}As$, 41
$Ga_{0.8}Al_{0.2}As/AlAs$, 38, 41
$Ga_{1-x}Al_xAs$, 36
GaAs, 24, 29, 31, 36, 47, 50, 61
substrate, 56
GaAs-like
optical phonon, 61
vibrations, 64
GaAs/AlAs, 24, 26, 41, 45, 75, 91
MQW, 92, 94
parameters, 27
quarter-wave layers, 49
Stokes and anti-Stokes spectra, 70

H_2
antibonding states, 42
bonding states, 42
molecule, 42
half-cavity
low-Q, 21
half-width at half maximum (HWHM), 51
Hamiltonian
exciton plus cavity photon, 88
of light–matter interaction, 34
hypersound, 17

$In_{0.14}Ga_{0.86}As$, 91
inelastic light scattering
polariton-mediated, 22
interband transitions, 18
interface modes, 59

Lamb-shift
atom-vacuum-field, 20
laser, 74
excitation
self-interference, 46
HeNe, 38
Ti-sapphire, 38
tunable, 38
lateral nanostructuring, 104
lifetime
detuning dependent, 90
light–matter interaction, 17, 74
absorption, 20
emission, 20
linewidths
cavity-photon, 99
exciton, 99
LO GaAs-like phonons
GaAlAs, 65
LO phonon, 51
in bulk GaAs, 99
Stokes frequency, 50
Lorentzian probability function, 90
luminescence, 20, 42

magnons, 18
matrix element
squared, 34
metallic tips, 20
microcavity
amplification power, 36
confinement, 35
confocal, 22
for Raman enhancement, 17
high-finesse, 23
II–VI, 96
III–V, 93

optical, 17
piezoelectrically controlled, 21
planar, 17, 22, 39, 90
Polaritons, 86
semiconductor, 29
strongly coupled, 19, 83, 86
microcavity modes
parity, 63
minigaps, 58
mirror
dispersive, 23
losses, 93
low-reflection, 32
superlattices, 81
mode anticrossing, 86
molecular beam epitaxy (MBE), 19, 36
multiple quantum well (MQW), 37

nanocavities
acoustic, 19, 74
nanoparticle
self-assembled, 19
nanostructures
metallic (SERS), 18
near-field optical microscopy, 20

optical phonon
double-cavity dispersion, 67
in finite superlattices, 64
optical reflectivity
for different angles, 40
optical resonance
double (DOR), 35
optical resonant, 22

perturbation theory
second-order, 18
third-order, 18
phonon, 18
cavity, 76
cavity design, 75
confined, 59
dispersion
bulk-like, 73
inplane wavevector, 62, 63
engineering, 74
forbidden and allowed, 82
interface, 50, 59
lasing, 74
longitudinal optical, 50
phonon
m confined, 66
mirror reflectivity, 75
odd and even, 81
optical, 47
photoelastic
constant, 69, 82
mechanism, 69
model, 72
photoluminescence, 38
photon
confined, 48
confinement, 17
photonic
bandgap materials, 21
molecule, 42
structures, 21
pillars, 104
plasmon, 18
polar dielectric materials, 58
polariton
cavity, 17, 19
damping, 89
damping factors, 102
dephasing, 89
dynamics, 85
exciton, 83
gap, 84
inplane dispersion, 37
low (LP), 88, 101
medium (MP), 101
radiatively stable, 86
upper, 88
polariton-mediated scattering
in bulk materials, 90
porous silicon, 21
process
Stokes, 33

Q-factor, 21, 35
quantum dots, 83
quantum wells, 83
quantum wires, 83

Rabi gap, 91, 96
large, 94
small, 96
Rabi oscillations

coherent, 86
Rabi splitting, 87, 89, 91, 93, 99
radiation pattern
azimuthal angle, 46
Raman
amplification, 17, 21, 48
by optical confinement, 33
cross section, 18, 20
total, 48
double-resonance, 55
efficiency, 103
amplifications, 21
is enhanced, 103
maximum for zero detuning, 98
photoelastic model, 68, 78
theoretical, 57
enhancement, 20
by optical confinement, 19
intensity
single-resonance scans, 55
luminescence, 77
optical resonances, 23
photons and luminescence, 93
scattering, 22
cavity-polariton-mediated, 17, 104
confined photons, 80
DOR, 41
double optical resonance, 91
double optical resonant, 21
double-resonance, 54
droplets, 21
efficiency, 96
efficiency enhancement, 30
forward, 74
in cavities, 17
in microcavities, 43
interference, 99
interference-enhanced (IERS), 21
lower polariton, 93
on droplets, 39
optical resonant, 17, 19
optically confined structures, 22
organic molecules, 21
planar cavities, 88
polariton-mediated, 85, 88
resonant (RRS), 17, 18, 22, 35
stimulated, 39, 103
surface-enhanced (SERS), 18
tensor, 35
under optical confinement, 17
upper polariton, 93
scattering amplitute
in microcavities, 35
spectroscopy
of single low-dimensional nanostructures, 103
optically confined, 57
reflection
cavity at boundaries, 29
refractive index
$n_{\rm eff}$, 29
contrast, 23
modulation, 25
resonance
morphology-dependent, 21
resonant
Raman
experiments outgoing, 96
Raman scattering
ingoing, 84
optical, 35
outgoing, 84
resonators
Fabry–Perot (FP), 27
Rytov's model, 70

scattering
deformation-potential, 63
probabilities, 18
volumes
small, 18
second-harmonic generation, 31
semiconductor
II–VI, 36, 85
III–V, 36, 85
SERS
chemical effect, 19
electromagnetic effect, 19
hot-spots, 19
Si/Ge, 75
spacer, 45
spectral density, 90
$SrTiO_3$, 76
standing optical vibrations, 64
standing waves
optical, 22
states
antibonding, 42

bonding, 42
stimulated
emission
of coherent phonons, 104
Stokes energy, 50, 52
Stokes peaks, 21
Stokes-shifted photon, 34
stop band, 23, 31, 42
edge, 79
reflectivity, 23
width, 23
stratified media, 45
strong coupling, 30
polaritonic, 34
superlattices
mirror-like, 68
surface
chemically roughened, 19
modes, 58
vibrations, 59

three-oscillator model, 101
transfer-matrix formalism, 44
transmission
factor
bulk, 89
triple spectrometer
additive mode, 79
subtractive mode, 80
tunability
of energies, 37
tunnel barriers, 32

ultrasound
THz, 23

vacuum-field fluctuations, 20
virtual process, 35

waves
evanescent, 45
leaky, 45
wavevector
partial conservation, 64
whispering-gallery modes, 21, 104

Raman Scattering in Carbon Nanotubes

Christian Thomsen[1] and Stephanie Reich[2]

[1] Institut für Festkörperphysik, PN5-4, Technische Universität Berlin, Hardenbergstraße 36, 10623 Berlin, Germany
thomsen@physik.tu-berlin.de

[2] Department of Materials Science and Engineering, Massachusetts Institute of Technology, 77 Massachusetts Avenue, Cambridge, MA 02139-4307, USA
sreich@mit.edu

Abstract. The vibrational properties of single-walled carbon nanotubes reflect the electron and phonon confinement as well as the cylindrical geometry of the tubes. Raman scattering is one of the prime techniques for studying the fundamental properties of carbon tubes and nanotube characterization. The most important phonon for sample characterization is the radial-breathing mode, an in-phase radial movement of all carbon atoms. In combination with resonant excitation it can be used to determine the nanotube microscopic structure.

Metallic and semiconducting tubes can be distinguished from the high-energy Raman spectra. The high-energy phonons are remarkable because of their strong electron–phonon coupling, which leads to phonon anomalies in metallic tubes. A common characteristic of the Raman spectra in nanotubes and graphite is the appearance of Raman peaks that correspond to phonons from inside the Brillouin zone, the defect-induced modes. In this Chapter we first introduce the vibrational, electronic, and optical properties of carbon tubes and explain important concepts such as the nanotubes' family behavior. We then discuss the Raman-active phonons of carbon tubes. Besides the vibrational frequencies and symmetries Raman spectroscopy also allows optical (excitonic) transitions, electron–phonon coupling and phase transitions in single-walled carbon nanotubes to be studied.

1 Introduction

Carbon nanotubes have attracted tremendous interest from the scientific community over the last few years since their discovery by *Iijima* in 1991 [1] and in 1993 in their single-walled form [2]. Over 15 000 publications have appeared up to the time of writing this article[1] with an ever-growing number of papers per year. The reason for this enthusiasm lies in their fascinating physical properties, which make them a model system for one-dimensional physics, their relative ease of preparation, and their large potential for applications. The combination of these features keeps the interest alive for new investigations in the research community as well as in large and small industrial companies.

[1] Institute of Scientific Information; general search with keywords "nanotube*".

M. Cardona, R. Merlin (Eds.): Light Scattering in Solid IX,
Topics Appl. Physics **108**, 115–236 (2007)

The one-dimensionality of carbon nanotubes leads to a number of easily observable, physically interesting effects, from absorption anisotropy, depolarization effects and one-dimensional vibrational confinement to possible Luttinger-liquid behavior of the electrons. The physical properties of the "theoretical" starting material for a single-walled carbon nanotube – a sheet of graphene – are relatively well understood. Still, nanotube research has given new impetus to the study of graphite e.g., its phonon dispersion [3] or the large shift of the so-called D-mode with laser excitation energy [4]. Theoretically, the electronic dispersion of nanotubes may be described by analytic expressions using the tight-binding parameters of graphite [5], which have given access to an initial understanding of the processes involved in absorption. Improvements of this formula taking into account more distant neighbors [6] allowed a quantitative comparison of the electronic properties of graphite with those of carbon nanotubes.

The nanotube wall has a curvature with profound effects on the electronic properties, in particular of small nanotubes. The effect of $\sigma - \pi$ hybridization has been described by a symmetry-adapted nonorthogonal tight-binding model [7] and, in a more sophisticated way, by ab-initio calculations [8–10] with increasing agreement with experiment. Most recently, the effect of excitons on the absorption properties has come into focus [10–13]; we will discuss this topic in Sect. 5.3.

A number of exciting applications have emerged for carbon nanotubes. Due to their extreme toughness-to-weight ratio they have become a prime choice for mechanical reinforcement of polymers and composites [14]. The fact that they can be both semiconducting or metallic (displaying even ballistic electron transport) for slightly different atomic structures opens up the field of applications in micro or nanoelectronics. Although it is not yet possible to control the growth of one specific type (chirality) of nanotube, successful procedures for separating metallic and semiconducting nanotubes have been reported [15]. We expect schemes for specific-chirality growth to appear in the near future [16]. Identifying the chiral index – i.e., their microscopic structure – of individual isolated nanotubes via Raman scattering has become routinely possible. The nonlinear absorption properties of carbon nanotubes have opened the field of saturable absorbers in laser and optical limiters [17]. Other research has focused on the application in fiber materials [18] or as artificial muscles [19].

Raman scattering has become one of the main characterization tools for carbon nanotubes. It tells us about the quality of the material, the microscopic structure of the tube, and phonon and electron quantum confinement. Other methods, such as electron microscopy, near-field optical techniques, or tunneling measurements can give similar information. Often, though, these methods require expert knowledge in handling and are quite time consuming. They are less suited for routine characterization.

Carbon nanotubes have advanced in the general interest and found their way into several textbooks, see the book by *Reich* et al. [20] for a recent overview of their fundamental physical properties. Older texts include those by *Dresselhaus* et al. [21, 22], *Saito* et al. [23], and *Harris* [24]. A number of review articles on specialized topics have been published recently, e. g., on the electronic states [25], Raman scattering [26], power applications of nanotubes [27], application as biosensors [28], and the behavior of nanotubes under high pressure [29]. Particularly useful collections of articles can be found in a special issue of The New Journal of Physics [30] (Eds. C. Thomsen and H. Kataura), and the Philosophical Transactions of the Royal Society [31] (Eds. A. C. Ferrari and J. Robertson).

In this Chapter, after a general introduction to structure and growth of carbon nanotubes (Sect. 2), we describe their lattice dynamics in Sect. 3, their electronic properties in Sect. 4, and their optical properties in Sect. 5. We then discuss some features of Raman scattering specific to one-dimensional systems such as the carbon nanotubes (Sect. 6). In Sect. 7 we show how, with resonant Raman spectroscopy, it is possible to find a unique assignment to the multitude of tubes with different chirality. The assignment is based on the chiralty dependence of both the bandgap energies and the vibrational frequency of the radial-breathing mode (RBM) of the nanotubes. In Sect. 8 we discuss the difference between the Raman spectra of metallic and semiconducting carbon nanotubes, which is important for the different envisioned technological applications of nanotubes, e.g., as electrical connectors or in devices with transistor action. Raman scattering offers a nondestructive, efficient way of identifying this property. Some aspects of electrochemical studies of carbon nanotubes are reviewed as well. Effects of electron–phonon coupling on the Raman spectra are presented in Sect. 9. This section also deals with interference effects on the resonant Raman intensities and Kohn anomalies. In Sect. 10 the role of defects in the high-energy region of the Raman spectra are discussed; they show a strong, curious shift of one of the defect-induced modes with excitation energy, which can be explained in terms of a Raman double resonance also found in graphite [4]. Throughout the Chapter we focus on single-walled carbon nanotubes.

2 Structure and Growth

Carbon nanotubes are tiny hollow graphite-based cylinders [20, 23]. Depending on the number of graphite layers of the wall they are called single-walled (one graphite layer), double-walled (two layers), triple-walled (three layers) and so forth. Tubes with a large number of concentric cylinders – ten or more – are known as multiwall nanotubes. Graphene, a single sheet of graphite, is thus the basic building block of carbon nanotubes. We obtain a carbon nanotube by cutting a tiny strip out of a graphene sheet and rolling it up into a cylinder. This procedure is shown in Fig. 1. In the laboratory, carbon nanotubes grow from carbon plasma by adding metal catalysts.

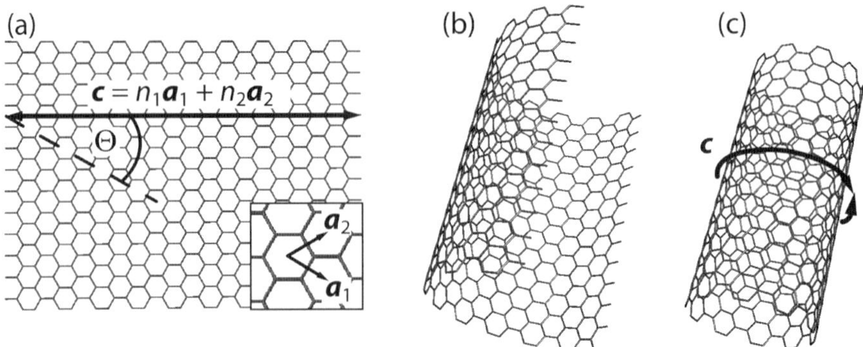

Fig. 1. Carbon nanotubes are tiny graphene cylinders. A strip is cut out of graphene (**a**) and then rolled up (**b**) to form a seamless cylinder (**c**). The chiral vector $\boldsymbol{c} = 10(\boldsymbol{a}_1 + \boldsymbol{a}_2)$ in (**a**) forms the circumference of the (10,10) nanotube in (**c**). Along the nanotube axis the unit cell is repeated periodically, i.e., a nanotube is a one-dimensional periodic system. The segment shown in (**c**) contains 13 unit cells along the nanotube axis. The *inset* in (**a**) shows the graphene lattice and the graphene unit cell vectors $\boldsymbol{a}_1$ and $\boldsymbol{a}_2$ on an enlarged scale

Before discussing the growth, we introduce the basic concepts and parameters that determine the properties of carbon nanotubes. As can be seen in Fig. 1a the cutting of graphene fixes the so-called chiral or roll-up vector $\boldsymbol{c}$. This vector goes around the circumference of the final tube in Fig. 1c. There are two parameters that control the microscopic structure of a nanotube, its diameter and its chiral angle or twist along the axis. Both are specified completely by $\boldsymbol{c}$, which is normally given in terms of the graphene lattice vectors $\boldsymbol{a}_1$ and $\boldsymbol{a}_2$ (see inset of Fig. 1a)

$$\boldsymbol{c} = n_1\boldsymbol{a}_1 + n_2\boldsymbol{a}_2 \,. \tag{1}$$

(n_1, n_2) are called the chiral index of a tube or, briefly, the chirality. A tube is characterized by (n_1, n_2). In Fig. 1c we show as an example the (10,10) nanotube.

The diameter of a tube is related to the chiral vector by

$$d = \frac{|\boldsymbol{c}|}{\pi} = \frac{a_0}{\pi}\sqrt{n_1^2 + n_1 n_2 + n_2^2}\,, \tag{2}$$

where $a_0 = 2.460$ Å is the graphene lattice constant [32–35]. For small tubes ($d < 0.8$ nm) the diameter is predicted to deviate from the geometrical diameter of a graphene cylinder in (2). Moreover, ab-initio calculations show that d becomes a function of the chiral angle below 0.8 nm [36]. Deviations from (2) are below 2 % for tube diameters $d \geq 5$ Å [36]. Note also that sometimes different lattice constants are used to calculate d. A popular value is $a_0 = 2.494$ Å obtained from a carbon–carbon distance $a_{\mathrm{CC}} = a_0/\sqrt{3} = 1.44$ Å; this value

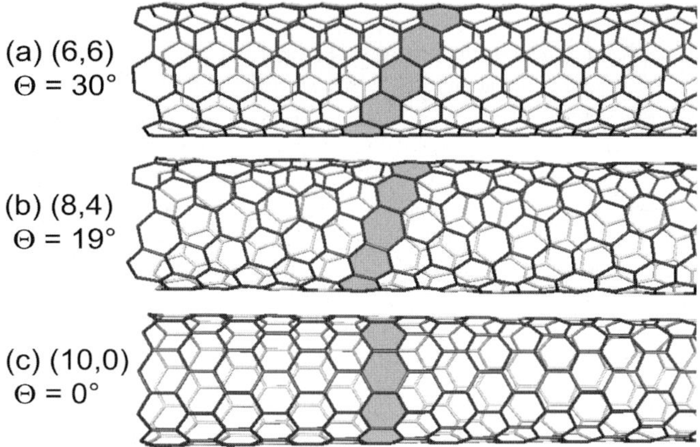

Fig. 2. Three carbon nanotubes with diameters around 0.8 nm: (**a**) (6,6) armchair, (**b**) (8,4) chiral tube and (**c**) (10,0) zigzag nanotube. The *gray hexagons* indicate the helix around the nanotube

for a_{CC} was most likely taken from fullerenes that have a larger carbon–carbon spacing than graphite and carbon nanotubes [37].

The second important parameter for carbon nanotubes is the chiral angle Θ, which is the angle between $\boldsymbol{a}_1$ and the chiral vector (see Fig. 1a). The chiral angle specifies the arrangement of the graphene hexagons on the wall of the tube. This is illustrated in Fig. 2, where we show three single-walled nanotubes with similar diameters but very different microscopic structure due to different chiral angles. Θ is related to the chiral index by

$$\Theta = \arccos\left[\frac{n_1 + n_2/2}{\sqrt{n_1^2 + n_1 n_2 + n_2^2}}\right],$$

or

$$= 30° - \arctan\left[\frac{1}{\sqrt{3}}\frac{n_1 - n_2}{n_1 + n_2}\right]. \tag{3}$$

The chiral angle is allowed to vary between $0° \leq \Theta \leq 30°$; all other ranges of Θ are equivalent to this interval because of the hexagonal symmetry of graphene (see Fig. 1a). A chiral angle of 0° and 30° corresponds to tubes with a particular high symmetry, as we discuss later. They are called *zigzag* ($\Theta = 0°$) and *armchair* tubes ($\Theta = 30°$).

The chiral vector not only determines the tube diameter and chiral angle, but all other structural parameters including the length of the unit cell and the number of carbon atoms in the unit cell. A compilation of these parameters can be found in Table 1.

A very useful illustration for the chiral vectors is to draw all possible nanotubes of a given diameter range onto a graphene lattice as in Fig. 3.

Table 1. Structural parameters, symmetry and reciprocal lattice vectors of single-walled carbon nanotubes. Armchair and zigzag tubes are special cases of the general expression given in the last column. n is the greatest common divisor of n_1 and n_2. n_c and q are the number of carbon atoms and the number of graphene hexagons in the nanotube unit cell, respectively; wv stands for wavevector. m is the z component of the angular momentum quantum number; it indexes electronic and vibrational bands of carbon nanotubes and takes only integer values

Parameter	Armchair	Zigzag	Chiral
	(n, n)	$(n, 0)$	(n_1, n_2)
Chiral vector $\boldsymbol{c}$	$n(\boldsymbol{a}_1 + \boldsymbol{a}_2)$	$n\boldsymbol{a}_1$	$n_1\boldsymbol{a}_1 + n_2\boldsymbol{a}_2$
Diameter d	$\sqrt{3}a_0 n/\pi$	$a_0 n/\pi$	$a_0\sqrt{n_1^2 + n_1 n_2 + n_2^2}/\pi$
Chiral angle Θ	$30°$	$0°$	$30° - \arctan \frac{1}{\sqrt{3}} \frac{n_1 - n_2}{n_1 + n_2}$
Lattice vector $\boldsymbol{a}$	$-\boldsymbol{a}_1 + \boldsymbol{a}_2$	$-\boldsymbol{a}_1 + 2\boldsymbol{a}_2$	$-\frac{2n_2 + n_1}{n\mathcal{R}}\boldsymbol{a}_1 + \frac{2n_1 + n_2}{n\mathcal{R}}\boldsymbol{a}_2$
Lattice const. a	a_0	$\sqrt{3}a_0$	$a_0\frac{\sqrt{3(n_1^2 + n_1 n_2 + n_2^2)}}{n\mathcal{R}}$
Point group	D_{2nh}	D_{2nh}	D_q
$n_c = 2q$	$4n$	$4n$	$4\frac{(n_1^2 + n_1 n_2 + n_2^2)}{n\mathcal{R}}$
$\mathcal{R}$	3	1	3 if $(n_1 - n_2)/3n =$ integer 1 if $(n_1 - n_2)/3n \neq$ integer
Quantized wv $\boldsymbol{k}_\perp$	$\frac{\boldsymbol{k}_1 + \boldsymbol{k}_2}{2n}$	$\frac{\boldsymbol{k}_1}{n} + \frac{\boldsymbol{k}_2}{2n}$	$\frac{2n_1 + n_2}{qn\mathcal{R}}\boldsymbol{k}_1 + \frac{2n_2 + n_1}{qn\mathcal{R}}\boldsymbol{k}_2$
Axial wv $\boldsymbol{k}_z$	$\frac{-\boldsymbol{k}_1 + \boldsymbol{k}_2}{2}$	$\frac{\boldsymbol{k}_2}{2}$	$\frac{-n_2\boldsymbol{k}_1}{q} + \frac{n_1\boldsymbol{k}_2}{q}$
Range of m	$-(n-1)\ldots n$	$-(n-1)\ldots n$	$-(q/2-1)\ldots q/2$

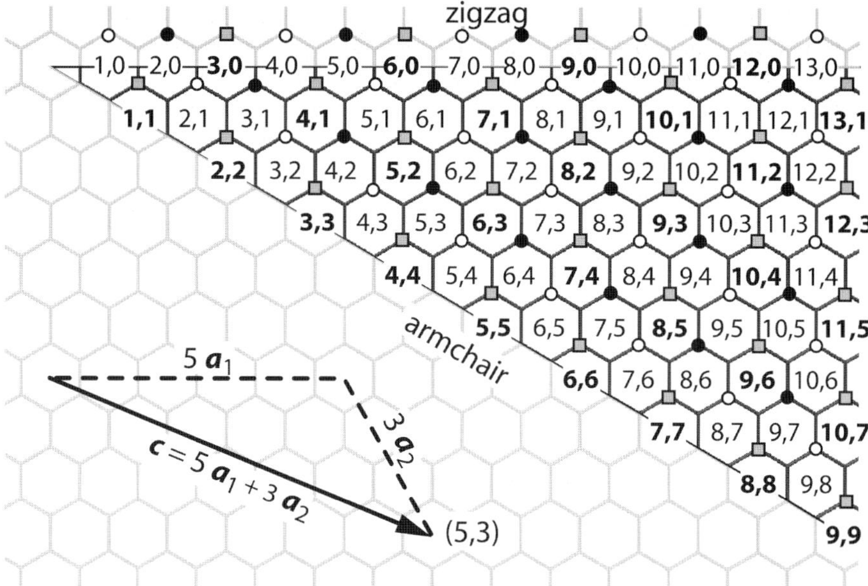

Fig. 3. Chiral indices of single-walled carbon nanotubes mapped onto a graphene sheet for tube diameters $d \leq 1.2\,\text{nm}$. The *small symbols* at the upper corner of a hexagon indicate the family of a nanotube. *Closed circles* are semiconducting tubes belonging to the $\nu = (n_1 - n_2) \bmod 3 = -1$ family; *open circles* belong to the $\nu = +1$ family, the second semiconducting family. *Squares* are tubes from the $\nu = 0$ family; their chiral indices are in boldface. These tubes are metallic or quasimetallic, which means that their electronic bandgap is 10 meV to 100 meV and induced by the curvature of the nanotube wall. Note that tubes from one family are connected by the three armchair-like directions. Nanotube indices that are connected by the vertical armchair-like direction belong to the same nanotube branch (see also Table 2)

This plot highlights systematics in the nanotube properties such as their metallic and semiconducting character, the family behavior, and nanotube branches. We will discuss these topics throughout this Chapter. In the lower left part of Fig. 3 we show in detail how to construct a (5,3) nanotube on a graphene sheet.

The series of $(n, 0)$ zigzag nanotubes that run horizontally in Fig. 3 have $\Theta = 0°$. The name derives from the zigzag chain that forms the edge of this high-symmetry type of tubes (see Fig. 2c). The other high-symmetry nanotube structures are the (n, n) armchair tubes with $\Theta = 30°$. The edge of their unit cell resembles a row of armchairs when viewed from above (Fig. 2a), hence the name. Zigzag and armchair tubes are both called *achiral* tubes. A general nanotube with lower symmetry, like the (8,4) tube in Fig. 2b, is referred to as *chiral*.

Table 2. Frequently used classifications for a general nanotube used in this work. Families and branches may be visualized as lines parallel to the armchair direction and as vertical lines in Fig. 3, respectively. "Neighbor" refers to the nanotube closest to a particular one on the same branch and with $d' \geq d$

	Symbol	Definition	Remark
Family index	ν	$(n_1 - n_2) \mod 3$	$\nu = 0, \pm 1$ (or $\nu = 0, 1, 2$)
			lines parallel to armchair
Semiconducting	S	$(n_1 - n_2) \mod 3 = \pm 1$	
Metallic	M	$(n_1 - n_2) \mod 3 = 0$	
Branch index	β	$2n_1 + n_2$	vertical lines in Fig. 3
Neighbor	(n_1', n_2')	$(n_1 - 1, n_2 + 2)$	$n_1' \geq n_2'; \beta = \text{const}$

Achiral tubes have a number of properties that distinguish them from chiral tubes. For example, armchair nanotubes are the only tubes where the electronic bandgap is strictly zero. From a practical point of view the number of carbon atoms in the unit cell n_c is much smaller in achiral than in chiral tubes of the same diameter, see Table 1. Most calculations of, e.g., the lattice dynamics or the electronic band structure have been performed for armchair and zigzag tubes [38, 39]. This should be kept in mind when comparing experiment and theory, since chiral tubes greatly outnumber the achiral species in real samples.

Carbon nanotubes are grown by three major methods, laser ablation, arc-discharge, and chemical vapor deposition (CVD) and CVD-related processes [1, 2, 40, 44, 45, 49–54]. All growth processes have in common that a carbon plasma is formed either by heating or by chemical decomposition. Tubes grow when metal catalysts are added to this plasma. The type of catalyst and the growth conditions determine whether single, double or multiwall tubes are obtained; the latter also grow in small quantities without any catalysts. Besides the number of walls, the yield, tube diameter and length can be controlled during growth. Table 3 lists some typical values for the tubes most widely used in optical studies.

Different methods used to grow carbon nanotubes have their advantages and disadvantages [52]. We are not going to discuss them in detail. The first processes that produced nanotubes with high yield were arc-discharge and laser ablation [40, 51].

A scanning electron microscopy (SEM) image of carbon nanotubes grown by arc-discharge is shown in Fig. 4a. The web-like structure consists of nanotube bundles. Every bundle or rope in Fig. 4a is formed by 20–100 single-walled nanotubes, as can be seen in the high-resolution transmission electron

Fig. 4. (**a**) SEM images of single-walled carbon-nanotube bundles. (**b**) High-resolution TEM image showing the cross section of a bundle of tubes. Every circle is the circumference of a single-walled carbon nanotube. The tubes are arranged in a hexagonal packing. (**c**) Water-assisted CVD-grown carbon nanotubes of high purity and length. The nanotube mat is shown next to the head of a matchstick. (**d**) A patterned substrate allows the growth of carbon nanotubes in predefined places as shown in this SEM picture. The *inset* displays an expanded view of one of the pillars. After [40] and [41]

microscope picture in part (b) of the figure. In such a bundle the tubes are packed in a two-dimensional hexagonal lattice.

The bundling results from the attractive van-der-Waals forces between nanotubes [55–58]. For more than ten years it had been a great challenge to break these bundles and prevent the isolated tubes from rebundling. This problem was solved in 2002 by a postprocessing routine [59], where first the bundles are broken apart by ultrasonification. The tubes are then coated by a surfactant to eliminate interactions between them. Finally, the remaining small bundles with 3–10 tubes are removed from the solution with a centrifuge [59]. In this way, solutions with isolated nanotubes were produced for a wide range of nanotubes [44, 59–62].

Finding a way to produce large quantities of isolated nanotubes in solution is particularly important for optical studies and Raman spectroscopy on single-walled carbon nanotubes [60, 63–65]. In contrast to bundled tubes, isolated nanotubes behave like truly one-dimensional systems from an optical point of view. They show sharp absorption peaks and Raman resonances [59, 63]; they also emit light in the near infrared [60]. Based on optical spectroscopy of isolated nanotubes in solution it was possible – for the first time – to determine which nanotube chiralities are present in a sample [44, 60–64]. These topics will be discussed in Sects. 5 and 7.

Table 3. Typical diameters for carbon nanotubes grown by different methods. σ is the standard deviation of a Gaussian distribution. The length is categorized as short (below 100 nm), medium (100 nm to 1000 nm) and long (above 1 μm). A special, ultralong case is the 4 cm nanotube reported in [42]. The "special" tubes types (HiPCo, CoMocat etc.) are all grown by CVD-related processes

Growth method	d (nm)	σ (nm)	Length	Remark
Laser ablation	1.5	0.1	medium	
Arc-discharge	1.5	0.1	medium	
CVD	strongly varying depending on catalyst and growth conditions			
HiPCo	1.0	0.2	medium	[43]
CoMoCat	0.7	0.1	medium	[44]
Alcohol	0.7	0.1	medium	[45]
Zeolites	0.4	–	short	[46]
cm tubes	1.4	–	ultralong	[42]
Supergrowth	2.0	1.0	long	[41]
Double-walled tubes (inner tube diameters)				
CVD	0.7	0.05		[47]
Peapods	0.75	0.05		[48]

Chemical vapor deposition recently evolved into the most widely used growth technique, because it is cheap, scalable and allows a high degree of control over the tubes. With CVD it is possible to grow nanotubes on well-defined places with patterned substrates and, e.g., to produce an individual tube that bridges two catalytic particles [66, 67]. Also, CVD tends to produce isolated rather than bundled tubes, which is interesting for optical spectroscopy [68, 69], and it works at lower growth temperatures [70]. The great disadvantage of CVD is the lower quality of the nanotubes obtained.

A major step forward in terms of quality, purity, yield, and length of CVD tubes was recently reported by *Hata* et al. [41]. By adding water during the CVD process they obtained mm-thick mats of single-walled nanotubes. Figure 4c compares such a mat with a matchstick. The nanotube forest is free of catalytic particles and amorphous carbon. These two contaminations are normally present in as-grown tube samples and need to be removed by subsequent purification. Figure 4d demonstrates how the places in which the tubes grow can be defined by patterning the substrate.

The growth of carbon nanotubes remains a rapidly evolving field. Especially on the theoretical side much needs to be done for a better understanding of the processes. The major challenge for the growth of carbon nanotubes is to grow tubes with a unique chirality. We mentioned that the tube diameter can

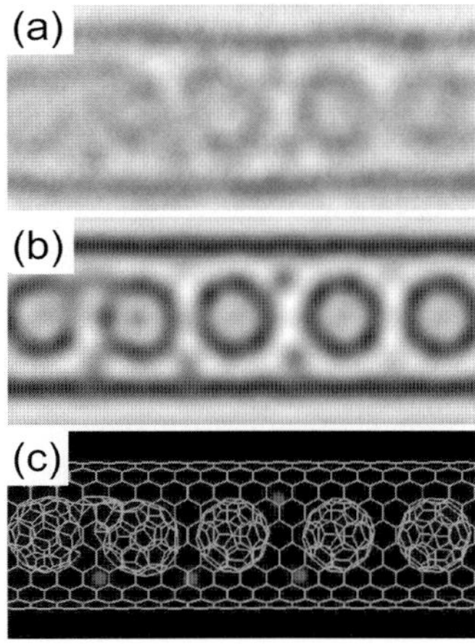

Fig. 5. (**a**) High-resolution transmission-electron-microscope image of a potassium-doped C_{60} peapod. (**b**) A simulation of the image. (**c**) Model of the doped peapods. Seen are the nanotube walls (*horizontal dark lines*), the fullerene molecules (large circular molecules), and the potassium doping atoms (*small black dots*). From [71]

be controlled to some extent by the growth parameters. What about the second important parameter, the chiral angles of the tubes? The general answer is that nanotube samples show a homogeneous distribution of chiral angles from armchair to zigzag [72, 73]. There seems to be some preference for specific chiralities when the tube diameters are very small ($d < 0.8\,\mathrm{nm}$) [44, 45]. This conclusion was drawn, however, from optical spectroscopy under the assumption that the cross section is independent of chirality. This assumption is most likely incorrect [74, 75]. For fundamental studies and Raman scattering, in particular, nanotube samples with a unique type of tube are not as critical as, e.g., for applications. By tuning the laser-excitation energy we can now select a specific nanotube chirality with resonant Raman scattering. This will be discussed in Sect. 7.

Carbon nanotubes can be prepared containing small molecules and one-dimensional crystals. We show in Fig. 5, as an example, an image of so-called peapods, that is, nanotubes containing C_{60} molecules. Such peapods were discovered by *Smith* et al. [76] and produced in large quantities by *Kataura* et al. [77]. The peapods used for the image in Fig. 5 were doped with potassium. One can readily see the K-atoms in the image. From a study of the electronic state of the potassium atoms *Guan* et al. [71] were able to show that the peapod was n-doped, suggesting that electrically conductive nanowires are technologically feasible. Doped peapods were also studied with Raman scattering by *Pichler* et al. [78]. For more details on the determination of the charge state of doped peapods, see [71, 78] and Sect. 8.3.

Another example of inorganic molecules or crystals prepared in nanotubes is shown in Fig. 6 [79]. The image shows two layers of a KI crystal in the carbon nanotube. It is remarkable that the atoms in the KI crystal all lie at the surface; the crystal has no “volume”. The structural and electronic properties of one-dimensional crystals of this type can be investigated only if the surface atoms are passivated, here by the surrounding nanotube. Possible

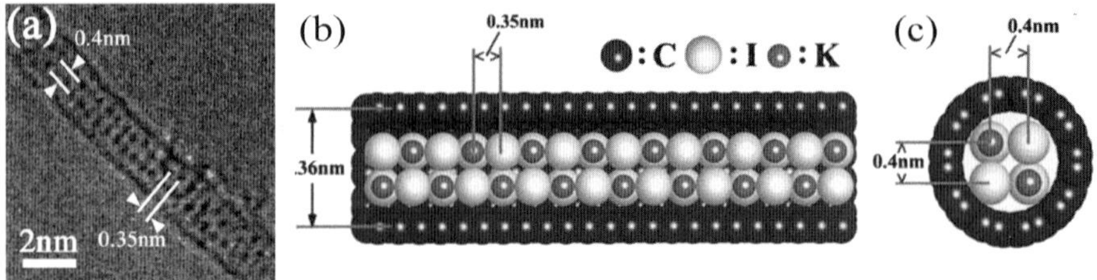

Fig. 6. (**a**) High-resolution transmission electron microscope (TEM) image of a bilayer crystal of potassium iodide grown inside a carbon nanotube with diameter of ≈ 1.36 Å. (**b**) Model of the KI crystal-in-nanotube image of (**a**). (**c**) Cross section of this nanotube. After [79]

structures of alkali halide clusters have been predicted by *Diefenbach* and *Martin* [80].

3 Lattice Dynamics

Carbon nanotubes have many vibrational degrees of freedom because of the large number of atoms in their unit cells. Achiral tubes of $d \approx 1\,\mathrm{nm}$ diameter have 100–150 phonon branches; in chiral nanotubes this number can be higher by one or two orders of magnitude. Only a very small fraction of these phonons is Raman or infrared active. The Raman-active modes fall into a low-energy range where radial vibrations are observed, a high-energy range with inplane carbon–carbon vibrations, and an intermediate-frequency range. The low- and high-energy phonons have received most attention; their Raman signal is very strong and they can be used for characterizing and studying nanotubes as they give information about the tube diameter and chirality, phonon confinement, the semiconducting or metallic character, optical transition energies and more. In this section we first study the number of nanotube vibrations and their symmetry from group theory. Then we discuss the distinct energy ranges of the phonons that can be studied by optical spectroscopy.

3.1 Symmetry

Carbon nanotubes are structures with particularly high symmetry. This comes from the underlying hexagonal lattice. The translations of graphene turn into rotational and helical symmetry operations for nanotubes, because we build the tube by rolling up graphene. The symmetry of carbon nanotubes was rigorously derived within the framework of line groups [81, 82]. Line groups deal with systems that are periodic in one dimension. They are the equivalent of crystal space groups for one-dimensional solids. The line-group treatment is very powerful, see *Reich* et al. [20] for an introduction

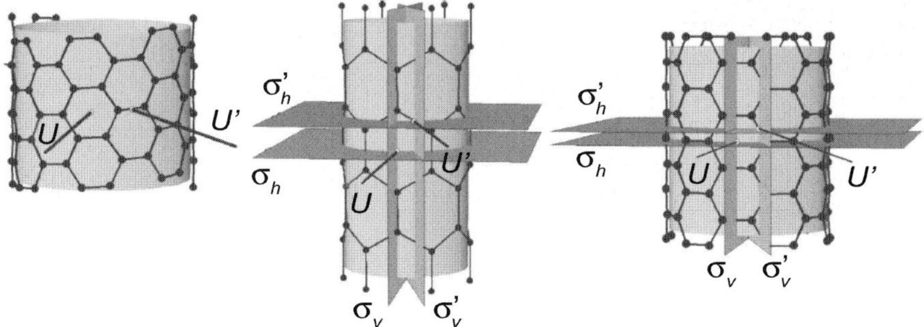

Fig. 7. Symmetry operations of chiral (*left*), zigzag (*middle*) and armchair (*right*) nanotubes. Besides the operations shown here all tube types also have a principal screw axis – achiral tubes also have a principal rotation axis. U and U' are frequently denoted by C_2' and C_2'', respectively. In all three panels, primed operations involve translations by a fraction of the nanotube lattice vector $\boldsymbol{a}$. From [81]

and [83–88] for applications of the method to carbon nanotubes. Here we will use the isogonal point groups of single-walled carbon nanotubes to discuss their symmetry. We assume that the reader is familiar with this aspect of group theory [89, 90].

Looking back at the (10,0) nanotube in Fig. 2c a number of symmetry operations are apparent: We can rotate the tube by $360°/10 = 36°$ about its z-axis, the nanotube axis. There is a series of so-called primed rotations, which are perpendicular to z. The most obvious one has its rotational axis normal to the paper. We can also reflect the lattice at a mirror plane normal to z going through the centers of the hexagons (horizontal mirror plane). And, finally, one can identify a set of vertical mirror planes including the nanotube axis and the carbon–carbon bonds parallel to it [81].

The symmetry operations mentioned so far do not require a translation of the tube by a fraction of its lattice vector $\boldsymbol{a}$. There is, however, a second set of operations in a (10,0) nanotube that preserve the lattice. They involve translations, i.e., screw axes and glide planes. The main screw axis is best seen for the (6,6) nanotube in Fig. 2a. The tube can be rotated by $360°/6 = 60°$, but we can also apply a rotation by 30° followed by a translation by $\boldsymbol{a}/2$. This generates the helix twisting around the tube's circumference (gray hexagons in Fig. 2a). Likewise, the (8,4) and (10,0) tubes have a screw axis. In addition, carbon nanotubes always have two sets of C_2 rotations perpendicular to z, see U and U' in Fig. 7. Achiral tubes possess two horizontal (σ_h, σ_h') and two sets of vertical (σ_v, σ_v') mirror planes (see Fig. 7).

The isogonal point group of a carbon nanotube depends on its chirality. (n, n) armchair and $(n, 0)$ zigzag tubes belong to D_{2nh}; chiral (n_1, n_2) nanotubes to D_q, where q is the number of hexagons in the unit cell of the tube [81], see Table 1. The character table for these groups is given in Table 3.1.

Table 4. Character table of the D_{qh} point groups with even q, which applied to all nanotubes (q is always even, see Table 1). $\alpha = 2\pi/q$, $n = q/2$. To obtain the table for D_q omit the inversion I plus all following symmetry operations and drop the subscripts g and u. The basis functions indicate the translations ($T_{x,y,z}$) and rotations ($R_{x,y,z}$). Phonons belonging to the translational representations are infrared active. The Raman activity is indicated in the last column, which also gives the Raman selection rules

D_{qh}	E	$2C_q$	$2C_q^2$	…	$C_q^n = C_2$	$\frac{q}{2}C_2' = U$	$\frac{q}{2}C_2'' = U'$	I	$2IC_q$	$2IC_q^2$	…	$IC_2 = \sigma_h$	$\frac{q}{2}\sigma_v$	$\frac{q}{2}\sigma_{v'}$	Basis functions	
A_{1g}	1	1	1	…	1	1	1	1	1	1	…	1	1	1		$\alpha_{xx}+\alpha_{yy}, \alpha_{zz}$
A_{1u}	1	1	1	…	1	1	1	−1	−1	−1	…	−1	−1	−1		
A_{2g}	1	1	1	…	1	−1	−1	1	1	1	…	1	−1	−1	R_z	
A_{2u}	1	1	1	…	1	−1	−1	−1	−1	−1	…	−1	1	1	T_z	
B_{1g}	1	−1	1	…	$(-1)^n$	1	−1	1	−1	1	…	$(-1)^n$	1	−1		
B_{1u}	1	−1	1	…	$(-1)^n$	1	−1	−1	1	−1	…	$-(-1)^n$	−1	1		
B_{2g}	1	−1	1	…	$(-1)^n$	−1	1	1	−1	1	…	$(-1)^n$	−1	1		
B_{2u}	1	−1	1	…	$(-1)^n$	−1	1	−1	1	−1	…	$-(-1)^n$	1	−1		
E_{1g}	2	$2\cos\alpha$	$2\cos 2\alpha$	…	−2	0	0	2	$2\cos\alpha$	$2\cos 2\alpha$	…	−2	0	0	(R_x, R_y)	$(\alpha_{yz}, \alpha_{zx})$
E_{1u}	2	$2\cos\alpha$	$2\cos 2\alpha$	…	−2	0	0	−2	$-2\cos\alpha$	$-2\cos 2\alpha$	…	2	0	0	(T_x, T_y)	
E_{2g}	2	$2\cos 2\alpha$	$2\cos 4\alpha$	…	2	0	0	2	$2\cos 2\alpha$	$2\cos 4\alpha$	…	2	0	0		$(\alpha_{xx}-\alpha_{yy}, \alpha_{xy})$
E_{2u}	2	$2\cos 2\alpha$	$2\cos 4\alpha$	…	2	0	0	−2	$-2\cos 2\alpha$	$-2\cos 4\alpha$	…	−2	0	0		
…	…	…	…	…	…	…	…	…	…	…	…	…	…	…		
$E_{(n-1)g}$	2	$2\cos(n-1)\alpha$	$2\cos 2(n-1)\alpha$	…	$2\cos(n-1)\pi$	0	0	2	$2\cos(n-1)\alpha$	$2\cos 2(n-1)\alpha$	…	$2\cos(n-1)\pi$	0	0		
$E_{(n-1)u}$	2	$2\cos(n-1)\alpha$	$2\cos 2(n-1)\alpha$	…	$2\cos(n-1)\pi$	0	0	−2	$-2\cos(n-1)\alpha$	$-2\cos 2(n-1)\alpha$	…	$-2\cos(n-1)\pi$	0	0		

Table 5. Phonon symmetries of single-walled carbon nanotubes

Tube	Phonon symmetry	m
(n,n) armchair		
n even	$2(A_{1g}+A_{2g}+B_{1g}+B_{2g})+A_{2u}+A_{1u}+B_{2u}+B_{1u}$ $+\sum_{m\,\mathrm{odd}}(4E_{mu}+2E_{mg})+\sum_{m\,\mathrm{even}}(4E_{mg}+2E_{mu})$	$[1,n-1]$
n odd	$2(A_{1g}+A_{2g}+B_{1u}+B_{2u})+A_{2u}+A_{1u}+B_{2g}+B_{1g}$ $+\sum_{m\,\mathrm{odd}}(4E_{mu}+2E_{mg})+\sum_{m\,\mathrm{even}}(4E_{mg}+2E_{mu})$	$[1,n-1]$
$(n,0)$ zigzag	$2(A_{1g}+A_{2u}+B_{1g}+B_{2u})+A_{2g}+A_{1u}+B_{2g}+B_{1u}$ $+3\sum_{m}(E_{mg}+E_{mu})$	$[1,n-1]$
(n_1,n_2) chiral	$3(A_1+A_2+B_1+B_2)+6\sum_{m}E_m$	$[1,\frac{q}{2}-1]$

In the literature, one often finds lower point group symmetries for both achiral and chiral nanotubes [21, 23, 91–94]. We want to comment on this, because the (incorrect) lower-symmetry groups strongly affect the phonon-selection rules. Essentially, they double the number of Raman-active vibrations in armchair and zigzag tubes [20, 93]. For chiral tubes the effect is less pronounced; we expect one totally symmetric A_1 phonon less when using the correct nanotube symmetry [82]. The screw axes and the glide planes of achiral nanotubes are often neglected in the literature [23]. These symmetry operations were most likely ignored because in the early days of nanotube research carbon nanotubes were modeled as extended fullerenes [91]. Typical carbon nanotubes, however, are at least several hundred of nanometers long. A description as a one-dimensional solid is therefore much more adequate than the molecular approach.[2] In chiral tubes the C_2 rotations perpendicular to the nanotube axis are often missing. This seems to be due to an error in the first publication about the symmetry of chiral tubes [95].

From the point group and the positions of the carbon atoms the number and symmetry of the phonon eigenvectors are found by one of the standard procedures [20, 81, 89, 90, 96]. The symmetries of the eigenmodes are given in Table 5. A and B modes are nondegenerate; E modes have a twofold degeneracy. The character table (Table 3.1) shows that the $A_{1(g)}$, $E_{1(g)}$ and $E_{2(g)}$

[2] The solid-state and the molecular picture must not be mixed. When using solid-state concepts such as band structures and phonon dispersions, translations need to be included among the symmetry operations.

representations are Raman active (the subscript g represents achiral tubes). We thus obtain a total of 8 Raman-active phonons in achiral and 15 in chiral nanotubes

$$\begin{array}{rl} \text{armchair} & 2A_{1g}\oplus 2E_{1g}\oplus 4E_{2g} \\ \text{zigzag} & 2A_{1g}\oplus 3E_{1g}\oplus 3E_{2g} \\ \text{chiral} & 3A_1 \oplus 6E_1 \oplus 6E_2\,. \end{array} \tag{4}$$

The totally symmetric modes in achiral tubes are the radial-breathing vibration, which is discussed in the next section, and a high-energy phonon. In armchair tubes the circumferential eigenvector is Raman active; in zigzag tubes the axial vibration. In chiral nanotubes, the three totally symmetric modes are the radial-breathing mode and two high-frequency vibrations. They resemble the circumferential and axial vibrations, but are, in general, of mixed character. This is typical for low-dimensional systems; it originates from the mechanical boundary conditions (around the circumference in the case of nanotubes) [38, 97, 98]. As we show in Sect. 6 the Raman spectrum of carbon nanotubes is dominated by scattering from totally symmetric phonons [99–102]. In the following we discuss the radial vibrations of carbon nanotubes and, in particular, the radial-breathing mode before turning to the high-energy frequencies.

3.2 Radial-Breathing Mode

The radial-breathing mode (RBM) is an important mode for the characterization and identification of specific nanotubes, in particular of their chirality. Its eigenvector is purely radial by symmetry for armchair tubes only; these have mirror planes perpendicular to the nanotube axis, and the radial-breathing mode is fully symmetric. For all other chiralities $n_1 \neq n_2$ the eigenvector has a small axial component, which is largest for zigzag tubes [36, 103, 104]. For most practical purposes, however, the radial-breathing mode may be considered purely radial. In Fig. 8 we show its eigenvector for an (8,4) nanotube.

The significance of the radial-breathing mode for the characterization of nanotubes comes from the inverse dependence of its frequency on the diameter of the nanotube

$$\omega_{\mathrm{RBM}} = \frac{c_1}{d} + c_2\,. \tag{5}$$

This quite general relationship may be derived, e.g., from the continuum mechanics of a small hollow cylinder as shown by *Mahan* [105]. From one of the solutions to the standard wave equation for sound in homogeneous solids, one obtains the frequency of the breathing mode[3]

$$\omega_{\mathrm{RBM}} = \frac{v_{\mathrm{B}}}{2\pi c d}\,, \tag{6}$$

[3] Note that in (6) the phonon frequency has units of cm^{-1}. This is the common definition in the nanotube community for the quantity "wave number", $\tilde{k} = \lambda^{-1}$, and differs from that of Mahan.

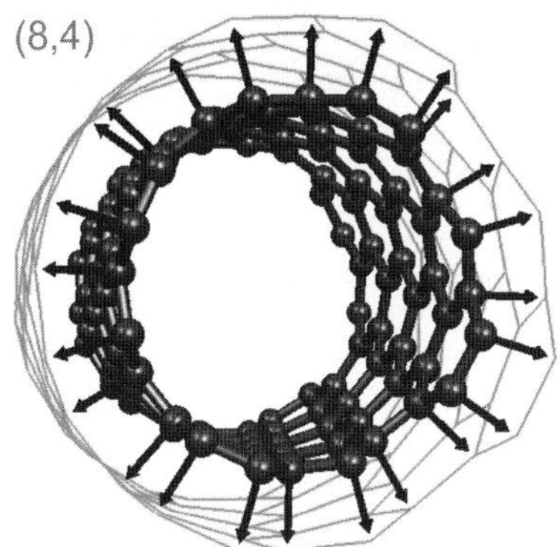

Fig. 8. Radial-breathing mode of an (8,4) nanotube. Shown are the displacement vectors of the atoms along the radius vector. In armchair nanotubes the eigenvectors are *exactly* radial, in all other tubes the eigenvector has a small axial component (on the order of 10^{-3} of the radial component) [103, 104]

with the velocity v_B given by

$$v_\mathrm{B} = 4c_\mathrm{t}\sqrt{1-\frac{c_\mathrm{t}}{c_\mathrm{l}}} \approx 42.8\,\mathrm{km/s}\,. \tag{7}$$

The constants $c_\mathrm{t} = (C_{66}/\rho)^{1/2} = 14.0\,\mathrm{km/s}$ and $c_\mathrm{l} = (C_{11}/\rho)^{1/2} = 21.7\,\mathrm{km/s}$ are the transverse and longitudinal sound velocities ($C_{66} = 44 \times 10^{11}\,\mathrm{dyn/cm^2}$, $C_{11} = 106 \times 10^{11}\,\mathrm{dyn/cm^2}$ [106–108], and $\rho = 2.66\,\mathrm{g/cm^3}$, the density of graphite). Inserting the velocity (7) in (6) yields

$$\omega_\mathrm{RBM} = \frac{227\,\mathrm{cm^{-1}} \cdot \mathrm{nm}}{d}\,, \tag{8}$$

which is in very good quantitative agreement with other calculations for isolated nanotubes and with experiment, as we shall see.

The constant c_2 in (5) is believed to describe the additional interaction in bundles of nanotubes. For a compilation of values for c_1 and c_2 both from experiment and theory (see Table 8.2 in Ref. [20]). Alternative ways of showing the inverse dependence of ω_RBM on diameter have been published by *Jishi* et al. [95] and *Wirtz* et al. [109].

The problem with using (5) to find the chiral index in practice is that many slightly different nanotubes lie in a small range of diameters. The experimental uncertainty of the constants in (5) is too large to find uniquely the chiral index of a specific tube. To give an example, for a diameter distribution centered around 0.95 nm with $\sigma \approx 0.2$ nm, as is typical for the HiPCO process, there are 40 chiral indices to be assigned. In laser-ablation-grown samples the mean diameters are larger and, correspondingly, many more chiralities fall into the range of tubes to be identified in a sample.

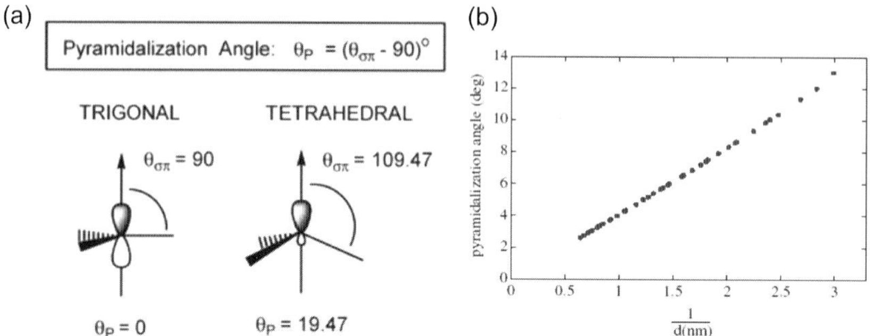

Fig. 9. (**a**) Definition of the pyramidalization angle θ_{p} [110]; $\theta_{\mathrm{p}} = 0°$ corresponds to sp^2- and $\theta_{\mathrm{p}} = 19.47°$ to sp^3-coordinated carbon. (**b**) Plot of θ_{p} as a function of inverse tube diameter showing how smaller tubes lead to a more sp^3-type coordination of the carbon atoms. After [36]

It turns out that only the simultaneous measurement of the radial-breathing mode frequency and the optical transition energies by resonant Raman spectroscopy yields sufficient information to identify uniquely the chiral index (n_1, n_2) of a nanotube. We will show how this works in Sect. 7.

3.2.1 Chirality Dependence

As the diameter of the nanotubes becomes smaller, the deviations from the analogy to graphene and the continuum-like models become larger. This concerns the geometric changes introduced by the curvature of the nanotube as well as the deviations of the radial-breathing mode frequency from the large-d behavior in (5). Aside from bond-length changes and bond-angle distortions *Niyogi* et al. [110] considered the chemical reactivity that arises from increasing sp^3 components in the bonds of rolled-up graphene. They defined the pyramidalization angle θ_{p} as the difference between 90° and the angle between the inplane σ and out-of-plane π bonds in the trigonal configuration (graphite or graphene). Graphene thus has $\theta_{\mathrm{p}} = 0°$; the other limit is $\theta_{\mathrm{p}} = 19.47°$, which corresponds to purely tetrahedrally bonded carbon (diamond) (see Fig. 9a).

Figure 9b shows the pyramidalization angle θ_{p} as function of inverse diameter for tubes in the diameter range 0.3 nm to 1.5 nm. While it is clear that larger tubes (say, diameter $\geq$ 1 nm) are not affected much, tubes with diameter below 0.5 nm have bonds with considerable sp^3 character. Fullerenes, by comparison, have $\theta_{\mathrm{p}} = 11.4°$ [110].

These geometric deviations lead naturally also to deviations of the radial-breathing mode frequency from the simple $1/d$ relationship. Because of the different symmetry of armchair and zigzag nanotubes, the zigzag tubes are more strongly affected by increasing curvature. This is shown in Fig. 10, where the deviation of ω_{RBM} from the value predicted by (5) is plotted vs. diameter

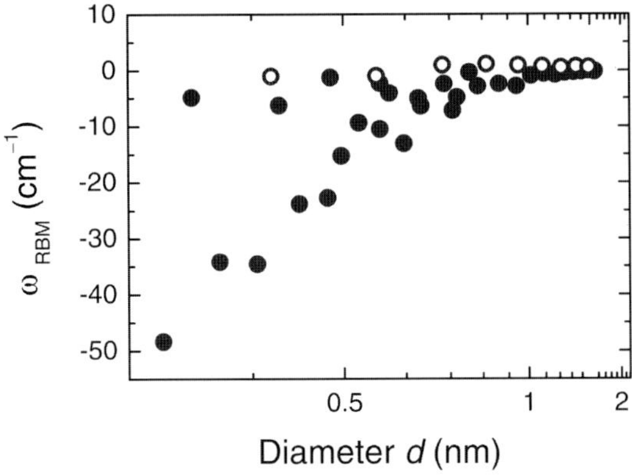

Fig. 10. Frequency of the radial-breathing mode vs. log d to illustrate the departure from the simple $1/d$ behavior predicted by (5). The deviations become larger for smaller nanotubes. Armchair nanotubes (*open symbols*) are not affected. After [36]

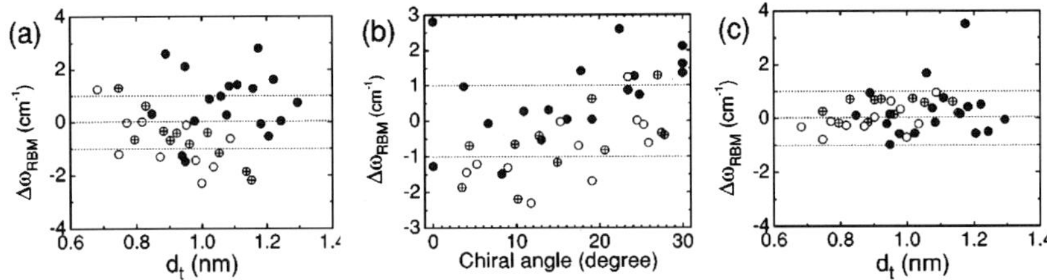

Fig. 11. Frequency differences $\Delta\omega_{\mathrm{RBM}}$ of the radial-breathing mode with respect to the simple $1/d$ behavior, as a function of nanotube diameter (**a**) and chiral angle (**b**). The *horizontal lines* indicate the $1/d$ approximation and the experimental error bars of the measurement. A higher parametrization, (9), leads to a much better description of the data (**c**). *Open circles*, *crossed circles* and *solid circles* correspond to $\nu = -1$, $\nu = +1$ semiconducting, and $\nu = 0$ metallic nanotubes, respectively. From [111]

of the nanotubes. *Kürti* et al. [36] observed that, in addition to a softening for smaller nanotubes, there is a dependence on chirality in the vibrational frequencies calculated ab initio. Again, armchair tubes are almost unaffected, even for quite small diameters. Alternatively, continuum mechanics becomes worse as an approximation, and the chirality of a nanotube should become more noticeable for very small nanotubes. For nanotubes around 1 nm the differences are small, but for even smaller tubes there is a systematic deviation, stronger for zigzag tubes, as expected.

Jorio et al. [111] did a systematic experimental study of the ω_{RBM} of small nanotubes. Their results are shown in Figs. 11a and b where the frequency deviations $\Delta\omega_{\mathrm{RBM}}$, compared to (5), are plotted as a function of diameter and chiral angle. There appear to be systematic deviations as a function of diameter for metallic tubes (closed circles) and the two semiconducting families (crossed and open circles in Fig. 11a). If we assume that the positive offset of the armchair tubes is an artifact of the fit and shift down the zero in Fig. 11a by, say, $2\,\mathrm{cm}^{-1}$, the agreement of the data to the theory of [36] turns out rather well.

The data in Fig. 11b show a systematic increase of $\Delta\omega_{\mathrm{RBM}}$ as a function of chiral angle Θ. Again, compared to the results of Kürti et al., who showed that on an absolute scale armchair tubes (large chiral angles) tend to follow the simple $1/d$ law, the data of Fig. 11b appear centered around their mean value at $\Theta \approx 15°$. Shifting them down by about $2\,\mathrm{cm}^{-1}$ to make their armchair shift zero shows that zigzag tubes deviate by about $3\,\mathrm{cm}^{-1}$ to $4\,\mathrm{cm}^{-1}$ from the $1/d$ expression. Thus, there is a good agreement between theory and experiment.

It is possible to parametrize the deviations from the simple $1/d$ form to include the effect of the threefold symmetry of the band structure [111]:

$$\omega_{\mathrm{RBM}} = \frac{A}{d} + B + \frac{C + D(\cos 3\Theta)^2}{d^2} \,. \tag{9}$$

Equation (9) was used in Fig. 11c providing a much better description of the data than a or b. However, the need to do this parametrization separately for metallic and semiconducting nanotubes limits somewhat the use of expression (9) with eight parameters. We list them in Table 6; for their discussion, see [111].

Table 6. Parameters for metallic and semiconducting nanotubes used in (9). These parameters describe the frequency of the radial-breathing mode in carbon nanotubes. From [111]

	A (cm^{-1} nm)	B (cm^{-1})	C (cm^{-1} nm)	D (cm^{-1})
Semicond.	227	7.3 ± 0.3	-1.1 ± 0.3	-0.9 ± 0.2
Metallic	227	11.8 ± 1.0	-2.7 ± 1.2	-2.7 ± 0.8

3.2.2 Double-Walled Nanotubes

The radial-breathing modes of double-walled nanotubes are interesting as well. The eigenmodes of the inner and outer tubes combine into inphase and out-of-phase modes in the double-walled tube, and their coupling is described best by the graphite interlayer coupling strength. *Popov* and *Henrard* showed using a valence-force field model that the coupled radial modes both increase in frequency compared to the same tube in single-walled form (see Fig. 12 [112]).

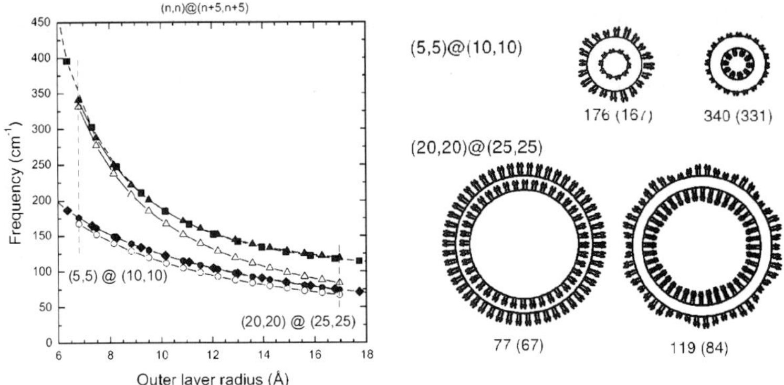

Fig. 12. (**left**) Frequencies of coupled radial-breathing modes in double-walled nanotubes. *Open symbols* belong to uncoupled tubes of the same diameter, *solid symbols* to double-walled nanotubes. The coupled frequencies are always higher than the uncoupled ones. (**right**) Eigenvectors and eigenfrequencies (in cm^{-1}) of the radial-breathing mode of two different diameters double-walled tubes. Inphase and corresponding out-of-phase modes are seen. After [112]

A simplified view of the coupled radial vibrations may be obtained by considering inner and outer tube oscillators connected by a spring. The inner and outer tubes have a diameter difference of 3.44 Å, the interplane spacing of graphite, and hence different eigenfrequencies $\omega_\mathrm{i} = \sqrt{k_\mathrm{i}/m_\mathrm{i}}$, as given by (8). From the solution of the coupled-oscillator equations

$$\begin{aligned} m_1\ddot{x}_1 &= -k_1 x_1 - k(x_1 - x_2) \quad \text{and} \\ m_2\ddot{x}_2 &= -k_2 x_2 - k(x_2 - x_1)\,. \end{aligned} \tag{10}$$

Dobardžić et al. [113] obtained the same dependence of inner and outer frequencies on diameter as [112]. In addition, the analytical solution to (10) yields the relative phases of inner and outer displacements and allows an approximate expression for the shift of both upper and lower modes:

$$\Delta\omega_\mathrm{RBM} \approx \frac{\kappa' d_\mathrm{i}}{2cc_1} \quad \text{for} \quad d_\mathrm{i} \leq 7\,\text{Å}\,, \tag{11}$$

where $\kappa' = 115\,927\,\mathrm{u}\cdot\mathrm{cm}^{-2}\cdot\text{Å}^{-1}$ is the effective spring constant between inner and outer shell, d_i is the diameter of the inner shell, $c = 14.375\,\mathrm{u}\cdot\text{Å}^{-1}$ is the linear mass density of a carbon nanotube, and c_1 is defined in (5). Note that the constant κ' is determined by the limiting case $d \to \infty$, where the out-of-phase mode becomes the interlayer mode of graphite [113]. For the diameters typically involved in the investigation of double-walled tubes, (11) is a useful expression for estimating the RBM frequencies. For further work based on this model, see [114].

Experimentally, the modes of double-walled tubes have not been conclusively identified. The measurements reported in [115, 116] have to be re-

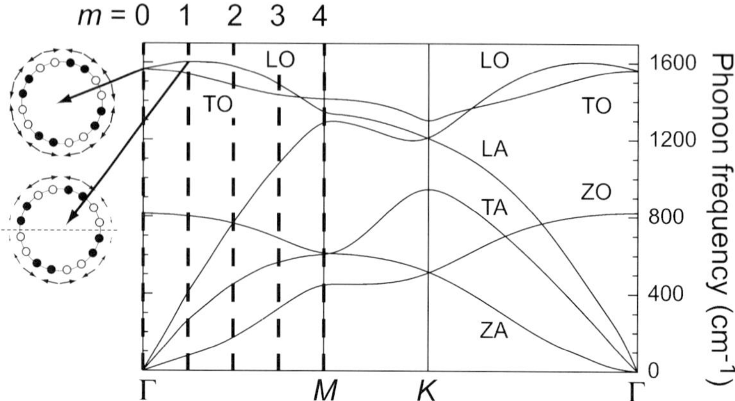

Fig. 13. Phonon dispersion of graphite and Γ-point vibrations of a (4,4) nanotube by zone folding. The nanotube contains several graphene unit cells around the circumference. Traveling waves of graphite that fulfill the periodic boundary conditions around the tube circumference are Γ-point vibrations of the nanotube (*dashed lines*). m labels the eigenvalues and is related to the z component of the angular momentum of a mode. Eigenvectors from the mth line have $2m$ nodes around the circumference, see the two circumferential eigenvectors for examples. The two eigenvectors are the A_{1g} (*upper*) and E_{1u} (*lower*) circumferential modes of the (4,4) nanotube. The tube cross section is schematically shown by the *large open circle* with the carbon atoms at $0 \cdot a$ (*closed symbols*) and $1/2 \cdot a$ (*open*) arranged around the circumference. The *arrows* next to the atoms indicate their displacement. The horizontal *dashed line* marks the nodal plane (zero displacement) of the lower $m = 1$ eigenvector. Data from [3]

evaluated in view of recent resonant Raman studies [63]. *Bacsa* et al. [47] identified several pairs of RBM peaks in tubes grown by a catalytic chemical vapor deposition (CVD) method. However, in their analysis they neglected the intertube interaction, i.e., they assumed $\Delta\omega_{\mathrm{RBM}} = 0$ in (11).

3.3 Tangential Modes

Tangential modes refer to all phonon bands of a nanotube originating from the optical phonons of graphite. Their eigenvectors are characterized by an out-of-phase displacement of two neighboring carbon atoms. The displacement is directed parallel to the nanotube wall, along the circumference, the axis, or a direction in between. The tangential modes involve predominantly the sp^2 inplane carbon–carbon bonds, which are extremely strong, even stronger than in the sp^3 diamond bond. Therefore, these modes have very high frequencies lying between $1100\,\mathrm{cm}^{-1}$ and $1600\,\mathrm{cm}^{-1}$ [20]. The Raman-active vibrations of the tangential modes fall into two groups, the high-energy modes (HEM) just below $1600\,\mathrm{cm}^{-1}$ and the D mode $\approx 1350\,\mathrm{cm}^{-1}$. The HEM is also called the G line in the nanotube literature; the "G" originally stood for

Table 7. Raman-active tangential modes. These phonons originate from the LO and TO branches of graphene; neighboring carbon atoms move out of phase. "circumferential" means parallel to the tube cirumference or chiral vector $\boldsymbol{c}$; "axial", parallel to the nanotube axis or its unit cell vector $\boldsymbol{a}$. In armchair and zigzag tubes there is only one phonon per irreducible respresentation, in chiral tubes there are two. The subscript g holds for achiral tubes. All Raman-active eigenvectors of archiral tubes can be found in [20]

	$A_{1(g)}$	$E_{1(g)}$	$E_{2(g)}$
$\lvert q_m \rvert$	0	$2/d$	$4/d$
(n, n) armchair	circumferential	axial	circumferential
$(n, 0)$ zigzag	axial	cirumferential	axial
(n_1, n_2) chiral	mixed (2 modes)	mixed (2modes)	mixed (2modes)

graphite and was taken over from the graphite Raman spectrum to carbon nanotubes.

The theoretical phonon dispersion of the tangential modes and their eigenvectors can be derived from the graphite phonon dispersion [8,20,23,117–119] using the concept of zone folding, which we explain in more detail in Sect. 4.1. The general idea is shown in Fig. 13 for a (4,4) armchair tube. In such small tubes, zone folding is not very accurate for the phonon frequencies on an absolute scale, but it helps to understand the systematics in the tangential modes.

The unit cell of a nanotube contains many more carbon atoms than that of graphene, but they are built from the same lattice. Hence we expect similar phonon frequencies in the two materials. The main difference between graphene and a tube is that a number of traveling waves in graphene ($q \neq 0$) correspond to the Γ point of the tube ($q = 0$). More precisely, the wave with a wavelength $\lambda_m = \pi d/m$ fulfills the periodic boundary conditions around the tube (m integer, see also Sect. 4.1). In an (n, n) armchair tube, several graphene unit cells are joined across the corners of the hexagons ($\boldsymbol{a}_1 + \boldsymbol{a}_2$ direction) to the nanotube unit cell. A wave with wavelength λ_m and traveling along the ΓM direction of graphene thus corresponds to the Γ point of the tube. This is shown in Fig. 13; the dashed lines indicate $\boldsymbol{k}$ vectors that are folded onto the Γ point of the (4,4) nanotube.

The Γ point of graphene gives rise to nanotube phonon eigenvectors with a constant displacement around the circumference, see the circumferential A_{1g} eigenvector in Fig. 13. These modes are always nondegenerate (A or B modes). There is a second tangential mode coming from the Γ-point optical mode of graphene. It is axial and Raman inactive in armchair tubes. Phonons with $m = 1, 2, \ldots, (n - 1)$ are doubly degenerate E modes. They have $2m$ nodes in their phonon displacements when going around the tube's circumference. The circumferential E_{1u} phonon in Fig. 13 has two nodes (dashed horizontal line) and is infrared active. Note that this phonon is higher

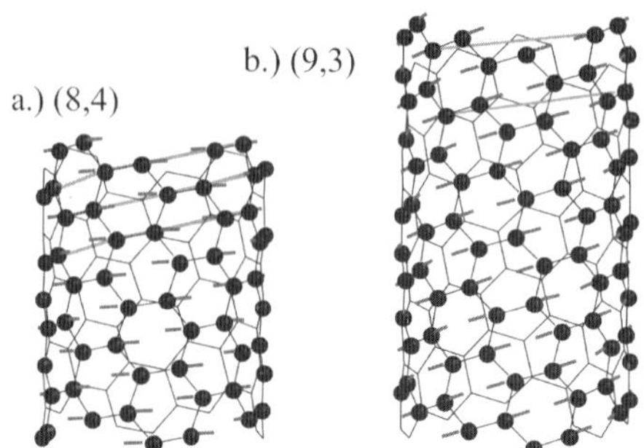

Fig. 14. (**a**) A_1 high-energy eigenvector of an (8,4) nanotube calculated from first principles. The atomic displacement is parallel to the circumference. (**b**) A_1 high-energy eigenvector of a (9,3) tube. The displacement is parallel to the carbon–carbon bonds. The direction of the helix in both tubes, obtained from the screw-axis operation, is indicated by the *gray lines*. From [20]

in frequency than the A_{1g} mode discussed previously. This comes from the overbending of the graphene LO branches (overbending means that the highest phonon frequency occurs inside the Brillouin zone, not at the Γ point, as in most materials; other examples of this phenomena besides graphite are diamond [120, 121] and hexagonal boron nitride [122, 123]). Phonons from the $m = 1$ and 2 line can be Raman active, see Table 7. Finally, phonons with $m = n$ originate from the M point of graphene. They are nondegenerate as are the modes originating from the graphene Γ point. These phonons are infrared- and Raman-inactive in all carbon nanotubes [20, 81].

To find the Γ-point vibrations of an $(n, 0)$ zigzag tube we need to divide the ΓKM direction in Fig. 13 into n segments. The degeneracy and the number of tangential high-energy modes follow as for armchair tubes. The Raman activity is reversed, see Table 7. An axial phonon that is Raman active in an armchair tube is inactive in a zigzag tube and vice versa [38, 81, 103].

In chiral tubes the chiral vector is along a low-symmetry direction of graphene. Then, also the confinement direction is a low-symmetry direction in reciprocal space. Once again, the degeneracy of the modes is the same as in armchair tubes (nondegenerate for $m = 0, q$, twofold degenerate for all other m and $2m$ nodes around the circumference). Because of the low symmetry of the tubes, the phonons are no longer axial and circumferential by symmetry [38, 124]. Examples are the A_1 eigenvectors of an (8,4) and a (9,3) tube in Fig. 14. Moreover, both A_1 ($m = 0$) phonons are Raman active. Their relative intensity depends on the chiral vector of the tube and the displacement direction of the eigenvector. For a nanotube with large Θ (close to armchair), the circumferential mode is expected to dominate, for a tube with small Θ (close to zigzag) the axial mode dominates.

Zone folding is a first approximation to the phonon frequencies of the high-energy tangential modes. As can be seen in Fig. 13 it predicts the axial

Table 8. Labeling used for the displacement direction (displ.) of the high-energy modes. tube: displacement direction is given as circumferential/axial; confinement: classification as LO/TO with respect to the confinement direction (tube circumference); propagation: classification as LO/TO with respect to the propagation direction (tube axes). symmetry: representation of the $m = 0$ mode in armchair (AC), zigzag (ZZ) and chiral (CH) tubes; graph.: graphene phonon branch from which a mode originates in the zone-folding scheme; overb. in tube indicates whether the $m = 0$ nanotube branches have an overbending or not. Selected references are given for each convention

Displ.	Symmetry	Graph.	Overb. in tube	Tube [38, 129, 130]	Confinement [8, 38, 117]	Propagation [126, 131, 132] [118, 128]
$\|\boldsymbol{c}$	A_{1g}(AC) A_{1u}(ZZ) A_1(CH)	LO	no	circumfer.	LO-like, LO_m	TO, T[a]
$\|\boldsymbol{a}$	A_{1u}(AC) A_{1g}(ZZ) A_1(CH)	TO	yes	axial	TO-like, TO_m	LO, L

[a] [133] The modes with a displacement along the nanotube circumference are called "axial" (A) in [118]; for the displacement along the tube axis "longitudinal" (L) is used in both [133] and [118]

and circumferential $m = 0$ eigenvector of a (4,4) tube to be degenerate, because they both arise from the doubly degenerate E_{2g} optical phonon of graphene (see [125] for a review of the lattice dynamics of graphite). There are two reasons why this is not the case for carbon nanotubes: curvature and, in the case of metallic tubes, the Kohn anomaly [3, 8, 38, 117, 118, 126–128].

Curvature softens the force constant in both semiconducting and metallic tubes. In Sect. 3.2 we discussed how the bonds along the nanotube circumference get an sp^3 component with decreasing diameter. Since the optical phonon of diamond is at a lower frequency ($1332\,\mathrm{cm}^{-1}$) than that of graphite ($1589\,\mathrm{cm}^{-1}$), curvature shifts the circumferential vibrations to lower frequencies. The axial vibrations are much less affected, see [8] and [118].

The Kohn anomaly affects the phonons in metallic tubes (see below). It strongly reduces the frequency of the axial A_1 mode. Before discussing this, we consider briefly the terms axial and circumferential versus transversal and longitudinal vibrations in connection with carbon nanotubes. These terms are used in a somewhat confusing way in the literature and are frequently intermixed.

We already pointed out that the tangential vibrations are, in general, of mixed axial and circumferential character in chiral carbon nanotubes [38], because of the mechanical boundary conditions and the low-symmetry confinement direction. Most of the theory of lattice dynamics, however, was developed for armchair and zigzag tubes. In achiral tubes the tangential modes have a definite parity and displacement direction with respect to the tube axis. Therefore, classifications of the tangential modes as axial/circumferential, LO/TO, and LO-like/TO-like are often used; see Table 8. The main confusion with the terms transverse and longitudinal arises because they can be referred to the confinement direction – as is normally done for low-dimensional systems [98, 134] – or with respect to the propagation direction – as is the standard convention for three-dimensional systems.

Take, for example, the two eigenvectors in Fig. 13. These circumferential vibrations should be called LO-like when referred to the confinement direction (the circumference), but TOs when referred to the only possible propagation direction, the z-axis of the nanotube. Labeling them as TOs leads to the result that the highest TO vibrations are higher in frequency than the LOs. Referring to the eigenvectors with respect to the confinement direction as LO-like or LO_0 and LO_1 in Fig. 13, on the other hand, we find that the nanotube LO modes have no overbending, whereas the TO modes do, see Table 8. While the "folded" nanotube Γ-point modes are higher in frequency than the $m = 0$ Γ-point mode due to overbending of the graphene LO mode, their propagation direction is perpendicular to the confinement direction, where the phonon dispersion has no overbending. This might lead to confusion compared to graphene, where the LO branches show a strong overbending. For these reasons we use the terms axial and circumferential in this work; they indicate the atomic displacement direction without making direct reference to the confinement or propagation direction of a phonon.

In armchair nanotubes, the valence and conduction bands cross at $\approx 2\pi/3a$. These tubes are always metallic and have a Fermi surface that is a single point, see Sect. 4. Chiral and zigzag tubes are metallic within the zone-folding approximation (neglecting curvature) if $(n_1 - n_2)/3$ is an integer. When curvature is included, these tubes have a small secondary gap at the Fermi level, on the order of 10 meV [6, 135–137] and are called quasimetallic tubes. A phonon that couples to the electronic states at the Fermi level opens a gap in armchair tubes and strongly modulates the gap in quasimetallic tubes [133]. This coupling reduces the total energy of the distorted system (nanotube plus phonon) and hence softens the force constants. One then finds in the phonon dispersion of the material a singular behavior for certain q vectors and phonon branches, referred to as a Kohn anomaly [138, 139].

In graphite and single-walled carbon nanotubes, a Kohn anomaly occurs at the Γ and the K point or at $2\pi/3a_0$ for the tubes[4] [3, 127, 140–142]. It reduces the Γ-point frequency of the axial high-energy vibration in metallic tubes. The softening by the electron–phonon coupling is larger than the curvature-induced softening [133]. Therefore, the axial $m = 0$ vibration of

[4] The Kohn anomaly at $q = 0$ originates from the special Fermi surface of graphene and nanotubes, which are single points. We will explain this in Sect. 9.3.

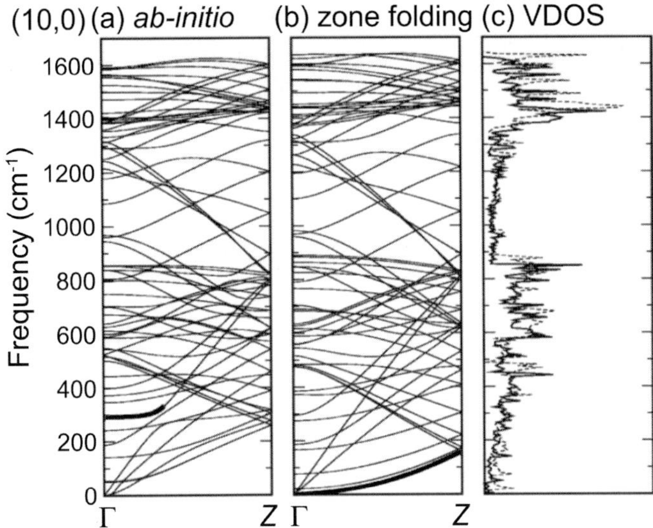

Fig. 15. Phonon dispersion of a (10,0) nanotube from (**a**) first principles and (**b**) zone folding of the graphene phonon dispersion. (**c**) Vibrational density of states from first principles (*full line*) and zone folding (*dashed*). From [118]

metallic nanotubes is lower in frequency than the circumferential phonon. This is opposite to semiconducting tubes, for which no Kohn anomaly occurs, but the circumferential phonon still feels the curvature of the nanotube wall.

We can obtain a good approximation for the phonon dispersion using zone folding. This works best for semiconducting tubes (no Kohn anomaly) with diameters $d \gtrsim 1\,\mathrm{nm}$ (little curvature effects). As an example, we compare the ab-initio phonon dispersion of a semiconducting (10,0) tube in Fig. 15a with the zone-folding approximation in Fig. 15b. In the low-energy range, zone folding fails to predict the radial-breathing mode (thick line at $300\,\mathrm{cm}^{-1}$ in Fig. 15a) [8,133,143]. This totally symmetric phonon has no simple equivalent among the Γ-point vibrations of graphene (upon zone folding $A_{1(g)}$ totally symmetric modes must originate from the graphene Γ point, see [20, 103]). Instead, the radial-breathing mode corresponds to an inplane uniaxial stress along the chiral vector [144].

Other low-energy vibrations typically show a frequency change when comparing Figs. 15a and b. This is mainly a curvature effect due to the rather small diameter of the (10,0) tube [9]. Another interesting difference is the parabolic acoustic dispersion upon zone folding shown by the thick line in Fig. 15b originating from the out-of-plane acoustic (ZA) branch of graphene (see Fig. 13). Because of the cylindrical geometry of the tube this peculiar dispersion disappears.

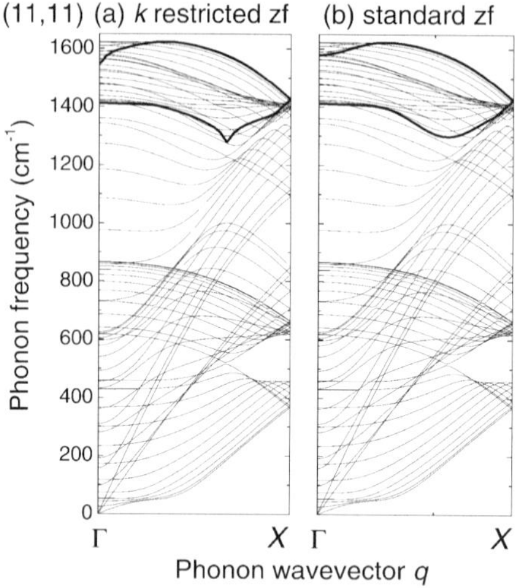

Fig. 16. Phonon dispersion of an (11,11) nanotube. (**a**) Calculated using linear response (ab-initio calculation) by restricting the *electronic* wavevectors to the allowed wavevectors of an (11,11) tube. This is an unconventional zone folding scheme, which takes into account the one-dimensional character of the tube, but neglects curvature. (**b**) Conventional zone folding of the graphene phonon dispersion obtained by restricting the *vibrational* states to the allowed wavevectors of the tube. In the one-dimensional nanotube Kohn anomalies are stronger (see *bold lines* in (**a**)) than in graphene (*bold lines* in (**b**)). From [141]

Zone folding is a better approximation for the high-energy tangential modes than for the low-energy vibrations. We can see some curvature induced effects in Fig. 15, for example, the softening of the highest-frequency vibrations in Fig. 15a compared to Fig. 15b. As discussed above, the highest phonons branches in carbon nanotubes are always circumferential because of the overbending in graphene. Circumferential modes soften due to curvature, a fact that explains the high-energy part of the phonon dispersion. This is also seen in the vibrational density of states (Fig. 15c). The softening of the vibrational DOS has been measured by two-phonon Raman scattering, see Sect. 10.3 and [145].

Figure 16 shows phonon-dispersion curves for a metallic (11,11) armachair tube [141]. Both dispersions were obtained by a zone-folding approach. Figure 16a represents an ab-initio calculation of graphene, where the calculated points in reciprocal space were restricted to the allowed $\boldsymbol{k}$ vectors of a (11,11) tube. Using this restricted Hamiltonian, the phonon bands were obtained from linear response theory [141]. This *unconventional zone folding* takes into account the one-dimensional character of carbon nanotubes, but not their curvature (for the (11,11) tube with $d = 1.5\,\mathrm{nm}$ curvature effects are small [9]). The dispersion in Fig. 16b corresponds to standard zone folding,

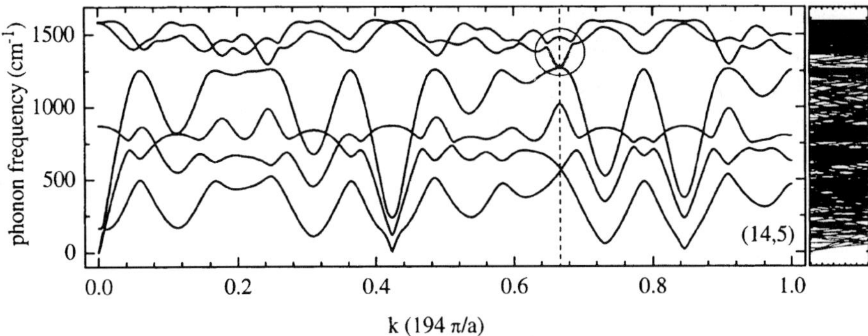

Fig. 17. Phonon dispersion of a (14,5) nanotube. **Left**: Dispersion as a function of the helical wavevector $\tilde{k}$. **Right**: Dispersion as a function of the linear wavevector k as in Figs. 15 and 16. The unit cell is very small in this chiral tube because the axial lattice vector $a = 29.5a_0$. From [85]

i.e., the graphene phonon dispersion was calculated from first principles and then used for finding the dispersion of the (11,11) tube (see Sect. 4.1 or [20]).

The axial $m = 0$ phonon band in Fig. 16a (thick line) drops sharply in frequency when approaching the Γ point [127, 140, 141, 146]. The phonon dispersion has a singularity at Γ, which reflects the Kohn anomaly of metallic nanotubes introduced above. The Γ-point axial frequency – A_{1u} symmetry – is at $1547\,\mathrm{cm}^{-1}$ as opposed to $1579\,\mathrm{cm}^{-1}$ in standard zone folding in Fig. 16b. A second singularity occurs for the back-folded ΓKM branch at $2\pi/3a$ ($m = n$ phonon of the (11,11) tube). This corresponds to the K point of graphene, see Sect. 4.1. Graphene also has Kohn anomalies at Γ and K [3, 127]. These two wavevectors couple electrons at the Fermi level. The Kohn anomalies are more pronounced in metallic nanotubes than in graphene, because of the one-dimensional character of the tubes [140–142].

In Fig. 16a only one phonon branch shows a singular behavior at Γ and K, respectively. Besides the correct q, a phonon also needs the correct symmetry to mediate an interaction between electrons. We discuss this topic in more detail in Sects. 8 and 9.3.

Carbon nanotubes have a large number of vibrational degrees of freedom and hence phonon bands. Figures 15 and 16 correspond to achiral tubes, where the number of atoms in the unit cell is small compared to the chiral species of similar diameter. In the right panel of Fig. 17 we show as a final example the dispersion of a (14,5) nanotube. The number of phonon bands in this case is 1164 [85]. A more transparent dispersion is obtained when using the so-called helical quantum numbers [85, 135]. In the left panel of Fig. 17 we show the phonon dispersion of the (14,5) as a function of the helical momentum instead of the linear momentum in the right panel. Details can be found in [20, 86, 135].

Out of the hundreds or thousands of phonon branches only a tiny fraction is Raman active (8 in achiral and 15 in chiral tubes, see Sect. 3.1). Moreover, the Raman spectra are dominated by totally symmetric phonons, see Sect. 6.4 [99, 101, 147]. Thus, only three branches are easily accessible experimentally, the radial-breathing mode and in the high-energy range the $m = 0$ tangential vibrations. Additionally, the K-point phonon of graphene gives rise to the D mode in the Raman spectra of nanotubes. The other vibrations are either very low in intensity or not observed at all. In the remainder of this work we mainly discuss the totally symmetric vibrations and the D mode, their overtones, resonances, and the corresponding electron–phonon coupling.

4 Electronic Properties

Single-walled carbon nanotubes are metals or semiconductors depending on their chirality. From the chiral index plot in Fig. 3 we see that every metallic tube (squares) is surrounded by six semiconducting neighbors (open and closed circles). Very subtle changes in the microscopic structure, e.g., a (10,5) and a (10,6) tube are almost indistinguishable concerning their atomic structure, result in drastic changes in the electronic properties. This is one of the reasons why carbon nanotubes evolved into a model system for one-dimensional nanostructures. They are a prime example of structure and shape being as important as, or even more than, chemical composition for very small systems.

A detailed review of the electronic and optical properties of carbon nanotubes is beyond the scope of this Chapter. Instead we introduce in this section the main concepts used for describing their physical properties: zone folding, family and branch behavior. We then concentrate on the optical range (visible and infrared transition energies) and exciton formation in nanotubes. This field has developed very rapidly over the last two years, after the first report of photoluminescence from carbon nanotubes [59, 60].

4.1 Zone Folding

Single-walled carbon nanotubes are one-dimensional solids [20]. Along the nanotube circumference electrons, phonons and other quasiparticles can only have certain, discrete wavelengths, because of the periodic boundary conditions [6, 20, 23, 83, 148–151]. On the other hand, quasiparticles can travel along the nanotube axis. An infinitely long tube has continuous electronic and vibrational states in this direction.

In the following we show how the quantization condition can be used to estimate the electronic band structure of carbon nanotubes from the electronic states of graphene. We do this by restricting the graphene eigenstates

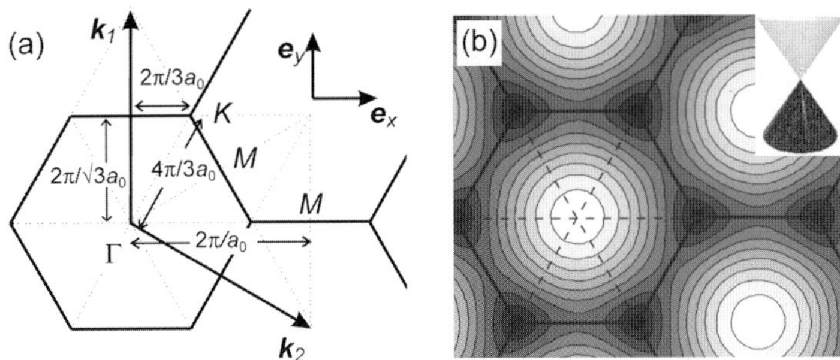

Fig. 18. (**a**) Hexagonal Brillouin zone of graphene. The reciprocal lattice vectors in terms of a Cartesian coordinate system are given by $\boldsymbol{k}_1 = 4\pi/\sqrt{3}a_0(1,0)$ and $\boldsymbol{k}_2 = 4\pi/\sqrt{3}a_0(\sqrt{3}/2, 1/2)$; their length is $4\pi/\sqrt{3}a_0$. The figure shows the high-symmetry Γ, M and K points. (**b**) Same as (**a**) but with a contour plot of the electronic band structure of graphene. Eigenstates at the Fermi level are shown in *black*; *white* marks energies far away from the Fermi level. The *inset* shows the valence (*dark*) and conduction (*bright*) band around the K points of the Brillouin zone. The two bands touch at a single point (K). After [20]

to the allowed wavevectors of a tube. We now introduce briefly the electronic properties of graphene.

Graphene is a semimetal. This means that its valence and conduction bands cross, but the electronic density of states is zero at the Fermi level (the crossing point). The most widely used description of the electronic band structure of graphene is an empirical tight-binding model [5, 6, 20, 23]. This model includes only the π states perpendicular to the graphene sheet and their interactions, because these states give rise to the electronic bands close to the Fermi level. We restrict the interaction between the carbon electrons to nearest neighbors in the graphene sheet and neglect the overlap between two π wavefunctions centered at different atoms. Under these approximations the valence E^- and conduction bands E^+ of graphene are given by the simple analytic expression [5, 20, 23]

$$E^{\pm}(\boldsymbol{k}) = \gamma_0 \sqrt{3 + 2\cos \boldsymbol{k}\cdot\boldsymbol{a}_1 + 2\cos \boldsymbol{k}\cdot\boldsymbol{a}_2 + 2\cos \boldsymbol{k}\cdot(\boldsymbol{a}_1 - \boldsymbol{a}_2)}\,. \tag{12}$$

γ_0 describes the interactions between two π electrons; typical values range from 2.7 eV to 3.1 eV. $\boldsymbol{k}$ is the electronic wavevector (see Fig. 18a) for the graphene Brillouin zone, and $\boldsymbol{a}_1$ and $\boldsymbol{a}_2$ are the lattice vectors.

The nearest-neighbor tight-binding model of (12) describes well the electrons close to the Fermi energy (within $\sim$ 0.2 eV of E_{F}). In the range of optical-transition energies, however, it deviates by 0.5 eV to 1 eV from ab-initio calculations and from measurements [6]. This can be fixed by including more neighbors; if up to third-neighbor interaction and the overlap between the π wavefunctions are included, there is virtually no difference between the semiempirical description and first-principles results. A discussion of the

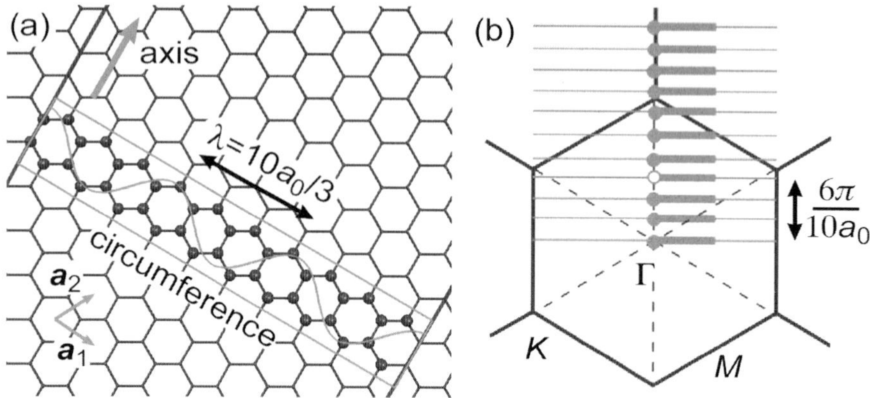

Fig. 19. Quantized wavevectors and zone folding of graphite. (**a**) a (10,0) nanotube unwrapped to a graphene sheet. The atoms shown are those inside the unit cell of the tube. The *gray line* is a wave with wavelength $\lambda = 10a_0/3$, which corresponds to an allowed state of the nanotube. (**b**) Allowed, discrete wavevectors of a (10,0) tube shown in the Brillouin zone of graphene. The wave in (**a**) corresponds to the *open circle* in (**b**). The *thick lines* form the one-dimensional Brillouin zone of the (10,0) tube

level of approximation and the derivation of (12) as well as a corresponding analytic expression for larger interaction ranges can be found in [6, 20].

Figure 18b shows a contour plot of the electronic band structure of graphene. For the π^* conduction band the electronic energies range from 12.2 eV at Γ to zero at the K point of the Brillouin zone; the π valence band varies between -8.8 eV and the Fermi level. The two Γ-point energies are from experiment [20,152]. To obtain nonsymmetric π and π^* bands within the tight-binding approximation we need to consider the finite overlap between electrons at different atoms or include more neighbors [5, 9, 20, 23]. Close to the six K points, where valence and conduction bands touch at a single point, the two bands are linear with k to a very good approximation. The inset of Fig. 18b shows the cones that represent the band structure very close to the Fermi energy.

Having obtained the electronic band structure of graphene, how can we use it to obtain the band structure of single-walled carbon nanotubes? Consider quantization in a (10,0) nanotube. An allowed wave for this tube is shown in Fig. 19a. It has three nodes along the circumference or a wavelength $\lambda = 10a_0/3$. For graphene this corresponds to wavevectors away from the Γ point of the hexagonal Brillouin zone (see Fig. 18a). The quantization condition for an (n_1, n_2) nanotube in reciprocal space can be derived from its chiral vector $\boldsymbol{c}$ and the axial lattice vector $\boldsymbol{a}$. The general expressions are given in Table 1.

For a (10,0) zigzag tube we find

$$\boldsymbol{k}_\perp \overset{(n,0)}{=} \frac{\boldsymbol{k}_1}{n} + \frac{\boldsymbol{k}_2}{2n} \overset{(10,0)}{=} \frac{\boldsymbol{k}_1}{10} + \frac{\boldsymbol{k}_2}{20}, \tag{13}$$

i.e., when starting from the Γ point and going along ΓKM ($\boldsymbol{k}_1 + \boldsymbol{k}_2/2$, see Fig. 18a) we find 10 equally spaced points that are allowed. All these points correspond to the Γ point of the (10,0) nanotube. They are shown as circles in Fig. 19b. The forth circle (open) represents the wave in Fig. 19a in reciprocal space. The distance between two allowed points, $2\pi/10a_0$ for the (10,0) tube, is given by the nanotube diameter $|\boldsymbol{k}_\perp| = 2/d$.

The reciprocal lattice vector along the nanotube axis is $\boldsymbol{k}_z = \boldsymbol{k}_2/2$ for the (10,0) and any other zigzag tube. The nanotube Brillouin zone is thus built as shown in Fig. 19b: The quantized circumferential vectors (circles) correspond to the Γ point of the tube (Γ_{nt}). Perpendicular to the quantization direction the allowed states are continuous. For the special case of zigzag tubes the general expression for allowed wavevectors (see Table 1) reduces to [20]

$$\boldsymbol{k}_{zz} = \frac{m}{n}\boldsymbol{k}_1 + \left(\frac{m}{2n} + \frac{k_z}{2}\right)\boldsymbol{k}_2 \quad \text{with } m = -(n-1)\ldots n; k_z = 0\ldots 0.5\,. \tag{14}$$

m (the z component of the angular momentum of a particle) indexes the bands of a nanotube. k_z is the magnitude of the axial wavevector in units of the reciprocal lattice vector. Inserting the allowed wavevectors in (12) we obtain the electronic band structure of zigzag carbon nanotubes within the tight-binding zone-folding approximation

$$E^{\pm}_{zz}(m, k_z) = \gamma_0 \sqrt{3 + 2\cos 2\pi\frac{m}{n} + 4\cos\pi\frac{m}{n} \cdot \cos\pi k_z}\,. \tag{15}$$

The corresponding expression for armchair tubes is

$$E^{\pm}_{\text{ac}}(m, k_z) = \gamma_0 \sqrt{3 + 4\cos\pi\frac{m}{n} \cdot \cos\pi k_z + 2\cos 2\pi k_z}\,. \tag{16}$$

For armchair as for zigzag tubes, the quantum number m satisfies $-(n-1) \leq m \leq n$. Finally, the band structure for a chiral (n_1, n_2) nanotube within the nearest-neighbor tight-binding model is given by

$$\begin{aligned} E^{\pm}_{\text{c}}(m, k_z) = & \gamma_0 \left[3 + 2\cos\left(m\frac{2n_1 + n_2}{qn\mathcal{R}} - \frac{n_2}{q}k_z\right)\right. \\ & + 2\cos\left(m\frac{2n_2 + n_1}{nq\mathcal{R}} + \frac{n_1}{q}k_z\right) \\ & \left. + 2\cos\left(m\frac{n_1 - n_2}{qn\mathcal{R}} - \frac{n_1 + n_2}{q}k_z\right)\right]^{1/2}, \end{aligned} \tag{17}$$

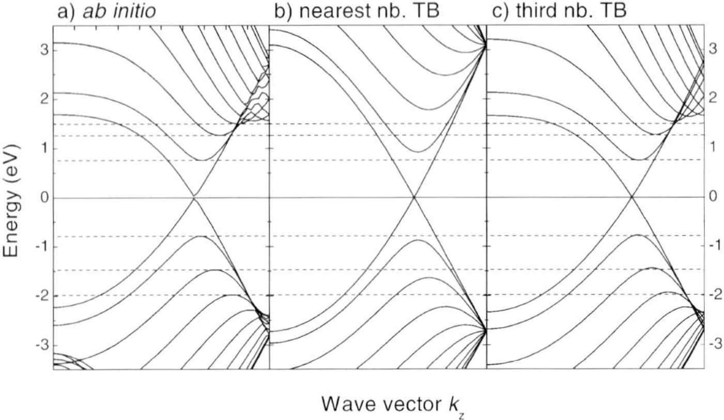

Fig. 20. Band structure of a (10,10) armchair nanotube with diameter $d = 1.4\,\text{nm}$. (**a**) Ab-initio calculation; (**b**) nearest-neighbor tight-binding calculation with $\gamma_0 = -2.7$ eV [(16) with $n = 10$]; (**c**) third-nearest neighbors tight-binding calculation. The *dashed lines* denote ab-initio calculated energies of the band extrema. The agreement of the energies in (**a**) and (**c**) is excellent. From [6]

where m is an integer running from $-(q/2-1)$ to $q/2$ [81]; see Table 1 for q, n and $\mathcal{R}$ as a function of n_1 and n_2.

The quantum number m is very useful to index bands and phonon branches and to derive selection rules, e.g., for Raman scattering, infrared vibronic and optical electronic absorption [20, 81, 83, 87]. The $m = 0$ electronic bands and phonon branches always contain the graphene Γ point (see Fig. 19b); $m = q/2$ ($= n$ for achiral tubes) is the M point of graphene for $k_z = 0$ [9]. These two bands are nondegenerate for any quasiparticle and any nanotube. In achiral tubes, all other bands are twofold degenerate in chiral tubes, none, see [20, 85, 86] for a discussion and examples.

4.2 Electronic Band Structure

Figures 20 and 21 show the electronic band structure of a (10,10) and (19,0) nanotube, respectively. Parts (a) in both figures are from first-principles calculations, while parts (b) were obtained with (16) and (15), and parts (c) are the results of the extended tight-binding model using up to third neighbors [6]. For all practical purposes, the extended tight-binding model is indistinguishable from the ab-initio calculations. The simple nearest-neighbors tight-binding scheme works reasonably well. There are certain systematics in Figs. 20 and 21 about which we will not comment in detail. For example, the band extrema are at the Γ point in the (19,0) tube, but at $2\pi/3a$ in the (10,10) tube. General rules for the overall shape of the band structure and the position of the band extrema as a function of n_1 and n_2 can be found in [20].

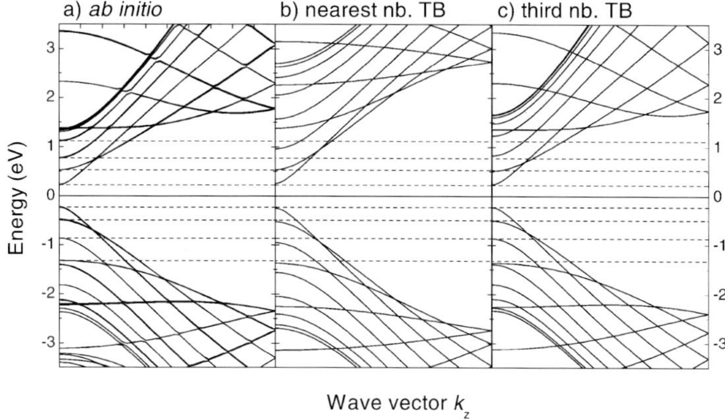

Fig. 21. Band structure of a (19,0) zigzag nanotube with a diameter of $d = 1.5\,$nm. (**a**) Ab-initio calculation; (**b**) nearest-neighbor tight-binding calculation with $\gamma_0 = -2.7$ eV [(15) with $n = 19$]; (**c**) third-nearest neighbors tight-binding calculation. The *dashed lines* denote ab-initio calculated energies of the band extrema. From [6]

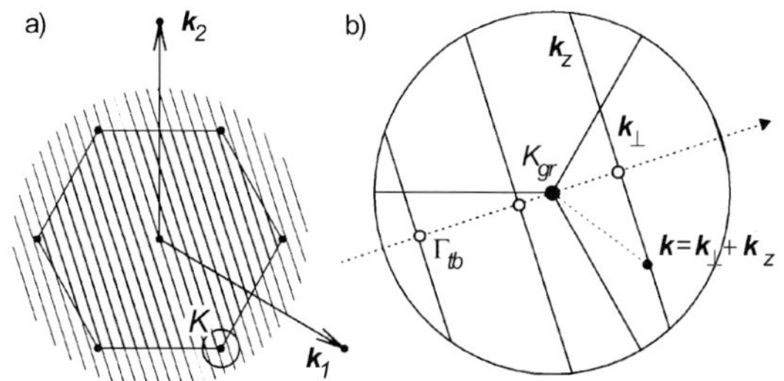

Fig. 22. (**a**) Allowed wavevectors of a nanotube in the Brillouin zone of graphene (see also Fig. 19). (**b**) Expanded view of the allowed wavevectors $\boldsymbol{k}$ around the K point. $\boldsymbol{k}_\perp$ is an allowed wavevector around the circumference of the tube; $\boldsymbol{k}_z$ is continuous. The *open dots* are points with $k_z = 0$; they all correspond to the Γ point of the tube. From [153]

As can be seen in Figs. 20 and 21 the (10,10) tube is metallic, whereas the (19,0) tube has a bandgap of 0.5 eV. The metallic and semiconducting character of the nanotubes can be explained by zone folding using the Fermi surface of graphene. In the last section we saw that the valence and conduction bands of graphene cross at the K point of the Brillouin zone. If the graphene K point is among the allowed states of a carbon nanotube, the tube is metallic. Otherwise, it is semiconducting with a moderate bandgap. To quantify this hand-waving argument let us consider an (n_1, n_2) nanotube. Figure 22a shows the quantized states of a general nanotube; the region around the K point is expanded in Fig. 22b. The electronic states are restricted to wavevectors that fulfill the condition $\boldsymbol{k} \cdot \boldsymbol{c} = 2\pi m$, where $\boldsymbol{c}$ is the chiral vector of the tube and m is an integer. For the tube in Fig. 22 the K point is not allowed and the tube is a semiconductor. The K point of graphene is at $\frac{1}{3}(\boldsymbol{k}_1 - \boldsymbol{k}_2)$; thus, a nanotube is a metal if

$$\boldsymbol{K} \cdot \boldsymbol{c} = 2\pi m = \tfrac{1}{3}(\boldsymbol{k}_1 - \boldsymbol{k}_2)(n_1\boldsymbol{a}_1 + n_2\boldsymbol{a}_2) = \tfrac{2\pi}{3}(n_1 - n_2)\,,$$

which implies

$$3m = n_1 - n_2\,. \tag{18}$$

This famous result, derived by *Hamada* et al. [148] and *Saito* et al. [149], means that a tube is a metal if $n_1 - n_2$ is a multiple of three.

We can now understand the pattern of metallic tubes in the chiral index plot in Fig. 3. (n, n) armchair nanotubes are always metallic, $(n, 0)$ zigzag tubes only if n is a multiple of three. The sets of metallic chiral indices parallel to the armchair-like directions arise, because for these sets $n_1 - n_2$ is either constant or changes by ± 3 from tube to tube. Take, for example, the (5,5) nanotube. Along the main armchair line (labeled armchair in Fig. 3) $n_1 - n_2$ is constant, since we change both n_1 and n_2 by the same amount, e.g., (5,5) to (6,6). Along the vertical direction n_1 decreases by 1 and n_2 increases by 2 when going from top to bottom, i.e., the difference between n_1 and n_2 changes by 3, such as for (5,5) to (6,3); similarly for the third armchair-like direction [(5,5) to (7,4)]. The same pattern is found for the open and closed symbols in Fig. 3. The reason is that metallic nanotubes are just a special case of the nanotube family, which we discuss in connection with the optical spectra of single-walled carbon nanotubes (Sect. 5.2).

The good agreement between first-principles and zone-folding calculations for the (10,10) and (19,0) nanotubes is also found for tubes with diameters ≥ 1.5 nm. For smaller-diameter tubes, the curvature of the nanotube wall can no longer be neglected. Curvature mixes the inplane σ and out-of-plane π states of graphene [9, 136]. It thereby shifts the electronic bands towards the Fermi level. Since this shift is larger for bands originating from the part between the K and the M point of the Brillouin zone than for bands from the part between Γ and K, curvature can change the order of the bands in single-walled nanotubes. It mainly affects the conduction bands and thus lifts the symmetry between the electron and the hole, which – to a very good approximation – is present in graphite, compare Figs. 20 and 21. For very small diameter nanotubes ($d \approx 0.5$ nm) the zone-folding picture fails completely [9, 10, 88, 136, 154–159]. For example, the (5,0) nanotube is metallic, although it is predicted to be semiconducting within the zone-folding approximation.

5 Optical Properties

The resonant Raman effect is intimately related to the optical properties of a material. Using Raman spectroscopy we can, therefore, study not only

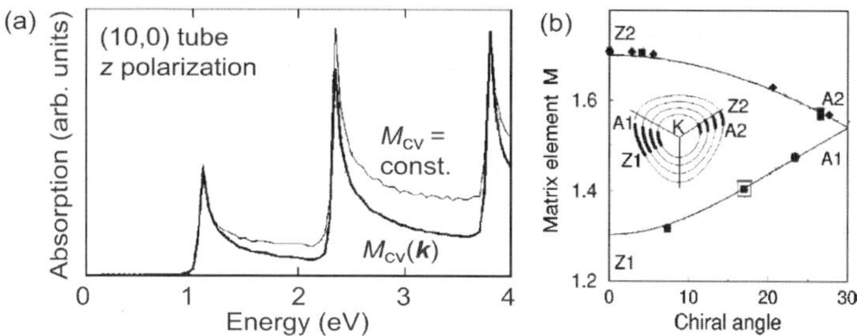

Fig. 23. (**a**) Band-to-band absorption spectra for a $(10, 0)$ nanotube calculated within tight binding. Note the sharp van-Hove singularities, typical for one-dimensional systems. The *thin line* assumes the optical matrix elements to be independent of the axial wavevector of the tube k_z; the *thick line* includes the dependence of the optical transition elements on k_z. (**b**) Chirality and family dependence of the matrix elements. A2 to Z2 refers to transitions originating close to the KM line of the graphene Brillouin zone. A1 to Z1 correspond to states between Γ and K (see *inset*). After [162]

the vibrations of a molecule or solid, but their electronic excitations and the optical response. This adds greatly to the power of Raman scattering for investigating and characterizing materials.

Resonances arise when the incoming or outgoing photon in the Raman process matches an optical singularity of the system (Sect. 6). They can be very strong in one-dimensional structures, because the electronic density of states diverges at the band extrema. In the following, we discuss the optical properties of carbon nanotubes. We start by considering the joint density of states (JDOS) in band-to-band transitions and then turn to exciton formation, which is important for understanding the optical properties.

5.1 Band-to-Band Transitions

In a band-to-band transition a photon excites an electron from the valence into the conduction band. The interaction between the electron and the hole it leaves behind is neglected, in contrast to an exciton formed by an interacting electron and hole (see Sect. 5.3). Band-to-band transitions have not been observed in carbon nanotubes, because the exciton binding energy is large (0.1 eV to 1 eV, much larger than $k_{\mathrm{B}}T$ at room temperature, k_{B} is the Boltzmann constant) [10,60,160,161]. We, nevertheless, discuss band-to-band transitions since they give a good insight into density-of-states arguments and the optical selection rules for carbon nanotubes. These arguments will remain valid when we later consider excitons.

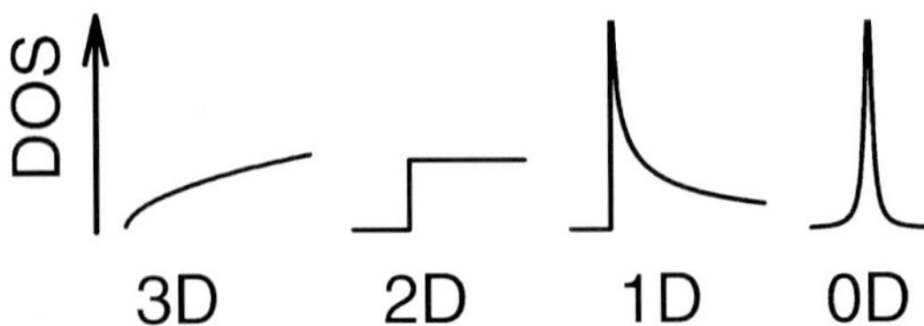

Fig. 24. Density of states in 3, 2, 1, and 0-dimensional systems. For one-dimensional systems the density of states follows a $1/\sqrt{E}$-behavior

In the dipole approximation the imaginary part of the dielectric function $\varepsilon_2(\omega)$ originating from band-to-band transitions is proportional to [134]

$$\varepsilon_2(\omega) = \frac{|M_{\mathrm{cv}}|^2}{\omega^2} \int \frac{dS_{\boldsymbol{k}}}{\nabla_{\boldsymbol{k}}(E_{\mathrm{c}}(\boldsymbol{k}) - E_{\mathrm{v}}(\boldsymbol{k}))}\,, \tag{19}$$

where ω is the photon frequency and M_{cv} are optical matrix elements. The integral corresponds to the joint density of electronic states (JDOS) of the valence E_{v} and conduction band E_{c}. Regions in the electronic band structure with parallel bands thus contribute strongly to band-to-band absorption.

For simplicity we assumed that the optical matrix elements M_{cv} are independent of the axial wavevector [87, 154, 162]. Within one pair of electronic subbands this is, in fact, quite accurate. Figure 23a compares a calculated band-to-band absorption spectrum for k-independent matrix elements (thin line) and including the k dependence of M_{cv} (thick). The peaks in the absorption are somewhat more pronounced in the latter, but the differences are small. For transitions between two different pairs of bands or transitions in tubes with different chirality, however, the matrix elements change. Figure 23b shows that the matrix elements depend on the chiral angle. M_{cv} also changes for electronic states originating from the KM (Z2, A2) and the ΓK (Z1, A1) part of the graphene Brillouin zone in the zone-folding approximation. This can be traced back to the optical properties of graphene [87,162].

The electronic density of states and also the joint density depend dramatically on the dimensionality of a system (see Fig. 24). For parabolic bands, found in most semiconductors, it rises as the square root of the energy above the bandgap in the three-dimensional case, exhibits a step-like dependence in two-dimensional solids, diverges as the inverse of a square root in one-dimensional systems, and, finally, is a δ-function in zero dimensions. Close to their minimum and maximum bands can always be approximated as parabolic. We thus expect a $1/\sqrt{E}$-behavior for the density of states in nanotubes. This result can be derived more rigorously using, e.g., the zone-folding approach [20, 153, 163]. The singular square-root dependence of the DOS on energy was observed by scanning tunneling spectroscopy [164, 165].

The second important term determining the strength of the optical absorption is the matrix element. Table 9 lists the selection rules for optical ab-

Table 9. Optical selection rules for carbon nanotubes [20, 87]. The (x, y) polarized transitions are strongly suppressed by the depolarization effect, see Sect. 6.4

Polarization	Representation	Δm	Remarks
z	A_{2u}	0	valence and conduction bands lying symmetrically to E_F
(x, y)	E_{1u}	± 1	only first pair of E_v and E_c in zigzag tubes

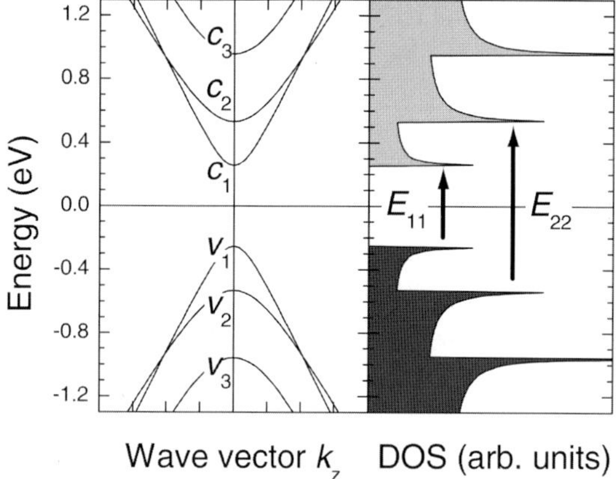

Fig. 25. Band structure and density of states of a zigzag nanotube. The band-to-band transition picture predicts that transitions between symmetrically lying pairs of valence and conduction bands contribute the most to the absorption. The transition energies are labeled E_{ii} where i labels the bands

sorption. For z-polarized light absorption is allowed between electronic bands with the same m, otherwise it is forbidden. Under perpendicular polarization the photon changes the band index by ± 1. In zigzag tubes (x, y) polarized absorption is allowed starting from the first two valence bands only. The perpendicularly polarized transitions are strongly suppressed by the anisotropic polarizability [166]. The tube acts as a Faraday cage for an electric field perpendicular to its axis. This makes carbon nanotubes almost transparent for inplane polarization (see Fig. 39b and the discussion in Sect. 6.4).

The overall picture for explaining the optical properties of carbon nanotubes within the band-to-band model is summarized in Fig. 25. The valence (v_i) and conduction bands (c_i) give rise to square-root singularities in the electronic density of states. Selection rules and the depolarization effect make the transitions for symmetrically lying pairs by far the dominant contribution, e.g., $v_1 \rightarrow c_1$. The absorption probability is particularly large when the density of states is high in the initial (valence) and final (con-

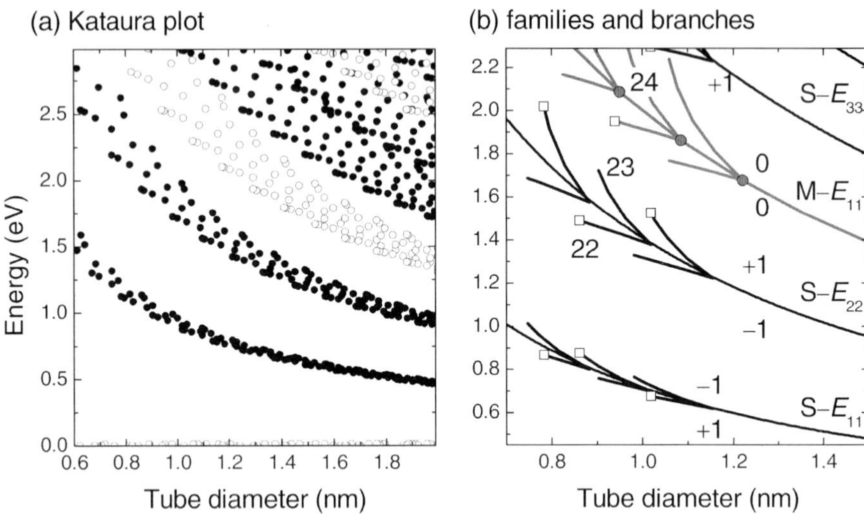

Fig. 26. (**a**) Kataura plot: transition energies of semiconducting (*filled symbols*) and metallic (*open*) nanotubes as a function of tube diameter. They were calculated from the van-Hove singularities in the joint density of states within the third-order tight-binding approximation [6]. (**b**) Expanded view of the Kataura plot highlighting the systematics in (**a**). The optical transition energies are roughly proportional to $1/d$ for semiconducting (*black*) and metallic nanotubes (*gray*). The V-shaped curves connect points from selected branches in (**b**), $\beta = 22, 23$ and 24; see text for details. We indicate whether the $\nu = -1$ or the $+1$ family is below or above the $1/d$ trend. *Squares* (*circles*) are transitions of zigzag (armchair) nanotubes

ducting) state. Thus, the absorption should show peaks corresponding to the transitions from the van-Hove singularity of the first valence band v_1 to the singularity of the first conduction band c_1 at an energy E_{11}, followed by $v_2 \rightarrow c_2$ at E_{22} and so forth (see Fig. 25). The absorption spectrum is often further approximated by considering only the van-Hove-related transition energies E_{ii} [167, 168]. Plotting them as a function of diameter one obtains the so-called Kataura plot, which we discuss in the next section.

The band-to-band transition picture was believed to be correct for almost ten years and most interpretations of the Raman and optical spectra relied on it. This view changed fundamentally after *Bachilo* et al. [60] measured photoluminescence and absorption from isolated nanotubes, followed by resonant Raman experiments on similar samples by *Telg* et al. [63] and *Fantini* et al. [64]: The experimental data could only be understood on the basis of excitonic transitions. Therefore, the literature has to be viewed with care because many studies relied on the band-to-band description.

5.2 Kataura Plot: Nanotube Families and Branches

When discussing the electronic band structure of carbon nanotubes we found two dependences on the nanotube chirality through zone folding: The separation between two lines of allowed wavevectors is given by $2/d$. On the other hand, the direction of the allowed lines with respect to the graphene Brillouin zone was connected with the chiral angle Θ (see Sect. 4.1). The optical transition energies depend, consequently, on these two parameters. This can be used to assign the chirality of a tube by Raman scattering and optical spectroscopy (see Sect. 7) [60, 61, 63, 64, 68].

Taking the maxima in the band-to-band absorption probability (the energy separation of the van-Hove singularities in Fig. 25) for an ensemble of nanotubes and plotting them as a function of the tube diameter d, we obtain Fig. 26a. The plot is named after *Hiromichi Kataura*, who first used it in connection with optical spectroscopy [167]. There are a number of systematics in the Kataura plot, which we discuss now. In the infrared and over most of the visible energy range an ensemble of nanotubes has well-separated ranges of transition energies for a given diameter range. The energies follow roughly $1/d$. The deviations from this trend reflect the chiral-angle dependence of the optical spectra. Figure 26b shows a part of the plot on an enlarged scale and concentrates on the systematics. The transition energies of metallic tubes (M–E_{11}, gray lines) are clearly separated from the semiconducting transitions (S–E_{ii}, black). Also, at first sight, semiconducting tubes seem to have more electronic transitions than metallic ones (however, this impression is not correct, in metallic tubes two transitions are always close in energy, see below). The first observation can be understood by the zone-folding approach.

Figure 27a–c shows the allowed wavevector lines of a metallic (a) and two semiconducting tubes (b,c) close to the K point of graphene (see Fig. 18). The electronic energies in the vicinity of this point give rise to transitions in the near infrared and the visible. In the metallic tube of Fig. 27a an allowed line goes through the K point (zero transition energy); the two neighboring lines, which are the origin of the M–E_{11} absorption, are at a distance of $2/d$ from K. In a semiconducting tube the first allowed line is $2/3d$ away from K (S–E_{11}), the next $4/3d$ (S–E_{22}), then $8/3d$ (S–E_{33}), $10/3d$ and so forth (see Fig. 27b). Close to the K point, the graphene band structure is approximately linear with wavevector. Therefore, the ordering of the optical energies will be S–E_{11}, S–E_{22}, M–E_{11}, S–E_{33}, ..., all of them well separated in energy for a given $d \approx 1\,\mathrm{nm}$ [153, 163].

The systematic deviations of the optical energies from the $1/d$ trend in Fig. 26 reflect the trigonal distortion of the graphene band structure, i.e., the fact that the energies depend on the direction of $(\boldsymbol{k} - \boldsymbol{K})$ [150, 169]. The trigonal distortion is clearly visible in the contours of Figs. 27a–c (see also Fig. 28): Without trigonal distortion the contour lines would be circles. The two M–E_{11} lines in Fig. 27a cross slightly different contours. Therefore,

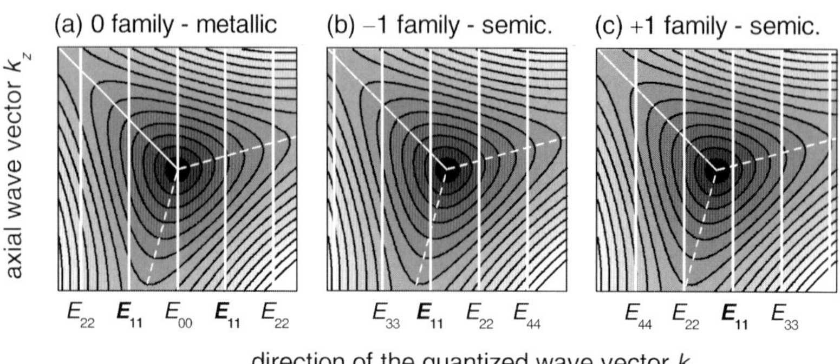

Fig. 27. Lines of allowed wavevectors for the three nanotube families on a gray-scale plot of the electronic band structure of graphene (K point at center). (**a**) Metallic nanotube belonging to the $\nu = 0$ family, (**b**) semiconducting -1 family tube, and (**c**) semiconducting $+1$ family tube. Below the allowed lines, the optical transition energies E_{ii} are indicated. Note how E_{ii} alternates between the left and the right of the K point in the two semiconducting tubes. The assumed chiral angle is 15° for all three tubes; the diameter was taken to be the same, i.e., the allowed lines do not correspond to realistic nanotubes. After [150]

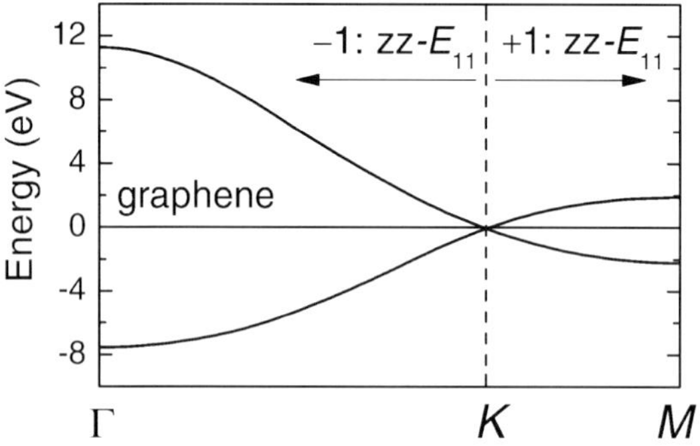

Fig. 28. Electronic band structure of the π and π^* states of graphene along the ΓKM line, which is the confinement direction for zigzag nanotubes. The electronic dispersion between Γ and K is stronger than between K and M. Zigzag tubes from the -1 family have a larger bandgap than those of the $+1$ family; see *arrows*

metallic nanotubes have two close-by transition energies that generate the V-shaped structure in Fig. 26b. A similar systematics applies to semiconducting nanotubes. Here, the splitting does not occur within one tube, but for two groups of semiconducting tubes, known as nanotube families [150].

The first allowed line in Figs. 27b and c are at the same distance from K. However, in Fig. 27b E_{11} is to the left of the K point (between Γ and K), whereas it is to the right in Fig. 27b (between K and M). Since the electronic dispersion is stronger between Γ and K than between K and M (see Fig. 28), the tube in Fig. 27b has a larger band gap and hence a larger transition energy E_{11} than the tube in Fig. 27c. For E_{22} the situation is reversed, because this transition originates from the opposite side of K. S–E_{33} behaves like S–E_{11}.

Whether S–E_{11} is to the right or to the left of the K point of graphene after the zone-folding scheme, is a function of n_1 and n_2 [150]. A nanotube belongs to the group or "family" in Figs. 27a,b and c if

$$\nu = (n_1 - n_2)\mathrm{mod}3 = \begin{cases} -1 & \text{semiconducting, large bandgap } E_{11}\text{, part b} \\ 0 & \text{metallic, part a} \\ +1 & \text{semiconducting, small bandgap } E_{11}\text{, part c.} \end{cases} \tag{20}$$

Nanotube families were first suggested by us [150], but picked up only two years later after optical spectroscopy on isolated nanotubes was reported for the first time [59, 60]. Optical spectroscopy was the first method sensitive enough to measure the family dependence of carbon nanotubes. The concept proved to be vital for the understanding of carbon nanotubes; many of their properties change with the family index ν. The three nanotube families are also indicated in the chiral-index plot in Fig. 3.

Nanotube families explain why some tubes fall above and some below the $1/d$ trend in Fig. 26. Where, however, does the peculiar V-shaped structure come from? The Vs reflect the "branches" in the Kataura plot. Within a branch β, by definition (Table 1),

$$2n_1 + n_2 = \beta = \text{const.} \tag{21}$$

Using (21) we immediately derive the neighbor rule, which gives the chiral index of a neighboring nanotube (n_1', n_2') with smaller diameter within a branch, i.e., with $\beta = $ const.

$$n_1' = n_1 - 1 \quad \text{and} \quad n_2' = n_2 + 2 \qquad \text{as long as } n_1 \geq n_2\,. \tag{22}$$

Looking back at Fig. 3 we see that $\beta = $ const. corresponds to vertical lines in the chiral-index plot. For such a line the diameter of a tube decreases with decreasing chiral angle. At the same time, trigonal warping increases with decreasing chiral angle [150, 169]. An armchair tube with $\Theta = 0°$ shows no splitting related to trigonal warping. Its allowed lines run perpendicular to the base of the triangle in Fig. 27, i.e., the left and the right of K are at the same energy. Armchair tubes are at the centers of the metallic V-shaped curves, see circles in Fig. 26b. In contrast, for zigzag tubes the trigonal distortion is strongest [150]. Their allowed lines touch the tip and the base of the triangle of Fig. 27. Hence, $(n, 0)$ zigzag and $(n_1, 1)$ chiral nanotubes are at the end of the Vs in Fig. 26b; see open squares. A summary of the nanotube classifications presented here is given in Table 2.

Nanotube families and nanotube branches play an important role in the interpretation of experiments. The key point is that the nanotube properties depend systematically on the family and branch they belong to and the position within one branch. The change in bandgap for different families is the most obvious example, further examples in connection with Raman and optical spectroscopy include the Raman, photoluminescence and absorption intensity, the softening of the high-energy phonons, electron–phonon coupling and many other properties [45, 60, 63, 64, 111, 144, 162, 170, 171].

5.3 Excitons

An electron and a hole in a crystal can form a bound state called an exciton [134]. This is mathematically equivalent to the hydrogen problem except for the existence of a center-of-mass k vector because of the translational invariance along the tube (see, e.g., [134]). The exciton binding energy in a three-dimensional solid assuming electron–hole symmetry is given by

$$E_\mathrm{b} = \frac{m^*}{2\varepsilon_\mathrm{cr}^2}\mathrm{Ry}\,. \tag{23}$$

This is much smaller than the ionization energy of hydrogen $1\,\mathrm{Ry} = 13.6$ eV, because the effective mass m^* is only a fraction of the electron mass and the Coulomb interaction is screened by the dielectric constant ε_cr. Inserting typical values for semiconductors into (23) $m^* \approx 0.01$ (in units of the electron mass) and $\varepsilon_\mathrm{cr} \approx 4$ we obtain binding energies on the order of a few meV. In many bulk crystals excitonic effects are only important at low temperature, where the thermal energy $k_\mathrm{B}T \lesssim E_\mathrm{b}$.

The exciton binding energy, however, depends on the dimension of the solid [172, 173]. For a 2D system E_b increases by a factor of four within the hydrogen model compared to 3D crystals. Although nanotubes are one-dimensional systems, a 2D model is most appropriate, since the electron and the hole move on a cylindrical surface [13]. Additionally, screening is less efficient in a low-dimensional solid. Assuming a 2D hydrogen model and no dielectric screening, the exciton binding energy increases to $E_\mathrm{b}(2\mathrm{D}) = 4E_\mathrm{b}(3\mathrm{D}) = m^* 4\,\mathrm{Ry}/2 \approx 300\,\mathrm{meV}$. More sophisticated estimates for the exciton binding energy in carbon nanotubes yield $E_\mathrm{b} = 1$ eV to 1.5 eV for diameters $d \approx 1\,\mathrm{nm}$ and tubes in vacuum [10, 160, 161, 174]. This is much larger than the thermal energy at room temperature $k_\mathrm{B}T \approx 25\,\mathrm{meV}$. Excitons in carbon nanotubes are thus observable at room temperature and below.

The particular importance of the excitonic interaction for spectroscopy is that it fundamentally changes the optical excitations. This is illustrated in Fig. 29. Instead of exciting electrons from the valence into the conduction band, as in the band-to-band model (see Fig. 25), the photon excites an exciton from the ground state into the optically active states eh_{ii} formed by

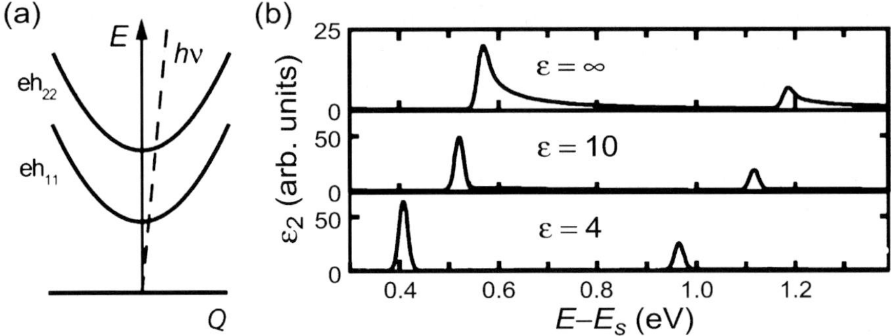

Fig. 29. (**a**) Schematic dispersion of the first eh_{11} and second eh_{22} subband excitons. Q is the exciton wavevector defined as the center-of-mass wavevector of electron and hole. $h\nu$ indicates the dispersion of photons. Absorption occurs at the crossing of the photonic and excitonic dispersion with a wavevector $Q_{\text{abs}} \approx 0$. After [134]. (**b**) Calculated absorption spectra neglecting electron–hole interaction (band-to-band transitions), $\varepsilon = \infty$ *top panel*, and for decreasing screening, *middle* and *bottom panels*. The shift in the main transition indicates the exciton binding energy as a function of ε. With decreasing screening, more spectral weight is transfered to the exciton. After [160]

electrons and holes in the ith subband (see Fig. 29a). The combined electron–hole wavevector $\boldsymbol{Q} = \boldsymbol{k}_{\text{e}} + \boldsymbol{k}_{\text{h}}$ has to be equal to the wavevector of the photon, $\boldsymbol{Q} \approx 0$ ($\boldsymbol{k}_{\text{e}}$ and $\boldsymbol{k}_{\text{h}}$ are the wavevectors of the electron and the hole, respectively).

Figure 29b shows how the optical absorption spectrum changes when electron–hole interaction is considered. The results were obtained for a (19,0) tube using a tight-binding model to solve the Bethe–Salpeter equation, a two-particle equation, which includes electron–hole interaction [160]. The characteristic $1/\sqrt{E}$ absorption predicted for band-to-band transitions (top panel and Fig. 25) changes into a δ-function when the spectral weight is transferred from the electron–hole continuum to the excitons. ε is the dielectric function of the medium surrounding the tube.

The two peaks in the three panels of Fig. 29b originate from transitions between the first two pairs of valence and conduction bands in the top panel and from exciting the first and second subband exciton in the middle and lower panels. The shift in the peak position gives the exciton binding energies for two dielectric constants ε of the medium surrounding the tube.

From what we discussed up to now, one would expect the experimental optical transitions to be lower in energy than predicted from the single-particle picture, because of the large exciton binding energies. Instead, photoluminescence and Raman spectroscopy found a blueshift of the transitions by some 100 meV when compared to a single-particle theory [44, 45, 60, 61, 63, 64, 68]. The origin of this shift is the increase of the bandgap due to electron–electron interactions [10,174]. As shown in Fig. 30, electron–electron

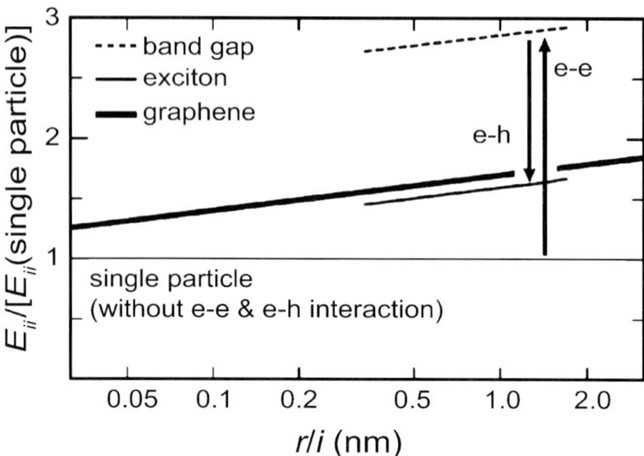

Fig. 30. Electron–electron interaction increases the separation between the ith valence and conduction band (*dashed lines*) by approximately 1.8 eV compared to the single-particle picture (*horizontal full line* at 1). The binding energy due to electron–hole interaction is around 1.3 eV. The excitonic transitions (*full thin lines*) are blueshifted compared to the band-to-band transitions. The *full thick line* was obtained by zone folding (see [174] for details). Note the log scale for x. After [174]

interaction (e–e) increases the electronic gap compared to the single-particle approximation. This increase is almost canceled by the electron–hole interaction (e–h). Nevertheless, the exciton in Fig. 30 is somewhat higher in energy than the single-particle states.

Figure 31 shows an experimental Kataura plot, i.e., the measured optical transition energies as a function of tube diameter. We will explain in Sect. 7 how the transition energies can be assigned to carbon nanotubes experimentally to plot them as in Fig. 31. The figure shows the first two transitions for semiconducting tubes measured by photoluminescence and Raman scattering and the first transition of metallic tubes from Raman spectroscopy [60, 63]. Only the lower parts of the metallic branches were observed experimentally; see Fig. 26. The energies of the optical transitions can be well understood, at least qualitatively, considering the single-particle states, electron–electron and electron–hole interaction, and the curvature of the nanotube walls [74, 160, 170, 174].

The exciton binding energies of single-walled carbon nanotubes were measured by two-photon absorption. These experiments probe the energy of excited excitonic states by the simultaneous absorption of two photons. The binding energy can be estimated from the energy difference between the exciton ground state (one-photon absorption) and the excited state (two-photon absorption). *Wang* et al. [12] and *Maultzsch* et al. [13] reported $E_\mathrm{b} \approx 0.4$ eV for nanotubes coated by surfactants. This is in reasonably good agreement with theory when the screening by the surfactant is taken into account.

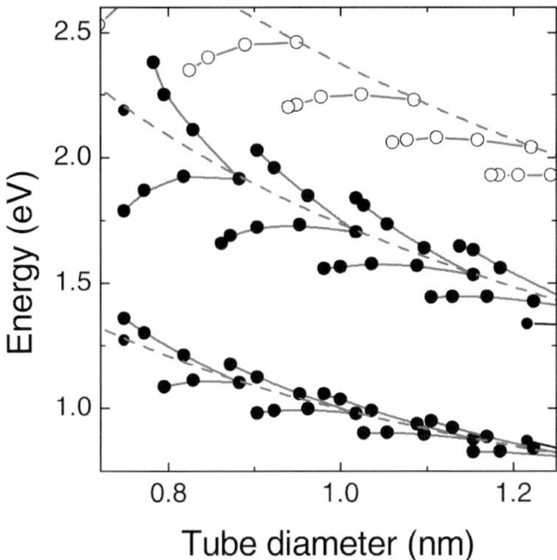

Fig. 31. Experimental Kataura plot for the first two semiconducting (*closed symbols*) and the first metallic (*open symbols*) transition. *Dashed lines* connect the (near-to) armchair tubes; *full lines* connect tubes in a branch, compare Fig. 26 and Table 2. Experimental transition energies were measured with photoluminescence [60] and resonant Raman scattering [63]

Optical spectroscopy of isolated nanotubes has been a rapidly evolving field over the last three years, but many aspects are still not understood. For example, side peaks in the photoluminescence excitation spectra of single-walled carbon nanotubes have been interpreted as phonon side bands of excitons [175, 176], coupled exciton–phonon excitations [11, 177], or excited excitonic states [13]. The variation in the nanotube absorption and emission intensity with chirality has been attributed either to electron–electron [74] or to electron–phonon interaction [75]. Nevertheless, the two-photon absorption experiments proved the optical excitations to be of excitonic character in single-walled carbon nanotubes [12, 13]; prior to the measurement of the exciton binding energy, time-resolved spectroscopy also suggested formation of excitons [178–182]. The band-to-band transition picture is thus not fully appropriate when modeling either photon absorption and emission or Raman scattering. On the other hand, a unified description of the Raman effect including excitonic effects is still missing. In this Chapter we will use mainly the band-to-band transition picture to model Raman scattering in carbon nanotubes pointing out important differences and consequences that should arise when electron–hole interaction is taken into account.

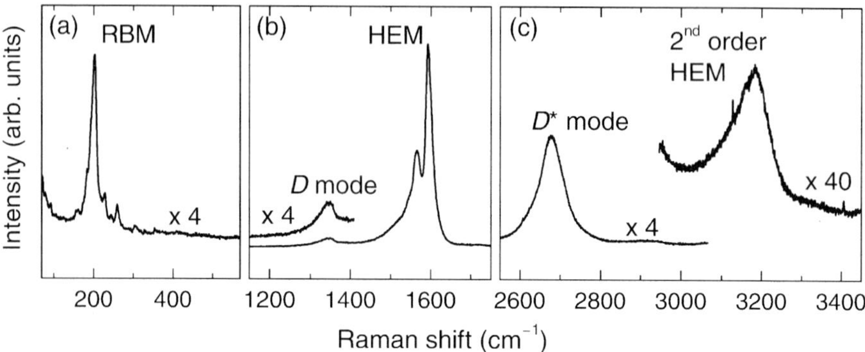

Fig. 32. Raman spectrum of bundles of single-walled carbon nanotubes excited with $\lambda_{\mathrm{L}} = 488\,\mathrm{nm}$. (**a**) Radial-breathing mode (RBM), (**b**) D mode and high-energy modes (HEM), (**c**) two-phonon Raman spectrum with the D^* (overtone of the D) and the overtone spectrum of the HEM. The spectra were multiplied by the factors given in the panels. From [186]

6 Raman Effect in Carbon Nanotubes

Resonance Raman scattering refers to a technique where one or more transitions participating in the scattering process involve real states. It has been widely used to study electronic states in solids (see other volumes of this series [183], in particular the seminal articles by *Cardona* [184] and by *Martin* and *Falicov* [185]).

After introducing the Raman spectrum of single-walled carbon nanotubes, we show in this section how the enhancement of the Raman-scattering intensity through resonances is involved in all the reported spectra of nanotubes: nonresonant Raman scattering is simply too weak to be observed in this material. Both single and double resonances are observed in carbon nanotubes; they have distinct features and yield different kinds of physical insight into the nanotubes. Double resonances are much more widely present in nanotubes than in 2- or 3-dimensional solids; they are, in fact, the dominant scattering mechanism for the high-energy modes.

6.1 Raman Spectra of Single-Walled Nanotubes

Figure 32 shows the three energy ranges where the most important Raman features of single-walled carbon nanotubes are observed [20,26,145,187–189]. In the low-energy range (Fig. 32a) there are several peaks in nanotube bundles or nanotubes in solutions resulting from the radial-breathing modes of the resonantly excited tubes (see Sect. 3.2). For a single tube, there would be only a single radial-breathing mode peak. Its frequency depends on diameter, these modes are typically found between $100\,\mathrm{cm}^{-1}$ and $400\,\mathrm{cm}^{-1}$. The radial-breathing mode is a vibration characteristic of carbon nanotubes and is often

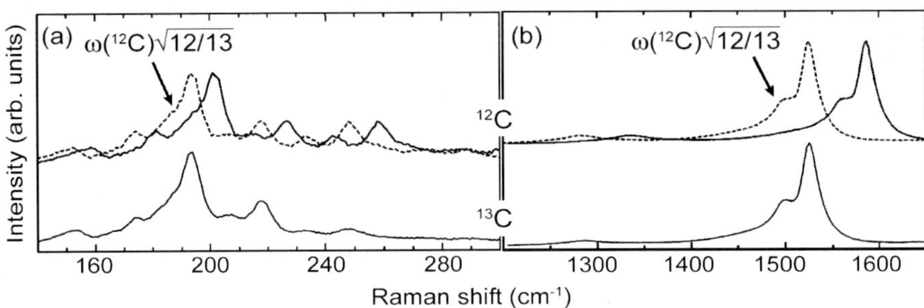

Fig. 33. Low (**a**) and high-energy (**b**) Raman spectra of ^{12}C (*upper traces, full lines*) and of ^{13}C (*lower traces*). The *dashed lines* are the ^{12}C spectra with the frequency multiplied by the expected isotope shift ($\sqrt{12/13}$). From [195]

considered as their fingerprint. However, since many materials have vibrational frequencies in this energy range, the presence of single-walled tubes in a sample cannot be surmised from the low-energy spectrum [190–193].

The D mode in Fig. 32b is an intriguing Raman peak that arises from double-resonant Raman scattering [4]. Its frequency depends on the energy of the exciting laser; we discuss this mode in Sect. 10.

The high-energy modes (HEM) of single-walled tubes originate from tangential vibrations of the carbon atoms (Sect. 3.3), i.e., along the nanotube walls [187, 194]. The double-peak structure close to 1600 cm^{-1} in Fig. 32b is a unique feature of single-walled carbon nanotubes. It is a more reliable Raman signature of single-walled tubes than the RBMs in the low-energy range. Typically, the most intense peak in the high-energy range occurs between 1592 cm^{-1} and 1596 cm^{-1} (depending on diameter and excitation energy); it is always higher in frequency than the corresponding peak in graphite (1589 cm^{-1} [145]). The smaller peak between 1560 cm^{-1} to 1590 cm^{-1} changes dramatically for resonant excitation of metallic tubes (Sect. 8 [168, 188]).

Between the RBM in Fig. 32a and the high-energy range with the D mode and the HEM a number of weak Raman peaks are observed [187, 196]. We will not review this so-called intermediate frequency range but refer the reader to [26].

Figure 32c shows the overtone Raman spectrum of single-walled carbon nanotubes. Note that D^*, the overtone of the D mode, is stronger in intensity than its fundamental. The two-phonon Raman spectrum of carbon nanotubes is similar to that of other sp^2-bonded carbons, but the overtone of the HEM is at lower frequency than in graphite (see Sect. 10).

In Fig. 33 we show Raman spectra of single-walled carbon nanotubes where ^{12}C (98.93 % natural abundance) was replaced by ^{13}C [195]. Phonon frequencies are proportional to $1/\sqrt{m}$ (spring model); hence we expect the frequencies to shift by $\sqrt{12/13} \approx 4\,\%$ when replacing naturally occurring carbon by the pure ^{13}C isotope. The full lines in the upper traces of Fig. 33

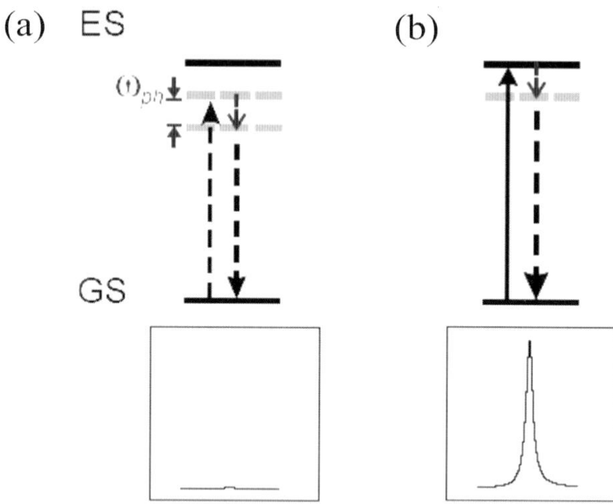

Fig. 34. Schematic of (**a**) nonresonant (Stokes) scattering and (**b**) a single (incoming) resonance. GS (ES) refer to the ground (excited) state of the material investigated and ω_{ph} to the phonon (or other elementary excitation) emitted in the process. The corresponding relative strengths in a Raman spectrum are indicated on the *bottom*

are the as-measured ^{12}C Raman spectra; for the broken lines the frequencies were multiplied by the expected isotope shift. There is excellent agreement with the ^{13}C Raman spectra shown in the lower traces of Fig. 33 [195, 197].

6.2 Single Resonances

The concept of a single resonance is illustrated in Fig. 34. Nonresonant Raman scattering corresponds to the excitation of an electron from the ground state (GS) to a virtual state (gray, dashed line) followed by emission (Stokes scattering) or absorption of a phonon (anti-Stokes, not shown) and subsequent recombination of the electron–hole pair (a). Increasing the excitation energy leads to the excitation of an electron into a real (empty) state (b).

Since the exciting laser light is in resonance, one speaks of an *incoming* resonance. Increasing the excitation energy further leads to an *outgoing* resonance and, finally, again to a nonresonant situation.

Whether incoming or outgoing resonances can be resolved experimentally depends on the width ($\sim$ inverse lifetime) of the excited electronic state. If it is too broad compared to the energy of the phonon (or other elementary excitation), they cannot be resolved. This is the situation for the resonances of the radial-breathing mode (ω_{RBM} in the range $150\,\mathrm{cm}^{-1}$ to $250\,\mathrm{cm}^{-1}$ for typical nanotubes), and an extraction of the precise transition energies from resonance spectra has to take this into account; see Sect. 7. In total, six different time-order diagrams contribute to first-order Raman scattering, of which we showed in Fig. 34 only one. In general, there are, of course, many excited states that have to be included for proper calculation of the Raman spectra; see, e.g., [185, 198].

The Raman efficiency of a bulk material for Stokes scattering, i.e., the number of photons per incident photon per unit length and solid angle $\mathrm{d}\Omega$ is given by [184, 185, 199]

$$\frac{\mathrm{d}S}{\mathrm{d}\Omega} = \frac{1}{\mathrm{V}} \frac{\mathrm{d}\sigma}{\mathrm{d}\Omega} = \frac{\omega_1 \omega_2^3}{4\pi^2} \frac{\eta_1 \eta_2^3}{c^4} \frac{V_\mathrm{c} N}{(\hbar\omega_1)^2} \sum_f |K_{2f,10}|^2 [N(\omega) + 1] , \tag{24}$$

where $K_{2f,10}$ is the Raman matrix element (see below) with the final state f with frequency ω. $\mathrm{d}\sigma/\mathrm{d}\Omega$ is called the differential Raman cross section. $V = V_\mathrm{c} N$ is the scattering volume and V_c is the volume of the unit cell. In the one-dimensional systems we want to consider here, the scattering volume is equivalent to the scattering length L. ω_1 and $\omega_2 = \omega_1 - \omega$ are the frequencies of the incoming and outgoing light, respectively. η_1 and η_2 are the indizes of refraction at ω_1 and ω_2. $N(\omega) = 1/[\exp(\hbar\omega/k_\mathrm{B}T) - 1]$ is the Bose–Einstein occupation factor and c is the speed of light in vacuum.

We can express the resonant process by using the Raman matrix element $K_{2f,10}$ [185]

$$K_{2f,10} = \sum_{a,b} \frac{\langle \omega_2, f, i | H_{\mathrm{eR},\rho} | 0, f, b \rangle \langle 0, f, b | H_{\mathrm{ep}} | 0, 0, a \rangle \langle 0, 0, a | H_{\mathrm{eR},\sigma} | \omega_1, 0, i \rangle}{(E_1 - E^e_{ai} - i\gamma)(E_1 - \hbar\omega - E^e_{bi} - i\gamma)} , \tag{25}$$

where $|\omega_1, 0, i$ denotes the state with an incoming photon of energy $E_1 = \hbar\omega_1$, the ground state 0 of the phonon (no phonon excited), and the ground electronic state i; the other states are labeled accordingly. Initial and final electronic states are assumed to be the same, the sum is over all possible intermediate electronic states a and b. The final state is denoted by f. The E^e_{ai} are the energy differences between the electronic states a and i; $\hbar\omega$ is the phonon energy, and the lifetimes of the various excited states are taken to be the same. The three terms in the numerator are the matrix elements for the coupling between electrons (e) and radiation (R, incoming and outgoing photons with polarization σ and ρ, respectively) and the coupling between electrons and phonons (p).

Figure 34b corresponds to the incoming photon of energy E_1 matching the energy of the electronic state E^e_{ai} in (25). As mentioned, Fig. 34b describes an incoming (single) resonance. In the outgoing resonance the photon emitted in the recombination process matches an eigenenergy of the system, $E_1 - \hbar\omega = E^e_{bi}$. The Raman intensity is given by the magnitude squared of $K_{2f,10}$, leading to the strong enhancements in the spectra when the incoming (or outgoing) photon energy matches a transition of the system. By using appropriate energies of tunable lasers the real electronic states of a material

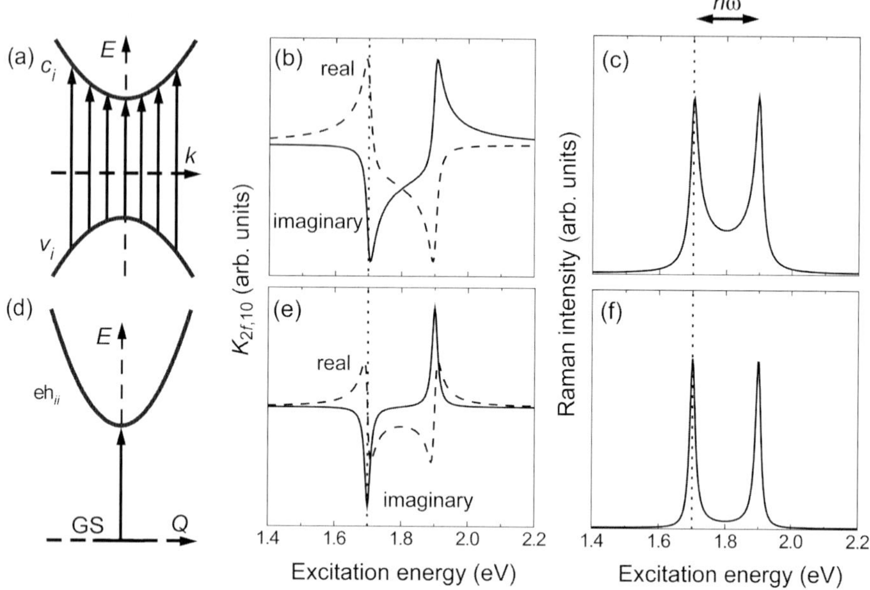

Fig. 35. Raman-scattering efficiency and resonance profiles for band-to-band transitions and excitons in one-dimensional systems. (**a**) Parabolic valence (v_i) and conduction (c_i) band with a gap E_{ii}. Optical excitations (*arrows*) are nearly vertical to conserve momentum. (**b**) Real (*dashed line*) and imaginary (*full*) part of the Raman matrix element $K_{2f,10}$ for band-to-band transitions and $E_{ii} = 1.7$ eV, $\hbar\omega = 0.2$ eV ($1600\,\text{cm}^{-1}$), and $\gamma = 0.01$ eV. (**c**) Raman efficiency $\propto |K_{2f,10}|^2$ calculated from (**b**). The profile is symmetric with an incoming and outgoing resonance. (**d**) Parabolic exciton dispersion eh_{ii} over the combined wavevector $Q = k_e + k_h$. Optical excitations (*arrow*) occur at $Q \approx 0$ only. (**e**) same as (**b**) but for excitonic intermediate states in the Raman process [the parameters are the same as in (**b**)]; (**f**) Raman profile calculated from (**e**)

can be studied systematically. The authoritative review article on solid-state systems using this technique is [184].

When calculating the matrix element, it is important to evaluate the sum in (25) *before* taking the absolute square in (24). Physically speaking, this corresponds to taking into account interferences between the different contributions to the scattering amplitudes. The simplification of replacing the sum by a density of states, i.e., ignoring interferences, is only correct if the scattering events are independent, say at very different energies of the elementary excitation, or in different tubes. In general, however, the sum must be evaluated explicitly. Although this is well known (see, e.g., *Martin* and *Falicov* [185]), it is frequently forgotten or ignored in the nanotube literature. *Bussi* and coworkers [200,201] recently calculated the sum in (25) and showed that for typical (RBM) phonon and electronic energies in metallic nanotubes such interference effects are indeed important for the identification of electronic transition energies (see Sect. 9.2). Because of its importance we show below for parabolic electronic bands and for transitions into excitonic states how the sum is evaluated, following the derivation in [185].

Abbreviating the numerator in (25) by $\mathcal{M}$ and taking the electronic bands to be described by

$$E^e_{ai}(k) = E^e_{bi}(k) = E_{ii} + \hbar^2 k^2/2\mu\,, \tag{26}$$

(see Fig. 35a) where k is the magnitude of the electron wavevector and $1/\mu = 1/m_\mathrm{e} + 1/m_\mathrm{h}$ is the reduced effective mass, we define

$$\zeta_0^2 = -\left(E_1 - E_{ii} - i\gamma\right) \quad \text{and} \quad \zeta_1^2 = -(E_1 - \hbar\omega - E_{ii} - i\gamma)\,. \tag{27}$$

For just one set of valence and conduction bands the sum in (25) can be converted into an integral over k, i.e., over all possible vertical transitions, we obtain

$$K_{2f,10} = \frac{\sqrt{2\mu}}{\hbar}\frac{L}{2\pi}\int_{-\infty}^{\infty}\frac{\mathcal{M}\mathrm{d}\zeta}{(\zeta^2+\zeta_0^2)(\zeta^2+\zeta_1^2)}\,, \tag{28}$$

where we have substituted k by $(\sqrt{2\mu}/\hbar)\zeta$, and L is the tube length (or scattering "volume").[5] We obtain for band-to-band transitions in the parabolic approximation

$$K_{2f,10} = \frac{L\sqrt{\mu/2}}{i\hbar^2\omega}\left(\frac{\mathcal{M}}{\sqrt{E_1 - E_{ii} - i\gamma}} - \frac{\mathcal{M}}{\sqrt{E_1 - \hbar\omega - E_{ii} - i\gamma}}\right)\,. \tag{29}$$

The first term in parentheses describes the incoming resonance, the second one the outgoing resonance.

When more than one electronic band is involved, the sum $\sum_{a,b}$ in (25) has to include different bands besides the different wavevectors. This leads to several similar terms in (29) with possibly different electron–phonon and electron–photon coupling strengths $\mathcal{M}'$. We will show an example where interferences between these terms lead to important differences in the determination of resonance energies in Sect. 9.2 [201].

In Fig. 35b we show the real and imaginary part of $K_{2f,10}$ in (29) as a function of the excitation energy E_1. $E_{ii} = 1.7$ eV is indicated by the dashed vertical line and $\hbar\omega = 0.2$ eV (see arrow in Fig. 35c). The imaginary part has the typical square root singularity $|E|/\sqrt{E^2 - E_{ii}^2}$ of the joint density of states in one-dimensional solids [153]. The real part, however, shows a different energy dependence: it is larger for energies below E_{ii} and $E_{ii}+\hbar\omega$. The resulting Raman efficiency, which is proportional to $|K_{2f,10}|^2$, is shown in Fig. 35c. The two maxima arise from the incoming resonance at E_{ii} and

[5] The integral in (28) can be solved by making ζ a complex variable and summing over the residues in, say, the upper half of the complex plane [202].

the outgoing resonance $E_{ii} + \hbar\omega$. The Raman profile does not resemble the joint density of electronic states, as erroneously assumed in, e.g., [94].

We now consider excitonic states as the intermediate states a, b in the Raman process. The evaluation of (25) is slightly different from the band-to-band case. For an exciton, the energies in (25) are (see Fig. 35d),

$$E^e_{ai} = E^e_{bi} = \mathrm{eh}_{ii}(Q \approx 0) , \tag{30}$$

i.e., there is only one transition, and no integration over the wavevector is needed because the exciton momentum $Q \approx 0$ for optical transitions (Q is the wavevector of the exciton $Q = k_\mathrm{e} + k_\mathrm{h}$, see Sect. 5.3). We obtain from (25) for excitonic transitions

$$K_{2f,10} = -\frac{1}{\hbar\omega}\left(\frac{\mathcal{M}}{E_1 - eh_{ii} - i\gamma} - \frac{\mathcal{M}}{E_1 - \hbar\omega - eh_{ii} - i\gamma}\right) . \tag{31}$$

Figure 35e shows the real and imaginary parts of $K_{2f,10}$ for an excitonic intermediate state. Both parts are different from the band-to-band case in Fig. 35b. The measurable quantity, however, the Raman efficiency (or intensity) in Fig. 35f is again a two-peak profile with an incoming resonance at eh_{ii} and an outgoing resonance at $eh_{ii} + \hbar\omega$ (we used $eh_{ii} = 1.7$ eV as for E_{ii} in the band-to-band transitions; in reality the two energies differ, see Sect. 5.3). Comparing Figs. 35c and f we find that the exciton resonance is somewhat sharper than a resonance involving band-to-band transitions ($\gamma = 0.01$ eV in both calculations). Using this difference to experimentally discriminate between the two cases is, however, rather difficult, because γ is not independently known.

As first pointed out by *Canonico* et al. [200] the joint density of states has been incorrectly used for modeling the Raman profile, e.g., in [26, 203–205]. This was recently corrected in a paper by the same group [75] but differences from earlier models and results were not discussed. The asymmetric experimental Raman profile in [203] that probably led to the JDOS-based theoretical model can be explained by interferences in the Raman matrix element as shown in [200, 201]. We also note that our paper [168] is somewhat misleading in deriving the Raman cross section (note that the JDOS in [168] corresponds to an ensemble of different tubes that are indeed independent scatterers). The model used for fitting the experimental data, however, is correct.

Graphite and metallic carbon nanotubes, among other materials, have strongly dispersive electronic bands near the Fermi surface; their bands cross the Fermi level with a nearly linear dispersion, see Sect. 4. This leads to real optical transitions at all energies, i.e., the Raman spectra are resonant regardless of the excitation energy, see Fig. 36.

The emitted phonons (gray dashed arrows) correspond to various possible $q \neq 0$ nonresonant carrier-scattering processes. The recombination between eigenstates occurs again such that quasimomentum is conserved, i.e.,

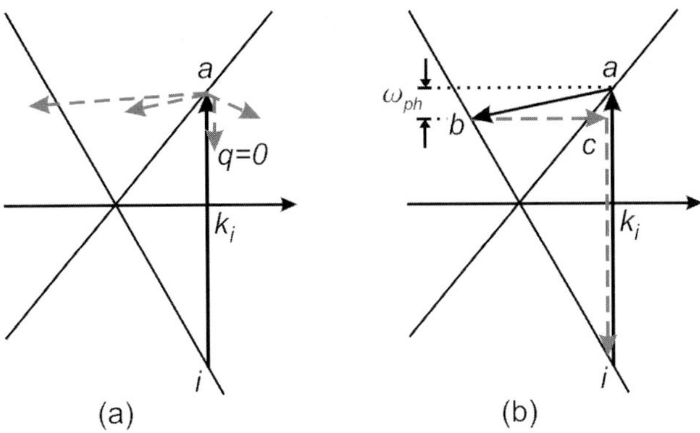

Fig. 36. In graphite the photon in the Raman process is resonant, regardless of the excitation energy. The transitions in both panels are in resonance for the incoming light ($i \rightarrow a$). (**a**) Single resonance: Of all possible phonon emissions (or absorptions) (*gray arrows*) into virtual states only those with ≈ 0 momentum contribute to the Raman-scattering process. There is no second resonant transition. (**b**) Double resonance: the emitted phonon with $q \neq 0$ makes a transition between eigenstates of the system ($a \rightarrow b$), the elastic (defect) scattering ($b \rightarrow c$) and the recombination ($c \rightarrow i$) do not. After [20]

only $q \approx 0$ phonons contribute to a single-resonant process, see Fig. 36a. In general, as long as the phonon emission is not another transition of an electron or hole to an eigenstate of the system, one speaks of a single-resonant Raman process [184]. Compared to nonresonant Raman spectra (Fig. 34a) the only essential new quality of single-resonant spectra (Fig. 34b) is the intensity enhancement due to the vanishing denominator in (25). There are qualitatively different effects in double-resonant Raman scattering that we discuss next.

6.3 Double-Resonant Raman Scattering

Double-resonant scattering processes play an important role in interpreting the Raman spectra of graphite and carbon nanotubes. Double and even triple resonances are known for semiconductors, although they are observed only under special conditions. In the carbon-based materials, though, the double resonance is the dominant process, and its discovery in graphite and carbon nanotubes has led to a big step forward in the understanding of these materials [4, 194, 206, 207].

In analogy to single-resonant scattering one speaks of a double resonance when *two* of the transitions to intermediate states of the system are real (see Fig. 36b). It is clear that both the photon and phonon involved must match quite well the energies in a system for multiple resonances to occur. In semiconductors, a double resonance was observed for a specific setting of parameters, e.g., by tuning the eigenenergies in a magnetic field or by applying pressure (see [208] or [209]).

Among the many possible nonresonantly emitted phonons in Fig. 36a there are some that scatter the electron from eigenstate a to another eigenstate b. Such a process is depicted in Fig. 36b and adds a second resonance to the Raman process. Note that this process is only allowed for a particular combination of energy and momentum of the phonon that scatters the electron. In this way, phonons with a general quasimomentum q participate in the double-resonant processes. Of course, the overall momentum is conserved in a double resonance as well; the recombination happens (for excitation with visible light) near the point in k-space where the initial absorption occurred. One way of scattering the electron back to k_i is to scatter it elastically off a defect or off a nearby surface. Another possibility is to scatter it inelastically with another phonon. The former process is indicated in Fig. 36b, (process $b \rightarrow c$ runs horizontally, i.e., elastically). The transition $b \rightarrow c$ is nonresonant, as c is not an eigenstate of the system. Recombination occurs from there, conserving quasimomentum. The second possibility, inelastic scattering with another phonon, leads to a Raman signal at twice the phonon energy and does not require a defect to conserve momentum.

The processes described lead to Raman matrix element with an additional resonant denominator as compared to (25) [4, 185, 207].

$$K_{2f,10} = \sum_{a,b,c} \frac{\mathcal{M}_{\mathrm{eR},\rho}\mathcal{M}_{\text{e-defect}}\mathcal{M}_{\mathrm{ep}}\mathcal{M}_{\mathrm{eR},\sigma}}{(E_1 - E^e_{ai} - i\gamma)(E_1 - \hbar\omega_{\mathrm{ph}} - E^e_{bi} - i\gamma)(E_1 - \hbar\omega_{\mathrm{ph}} - E^e_{ci} - i\gamma)}, \quad (32)$$

where we have abbreviated the matrix elements in the numerator by $\mathcal{M}_i$. Specifically, $\mathcal{M}_{\text{e-defect}}$ refers to the elastic interaction of the defect and the scattered electron. Not much is known about this interaction and the assumption that it is elastic and symmetry conserving for the scattered carrier is the simplest but not necessarily the only one. In analogy to the single-resonant term the Raman intensity is strongly enhanced when the denominator vanishes, i.e., when transitions between eigenstates take place in the process. As usual, the processes can occur in different time order and also for the hole instead of the electron. A full description takes all time orders and both carrier types into account.

The fascinating point about the double-resonant process is this: a different incoming photon energy (say, $i' \rightarrow a'$) leads to an excited electron with different momentum. To fulfill the second resonant transition a phonon with different momentum and energy is required for the second resonance. A larger incoming photon energy requires a larger phonon wavevector and – depending on the phonon dispersion – involves a higher or lower phonon energy. Scanning the incident photon energy thus corresponds to scanning the phonon energy in k-space. We will show this explicitly later in Fig. 53, where

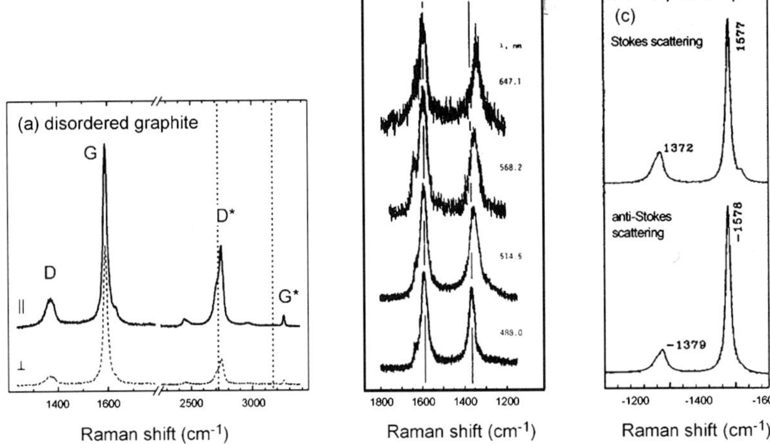

Fig. 37. (**a**) The Raman spectrum of a disordered graphite crystallite in parallel and crossed polarization of incident and scattered light. The defect-induced D mode, the Γ point or G mode, and their overtones D^* and G^*. From [125]. (**b**) Raman spectra excited at different excitation energies in the visible. Clearly seen is the curious shift of the D mode to higher energies for higher excitation energies. The G mode, in this sense, behaves like a "normal" excitation. After [210]. (**c**) Stokes and anti-Stokes spectra of the D and G mode with different frequencies for the D mode. After [211]

an experimental energy scan is compared to a calculated wavevector scan of an isolated nanotube.

The (approximate) phonon wavevector in a double-resonant process is related to the incoming photon energy. Neglecting the phonon energy compared to the much larger laser energy the double-resonant phonon wavevector may be approximated by $q = 2k_i$. This relationship is often used for a quick evaluation of the double-resonance conditions at different laser energies. For linear electronic bands, as is the case for graphite not too far from the Fermi level, (32) can be evaluated analytically, resulting in a linear relation between the incoming photon energy E_1 and phonon wavevector q, see [4]

$$q = \frac{E_1 - \hbar\omega_{\text{ph}}(q)}{v_2} \quad \text{or} \quad \frac{E_1 - \hbar\omega_{\text{ph}}(q)}{-v_1}, \tag{33}$$

where v_i are the Fermi velocities of the linear electronic bands. If the phonon energy is taken to depend on wavevector, (33) may be solved iteratively.

By scanning the incident laser energy it is thus possible to map the phonon frequencies in large parts of the Brillouin zone. Based on the double-resonant Raman interpretation *Reich* and *Thomsen* re-evaluated literature data on graphite and plotted them onto the phonon-dispersion curves [125] (see also Sect. 10.4 and [207]). The phonon dispersion of an isolated nanotube was

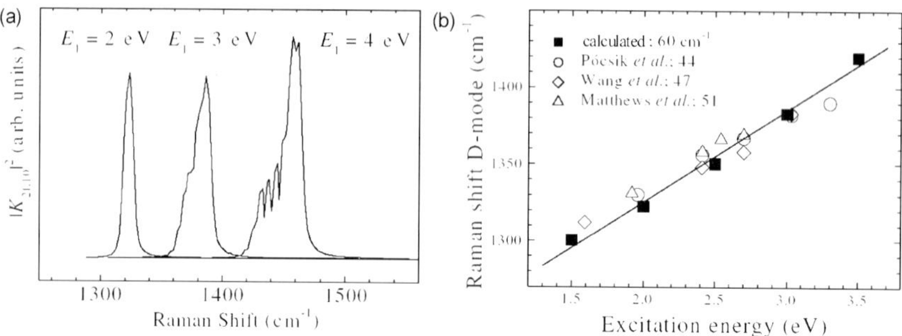

Fig. 38. (**a**) Calculated Raman spectra for the D mode in graphite for three different laser energies. (**b**) Calculated (*full squares*) and measured (*open symbols*) frequencies of the D mode as a function of excitation energy. From [4]; the measurements were taken from [215, 216] and [217]

measured and evaluated using this technique by *Maultzsch* et al. [129] (see Sect. 8.2).

Ge is another material where double resonance could explain an additional, excitation-energy dependent vibrational mode [212]. *Mowbray* et al. [213] invoked scattering by the surface as the process that conserved momentum in the double resonance. The essential evidence for this interpretation was the dependence of the observed Raman-frequency shifts on surface orientation. A detailed investigation of Raman double resonance in Ge was performed more recently by *Mohr* et al. [214].

We turn now to the double-resonant scattering process in graphite, which is closely related to that in carbon nanotubes. *Thomsen* and *Reich* first calculated the Raman spectra of the D mode using the concept of double resonances [4]. Their reason for re-examining the D mode was its curious and unresolved dependence on excitation energy. We show the characteristic features of the Raman spectrum of graphite in Fig. 37. The high-energy part of the spectrum is shown in Fig. 37a. The peak labeled D is a defect-induced mode – as *Tuinstra* and *Koenig* showed in a systematic study of graphite – and D^* its overtone. The shift with excitation energy reported by *Vidano* et al. is shown in Fig. 37b; it amounts to about $50\,\mathrm{cm}^{-1}/\,\mathrm{eV}$ [210, 215–217]. D^* shifts at twice the rate of D [218]. Finally, the D mode has different Stokes and anti-Stokes frequencies, see Fig. 37c, which is very unusual for Raman spectroscopy. In fact, it is impossible in first-order scattering.

The results of Thomsen and Reich, who based their interpretation on a double-resonant process, are reproduced in Fig. 38a and are the numerical evaluation of the Raman matrix element of (32). It is nicely seen that a Raman line appears by defect-induced resonances and shifts to higher frequencies when the excitation energy is increased. Figure 38b compares the experimental and theoretical frequency slopes showing an excellent agreement. The D^* mode is an overtone of the D peak where the electron is backscattered by a second phonon instead of a defect. The differences between Stokes and anti-Stokes scattering arise because the double-resonant condition is slightly different for the creation and destruction of a phonon, see also the discussion in Sect. 10 and [125, 219].

There is an alternative approach to understand the appearance of the D mode and its excitation-energy dependence; it is based on the lattice dynamics of small aromatic molecules [220, 221]. In these molecules the D mode has a Raman-active eigenvector; its frequency depends on the actual size and shape of the molecule. The shift with excitation energy results from a resonant selection of a particular molecule by the incoming laser. For defect-induced scattering the solid-state approach presented above and the molecular approach can be shown to be the same. The latter fails, however, in explaining the D^* mode properties in perfect graphite. On the other hand, it has been argued [222] that double-resonant Raman scattering cannot explain why only one phonon branch shows strong double resonances. The selectivity of the process can be understood when considering the matrix elements in the Raman cross section (see [125, 207]).

The frequency difference between Stokes and anti-Stokes scattering was shown in Fig. 37c: the D mode in graphite differs by $\Delta\hbar\omega = 7\,\mathrm{cm}^{-1}$ [211]. In double resonance the resonance conditions differ for the creation and destruction of a phonon [4]. The double-resonance process shown in Fig. 36b is for Stokes scattering. We obtain an anti-Stokes process by inverting all arrows in the picture. This process is also double resonant, but at a different excitation energy ($i \to a$) [219, 223]. In general, the Raman cross section for Stokes scattering K_{S} is the same as the cross section for anti-Stokes scattering K_{aS} if we interchange the incoming and outgoing photon (time inversion), i.e.,

$$\begin{aligned} K_{\mathrm{aS}}[E_2, \hbar\omega_{\mathrm{aS}}(E_1), E_1] &= K_{\mathrm{S}}[E_1, \hbar\omega_{\mathrm{S}}(E_2), E_2] \\ &= K_{\mathrm{S}}[E_1, \hbar\omega_{\mathrm{S}}(E_2), E_1 + \hbar\omega_{\mathrm{aS}}(E_1)] \,. \end{aligned} \tag{34}$$

In other words $\hbar\omega_{\mathrm{aS}}(E_1) = \hbar\omega_{\mathrm{S}}(E_2)$. Since we know the dependence of the Stokes frequency on excitation energy we can calculate the frequency difference

$$\Delta\hbar\omega = \hbar\omega_{\mathrm{aS}}(E_1) - \hbar\omega_{\mathrm{S}}(E_1) = \frac{\partial\hbar\omega_{\mathrm{S}}}{\partial E}\hbar\omega_{\mathrm{aS}}(E_1) \,. \tag{35}$$

Tan et al. [211] measured an anti-Stokes frequency of $1379\,\mathrm{cm}^{-1}$ and a D-mode slope of $43\,\mathrm{cm}^{-1}/\mathrm{eV}$. We predict $\Delta\hbar\omega = 7.4\,\mathrm{cm}^{-1}$ from (35), which is in excellent agreement with the reported value ($7\,\mathrm{cm}^{-1}$), see Fig. 37c. As discussed in [219] and [223] the frequency difference of Stokes and anti-Stokes scattering for the D^* overtone is predicted to be four times larger than for the D mode, because both the energy difference between E_1 and E_2 and the slope of the Raman peak with laser energy are doubled. This also agrees very well with the experimentally reported differences. In turn, (35) can be used to determine – at least approximately – the slope of a disorder band.

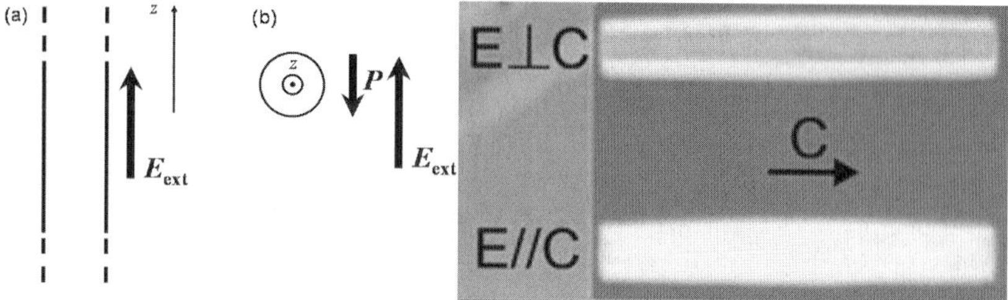

Fig. 39. Schematic of the shape depolarization in a nanotube. (**a**) An external field parallel to the nanotube axis will not induce any charge in a nanotube (except at ∞). (**b**) Fields perpendicular to the axis will induce charges in the walls and hence a perpendicular polarization that will tend to cancel the external field. Absorption is suppressed for $E \perp z$. From [20]. (**c**) Absortion of light polarized perpendicular (*upper*) and parallel (*lower*) to the nanotube axis, here for small tubes grown in the channels of a zeolite crystal. From [46]

6.4 Anisotropic Polarizability

The one-dimensional nature of carbon nanotubes affects their light-absorption properties.[6] Anisotropic polarizabilities in spheres and ellipsoids are treated in advanced textbooks, e.g., in [224], and nanotubes constitute a limiting case of these treatments. The effect of the shape anisotropy has to be taken into account when discussing the optical properties of nanotubes. In the direction parallel to the tube axis there is charge build up only at the ends of a nanotube (at $\pm\infty$). In the static limit this has no effect on the absorption. Perpendicular to the axis, as *Ajiki* and *Ando* pointed out [166], the induced polarization will reduce the external field (see Fig. 39).

The effect of the anisotropic polarizability on light absorption was studied in detail by several authors [166, 225, 226]. The essence is that in one-dimensional systems – like the single-walled carbon nanotubes – there is practically no absorption for polarizations perpendicular to the nanotubes axis. For multiwalled tubes the effect of anisotropic polarizability becomes smaller as the shape of the nanotubes becomes more isotropic. *Benedict* et al. [225] derived the following expression for the diameter dependence of the depolarization effect, i.e., the screened polarizability per unit length of a cylinder is

$$\alpha_{\perp}(\omega) = \frac{\alpha_{0,\perp}(\omega)}{1 + 8\alpha_{0,\perp}(\omega)/d^2}, \tag{36}$$

where d is the diameter of the cylinder, and $\alpha_{0,\perp}$ is the unscreened polarizability. See also *Tasaki* et al. [226] for a description at optical frequencies, or [20] for more details. Figure 39c shows an experimental verification of the depolarization effect in aligned nanotubes (within a zeolite crystal). For light

[6] In the carbon nanotube literature this is sometimes called the *antenna effect*.

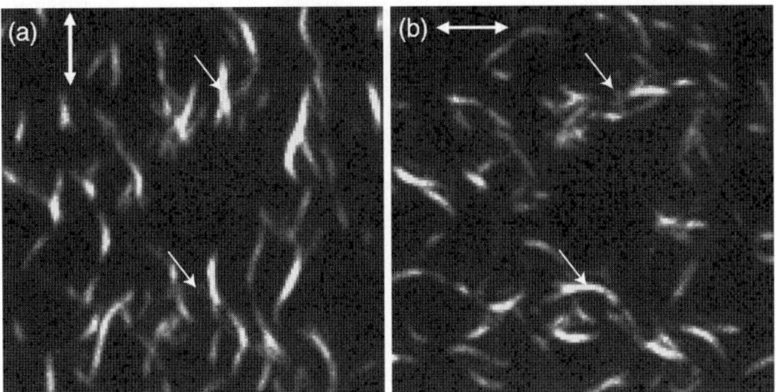

Fig. 40. The depolarization effect in bundles of single-walled nanotubes recorded on an area of $15 \times 15\,\mu m$. The polarizations of incident and scattered light in (**a**) and (**b**) are along the direction of the *double arrow*. The intensity shown is the integrated signal of the D^* mode at 2615 cm, the overtone of a defect-induced mode (Sect. 10.3). The *single arrows* highlight nanotube bundles parallel and perpendicular to the light polarizations. Raman scattering is completely suppressed for light perpendicular to the nanotube axis. From *Hartschuh* et al. [229]

polarized perpendicular to the nanotube's axes the sample is transparent (upper picture); for parallel polarization it is completely black.

The influence of the anisotropic polarizability is seen also in the Raman spectra. Since the absorption and emission of light is nearly zero for perpendicular polarization, only the zz component of the Raman tensor gives rise to scattering. This was observed in many experimental studies [99, 101, 147, 227, 228]. We show a nice confirmation in Fig. 40, where the Raman spectra of several bundles of carbon nanotubes were spatially resolved by a confocal Raman arrangement [229]. Nanotubes aligned with the light polarization are seen as bright short lines in the figure. Part (b) is taken on the same small area of the sample as part (a), but the polarization of the incoming light is rotated by 90°. Now other tubes light up. Tubes at a finite angle to the polarization show up in both panels.

An important consequence of the depolarization effect is that only $A_{1(g)}$ phonons have a reasonably strong Raman signal, because they are the only modes with nonzero zz component in the Raman tensor in the nanotube point groups (see Table 3.1). When modeling the Raman spectrum of single-walled carbon nanotubes, one of the challenges is to explain the complex spectrum on the basis of totally symmetric phonons alone [194, 230].

A nice example of where the Raman depolarization ratio of a spherical molecule is affected by the anisotropic polarizability of the nanotube was shown by *Pfeiffer* et al. [232] in peapods. Peapods are C_{60} molecules embedded in carbon nanotubes, see Sect. 2 for a TEM image and Fig. 41 for a Raman spectrum. The depolarization ratios, the ratio of the Raman intensity of a peak in perpendicular and parallel polarization in ran-

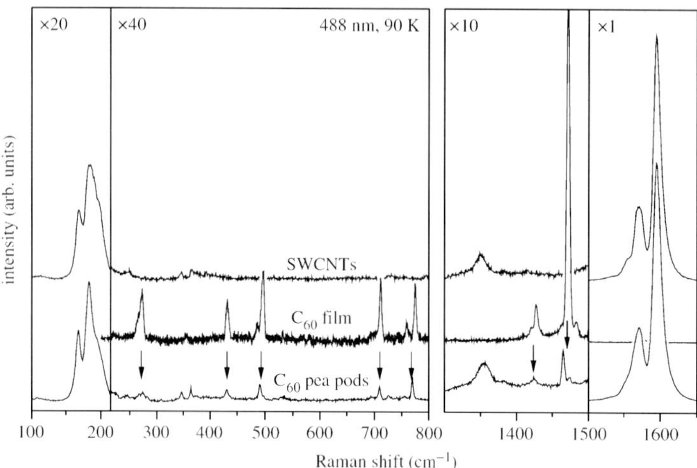

Fig. 41. (*upper*) Raman spectra of single-walled carbon nanotubes, (*middle*) of a C_{60} film, and (*lower*) of C_{60} peapods. The peapods, which are nanotubes containing C_{60} molecules, show a superposition of the spectra of the upper and middle traces. The *arrows* highlight the C_{60}-peak positions in the peapod. After [231]

domly oriented scatterers, is characteristic of the symmetry of a molecule or solid. For example, the two fully symmetric A_{1g} modes of polycrystalline C_{60} at 494 cm^{-1} and 1469 cm^{-1} have depolarization ratios of ≈ 0.1 at room temperature [232, 233]. For $\alpha_{xx} = \alpha_{yy} = \alpha_{zz} \neq 0$ and all other $\alpha_{ij} = 0$, the depolarization ratio should be zero. In a peapod, however, only the z component of the incident light "sees" the C_{60} molecule (z is along the nanotube axis). The other components are completely screened. For the Raman-scattering tensors of C_{60} this implies that only the zz components are effectively nonzero. The depolarization ratio of an A_{1g} tensor with only $\alpha_{zz} \neq 0$ is equal to 1/3. A depolarization ratio of 1/3 was indeed observed experimentally for the peapods [232], confirming the importance of the anisotropy of the polarizability. See, e.g., the books by *Hayes* and *Loudon* [234] or *Reich* et al. [20] how to obtain depolarization ratios for a given Raman tensor.

The depolarization effect was recently found in nanotubes of WS_2 (see Fig. 42 [235]). The intensity of the observed Raman signal of WS_2 is plotted as a function of angle in the polar plot Fig. 42b. The observed Raman frequencies in the nanotube were the same as in bulk WS_2 to within experimental error because the nanotubes were rather thick (compared to the usual carbon nanotubes), $d(WS_2) \approx 20$ nm. However, the Raman intensities of both the mode at 351 cm^{-1} and 417 cm^{-1} followed the expected angular dependence for the anisotropic polarizability in nanotubes. For light polarized along the tube axis, the modes were strong, perpendicular to it they nearly disappeared, see Fig. 42b, where we plot the intensity as a function of angle between the polarization and the WS_2 nanotube axis.

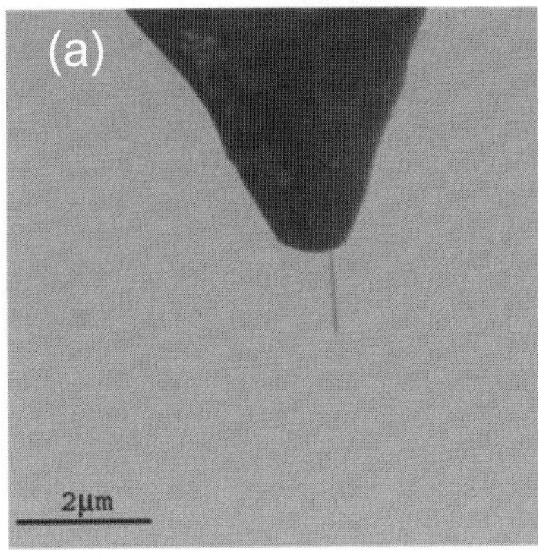

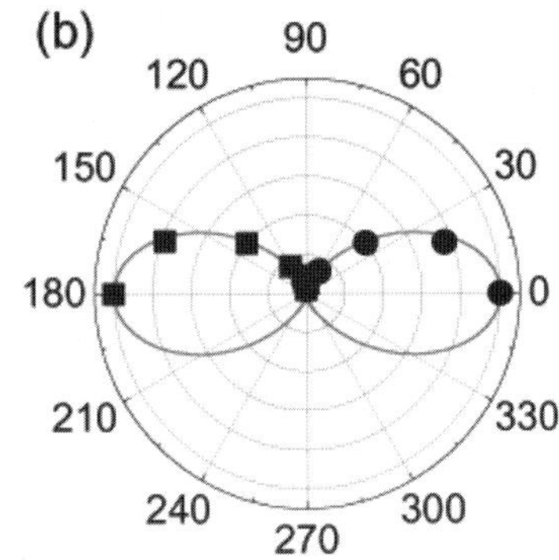

Fig. 42. (**a**) A free-standing WS_2 nanotube mounted on an AFM tip. (**b**) The anisotropic polarizability of the nanotube is observed in the intensity of the two Raman modes at $351\,\mathrm{cm}^{-1}$ (*circles*) and $417\,\mathrm{cm}^{-1}$ (*squares*) of the nanotube in (**a**). For clarity, the two modes are plotted in different quadrants; 0° and 180° correspond to light polarized parallel to the axis of the nanotube. After [235]

7 Assignment

Identifying the chirality of a particular nanotube or set of nanotubes has been an issue in the literature ever since the discovery of nanotubes. While it is possible to do so with electron microscopy and with scanning tunneling spectroscopy [164, 165] both techniques are rather time consuming and difficult to do without major sample manipulation. Using Raman spectroscopy, and the radial-breathing mode in particular, there have been many attempts to use the relationship between inverse diameter and the breathing-mode frequency (5) to determine the chiral index of a tube [47, 115, 236]. First, the ω_{RBM} is measured with high accuracy and then, taking your favourite constants c_1 and c_2, the diameter of a nanotube is calculated. The difficulties with this procedure lie in considerable uncertainties in c_1 and c_2 and the comparatively many different chiralties in even a small diameter range of a typical sample. Attempts to correlate the radial-breathing mode frequencies with the Raman excitation energy, in principle, improved the quality of such an assignment [236]. However, using the simple tight-binding expression for the electronic energies, as was done, e.g., in [115, 236], is not sufficient to identify a particular nanotube, in spite of persistent attempts to do so in the literature. We consider such assignments unreliable because they imply an unrealistic absolute accuracy of the calculated electronic energies [115, 116, 205, 236, 237].

Substantial progress in a Raman-based chirality assignment was only made after *Bachilo* et al. published their two-dimensional plot of excitation vs. luminescence energy of nanotubes. The tubes were wrapped by sodium dodecyl sulfate (SDS) molecules and thus unbundled in the solution [59]. *Bachilo* et al. were the first to observe luminescence from many nanotubes,

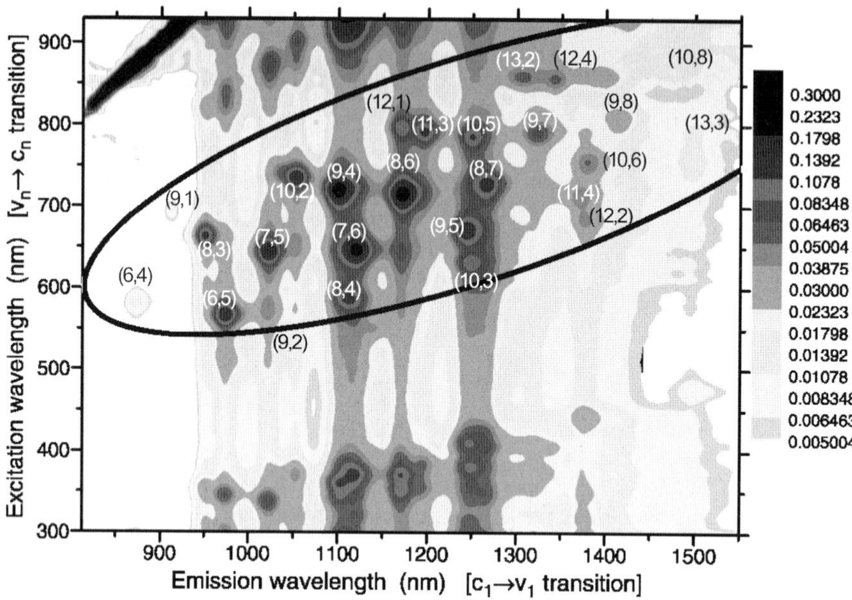

Fig. 43. Two-dimensional plot of the luminescence intensity (x-axis) of individual carbon nanotubes in solution as a function of excitation energy (y-axis). A cut parallel to x is a single luminescence spectrum, one parallel to y is a photoluminescence excitation or absorption spectrum. Every chirality corresponds to a single point in this plot. The assignment is indicated by (n_1, n_2). After [60]

and they showed that from the luminescence maxima observed in such a plot, a pattern related to the Kataura plot could be identified [60]. They still needed a least-square fit to "anchor" their experiment to theory and could not investigate metallic nanotubes, but the essential path to identifying individual chiral indices was found.

In the two-dimensional plot of *Bachilo* et al. (Fig. 43) it became possible to investigate simultaneously the second (or higher) optical transition of carbon nanotubes and their lowest excitation above the ground state.[7] The excitation energy is plotted on the y-axis and a horizontal line corresponds to a luminescence spectrum excited at one particular excitation energy. Maxima along the y-direction of the plot occur at energies where the absorption – as given by the intensity of the photoluminescence – is maximal. In addition to seeing the luminescence from individual nanotubes the result prompted a discussion about the ratio of E_{22} to E_{11} (or eh_{22} to eh_{11}), which turned out to be $\neq 2$ in the large-diameter limit as one expects for uncorrelated

[7] These transitions were generally identified as energies between the van-Hove singularities E_{ii}. In view of the experimental and theoretical progress concerning the importance of excitonic energies these should be rather referred to eh_{ii}, see Sect. 5.3.

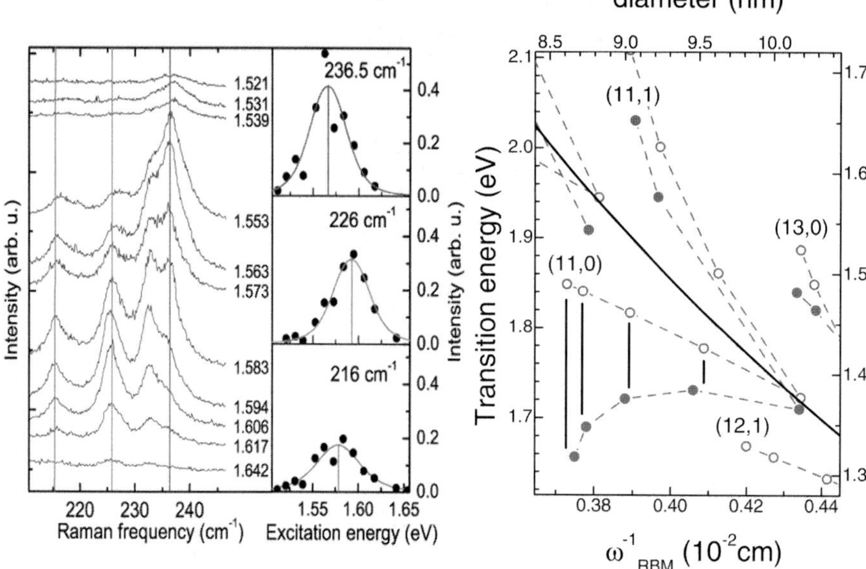

Fig. 44. (*left*) Resonant Raman spectra of a solution of sodium dodecyl sulfate wrapped around carbon nanotubes. Shown is a narrow interval of excitation energies with several radial-breathing mode peaks going in and out of resonance. Resonance profiles extracted along the *vertical lines* on the *left*; the maxima occur at different energies. (*right*) Resonance energies of the (11,0) branch, experimental (*closed symbols*) and theory (*open*). The vertical alignment, i.e., the diameter to ω_{RBM} relationship is nearly perfectly matched. For a larger range of diameters, see Fig. 46. After [242]

electrons and holes (band-to-band transitions). The so-called *ratio problem* was discussed by a number of authors, see, e.g., [20, 60, 74, 161, 174, 238, 239], and we refer the reader to the literature about this topic.

A step beyond the two-dimensional luminescence graph was made by plotting the Raman spectra as a function of excitation energy [63–65, 111, 240–242]. The y-axis corresponds again to the transition energies. However, instead of plotting the Raman spectra straight along the x-axis, ω_{RBM} is converted to nanotube diameter using (5). Now the data are in the form of the well-known Kataura plot (Fig. 26). The essence of the two-dimensional Raman plot lies in the fact that both electronic transitions (through Raman resonances E_{ii}) and vibrational properties (through the radial-breathing mode frequency ω_{RBM}) depend on chirality and are measured simultaneously. An assignment of chiralities may be obtained from such a plot without any prior assumptions about c_1 and c_2, as we will show in the following. Such a plot is obtained by using a variety of discrete and tunable laser sources with energies separated by less than the width of a typical excitation, say about 10 meV.

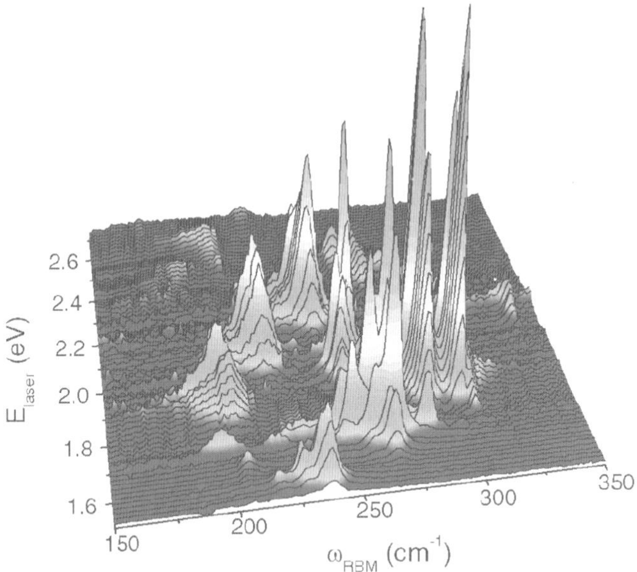

Fig. 45. Two-dimensional plot of the radial-breathing-mode range vs. laser excitation energy. Note the various la-ola-like resonance enhancements, from which we can determine both the optical transition energies and the approximate diameter of the nanotubes. The spectra were each calibrated against the Raman spectrum of CCl_4. From [64]

In Fig. 44 (left) we show the resonance profiles of several nanotubes with nearby resonance maxima. The largest intensity is seen to move through the peaks forming what looks like a *la-ola* wave [243]. There are several such la-ola-type resonances when exciting the nanotubes with different laser lines, each occurring over a fairly narrow energy and ω_{RBM} range. As we will see, each la-ola series of resonances corresponds to a particular branch in the Kataura plot of transition energies versus diameter. Within each branch the fixed relationship between the indices – n_1 decreases by one and n_2 increases by 2 as long as $n_1 > n_2$ – gives a fixed assignment once a particular branch is identified (see Table 2 for the definition of a branch). The tube with a smallest diameter in a branch is always a zigzag or near-zigzag tube, the one with the largest diameter an armchair or near-armchair one. (See Sects. 2 and 5.2 for more details.)

Figure 45 shows the two-dimensional resonance Raman plot where many of the branches can be seen, their intensity is plotted in the third direction. (Here, ω_{RBM} and not $1/d$ is plotted.) Each small group of peaks corresponds to one branch of the Kataura plot (Figs. 26 and 44).

We are now able to perform an (n_1, n_2) assignment. This is done by stretching and shifting the theoretical Kataura plot with respect to the experimental one [63, 242].

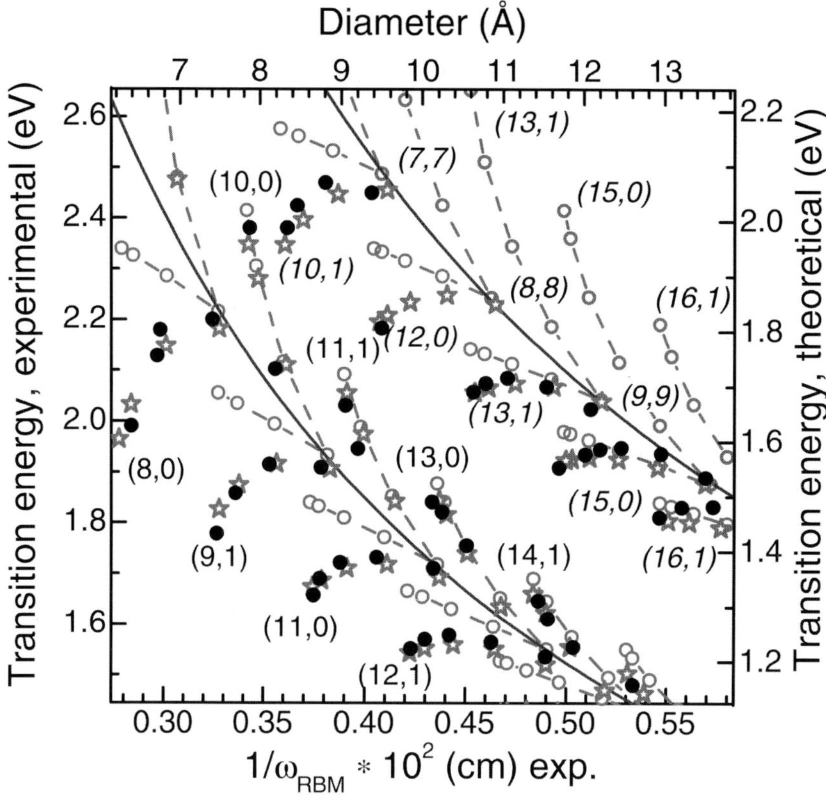

Fig. 46. Experimental Kataura plot. Shown is the part of the full plot (Fig. 26) where experimental data are available from a sample with isolated, SDS-wrapped nanotubes in solution. The two *solid* $1/d$ *lines* are the E_{22}^{S} and E_{11}^{M} average diameter dependence (isotropic approximation of the graphite band structure). The *open symbols* are calculated transition energies from a tight-binding approximation [6] (*circles*) and the empirical expressions (38) and (39) (*stars*) [242], which describe the transition energies rather well. *Closed circles* refer to experimental data points. The chiral index labels refer to the tube with largest chiral angle within a branch (zigzag or near-zigzag tubes). The index of neighboring tubes may be obtained from the $(n_1', n_2') = (n_1 - 1, n_2 + 2)$ rule (22), the branch index is obtained from $\beta = 2n_1 + n_2$ (21); armchair tubes lie on the $1/d$ line and are labeled as well. See also Table 2. Note that the branch index is the same in the upper and lower branches of the metallic tubes, whereas it differs in semiconducting tubes. Metallic nanotubes thus have two close-by optical transitions [150]. The region of the (11,0) branch is enlarged in Fig. 44 (*right*). After *Telg* et al. [63]

What is the physical meaning of such stretching and shifting? Along the x-axis it means changing the constants c_1 and c_2, respectively, as can be seen in (5). Along the y-axis it means adjusting the electronic energies between experiment and theory, a simple shift corresponds to a scissors operator. It is important to note that bringing theory and experiment into coincidence is not a fit of experiment to theory in the usual sense. Instead the best match relies on recognizing the patterns of the branches and families in such a plot. The stringent criterion is that in the vertical direction there is a full correspondence between theory and experiment as in Fig. 44 (right). We mean by this, e.g., that the x-axis cannot be shifted by just a little, say one diameter difference of neighboring tubes. For example, were we to shift the assignment such that the experimental (11,0) branch in Fig. 44 (right) were to lie a little more to the left, the outermost tube [the (11,0)] would no longer have a theoretical counterpart. Such a small shift corresponds to a change of $\Delta c_1 \approx 2.5\,\mathrm{cm}^{-1} \cdot \mathrm{nm}$ in (5). As the theoretical Kataura plot, however, is exhaustive (all possible combinations of n_1 and n_2 are contained in the plot) such a shift invalidates the overall pattern and is impossible [63, 242].

What about shifting the experiment by one entire branch? Take, e.g., all experimental values to smaller diameters so that the (11,0) ends up at a diameter of 7.5 Å, the value of the (9,1) in Fig. 46. Then one would find for the (9,1) branch more observed chiral indices, namely five, than exist for this branch, namely four [242] (see also Fig. 3). Again this is impossible, and so is this assignment, which corresponds to $\Delta c_1 \approx 30\,\mathrm{cm}^{-1} \cdot \mathrm{nm}$, i.e., a 15 % decrease in c_1.

Note also that the diameter difference between neighboring tubes is not constant. For instance, $d[(11,0)] = 8.6$ Å and $d[(10,2)] = 8.7$ Å, whereas the next tube in the same branch is at a larger separation, $d[(9,4)] = 9.0$ Å. These subtle differences are observed in the experiment as well. In fact, for the four branches with the smallest tubes (10,1), (11,0), (12,1) and (13,1) we observed all possible nanotubes and diameter–distance irregularities putting our assignment on firm ground [63,242]. In Table 10 we list, for about 50 nanotubes, the RBM frequencies and optical transition energies together with d, Θ, the family ν, the branch index β, and the magnitude squared of the experimental Raman susceptibility of a set of HiPCo-grown nanotubes. The susceptibility values were obtained by correcting the Raman intensity for the spectrometer response and ω^4 but not for the distribution of nanotube chiralities in the sample.

Table 10: RBM frequency ω_{RBM} and optical transition energies E_{ii} for 47 single-walled carbon nanotubes. The table lists, additionally, the tube diameter d, the chiral angle Θ, the nanotube family $\nu = (n_1 - n_2) \bmod 3$, and the Kataura branch $\beta = 2n_1 + n_2$ of each nanotube. The last column gives the square of the Raman susceptibility of a HiPCo-grown sample in arbitrary units (RS^2). The transition energies E_{ii} are the second transition for semiconducting tubes S–E_{22} ($\nu = \pm 1$) and the first transition for metallic ($\nu = 0$) tubes M–E_{11} (see Fig. 26). Data from [63, 242]

Tube	ω_{RBM} (cm^{-1})	E_{ii} (eV)	d (Å)	Θ (°)	ν	β	RS^2
(14, 5)	174.5	1.83	13.36	14.7	0	33	1.9
(10, 10)	175.7	1.889	13.57	30.0	0	30	0.9
(15, 3)	179.0	1.83	13.08	8.9	0	33	1.8
(16, 1)	182.0	1.81	12.94	3.0	0	33	0.5
(11, 8)	182.7	1.936	12.94	24.8	0	30	1.9
(12, 6)	189.6	1.948	12.44	19.1	0	30	1.9
(13, 4)	195.3	1.944	12.06	13.0	0	30	2.5
(9, 9)	195.3	2.02	12.21	30.0	0	27	0.5
(14, 2)	196.3	1.934	11.83	6.6	−1	29	3.6
(12, 5)	198.5	1.554	11.85	16.6	−1	29	0.1
(15, 0)	201.5	1.908	11.75	0.0	0	30	0.7
(13, 3)	203.3	1.610	11.54	10.2	−1	29	0.6
(10, 7)	204.0	2.067	11.59	24.2	0	27	0.9
(9, 8)	204.0	1.535	11.54	28.1	+1	26	0.5
(14, 1)	205.8	1.646	11.38	3.4	−1	29	0.3
(11, 5)	212.4	2.084	11.11	17.8	0	27	1.7
(9, 7)	216.0	1.564	10.88	25.9	−1	25	1.3
(12, 3)	217.4	2.075	10.77	10.9	0	27	2.6
(13, 1)	220.3	2.057	10.60	3.7	0	27	1.5
(11, 4)	221.8	1.76	10.54	14.9	+1	26	
(10, 5)	226.1	1.578	10.36	19.1	−1	25	2.3
(12, 2)	228.1	1.82	10.27	7.6	+1	26	12.2
(8, 7)	230.4	1.710	10.18	27.8	+1	23	0.4
(13, 0)	230.8	1.84	10.18	0.0	+1	26	0.2
(11, 3)	232.6	1.570	10.00	11.7	−1	25	2.0
(12, 1)	236.4	1.551	9.82	4.0	−1	25	4.1
(12, 0)	244.9	2.18	9.40	0.0	0	24	

Table 10: continued

Tube	ω_{RBM} (cm^{-1})	E_{ii} (eV)	d (Å)	Θ (°)	ν	β	RS[2]
(8,6)	246.4	1.73	9.53	25.3	−1	22	1.4
(7,7)	247.8	2.45	9.50	30	0	21	
(10,3)	252.1	1.945	9.24	12.7	+1	23	3.6
(11,1)	256.0	2.031	9.03	4.3	+1	23	9.8
(9,4)	257.5	1.72	9.03	17.5	−1	22	2.5
(8,5)	262.7	2.47	8.90	22.4	0	21	
(7,6)	264.2	1.909	8.83	27.5	+1	20	2.6
(10,2)	264.6	1.690	8.72	8.9	−1	22	2.3
(11,0)	266.7	1.657	8.62	0.0	−1	22	1.7
(9,3)	272.7	2.43	8.47	13.9	0	21	
(10,1)	276.3	2.38	8.25	4.7	0	21	
(8,4)	280.9	2.10	8.29	19.1	+1	20	0.3
(7,5)	283.3	1.915	8.18	24.5	−1	19	18.3
(10,0)	291.4	2.38	7.83	0.0	+1	20	
(8,3)	297.5	1.857	7.72	15.3	−1	19	35.8
(7,4)	305.4	2.63	7.55	21.1	0	18	
(9,1)	306.2	1.78	7.47	5.2	−1	19	9.1
(6,5)	308.6	2.20	7.47	27.0	+1	17	
(8,2)	315.5	2.52	7.18	10.9	0	18	
(8,0)	352.2	1.99	6.27	0.0	−1	16	0.1

From a least-square fit of our so-assigned nanotubes it is possible to find an experimental value for the constants c_1 and c_2 in (5)

$$\omega_{\mathrm{RBM}} = \frac{(215 \pm 2)\,\mathrm{cm}^{-1} \cdot \mathrm{nm}}{d} + 18\,\mathrm{cm}^{-1} . \tag{37}$$

We stress, however, that the precise values of these constants are not critically important in view of the concept of families and, in particular, branches introduced. Nanotubes within the same branch have similar physical properties. Small variations, e.g., in c_2 due to environmental effects on the nanotubes, may be sample specific. *Fantini* et al. [64], who performed a related study fitted separately metallic ($c_1 = 218\,\mathrm{cm}^{-1} \cdot \mathrm{nm}$, $c_2 = 10\,\mathrm{cm}^{-1}$) and semiconducting ($223\,\mathrm{cm}^{-1} \cdot \mathrm{nm}$, $17\,\mathrm{cm}^{-1}$) tubes to (5), which is slightly different from the values in (37). Nevertheless, their most recent assignment of RBM peaks to nanotube chiralities agrees with the one presented here. Note also that we

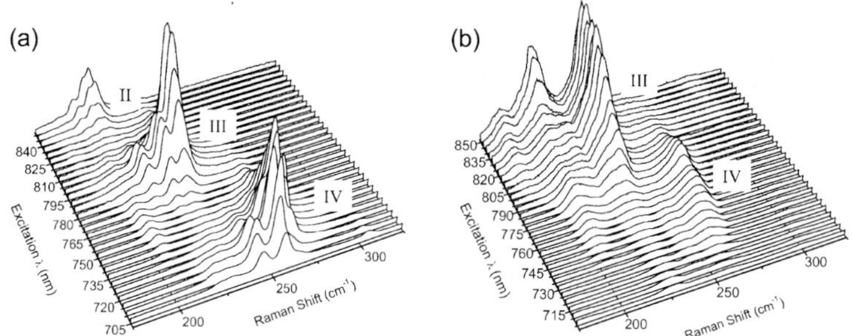

Fig. 47. Raman spectra of (**a**) individual nanotubes in solution and (**b**) ropes of nanotubes. From *O'Connell* et al. [245]

directly assigned a Raman peak to a nanotube chirality (n_1, n_2). We did not make use of (37) or similar relations to find the assignment in Table 10.

The typical bundling of nanotubes presents a form of environmental influence on the RBM spectra. Both transition energies and radial-breathing mode frequencies are shifted by the van-der-Waals interaction [244]. Predictions by *Reich* et al. based on ab-initio calculations of bundles showed that we should expect a shift of the electronic energies and observe a dispersion perpendicular to the nanotube axis [9]. With data based on the idea of an experimental Kataura plot, *O'Connell* et al. [245] recently showed that this is indeed the case. In Fig. 47 we reproduce part of their data. Figure 47a shows a sample of carbon nanotubes in solution, and three sets of la-ola (labeled II, III, and IV) are easily identified. For the same sample in bundled form, Fig. 47b shows the corresponding spectra. All three resonances have shifted down in transition energy upon bundling, their ω_{RBM} stay approximately the same. This is consistent with the prediction of *Reich* et al. [9] and confirms the concept and importance of the experimental Kataura plot.

While the experimental radial-breathing frequencies in Fig. 46 match theory nearly perfectly, there are systematic deviations of the transition energies from the tight-binding transition energies, see also right panel in Fig. 44. Within each branch – in particular for the -1 branch of E_{22} – the energies fall below the theoretical ones, the more so the smaller the chiral angle. In other words, the more "zigzag"-type the nanotube is, the less its electronic transition energies are described by zone folding the graphite ab-initio band structure, which is the basis for the theoretical points in Fig. 46. Furthermore, the more the (near-) zigzag nanotubes drop from the theoretical values, the smaller their diameter. Apparently, this is an effect of the increasing curvature on the band structure, which also affects the excitonic energies increasingly. Several authors have attempted to describe the curvature effect by empirical formulas [7, 60, 65, 111, 242, 246, 247]. They essentially agree in that they include a chirality-dependent correction, which increases with smaller diameter.

We show here an empirical formula parameterized by a fit to the data in Fig. 46 by *Maultzsch* et al. [242]. These are for the E_{22}^{S} transition of semiconducting tubes

$$E_{22}^{\mathrm{S}} = 4\gamma_0 \frac{a_{\mathrm{C-C}}}{d}\left(1+\gamma_1 \frac{a_{\mathrm{C-C}}}{4d}\right) + \nu\gamma_2 \frac{a_{\mathrm{C-C}}^2}{d^2}\cos 3\Theta\,, \tag{38}$$

with the parameters $\gamma_0 = 3.53$ eV, $\gamma_1 = -4.32$, $\gamma_2 = 8.81$ eV, the family index ν, the diameter d of the nanotube, and the carbon–carbon distance of graphene $a_{\mathrm{C-C}} = \sqrt{3}a_0 = 1.421$ Å. For the first set of metallic E_{11}^{M} transitions

$$E_{11}^{\mathrm{M}} = 6\gamma_0 \frac{a_{\mathrm{C-C}}}{d}\left(1+\gamma_1 \frac{a_{\mathrm{C-C}}}{6d}\right) - \gamma_2 \frac{a_{\mathrm{C-C}}^2}{d^2}\cos 3\Theta\,, \tag{39}$$

with the parameters $\gamma_0 = 3.60$ eV, $\gamma_1 = -9.65$, and $\gamma_2 = 11.7$ eV. When applying the empirical expressions to find the transition energy of a particular nanotube, d, ν, and Θ are computed from the chiral index of the nanotube (see Table 2).

Expressions (38) and (39) are quite useful; together with (37) they describe the position of a large number of nanotubes in the Kataura plot. Conversely, if ω_{RBM} and the approximate transition energy are known from a Raman measurement, the branch and the chiral index of a nanotube can be easily identified.

We describe now how one can, without an indepth understanding of the physics behind the idea, identify the chiral index of a nanotube or a set of nanotubes by going through the following steps.

- Identify the average diameter range of nanotubes you have; often this will depend on the method used to grow your tubes.
- Draw a vertical line in the experimental Kataura plot at your average diameter. You may take an existing plot like the one in Fig. 46, or generate one with an empirical formula, e.g., (38) and (39).
- Take a Raman spectrum with an excitation energy that runs through a branch with nanotubes that have your average diameter. This may be a branch belonging to any E_{ii}; choose one where you have a laser energy for excitation available in your lab.
- In your spectra, look at the region of expected ω_{RBM} and identify the branch, which may look like the resonance in Fig. 44 or like the one in Fig. 48. Several branches of tubes may appear in the spectra.
- Identify the maxima and use either Table 10 directly, or identify the highest ω_{RBM} within the la-ola and, using the simple $(n_1 - 1, n_2 + 2)$ rule (22), find the index for the peaks in the la-ola. (See also Sects. 2 and 5.2.)
- If in doubt, take a second laser line slightly shifted from your first. From the relative changes in intensity try to find out on which side of the la-ola you are on. This will give you additional confidence in your assignment.

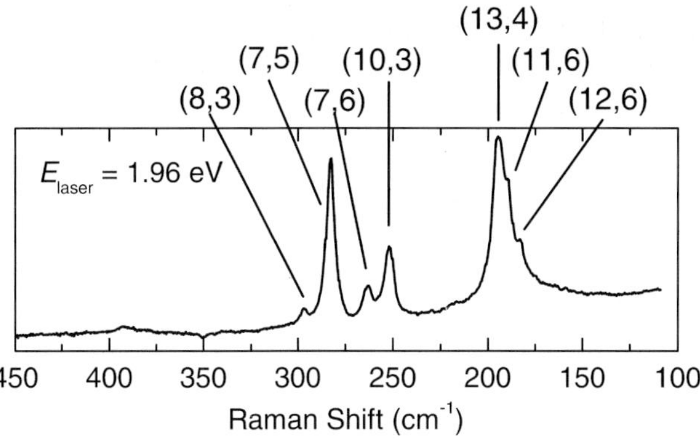

Fig. 48. Raman spectrum at $E_{\mathrm{exc}} = 1.96$ eV, an energy chosen specifically to identify the chirality of individual nanotubes; see text for details. After *Maultzsch* et al. [242]

In principle, if your tubes are in a different environment (different solution, wrapping, substrate, etc.) the Kataura plot may be shifted slightly as a whole. This concerns both Kataura axes, i.e., the constants c_1 and c_2 may change or the precise position of the electronic transition energies and the plot could shift up or down slightly. We can get an idea of how little the transition energies shift from studies using either sodium dodecyl sulfate (SDS) or sodium dodecylbenzene sulfonate (SDBS) surfactants as solvent for the same type of tubes [242], where the maximal shift observed was about 10 m eV. The frequency of the radial-breathing modes did not change within the experimental error. We emphasize that for the assigment these effects are expected to play a minor role, only. Once an entire branch is identified, the highest ω_{RBM} frequency is the one closest to the zigzag direction (see Sect. 5.2). Identifying a branch can be accomplished with a single suitably chosen excitation energy.

8 Metallic and Semiconducting Nanotubes

Metallic nanotubes – one third of all nanotubes in a random ensemble of nanotubes – are of special interest for applications as thin wires. They are not detectable by luminescence, and identifying them poses a challenge to Raman spectroscopy. To the extent that a full experimental Kataura map for a set of samples has become available (Sect. 7), the identification of metallic nanotubes, of course, is not an issue any longer. Given n_1 and n_2, if their difference is divisible by 3, a nanotube is metallic (see Sect. 2). Or, in other words, for one laser excitation energy metallic tubes have different diameters from semiconducting tubes. Based on the frequency of the radial-breathing mode a distinction becomes easily possible.

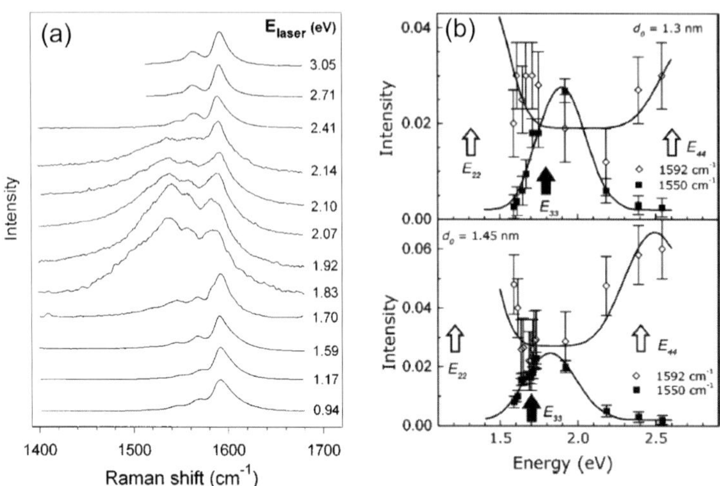

Fig. 49. (**a**) Resonant Raman spectra of the high-energy mode between 1 eV and 3 eV excitation energy. The strong enhancement of the mode at $\approx 1540\,\mathrm{cm}^{-1}$ at an excitation energy near 2 eV is clearly seen. After [188]. (**b**) Resonant Raman-excitation profiles from metallic (*open symbols*) and semiconducting (*closed symbols*) nanotubes. The *lines* are fits to (40). The *upper panel* refers to smaller-diameter nanotubes, the *lower* one to tubes with larger diameter. After [168]

8.1 Phonon–Plasmon Coupled Modes

Before the availability of isolated single-walled carbon nanotubes, one focus of Raman spectroscopy has been the shape of the high-energy mode, which appears very different in metallic and semiconducting nanotubes. The essential observation was that the shape of the high-energy peak changed at an excitation energy believed to match optical transitions in metallic nanotubes, enhancing those tubes systematically [188]. We show in Fig. 49 how a peak at $\approx 1540\,\mathrm{cm}^{-1}$ dominates the spectra at an excitation energy of 1.8 eV to 1.9 eV, while it nearly disappears at higher and at lower excitation energies. This peak, which had already been seen in an earlier paper [187], was interpreted as being due to an enhancement of optical transitions in the region of metallic tubes [188]. Because of the $1/d$ dependence of the electronic transition energies this "metallic resonance" should depend on the average diameter of the nanotube. Indeed it was shown that the resonance energies E_{ii} shift to larger energies for smaller nanotubes [168], see Fig. 49b. This analysis takes into account the average diameter d_0, the diameter distribution σ, considers δ-functions for the resonant transitions in every distinct nanotube chirality, and assumes constant Raman matrix elements when calculating the

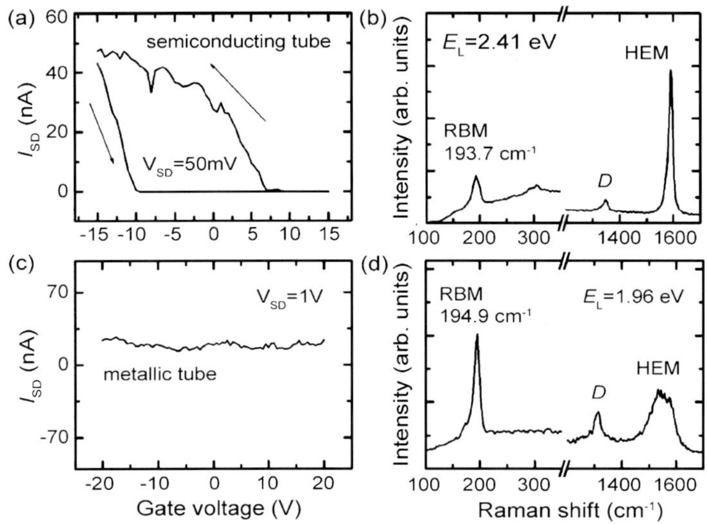

Fig. 50. Combined electric transport and Raman measurements on a single tube. (**a**) Transport characteristic of a semiconducting nanotube; (**b**) Raman spectrum of the tube in (**a**) showing the RBM, a weak D mode and a sharp Lorentzian-like high-energy mode. (**c**) Transport characteristic of a metallic nanotube; (**d**) Raman spectrum of the tube in (**c**) showing the RBM, a stronger D mode and a broadened asymmetric high-energy mode. This is the characteristic spectrum of a metallic tube. We thank R. Krupke for providing this figure

Raman intensity I as a function of the laser excitation energy $\hbar\omega_{\mathrm{exc}}$ according to

$$I(\hbar\omega_{\mathrm{exc}}) = \sum_d \frac{1}{d^2} \frac{A e^{-\frac{1}{2}[(d-d_0)/\sigma)]^2}}{[(E_{ii} - \hbar\omega_{\mathrm{exc}})^2 + \hbar^2\gamma^2][(E_{ii} - \hbar\omega_{\mathrm{exc}} + \hbar\omega_{\mathrm{ph}})^2 + \hbar^2\gamma^2]} \,. \tag{40}$$

The phonon energy used in the fit to data presented in Fig. 49b was 0.199 eV (semiconducting) and 0.196 eV (metallic nanotubes), resulting in energies of 1.2 eV and 1.3 eV for the second semiconducting transition and 2.4 eV and 2.6 eV for the third semiconducting transition. The transition energies for metallic nanotubes were at 1.8 eV and 1.9 eV in the two sets of tubes. The first value refers to the nanotubes with larger average diameter (d_0 = 1.45 nm), the second to those with smaller diameter (1.3 nm) [168]. Note that this analysis did not take into account excitonic effects, which are less important for nanotube bundles.

An independent proof of the connection of the semiconducting or metallic nature of a nanotube with the broadening of the high-energy Raman mode was given by *Krupke* et al. [248]. Their measurements of the transport properties and the Raman-scattering signal on the same tube specimen is shown

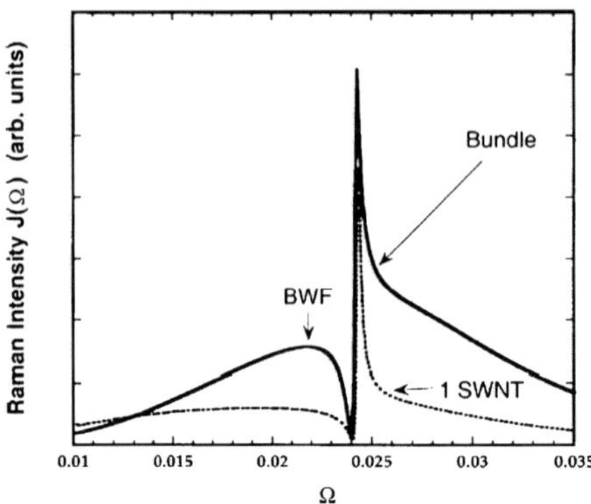

Fig. 51. Calculated lineshapes of the high-energy mode in nanotubes. An isolated nanotube, because of the small q-vector involved in light scattering, shows only a weak Fano resonance. Bundles of nanotubes have a much more prominent resonance. See also Fig. 52. From [132]

in Fig. 50. In the upper half of the figure a semiconducting nanotube was measured, the lower half shows the behavior of a metallic nanotube. The excitation energies, 2.41 eV and 1.96 eV corroborate the finding in Fig. 49 and the generally accepted treatment of nanotubes excited around 1.8 eV to 1.9 eV as predominantly metallic tubes.

Kempa [132] studied theoretically the origin of the broad line below the highest mode in the Raman spectrum. He worked out the proposal by *Brown* et al. [131] that the coupling of the tangential modes to the electronic continuum present in metallic nanotubes is the origin of the broad Raman line in the spectra. Semiconducting nanotubes, on the other hand, were responsible for uncoupled, Lorentzian-like peaks in the spectra. The plasmons in metallic tubes, according to Kempa, couple to the optical phonons and appear as phonon–plasmon coupled modes in the spectra. The small wavevector of the Raman excitation, however, produces only a narrow peak, see Fig. 51, for a single nanotube. Kempa invoked defects to incorporate phonon–plasmon scattering with larger q-vectors. Alternatively, bundles of nanotubes form a plasmon continuum, similarly as electronic bands in a solid form from the discrete states of atoms. The calculated phonon–plasmon coupled lineshape for bundles is also shown in Fig. 51 and displays a broad Fano lineshape, more similar to experiment.

An alternative explanation for the low frequency of one of the tangential modes in metallic nanotubes was put forward by *Dubay* et al. [133]. They found from first-principles calculations that in metallic tubes the Γ-point frequency of the longitudinal optical phonon is softened by what is called a Peierls-like mechanism (it is a Kohn anomaly of the axial transversal branch). In Fig. 52a their calculated optical-phonon frequencies of armchair, zig-zag, and one chiral tube are shown as a function of tube diame-

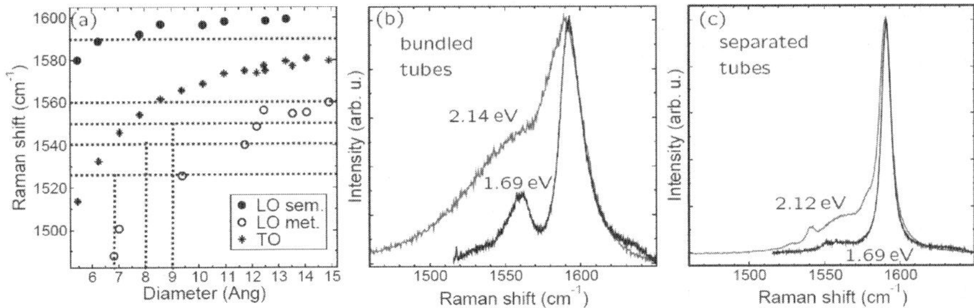

Fig. 52. (**a**) Ab-initio calculated frequencies of the high-energy optically active modes in nanotubes. Compared to semiconducting nanotubes the axial phonon in metallic tubes shows a large softening with decreasing diameter [133]. (**b**) The (softened) 1540 cm^{-1} mode in nanotube bundles is very broad. The origin of the larger half-width in bundled than in isolated nanotubes needs to be fully clarified [132, 250]. (**c**) In separated tubes, where the tubes are known from the experimental Kataura plot to belong to the metallic (13,1) branch, the metallic lineshape, although considerably smaller, is still visible in the spectra. From [251]

ter. With decreasing diameter, the phonon frequencies decrease, which was qualitatively found in force-constant calculations as well, although not for all optical modes [85, 103, 249]. The longitudinal and transverse modes in semiconducting nanotubes roughly follow the diameter dependence. However, the axial frequency of metallic nanotubes drops substantially below the other high-energy modes (see Fig. 52a). *Dubay* et al. [133] showed that this phonon periodically opens a gap at the Fermi level, which results in a strong electron–phonon coupling and softens the axial force constants (see Sects. 3.3 and 9.3).

We compare the high-energy mode of bundled and isolated nanotubes by *Telg* et al. [251] in Figs. 52b and c. The phonon frequencies, which are lower in the metallic nanotubes, explain the lower frequencies observed for peaks excited in resonance with transitions in metallic tubes. In semiconducting tubes, the curvature of the nanotube wall causes only a small decrease in frequency (see Fig. 52a). In Figs. 52b and c bundled tubes are compared to separated tubes at the same excitation energies. The latter were identified to belong to the (13,1) branch, i.e., to tubes with the chiralities (13,1), (12,3), (11,5), (10,7) and (9,9). The resonance of the metallic tubes is apparently much weaker in the isolated tubes compared to the bundled ones. The observations in Fig. 52 seem consistent with *Kempa*'s [132] suggestions, although the presence of a small Fano peak in separated tubes has remained controversial [252, 253]. In his model, however, the plasmon couples to the *circumferential* vibration, whereas *Dubay* and other groups predict a strong softening of the *axial* mode in metallic nanotubes [118, 140–142]. The two models are thus not compatible. This issue needs to be resolved by further experimental and theoretical work.

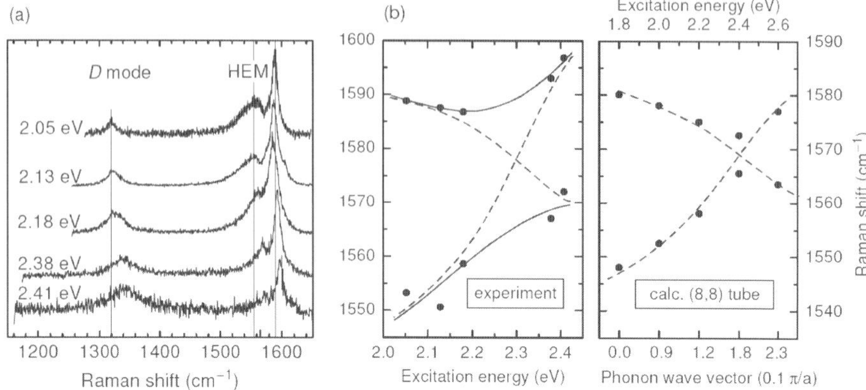

Fig. 53. (**a**) Raman spectra of an individual carbon nanotube at various excitation energies. Note the shift of both the D mode and the smaller of the high-energy modes. The *vertical lines* mark the peak frequencies in the uppermost trace. (**b**) Phonon frequencies from the two high-energy modes on the *left*, plotted against the energy of the exciting laser (experiment). The *right panel* shows a plot of the phonon frequencies taken from a model dispersion relation as corresponds to the excitation energies in the double-resonance model used to describe the Raman spectra in carbon nanotubes (calc. (8,8) tube). From [129]

8.2 Dispersion of Nanotube Phonons

The analysis of the high-energy phonons was pushed forward in a study of an individual metallic nanotube by *Maultzsch* et al. [129]. Take the uppermost trace in Fig. 53, where an (8,5) nanotube – as identified by its ω_{RBM} and its excitation energy in an experimental Kataura plot – is excited in optical resonance. The mode at $\approx 1560\,\mathrm{cm}^{-1}$ is broad and softened, similarly to that of the metallic tube in Fig. 52c. Experimentally, the difference is that in Fig. 53 *an individual* tube or a very thin isolated bundle with $\approx$ 3–5 tubes deposited on a silicon substrate was studied with several different excitation energies. As the excitation energy is increased beyond the resonance energy to 2.41 eV the signal becomes weaker by over one order of magnitude, the metallic resonance disappears, and the smaller of the two high-energy peaks moves up in frequency. In fact, for higher excitation energies the spectrum begins to look like that of a semiconducting nanotube. This is a characteristic of the double-resonance process, which dominates the Raman spectra of carbon nanotubes [194].

The details of the lineshape have been calculated within the double-resonance model and agree with the results in Fig. 53a [129]. Essentially, changing the excitation energy corresponds to changing the wavevector along the nanotube axis and mapping the phonon dispersion within the Brillouin zone of the nanotube. The phonon branch derived from the measurement is compared to a calculated branch of an (8,8) nanotube in Fig. 53b. The axial and circumferential modes are seen to cross in the achiral (8,8) armchair tube, where they have different symmetries (right panel). In the experiment (left panel) the two branches anticross, as expected for the lower-symmetry (8,5) chiral nanotube. We find a good agreement of these two dispersion curves.

An interesting effect follows from the interpretation within the double-resonance model and that cannot be neglected when using the appearance of the high-energy mode for the identification of metallic nanotubes. The broad peak due to the softened phonon branch in Fig. 52a becomes narrower when moving to larger phonon wavevectors. Larger wavevectors dominate the scattering at energies above the Γ-point resonance, see Fig. 53b. Consequently, the lineshape of the same individual nanotube changes continuously from metallic-like to semiconducting-like, see Fig. 53a. An identification of a metallic tube from the shape of its high-energy mode must therefore be performed in resonance. In other words, a broadened metallic-like high-energy peak is always evidence of a metallic nanotube. A narrower semiconducting-like one may be either semiconducting or, if excited above resonance, may also be a metallic nanotube. Note that the interpretation developed in [129] relied on band-to-band transitions. For excitation above the excitonic transition an outgoing resonance can be similarly double resonant for excitonic transitions if k conservation is relaxed by defects, see [254]. Screening by the substrate and other tubes in the bundle might also reduce the importance of excitonic effects.

8.3 Electrochemical Doping of Carbon Nanotubes

The excellent chemical stability and their large surface-to-volume ratio suggest the application of single-walled carbon nanotubes as supercapacitors [255], batteries [256], actuators [19], and other electronic devices. The change in resistivity of a semiconducting nanotube by many orders of magnitude upon electrostatic gating or gas absorption has a potential for applications such as nanotube field-effect transistors [257] or electrochemical sensors [258].

As all of these applications are based on the ability to dope nanotubes, it becomes of interest to examine the effect of charge transfer or doping on the Raman spectra. The effects of intercalation of ions, say alkali metals, iodine or bromine, on the optical and Raman spectra have been discussed in [259–264]. An alternative method with different applications is electrochemical doping of carbon nanotubes, which we do not intend to review here (see, e.g., [19, 265–268]). We focus on an aspect of spectroelectrochemistry related to the high-energy Raman mode in metallic nanotubes.

The electronic structure of bundled single-walled nanotubes can be changed electrochemically by varying the potential at their interface with an electrolytic solution [19, 265]. Although the achievable doping level is lower, because the electrolyte ions form a charged double layer only with the external surface of the ropes or bundles [269–272], it is possible to obtain quantitative information about the doping-induced strain. The strain is related to the transfered charge, which in the double-layer model can be calculated from the applied voltage. The result of such an experiment is independent of the electrolyte used and the basis for applying a linear relationship of the phonon frequency with applied voltage [269, 271]. In turn, this makes Raman spectroscopy a useful tool for analysis in spectroelectrochemical doping processes.

How much doping can be achieved by electrochemical doping? Given that one is certain to be in the double-layer charging regime, i.e., that there is no intercalation of ions from the electrolytic solution, the doping level f may be estimated from [267, 269]

$$f = \frac{M_{\mathrm{C}} C U}{F} , \tag{41}$$

where M_{C} is the atomic weight of carbon, $C \approx 35\,\mathrm{F/g}$ is the capacitance of the working electrode as determined by cyclic voltammetry, U is the applied potential, and F is the Faraday constant, the amount of charge of one mole of electrons ($F \approx 9.6 \times 10^4\,\mathrm{C/mol}$). In the experiment described here a mat of randomly interwoven nanotubes, so-called "buckypaper", was used as the working electrode. Intercalation effects were further reduced by using electrolytes with larger size ions, e.g., NH_4Cl, where the entire stability range of water can be explored ($\pm 1\,\mathrm{V}$). Under these conditions (41) yields a fraction not larger than $0.005 e^-$(holes)/(C-atom $\cdot$ V). Note that this is considerably smaller than what can be achieved by intercalation. In particular, the Fermi level in nanotubes of typical diameters cannot be shifted across one or several density-of-states singularities by electrochemical doping.

In Fig. 54 we show what happens when a voltage is applied to the electrode at an excitation energy of $E_{\mathrm{exc}} = 1.95$ eV. As expected from the strong phonon–plasmon coupled modes (Sect. 8.1) there is a broad peak at $\approx 1540\,\mathrm{nm}^{-1}$ at zero applied voltage, the spectrum looks "metallic". However, with both a positive and a negative voltage applied to the nanotube electrode, there is a qualitative change in the spectra. The characteristic broadening is reduced until near 1 V the spectra resemble much more semiconducting-like spectra. In fact, phenomenologically speaking, the upper- and lower-most traces in Fig. 54 are very similar to the corresponding ones in Fig. 49a. In other words, changing the excitation energy (within a certain range) appears to have the same effect as applying an electrochemical potential.

In order to understand how this can happen, let us analyze the electrochemical Raman spectra in more detail. Figure 54b shows the frequency of the largest peak in the high-energy region as a function of applied potential. For small applied voltages it shifts linearly with a rate of $1\,\mathrm{cm}^{-1}/\mathrm{V}$ to $1.5\,\mathrm{cm}^{-1}/\mathrm{V}$ and follows the sign of the applied voltage. For larger voltages the frequency becomes $\sim V^3$. The linear shift with applied voltage, which is largely independent of excitation energy, is a well-known property of the high-energy mode [273]. The behavior of the metallic modes P3 and P4 is quite

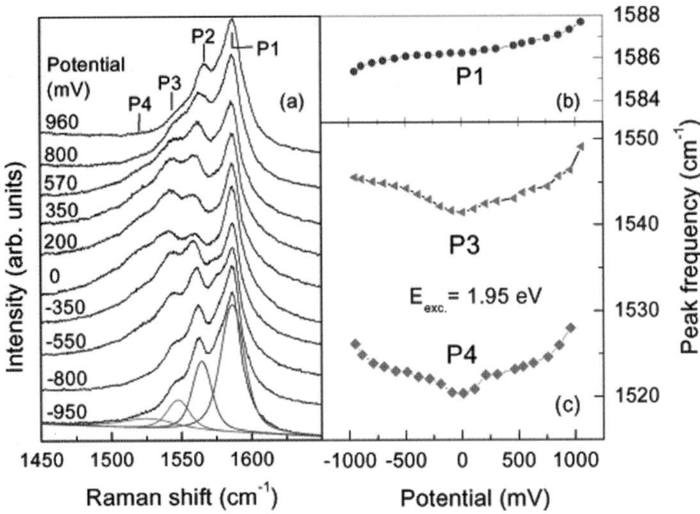

Fig. 54. (**a**) Raman spectra of a mat of single-walled nanotubes (buckypaper) at various electrochemical doping levels and an excitation energy of 1.95 eV, where metallic nanotubes are enhanced. While the 0 V potential indeed looks like the Raman spectrum of metallic tubes, see Fig. 49, the application of up to 1 V to the nanotube electrode in 1M NH_4Cl aqueous solution changes the lineshape drastically towards one that is semiconducting-like. (**b**) The frequency of the main peak, labeled P1 in (**a**), as a function of applied voltage. (**c**) Same as (**b**) but for the peaks labeled P3 (*triangles*) and P4 (*diamonds*). From [267]

different, however, see Fig. 54c. Both frequencies are $\sim V^2$, and for either sign of applied voltage they shift nonlinearly by up to $6\,\mathrm{cm}^{-1}/\mathrm{V}$ to $8\,\mathrm{cm}^{-1}/\mathrm{V}$ for the range of voltages applied (note the different frequency scales in b and c). Compared to the main peak P1 their intensity decreases drastically with applied voltage, causing the change in appearance of the high-energy mode in the Raman spectra.

An explanation offered by the authors of [274], who observed similar effects, is that the Fermi level, upon chemical doping, shifts beyond the next first valence- or conduction-band van-Hove singularity, like in intercalation experiments. As (41) shows, the electrochemically induced doping is far too small to shift the Fermi-level to the next singularity. From studies of the Fermi-level shift for various doping levels we conclude that the Fermi-level shift at $|f| = 0.005$ amounts to between 0.3 eV and 0.6 eV. However, the first metallic van-Hove singularity (≈ 0.9 eV) is not depleted below $U = 1\,\mathrm{V}$ for the vast majority of metallic SWNTs in the sample of buckypaper. For applied potentials between 300 mV and 500 mV, where the metallic modes in the Raman spectra in Fig. 54a have already changed considerably, we estimate ΔE_{F} to be between 0.1 eV and 0.2 eV [260, 265, 269, 275]. There must be another reason for this change for small Fermi-level shifts $\Delta(E_{\mathrm{F}}) < (E_{11} - E_{\mathrm{F}})$.

Let us see what follows from the interpretation within the double-resonance model [194] of Sect. 8.1 and the effect of electron–phonon coupling (Sect. 9.3). At an excitation energy of 1.95 eV in the spectra of Fig. 54a the metallic nanotubes are excited near or slightly above the first metallic resonance. Within the double-resonance model this implies a phonon wavevector near the Γ point, and the softening of the phonon branches due to the Peierls transition [133] is apparent in the low frequencies ($\approx 1540\,\mathrm{cm}^{-1}$), see also Fig. 52a. The physical reason for the softening of the phonon is a strong coupling of this mode to the electronic system, when the Fermi level lies where conduction and valence cross, i.e., at half-filling of the bands (Peierls-like softening, see Sect. 9.3). As the electrochemical doping sets in, the Fermi level moves away from the crossing point of the bands. The gap opening is reduced and disappears for large enough energies. At the same time, the phonon wavevector fulfilling the double-resonance condition increases, shifting the phonon frequency up, as in Fig. 53 for increasing excitation energies. The Peierls instability is thus switched off by applying a potential, regardless of the sign of the potential [267].

9 Electron–Phonon Interaction

Electron–phonon coupling enters critically many solid-state phenomena – the temperature dependence of the bandgap, thermalization of excited carriers, superconductivity, thermal and electrical transport among others [134, 276]. The interaction between electrons and phonons is also one step in the Raman process and thus determines the scattering cross section or Raman intensity. In turn, Raman scattering can measure the strength of the corresponding electron–phonon coupling given the optical properties of a material are known (see *Cardona* [184] and [277–279] for reviews and examples). On the other hand, electron–phonon coupling can also affect the phonon frequencies and drive structural phase transitions in solids.

In this section we first review electron–phonon coupling in carbon nanotubes and its effect on the radial-breathing mode Raman spectrum. We then discuss the Kohn anomaly (also called Peierls-like phonon softening) in metallic tubes.

9.1 Electron–Phonon Coupling of the Radial-Breathing Mode

An RBM Raman spectrum of ensembles of single-walled carbon nanotubes (bundled tubes or nanotubes in solution) typically shows several close-by peaks for a given excitation energy. We showed several examples throughout this work, in particular, in Sect. 7 (Fig. 44 on page 179). Figure 55 is another experimental spectrum (full line) obtained on nanotubes in solution [241]. The peaks within one la-ola group have different intensities. This reflects possibly a nonhomogeneous chirality distribution in the samples and the varying

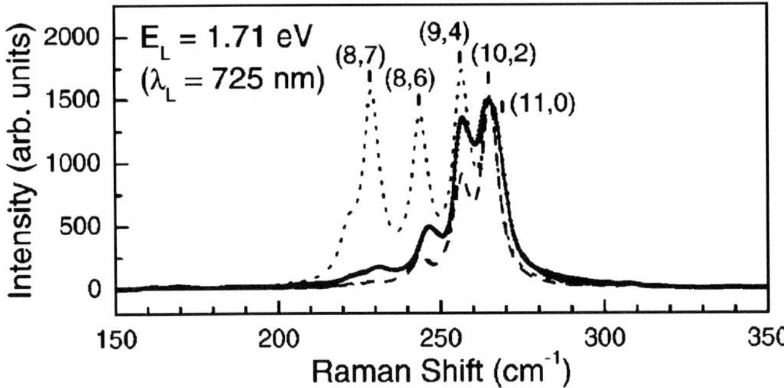

Fig. 55. Measured and calculated Raman spectra. *Full line*: Raman spectrum of isolated HiPCo tubes in solution excited with a laser energy $E_{\mathrm{L}} = 1.71$ eV. The chiral indices of the nanotubes are indicated. *Dotted*: Calculated spectrum assuming *constant* electron–phonon coupling for tubes of different chirality. *Dashed*: Calculated spectrum including the electron–phonon matrix elements as obtained from nonorthogonal tight binding. After [170]; experimental data from [241]

resonance enhancement for an excitation energy due to the differences in the optical transition energies. It can further be due to the differences in the optical properties and in electron–phonon coupling of specific nanotube chiralities.

Popov et al. [170] calculated the Raman spectra of ensembles of tubes within a tight-binding approximation (single-electron picture) to model electron–photon and electron–phonon interaction. Using an expression similar to (25) (p. 165) with *constant* electron–photon and electron–phonon matrix elements, they obtained the theoretical spectrum shown by the dotted lines in Fig. 55. This approximation describes rather poorly the experimental spectrum, although it accounts for the chirality distribution and the varying resonance energy for different chiralities. The dashed line in Fig. 55 was calculated including the tight-binding matrix elements in the Raman cross section. As can be seen, the experimental data are very well described by this approach, thus emphasizing the importance of the matrix elements in such an evaluation [144, 170, 280].

Figure 56 summarizes the maximum resonant Raman intensities observed on single-walled nanotubes in an experimental Kataura plot. Similar to what is seen in the experimental spectrum of Fig. 55 the Raman intensity is usually highest for tubes close to the zigzag direction (first or second outermost point in the V shapes branches). There is also a family dependence. $+1$ semiconducting tubes such as the (11,1) branch are weaker in intensity than -1 semiconducting tubes such as the (9,1) branch. Finally, the Raman intensity within one family decreases with increasing tube diameter [63–65, 111, 241, 242].

Machón et al. [144] calculated the electron–phonon matrix element of the RBM $\mathcal{M}_{\mathrm{ep}}$ using a frozen-phonon approximation and ab-initio techniques for

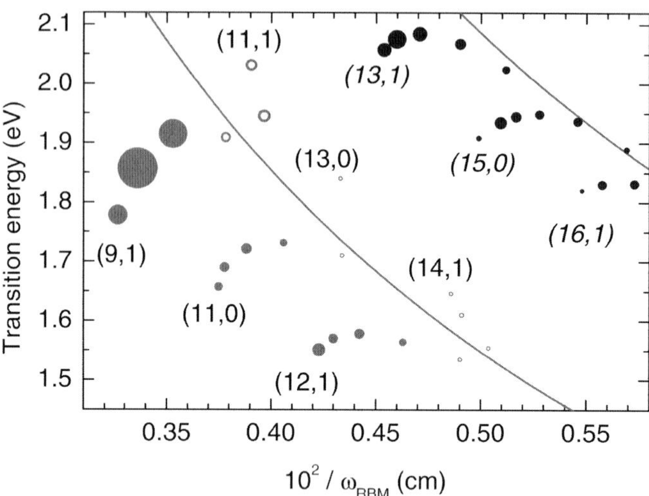

Fig. 56. Resonant Raman intensity of single-walled carbon nanotubes in a Kataura plot. The size of the symbols is proportional to the experimentally observed intensities (in arbitrary units). Intensities were obtained from the spectra normalized to the scattering intensity of BaF_2. The chiral indices indicate the branches. *Full lines* are proportional to $1/d$. The transition energies correspond to the second transition in semiconducting and the first transition in metallic tubes. The tubes were HiPCo tubes in solution (SDS coating). From [242]

the electronic structure [281]. Within this approach, the diagonal elements of the electron–phonon coupling Hamiltonian are found by imposing a deformation that mimics the phonon eigenvector and finding the shift this introduces in the electronic structure

$$\mathcal{M}_{\text{ep}} = \sqrt{\frac{\hbar}{2MN\omega_{\text{RBM}}}} \sum_a \boldsymbol{\epsilon}_a \frac{\partial E_{\text{b}}(\boldsymbol{k})}{\partial \boldsymbol{u}_{\text{a}}}\,, \tag{42}$$

where M is the atomic mass, N the number of unit cells, ω_{RBM} the RBM frequency, and a indexes the atoms in the unit cell of the tube. $\boldsymbol{\epsilon}_a$ is the normalized phonon eigenvector and $\partial E_{\text{b}}(\boldsymbol{k})/\partial \boldsymbol{u}_{\text{a}}$ describes the change in the electronic eigenenergies E_{b} due to the atomic displacement $\boldsymbol{u}_{\text{a}}$. *Machón* et al. [144] showed that the electron–phonon coupling in carbon nanotubes depends on chirality through $\partial E_{\text{b}}(\boldsymbol{k})/\partial \boldsymbol{u}_{\text{a}}$ in four ways: 1. a diameter dependence $1/d$, 2. a dependence on chiral angle, 3. a dependence on the nanotube family, and 4. a dependence on the optical band-to-band transition E_{ii} (related to eh_{ii} for excitons).

Table 11 summarizes $\mathcal{M}_{\text{ep}}$ for six zigzag nanotubes found from first principles.[8] Concentrating on the first transition, electron–phonon coupling de-

[8] The relative sign of the optical matrix elements will be important in the following. We therefore define $\mathcal{M}_{11}$ of the (10,0) tube as negative to facilitate the discussion.

Table 11. Electron–phonon coupling in zigzag carbon nanotubes. The table lists the matrix elements $\mathcal{M}_{ii}$ for the first three band-to-band transitions together with the transition energies E_{ii}. ν is the family index (see Table 2), d the tube diameter and ω_{RBM} the RBM frequency. Note the alternating signs between $\mathcal{M}_{11}$ and $\mathcal{M}_{22}$ as well as for one and the same transition i between +1 and −1 tubes. From [144]

Tube	ν	d	ω_{RBM}	$\mathcal{M}_{11}$	E_{11}	$\mathcal{M}_{22}$	E_{22}	$\mathcal{M}_{33}$	E_{33}
		Å	cm^{-1}	meV	eV	meV	eV	meV	eV
(10,0)	+1	7.9	287	−28	0.8	17	2.0	−30	2.4
(11,0)	−1	8.7	257	21	0.9	-28	1.3	−28	2.6
(14,0)	−1	11.0	203	16	0.7	-20	1.1	−21	2.4
(16,0)	+1	12.6	179	−17	0.6	13	1.2	−18	1.9
(17,0)	−1	13.4	170	14	0.6	-16	1.0	−17	2.1
(19,0)	+1	15.0	149	−15	0.5	12	1.0	−16	1.6

pends roughly on $1/d$ and is negative in +1 tubes, but positive in −1 tubes. At the same time, the absolute magnitude of $\mathcal{M}_{11}$ is larger in +1 tubes with $\mathcal{M}_{11}d = (220 \pm 4)$ eVÅ (root mean square deviations) than in −1 tubes (182 ± 5) eVÅ. For the second transition energy the family dependence is reversed, see Table 11; −1 tubes have a larger $\mathcal{M}_{22}$ with a negative sign, in +1 tubes $\mathcal{M}_{22}$ is smaller and positive. The relative magnitude of the squared matrix elements matches qualitatively the RBM intensity measurements in Fig. 56. The branches below the $1/d$ line (−1 tubes for eh_{22}) are stronger in intensity than the branches above $1/d$ (+1 tubes).

All four dependences can be explained on the basis of zone folding. As an example we discuss here the changes in electron–phonon coupling with nanotube family ν and transition index i in semiconducting zigzag tubes. Further details were given in [144], see also [170, 201, 280, 282].

To understand the dependence of $\mathcal{M}_{\mathrm{ep}}$ on family and transition index, we turn to Fig. 57. Here, the electron–phonon matrix elements are modeled by zone folding (see Sect. 4.1). Figure 57a shows the allowed lines of a (19,0) tube in the graphene Brillouin zone. In a zigzag tube the allowed lines run parallel to ΓM; the Γ point of the tube corresponds to the equidistant points along the ΓKM line. These are the wavevectors that contribute most to the optical spectra and hence Raman intensity in zigzag tubes; they give rise to the minima and maxima in the electronic band structure [20]. The full lines in Fig. 57b show the π band structure of graphene along ΓKM.

In the spirit of the frozen-phonon approximation to calculate $\mathcal{M}_{\mathrm{ep}}$ (42) and the zone-folding approach (Sect. 4.1) we now want to know how the π-band structure changes when imposing an RBM phonon eigenvector of a (19,0) tube [144]. For an unwrapped piece of graphene corresponding to the (19,0) tube, the RBM motion is a periodically alternating stress along $\mathbf{c}$, i.e., a uniaxial stress along $\boldsymbol{a}_1$ for $(n, 0)$ zigzag tubes. Figure 57b shows the

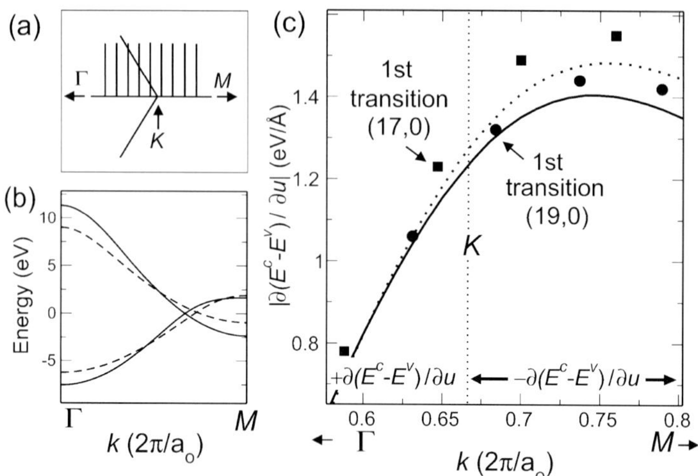

Fig. 57. Comparison between electron–phonon coupling of the RBM from ab-initio calculations and using zone folding. (**a**) Allowed k vectors of a (19,0) nanotube inside the graphene Brillouin zone. (**b**) Electronic band structure of graphene in equilibrium (*full lines*) and under the distortion that is equivalent to the RBM of a (19,0) nanotube (*dashed*). Only the π bands are shown for simplicity. (**c**) The *full line* and the *circles* give the electron–phonon coupling of the RBM in a (19,0) tube from zone folding and first principles, respectively. The *dashed line* and the *squares* are the corresponding data for a (17,0) tube. From [144]

π bands under a uniaxial stress along $\boldsymbol{a}_1$. The crossing point between the valence and conduction band (K point in unstressed graphene) moves along the ΓKM line. Let us first concentrate on the sign of ΔE_b of the conduction bands, i.e., the band above E_F regardless of the k vector. ΔE_b is negative at Γ (the dashed line is below the full line), goes through zero at roughly $1/2\Gamma M$ and is positive between the node and the valence and conduction-band crossing in the strained structure (dashed line above the full line). Then, at the crossing between the two dashed bands, ΔE_b changes discontinuously to negative. Finally, at $M \Delta E_\mathrm{b} = 0$.

How will the change in sign of ΔE_b at $\approx K$ affect the electron–phonon matrix elements in zigzag tubes? In Sect. 4.1 we showed that the first optical transition of +1 tubes originates from the KM line of the graphene Brillouin zone after zone folding. From Fig. 57b we expect $\mathcal{M}_{11}$ to be negative, because $\partial E/\partial u$ is negative. For −1 tubes we expect a positive sign, since the transition comes from the part between ΓK, where $\partial E/\partial u$ is positive. This is in excellent agreement with the ab-initio matrix elements in Table 11. For the second transition energy the signs are reversed. The second transition always comes from the opposite side of K to the first transition, hence the change in sign. The $\mathcal{M}_{33}$ are all negative, because the electronic dispersion is smaller between KM than between ΓK. Enumerating the transition energies by increasing energy (as we did in Table 11) E_{33} always originates from the KM line for the tubes given in the table.

After discussing the sign, let us turn to the magnitude of $\mathcal{M}_{\mathrm{ep}}$. The full line in Fig. 57c is the absolute value of the combined energy derivative of the valence and conduction band $|\partial(E^{\mathrm{c}} - E^{\mathrm{v}})/\partial u|$. We indicated the change in sign when crossing the K point close to the x-axis of the figure. Starting from the left the absolute energy derivative first increases (starting from zero at $1/2\Gamma M$, see Fig. 57b), reaches a maximum at $1/2KM$ and then decreases due to the vanishing coupling of the conduction band at M. Thus, the first two transitions of the (19,0) $\nu = +1$ tube are predicted as $|\mathcal{M}_{11}| > |\mathcal{M}_{22}|$ in excellent agreement with Table 11. The quantitative agreement between $\mathcal{M}_{ii}$ within zone folding (full line in Fig. 57c) and the ab-initio calculations (closed symbols) is also very good. The ab-initio results are slightly larger than those predicted from zone folding, because the RBM has a small nonradial component (Sect. 3.2), whereas the stress in the zone-folding calculation is uniaxial (purely radial RBM for the tube).

The squares in Fig. 57c show the ab-initio energy derivatives for the (17,0) -1 nanotube; the dashed line is the zone-folding result. Again, there is excellent agreement between the two calculations. Note also the change in sign and magnitude between $\mathcal{M}_{11}$ for the (19,0) and (17,0) tube (arrows) as explained before.

Within the zone-folding scheme we can also understand why $\mathcal{M}_{ii}$ depends on $1/d$. There are two reasons for this. First, for increasing diameter the same radial displacement of an atom corresponds to a smaller stress within zone folding (there are simply more bonds around the circumference). This explains why the full line in Fig. 57c is below the dashed line, $d(19,0) > d(17,0)$. Second, with increasing diameter the allowed k points for nanotubes move towards K. Considered alone, this second effect will decrease $\partial E/\partial u$ for the first transition in $+1$ zigzag tubes and increase it for the first transition in -1 tubes, see Fig. 57c. In $+1$ tubes (first transition) the two effects thus add up, whereas they have opposite signs in -1 tubes (the first effect dominates). Therefore, $\mathcal{M}_{11}d$ was larger in $+1$ tubes (220 eV $\cdot$ Å) than in -1 tubes (180 eV $\cdot$ Å).

The systematics in the electron–phonon coupling are quite interesting, because they potentially allow us to distinguish tube families based on their relative Raman intensities. Presently, we are not yet in a position to do this straightforwardly, because the optical part of the Raman process is poorly understood. A systematic confirmation of the alternating magnitude and sign of $\mathcal{M}_{ii}$ for one and the same nanotube, but different optical transitions, is still missing. *Reich* et al. [74] recently predicted other interesting differences between the Raman spectra in resonance with the first and second subband. They originate from the optical matrix elements in the expression for the Raman cross section and, in fact, enhance the expected differences between $+1$ and -1 tubes for eh_{11} resonant scattering as compared to eh_{22} (see [74] for details).

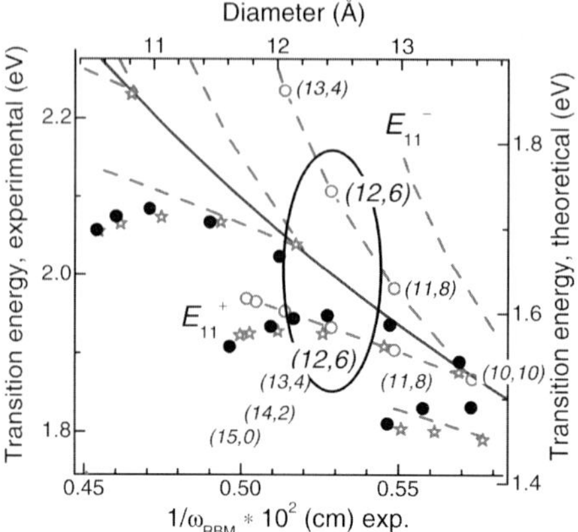

Fig. 58. Enlarged region of the experimental Kataura plot in Fig. 46. Highlighted is the (12,6) metallic nanotube, which like all metallic nanotubes, has two close-by optical transition energies. When calculating Raman matrix elements, interferences from these two transitions have to be taken into account [201]. In armchair tubes the two transitions energies are degenerate. See also [150]

9.2 Interference Effects

Turning back to Fig. 56 we note that in metallic tubes [(13,1), (15,0) and (16,1) branches] only the lower parts of the V-shaped branches have been observed experimentally [63, 64, 111, 242]. In contrast to semiconducting tubes, where the upper and lower part of a V have a different branch index β (they originate from different tubes), the two parts come from two close-by transitions in the same tubes in metallic nanotubes, i.e., the two parts of the V have the same branch index. Thus, the missing resonances cannot be due to differences in the abundance of certain chiralities in the sample.

In metallic nanotubes, trigonal warping of the graphene band structure always leads to two close-by transition energies $M\text{–}E_{ii}^{+}$ and $M\text{–}E_{ii}^{-}$. This is illustrated in the enlarged Kataura plot in Fig. 58. The (12,6), for example, has $M\text{–}E_{11}^{+} = 1.95$ and $M\text{–}E_{ii}^{-} = 2.1$ eV.[9] The first energy was measured, e.g., by *Telg* et al. [63], whereas the second one is obtained from the theoretical splitting between E_{11}^{-} and E_{11}^{+} [6], see Fig. 58 (we dropped the M in front of the energies for simplicity). *Bussi* et al. [201] showed that quantum-interference effects arise in the resonant excitation profiles of metallic nanotubes. The interferences occur between the two close-by transitions in metallic nanotubes. As we will see, this, unfortunately, does not explain the vanishing Raman intensity for the upper metallic branches in Fig. 56. Nevertheless, it is an intriguing feature and affects the transition energies extracted from a resonance profile.

[9] +/− is chosen in analogy to the +/−1 semiconducting families; therefore the + branch has the lower transition energy for E_{11}.

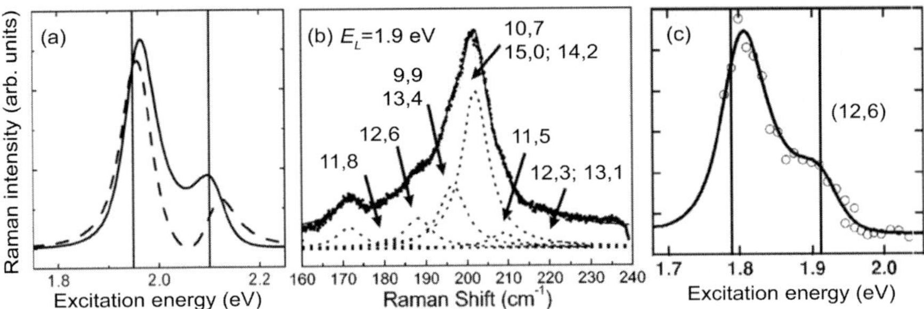

Fig. 59. Raman intereferences in metallic nanotubes. (**a**) Calculated resonance profile for two close-by transition energies $E_{11}^{+} = 1.94$ eV and $E_{11}^{-} = 2.1$ eV. *Full line*: negative ratio between the electron–phonon matrix elements for the two transitions, $\mathcal{M}_{11}^{+}/\mathcal{M}_{11}^{-} < 0$, as found from ab-initio calculations. *Dashed line*: $\mathcal{M}_{11}^{+}/\mathcal{M}_{11}^{-} > 0$. This plot proves that Raman-interference effects are present in the Raman resonances of metallic tubes, because the calculated profiles differ (*full* and *dashed lines*). (**b**) Raman spectrum of bundled carbon nanotubes produced by pulsed-laser vaporization; the laser energy was $E_\mathrm{L} = 1.9$ eV. (**c**) Raman resonance profile for the mode at 188 cm^{-1} corresponding to the (12,6) nanotube. After [200] and [201]

Bussi et al. [201] modeled the Raman profiles using the single-resonance expression for band-to-band transitions (29) (p. 167). The sum in the Raman matrix element $K_{2f,10}$ runs over *all* intermediate electronic states including contributions from different excited states, i.e., different optical transitions. We obtain for two transitions E_{11}^{+} and E_{11}^{-}

$$K_{2f,10} = \frac{L}{i\sqrt{2}\hbar^2\omega}\left[\frac{\mathcal{M}_{11}^{+}\mu_{11}^{+\,1/2}}{(E_1 - E_{11}^{+} - i\gamma)^{1/2}} - \frac{\mathcal{M}_{11}^{+}\mu_{11}^{+\,1/2}}{(E_1 - \hbar\omega - E_{11}^{+} - i\gamma)^{1/2}} + \frac{\mathcal{M}_{11}^{-}\mu_{11}^{-\,1/2}}{(E_1 - E_{11}^{-} - i\gamma)^{1/2}} - \frac{\mathcal{M}_{11}^{-}\mu_{11}^{-\,1/2}}{(E_1 - \hbar\omega - E_{11}^{-} - i\gamma)^{1/2}}\right], \quad (43)$$

where $\mathcal{M}_{11}^{+}$ are the combined electron–photon and electron–phonon matrix elements for E_{11}^{+}; correspondingly for $\mathcal{M}_{11}^{-}$. $\mu_{11}^{+/-}$ are the reduced effective masses for the $v_1^{+/-}$ valence and the $c_1^{+/-}$ conduction band. Depending on the relative sign of $\mathcal{M}_{11}^{+}$ and $\mathcal{M}_{11}^{-}$ the two resonances in (43) will add up constructively or destructively, i.e., we obtain an interference in the Raman cross section as a function of E_1.

Figure 59a shows the Raman profile obtained from (43) for $\mathcal{M}_{11}^{+}/\mathcal{M}_{11}^{-} = -2$ by the full line and for $+2$ by the dashed line. The two profiles differ, which is the manifestation of quantum interference [201]. Let us now use (43) for fitting the resonance profile of a metallic nanotube. The matrix elements were obtained from first principles in [201]. They showed similar systematics as in semiconducting tubes. In particular, $\mathcal{M}_{11}/\mathcal{M}_{22}$ is always negative (see also [144]). Figure 59b shows the Raman spectrum of bundled tubes excited at $E_{\mathrm{L}} = 1.9$ eV [200]. We assigned the peaks based on the experimental Kataura plot as explained in Sect. 7 taking into account the redshift of $\approx$ 0.1 eV induced by the bundling [9, 245]. The experimental resonance profile of the (12,6) tube with $\omega_{\mathrm{RBM}} = 188\,\mathrm{cm}^{-1}$ (symbols in Fig. 59c) was fitted by two resonances at 1.79 eV and 1.91 eV (full line for the fit, vertical lines for the energies) [201]. As pointed out by *Bussi* et al. a Raman assignment based on the splitting between the two metallic resonances can be affected quite severely by Raman interferences (see [201] for further details; similar results were obtained also in [282]).

A remark about the experimental data in Fig. 59c as well as other resonance Raman profiles in the literature are necessary. First, there are a number of papers using joint density of electronic states to model the Raman intensity of carbon nanotubes, which is incorrect. Regardless of the model (band-to-band transitions, excitons) the incoming and outgoing resonant Raman profiles are always symmetric peaks (as a function of excitation energy) and never expected to follow a $1/\sqrt{E}$ dependence [201]. We have already discussed this in Sect. 6.2. The assignment based on $1/\sqrt{E}$ Raman profiles and the experimental values, e.g., for the broadening parameter γ need to be re-evaluated. The experimental data in Fig. 59c show a strong high-energy shoulder assigned to the second transition E_{11}^{-}, which was not observed in isolated nanotubes in solution. On the other hand, the Raman data on semiconducting bundled tubes have a similar asymmetric profile as observed for the (12,6) in [200], although interferences are not expected in semiconducting tubes where $E_{22} \gg E_{11}$. Comparing experimental data on nanotube bundles with a theory for isolated tubes is always a little delicate. The shift in the resonance energy for E_{11}^{+} between (12,6) tubes in bundles and isolated (12,6) tubes in solution (0.1 eV), compare Fig. 59c and Table 10, is in rather nice agreement with ab-initio calculations and experiment [9, 245]. It needs to be clarified whether asymmetric profiles are really intrinsic to metallic tubes and why the surfactant-coated tubes do not show a second resonance.

In Fig. 59a the resonance due to E_{11}^{+} is stronger than the E_{11}^{-} resonance. The ratio between the two intensities is on the same order as the squared ratio between the combined matrix elements for electron–photon and electron–phonon interaction (four). The absence of the upper resonance in metallic tubes in solution might be caused by larger differences between the $\mathcal{M}_{11}^{+}$ and $\mathcal{M}_{11}^{-}$ or by an additional mechanism not considered so far.

9.3 Kohn Anomaly and Peierls Distortion

Kohn showed in a seminal paper that metals have anomalies in their phonon dispersion with $|\nabla_{\boldsymbol{q}}\omega(\boldsymbol{q})| = \infty$ [138]. For a simply connected Fermi surface, e.g., a circle, these anomalies occur at $q = 2k_{\mathrm{F}}$, where $+k_{\mathrm{F}}$ is the Fermi wavevector. For these qs and phonon symmetries that couple electrons at E_{F} the electron–phonon coupling is particularly large. This can even drive a static distortion of the entire lattice, which is known as a Peierls distortion for

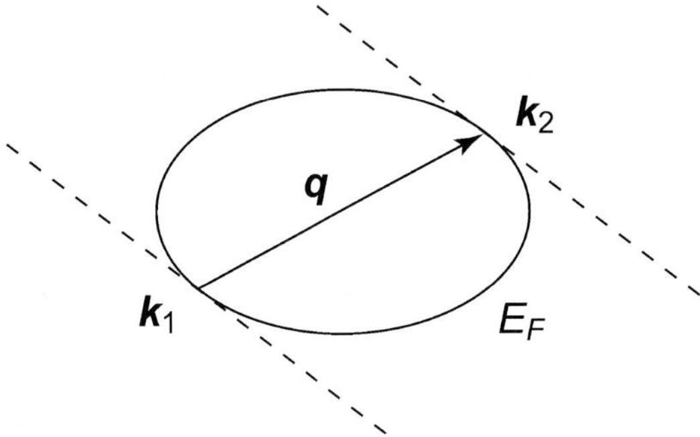

Fig. 60. Two parallel lines (*dashed lines*) touch the Fermi surface E_F in the points $\boldsymbol{k}_1$ and $\boldsymbol{k}_2$. At the wavevector $\boldsymbol{q} = \boldsymbol{k}_2 - \boldsymbol{k}_1$ the phonon dispersion becomes singular. Behind this Kohn anomaly is an abrupt change in the ability of the electrons to screen the lattice vibrations. After [138]

a one-dimensional system [283]. An atomic chain will dimerize and double its periodicity. The distortion opens up a gap at the Fermi energy and thus lowers the total energy of the system. Kohn anomalies and Peierls transitions are currently widely discussed for carbon nanotubes [127, 140–142].

Figure 60 shows the condition for the phonon wavevectors $\boldsymbol{q}$ at which $\omega(\boldsymbol{q})$ becomes singular [138]. Two parallel lines (dashed lines) touch the Fermi surface (full line) in $\boldsymbol{k}_1$ and $\boldsymbol{k}_2$. Then a Kohn anomaly occurs at $\boldsymbol{q} = \boldsymbol{k}_2 - \boldsymbol{k}_1$ given the phonon has the right symmetry to couple electrons at E_F. The Fermi surface is a single point in graphene; hence, in the limit $\boldsymbol{k}_2 \to \boldsymbol{k}_1$ we expect a Kohn anomaly at $\boldsymbol{q} = 0$. A second anomaly occurs for the wavevector connecting the K and K' point, i.e., $\boldsymbol{q} = K$. Only the E_{2g} optical branch at Γ and the TO (A') branch at K become singular in graphene for symmetry reasons [127].

Metallic nanotubes with $d \approx 1\,\text{nm}$ and above show two Kohn anomalies. One for the axial modes at Γ originating from the E_{2g} phonon of graphene and one for the backfolded branch at $2\pi/3a$ originating from the TO branch at K [3, 127]. These phonons are anomalous also in graphite, the $d \to \infty$ limit of carbon nanotubes. When the diameter of the tubes decreases the number of anomalous branches increases to five, see Fig. 61a [140, 142]. This is an interesting example of symmetry lowering and structural distortions in low-dimensional structures. In graphite, only the E_{2g} phonon at Γ and the TO branch at K can couple the π and π^* bands close to E_F [125, 207]. In carbon nanotubes we find, e.g., for armchair tubes that the coupling is allowed by symmetry for phonon branches correlated with the A_{1u}, A_{2g}, B_{1u} and B_{2g} representations at Γ; these modes are shown in Fig. 61b. Five optical modes and the rotation around z belong to these representations, see Table 5. A_{1u} and B_{2g} correspond to the anomalous graphite branches at Γ and K, respectively. Although four other modes are allowed to couple electrons at E_F as well, this will not have a strong effect until the atomic

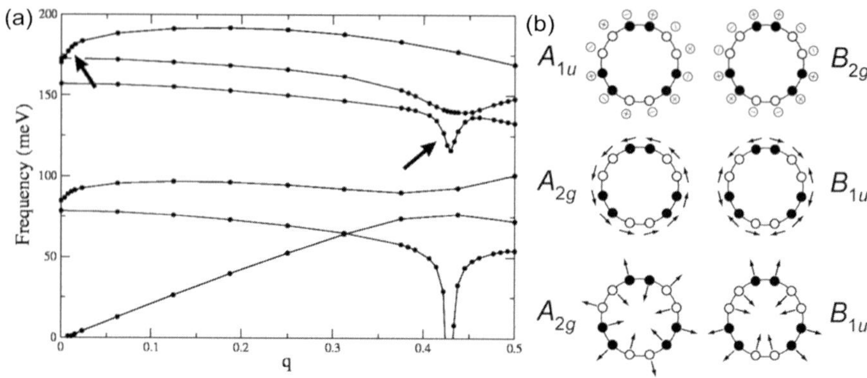

Fig. 61. Kohn anomaly and Peierls distortion in a (3,3) armchair nanotube. (**a**) Anomalous phonon branches calculated from first principles showing giant Kohn anomalies. The two modes indicated by *arrows* behave singular for larger-diameter tubes and graphene as well, see [127] and [141]. (**b**) Γ-point eigenvectors of the anomalous branches. The two axial modes belong to the arrowed branches in (**a**). The A_{2g} mode with the displacement along the circumference is the rotation around z (forth acoustic branch); the radial A_{2g} mode corresponds to the out-of-plane optical phonon of graphite (ZO mode). The radial B_{1u} mode drops to zero in (**a**); it induces the Peierls transition in the (3,3) tubes

structure becomes significantly different from the graphene cylinder limit. This only happens for tubes with $d \ll 1\,\text{nm}$ as for the (3,3) tube in Fig. 61a with $d = 4\,\text{Å}$ [154].

The Kohn anomaly for the radial branch with B_{1u} symmetry in a (3,3) nanotube is so strong that the phonon frequency was found to be imaginary in ab-initio calculations [140, 142]. This means the tube will undergo a Peierls transition; the transition energy was predicted between 240 K and 400 K (see [140] and [142] for details). Note that the phonon q vector inducing the Peierls transition is incommensurable with the unit cell. Thus, straightforward total-energy calculations of a (3,3) nanotube will not predict this transition.

A Kohn anomaly always implies a particularly strong electron–phonon coupling for the qs where $\omega(q)$ turns singular. In metals, the Raman spectrum, therefore, often reflects the strength of the electron–phonon coupling as a function of ω (note that in metals wavevector conservation is relaxed due to the small penetration length of the light) [139]. A strong experimental indication for the presence of the Kohn anomaly in graphite and carbon nanotubes is thus the high-energy range of the Raman spectrum. The spectra are dominated by scattering from the tangential A_1 high-energy vibrations (the axial mode of metallic tubes is singular at Γ) and the D mode (singular at $2\pi/3a$). Metallic isolated nanotubes often show a strong and sharp D line (see [99]).

According to [133] and [141] phonon softening due to Kohn anomalies is stronger in metallic tubes than in graphite. This results from the one-dimensional character of single-walled carbon nanotubes. The axial tangential mode periodically opens and closes a gap in metallic tubes, whereas it only shifts the Fermi point in graphene. Because of this oscillation of the bandgap *Dubay* et al. [133] used the term Peierls-like distortion when discussing lattice dynamics of metallic tubes. Experimentally, the D mode is indeed at lower frequencies in carbon nanotubes than in graphite, although a detailed study of this phenomenon is missing. Also, the excitation-energy dependence of the high-energy phonons in carbon nanotubes confirmed the extremely strong overbending for the softened axial phonon at Γ (see Sect. 8).

The calculated electron–phonon matrix element for the graphite optical modes is 200 meV at Γ ($\omega = 1540\,\mathrm{cm}^{-1}$) and 300 meV ($1250\,\mathrm{cm}^{-1}$) for the transverse branch at K [127]. In carbon nanotubes the coupling of these modes to the electronic system is thus at least an order of magnitude larger than for the RBM (see Table 11), in contrast to the predictions of [75]. Since electron–phonon interaction is also crucial for the relaxation of photoexcited carriers and in connection with the breakdown of ballistic transport in carbon nanotubes, we expect more studies of electron–phonon coupling to appear in the near future.

10 *D* Mode, High-Energy Modes, and Overtones

The D mode in graphite and carbon nanotubes is characterized by the dominance of the Raman double resonance in the spectra. The double resonance for the high-energy mode in nanotubes has similar experimental characteristics, but they are not as prominent. A similar statement holds for the overtones of the D and G modes. The detailed mechanism of this process was explained in Sect. 6.3. We summarize here the experimental evidence most frequently found for a double resonance [20, 125, 207, 230].

1. In the first-order spectrum the strength of the Raman signal is related to the number of defects or other symmetry-breaking elements, like the surface in Ge where visible light is strongly absorbed [213, 214] or the end of a nanotube [4, 218].
2. Excitation-energy dependence of the corresponding Raman peak, as best seen for the D band in graphite or in carbon nanotubes, see Fig. 38. The excitation-energy dependence is particularly large for the D band because the electronic and phononic bands are strongly dispersive for the double-resonant k and q vectors. In Ge, the excitation-energy dependence is less pronounced due to the flat optical-phonon bands [4, 210].
3. A large Raman intensity as seen, e.g., for individual carbon nanotubes, see Fig. 53 [4, 99, 284].
4. Frequency and lineshape differences in Stokes and anti-Stokes spectra, see [20, 125, 205, 211, 219, 230], or Fig. 37c.

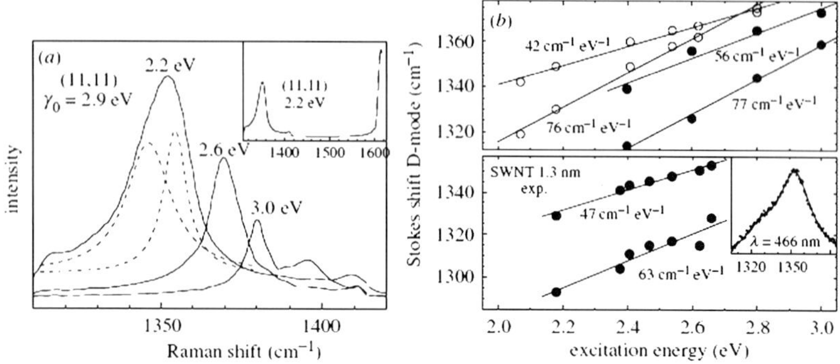

Fig. 62. (**a**) Calculated D-mode Raman spectrum of an (11,11) nanotube for different excitation energies (*inset*) extended scale. (**b**) Frequencies of the two D-mode components as a function of excitation energy. *Open symbols* refer to a (10,10) nanotube, *closed* ones to an (8,8). Experimental values are shown for comparison in the *lower part* of the figure. From [230]

5. Deviations from Lorentzian lineshapes, as observed for the high-energy mode in carbon nanotubes [194].

Not all of these observations are sufficient to identify double resonance; two-phonon scattering, e.g., also leads to deviations from a Lorentzian lineshape in the spectra. In the following sections we discuss several aspects of double resonance in the high-energy region ($> 1000\,\text{cm}^{-1}$).

10.1 *D* Mode and Defects

The D mode in the Raman spectrum of graphite was discussed extensively in connection with the double-resonance process in Sect. 6.3, and we focus here on some special aspects in nanotubes. The similarity of the excitation-energy-dependent shift found in the nanotubes suggested a physical process similar to that found in graphite. *Maultzsch* et al. [206] showed that there is a selection rule for the D mode related to the chirality of the nanotube. Namely, only when $(n_1 - n_2)/3n$ = integer (the so-called $\mathcal{R} = 3$ tubes) is D-mode scattering allowed. All other tubes ($\mathcal{R} = 1$) can also have double resonances, but the phonons involved are near-Γ-point phonons and hence not in the energy region of the typical D mode (see also Table 1).

In Fig. 62 we show the calculated D mode of an (11,11), a (10,10), and an (8,8) nanotube based on the double-resonance model in comparison to an experiment on nanotube bundles. The observed shift of the Raman frequency with excitation energy is well reproduced, which is the most prominent criterion for the observation of double resonance. The D mode is also observed in individual nanotubes, see Fig. 53, which implies that its Raman cross section is unusually large; the lineshape is non-Lorentzian. Differences in the Stokes

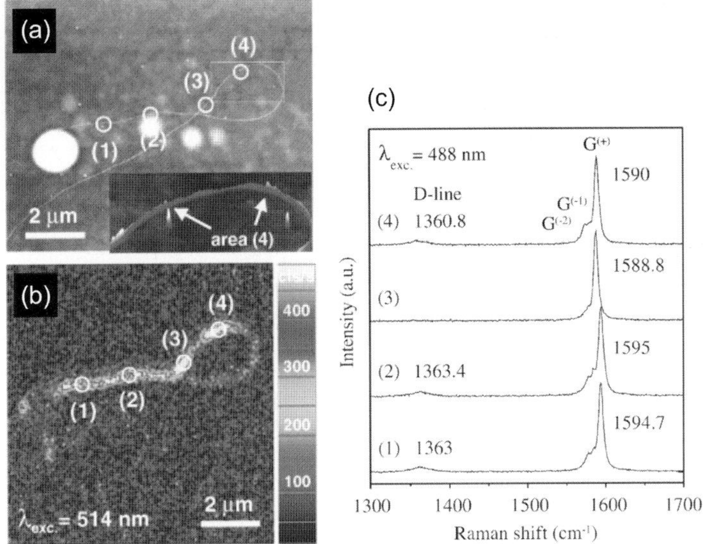

Fig. 63. (**a**) Atomic-microscope image of a carbon nanotube grown by a CVD method. The numbers mark four spots along the nanotube, for which the Raman spectra are shown in (**c**). (**b**) Raman image obtained by imaging the energy region of the G mode. (**c**) Raman spectra recorded at the specified points of the image in (**a**). Note the difference in D-mode intensity for the different spots, which reflects structural inhomogeneities along the nanotube. After [286]

and anti-Stokes spectra of the D mode have been reported also. They support the origin of the D mode in nanotubes as a double resonance [230].

The signal in Raman double resonance depends on a convolution of the two resonances. In carbon nanotubes the intensity therefore increases when the laser energy matches the separation in energy between a maximum in a valence band and a minimum in a conduction band in the band-to-band transition picture (only this model has been used, so far, for a calculation of the D-mode spectra). This can be seen in the variation of the Raman signal with excitation energy in Fig. 62a. The signal increases when the laser energy approaches the energies of the critical points in the electronic band structure, which was further discussed by *Kürti* et al. [285]. Note, however, that *Kürti* et al. disarded the selection rules for elastic scattering developed by *Maultzsch* et al. [206]. Conclusions about D-mode intensities for varying chirality, therefore, differ in [285] and [206].

Confocal Raman spectroscopy combined with atomic-force microscopy allows a space-resolved imaging of individual carbon nanotubes. We show in Fig. 63a combined measurement by *Jiang* et al. [286]. In Fig. 63a the atomic-force microcope picture of a curved nanotube is shown. Figure 63b shows the Raman image of the same tube taken by recording the spatially resolved G-

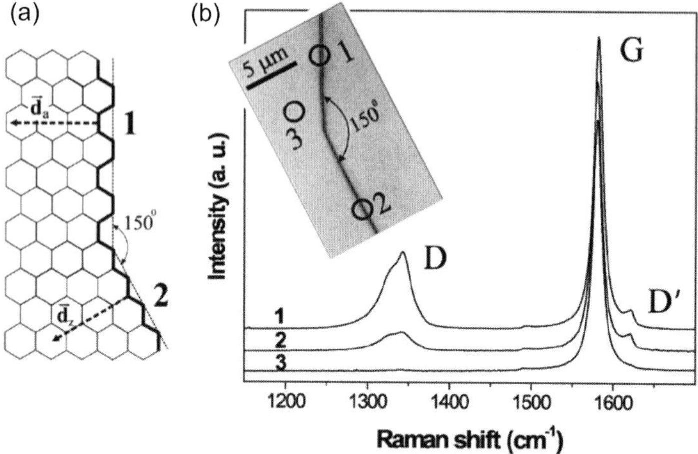

Fig. 64. (**a**) Orientation of a small HOPG crystallite as determined by atomic-force microscopy. Armchair and zigzag directions are indicated. (**b**) Raman spectra taken off the edge of a small HOPG crystallite at the numbered points 1,2 and 3. The D mode is largest at point 1, which corresponds to the "armchair" edge of the crystallite. At point 2, the "zigzag" edge, the D mode is nearly absent. These spectra verify the selection rules predicted in [206] for D-mode scattering. After [287]

mode intensity. A bright spot corresponds to a high G-mode intensity; it is by no means as homogeneous as one might expect for a first-order Raman process. Instead, the Raman intensity varies considerably along the tube, which is reminiscent of double-resonance scattering. Unfortunately, the authors of [286] did not show the intensity of the overtone spectra, which should be independent of defect density and would clarify the type of Raman process (see Sect. 10.3). As it stands, the variations in intensity of the D mode could be due to different defect concentrations or selection rules ($\mathcal{R} = 3$ vs. $\mathcal{R} = 1$) in the different tube sections or to a variation of the defect type and density.

A confirmation of the $\mathcal{R}=3$ selection rule was given by *Cançado* et al. [287]. These authors recorded Raman spectra of different types of edges of HOPG graphite crystallites that they also investigated with an atomic-force microscope (AFM). While there was no D mode far from a crystallite edge (position 3 in Fig. 64), they found a large D mode on one type of edge (position 1) and an ≈ 4 times smaller one on another type of edge (position 2), see Fig. 64b. Here the edge of the crystallite serves to break the translational symmetry, as is the case of the surface of the Ge crystal [213, 214]. From the AFM analysis the authors identified the graphite edge with a large D mode with an armchair-type edge, while the smaller D mode originated from a zigzag-type edge, see inset of Fig. 64b and a. Since all zigzag tubes are $\mathcal{R} = 1$ tubes, the D-mode scattering from the zigzag edge is forbidden by the selection rule derived in [206]. The small residual peak observed in the spectra is probably due to imperfections. Note also that the intensity of the

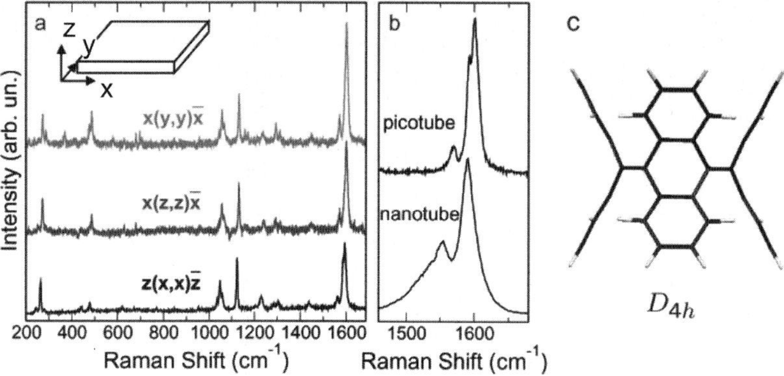

Fig. 65. (**a**) Raman spectra of a macroscopic picotube crystal for three different orientations along the crystal edges as shown in the *inset.* (**b**) a blowup of the high-energy Raman spectrum of the picotube crystal compared to that of a bundle of nanotubes at the same excitation energy (1.96 eV). The picotube spectra resemble those of nanotubes only for the HEM region. (**c**) Side view of the (4,4) picotube molecule. Shown is the idealized D_{4h} structure with inversion symmetry. The center ring of hexagons is the same as that of a (4,4) armchair nanotube. In the real molecule upper and lower wings bend away by different amounts from the z-axis of the tube. Going around the tube, larger and smaller bending angles alternate, resulting in a D_{2d} symmetry of the real molecule. After [288]

D' mode is constant in position 1 and 2. The double-resonant process giving rise to D' in graphite is the one responsible for the high-energy Raman spectra (G mode) in single-walled carbon nanotubes. This mode was predicted to be double resonant for $\mathcal{R} = 1$ and 3 tubes alike [206], in agreement with the edge Raman spectra in Fig. 64b [287].

10.2 Picotubes

A nice example of a material where a Raman double resonance does *not* occur, although the atomic structure is quite similar to those of nanotubes, are the recently investigated carbon picotubes [288]. They are molecules with the chemical formula $C_{56}H_{32}$ and D_{2d} symmetry and resemble short sections of (4,4) carbon nanotubes. They crystallize in macroscopic molecular crystals. See Fig. 65c for the structure of a single molecule; the crystal structure as deduced from X-ray diffraction can be found in [288]. These molecules have carbon-bond-related vibrational frequencies and Raman spectra quite similar to that of the corresponding nanotube. The interactions between the molecules are so weak and the picotubes so short that there is no dispersion in the vibrations. Consequently, the vibrational modes in picotube crystals do not show an excitation-energy dependence (also, the bandgap of picotubes is in the UV [289, 290]).

In Fig. 65a we show the Raman spectrum of a tiny picotube crystal for three different crystal orientations; the corresponding Raman tensors and mode symmetries were analyzed in [288]. Note the similarity of the high-energy mode in picotube crystals (upper trace) and in bundles of nanotubes (lower) in the enlarged spectrum in Fig. 65b. The trace of the picotube crystal is a simple superposition of three or more peaks with Lorentzian lineshape. In the metallic nanotubes, however, at $E_{\mathrm{exc}} = 1.96$ eV the spectrum at $\approx 1540\,\mathrm{cm}^{-1}$ in bundled nanotubes is asymmetric, as is typical for metallic nanotubes (see Sect. 8.1). At the same time, the high-energy shoulder of the largest peak in the nanotube bundle has broadened into a non-Lorentzian lineshape, which is characteristic of the double-resonance process. The lineshape of the high-energy mode in double resonance was calculated by *Maultzsch* et al. [194, 207]. In isolated nanotubes (i.e., unbundled ones) the high-energy mode becomes more symmetric and hence more like the spectrum in picotubes, but the asymmetry characteristic for double resonance remains [129]. The details of the asymmetry of this peak in metallic tubes are subject to ongoing discussion [252, 253]. Nevertheless, it is clear that in picotubes, for lack of appropriate transitions and the necessary phonon dispersion, the high-energy mode cannot be attributed to a double-resonance process.

Another important difference between picotubes and nanotubes is the absence of the shape-depolarization effect. As explained in Sect. 6.4 this effect in nanotubes is related to their large aspect ratio $\approx 10^2$ to 10^7. They exhibit a nonzero Raman intensity only if both the incoming and outgoing light are polarized along the nanotubes axis (zz configuration). Picotubes do not have a depolarization due to their shape, because they are very short (aspect ratio ≈ 1). Therefore, a Raman signal is observed in all three scattering configurations in Fig. 65a.

10.3 Two-Phonon Modes

The two-phonon Raman spectrum of graphite and carbon nanotubes extends from just above the high-energy mode to twice that frequency, say $1600\,\mathrm{cm}^{-1}$ to about $3200\,\mathrm{cm}^{-1}$ [145, 210]. This region is interesting not simply because overtones and combinations of the first-order modes can be found. Wavevector conservation plays an important role in two-phonon scattering, which we shall discuss in this section.

For the Raman double-resonance process we showed that for reasons of wavevector conservation a defect or a surface has to be involved (Sect. 6.3). This is best seen Fig. 36b, where the incident photon with ≈ 0 momentum (compared to the quasimomentum of the Brillouin zone) resonantly excites an electron–hole pair. The second resonant transition is scattering of the electron (or hole) by emission (or absorption) of a phonon (transition $a \rightarrow b$). In order to conserve quasimomentum the electron has to be scattered back to a point where its momentum is near that of the initial hole (or the hole has to be

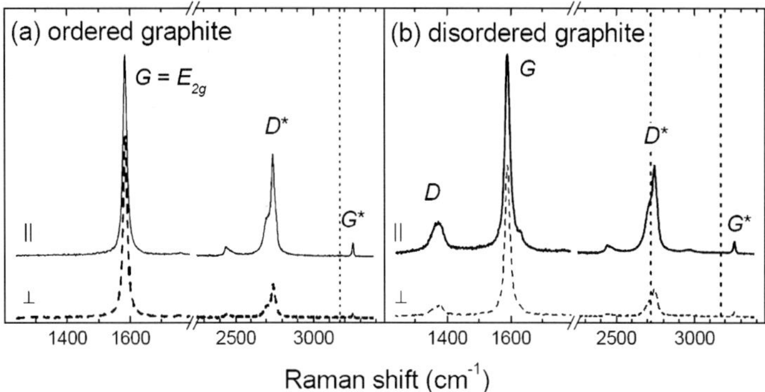

Fig. 66. The Raman spectrum of different spots of natural single-crystalline graphite (samples are $\sim 100\,\mu\text{m}$ in each dimension). (**a**) Ordered graphite shows the allowed first-order mode at $1582\,\text{cm}^{-1}$ and a second-order D^* mode at $2750\,\text{cm}^{-1}$. Note the absence of the D mode in the spectra. (**b**) Disordered graphite also shows the well-known D mode, here at $1370\,\text{cm}^{-1}$. It is apparently induced by defects in the lattice structure. Note also the overtone of the G mode at $3250\,\text{cm}^{-1}$ in both ordered and disordered graphite. Its frequency is more than twice the first-order frequency because of the overbending of the corresponding phonon branch, see text. $||$ and $\perp$ refer to parallel and perpendicular relative polarizations of the incident and scattered beams. From [125]

scattered to near the electron). The simplest form of scattering the electron back is inelastic scattering by a generic defect, which is shown as a horizontal dashed line in the figure. This is the double-resonance process that appears in the first-order Raman spectra [4, 125, 207].

It is also possible to conserve momentum by emitting (or absorbing) a second phonon. The essential difference as far as the Raman double resonance is concerned is that a defect is no longer necessary for quasimomentum conservation. Instead, the second phonon is emitted with equal and opposite momentum thus conserving momentum in the overall Raman process. It follows that an overtone in double resonance should shift with twice the rate as compared to the corresponding first-order mode. It should also have the additional characteristics of a double resonance, which are a large (relative) intensity, differences in Stokes and anti-Stokes frequencies, and non-Lorentzian lineshapes. All of these properties are indeed fulfilled for the overtone of the D mode in graphite, single and multiwalled nanotubes.

We show in Fig. 66 the role of defects in first-order and two-phonon scattering [125]. In ordered graphite (a) we see for both polarizations the strong G mode, which is the allowed mode (E_{2g} symmetry in graphite) [218]. At a frequency of $2750\,\text{cm}^{-1}$ we see the D^* mode, and at even higher frequencies the second-order peak of the G mode, commonly referred to as the G^* mode, at $3250\,\text{cm}^{-1}$. In disordered graphite, Fig. 66b, we see, in addition, the defect-induced D-mode at $1350\,\text{cm}^{-1}$ [218]. D^* can be regarded as the overtone of the D mode, which is only seen in the Raman spectra when induced by defects or when quasimomentum is conserved by a second emitted phonon. The D^* peak has approximately the same intensity in the two Raman spectra of Figs. 66a and b, whereas the first-order mode is completely absent in Fig. 66a. Hence the D^* mode is *not* induced by defects.

Owing to the lack of correlation of many of the defects the overtone D^* mode is ideally suited for a quantitative assessment of the defect density in graphite and graphite-based materials. Traditionally, this is done from Raman spectra by comparing the amplitude of the D mode to that of the G mode, which – as a first-order allowed mode – is supposed to be independent of defect density. In carbon nanotubes, however, there is still a large controversy as to whether the high-energy mode (the equivalence of the G mode in graphite) is truly a first-order mode [194, 291]; in fact, there is ample evidence that it is not. Even if it is only partially double resonant, it is not suited as normalization of the defect density, as its intensity is then partially dependent on defect density as well, which would result in a reduced defect-density dependence of the intensity ratio of the two peaks. Measurements on multiwalled carbon nanotubes with different concentrations of boron doping clearly show this effect [292]; see also Fig. 63, where the D and G-mode intensities vary along a curved nanotube.

The overtone spectra give us information about the phonon dispersion in a crystal. Because the quasimomentum conservation is fulfilled by the emission of the second phonon, the overtone spectra in a crystal reflect the density of phonon states. This effect is well known and, e.g., has led to an important conclusion as regards the phonon dispersion of graphite. In 1979 *Nemanich* and *Solin* [293] were the first to realize that the sharp peak at $3247\,\mathrm{cm}^{-1}$ in the second-order spectra of graphite was due to the overbending of the graphite phonon dispersion (see Sect. 3.3). The overbending leads to a peak in the overtone spectra at the highest phonon energies in the branch and can be estimated by $\omega_{2\mathrm{ph}} - 2\omega_{\mathrm{ph}} \approx 40\,\mathrm{cm}^{-1}$, see [125, 145, 293]. In Fig. 66 the positions of $2\omega_{\mathrm{ph}}$ are indicated by vertical dashed lines. Raman spectra of single-walled carbon nanotubes showed that the highest overtone phonon frequencies is not more than twice the frequency at the corresponding Γ point [145]. This observation is in accordance with the softening calculated from ab-initio work as compared to the zone-folding methods, see also Sect. 3.3.

In Fig. 67a we show the excitation-energy dependence of the second-order D^*-mode of single-walled nanotubes [145]. It shows a roughly twice as large shift when compared to the first-order mode, as expected from double-resonance theory [4, 219]. Again, the phenomenon is not unique to carbon nanotubes, but was shown, e.g., by *Reedyk* et al. in $Pb_2Sr_2PrCu_3O_8$, a semiconducting compound structurally related to a high-T_c superconductor. For a detailed discussion of the Raman features of this compound, see [294].

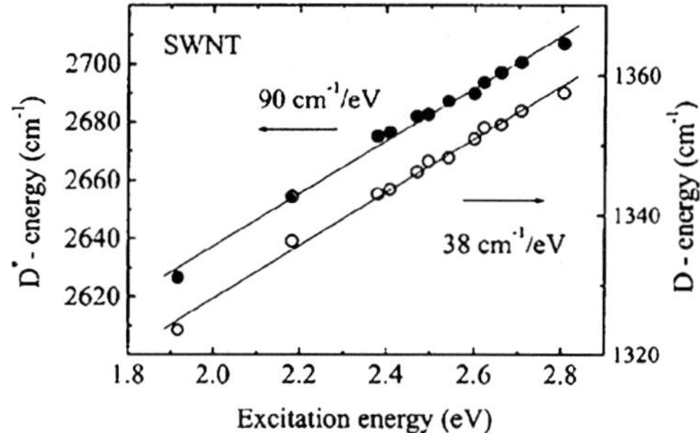

Fig. 67. The excitation-energy dependence of the overtone D^* in single-walled carbon nanotubes (*closed symbols*) as compared to that of the first-order mode (*open*). The overtone shifts with excitation energy roughly twice as fast as the first-order mode. From [145]

10.4 Phonon Dispersion

Double-resonant scattering involving large phonon q vectors may be utilized to map the phonon-dispersion relation as pointed out by *Thomsen* and *Reich* in [4]. This concept was first applied to graphite by *Saito* et al. [295], and later more completely by *Reich* and *Thomsen* [125]. The idea is based on the linear relation of the phonon wavevector q and the excitation energy in the double-resonance process (33). The (linearized) electronic dispersion is given by the Fermi velocities v_1 and v_2 and the phonon frequency $\hbar\omega_{\text{ph}}(q)$. To a simplest approximation the phonon energy may be neglected compared to excitation energy and q evaluated straightforwardly. If $\hbar\omega_{\text{ph}}(q)$ is to be taken into account, q involved in the double-resonance process may be obtained iteratively [4, 207].

Mapping the phonon dispersion in the simplest approximation corresponds to taking peaks identified as doubly resonant from Raman spectra (from their excitation-energy dependence) determining $q \approx E_1/v_i$ and plotting the so-obtained pairs of $[q, \hbar\omega_{\text{ph}}(q)]$. Figure 68 shows such a plot obtained from various experimental data extant in the literature [215,216,223,296,297]. The experimental phonon dispersion is seen to fall nicely on the calculated dispersion curves, which are ab-initio calculated curves. They, in turn, describe very well recent inelastic X-ray data on graphite [3]. For more details of the evaluation, see [125].

11 Conclusions

Raman spectroscopy is a very important and useful tool for the study and characterization of carbon nanotubes. It has contributed tremendously to the understanding of these fascinating novel carbon allotropes. We have given a detailed overview of the effects of reducing graphite to one dimension,

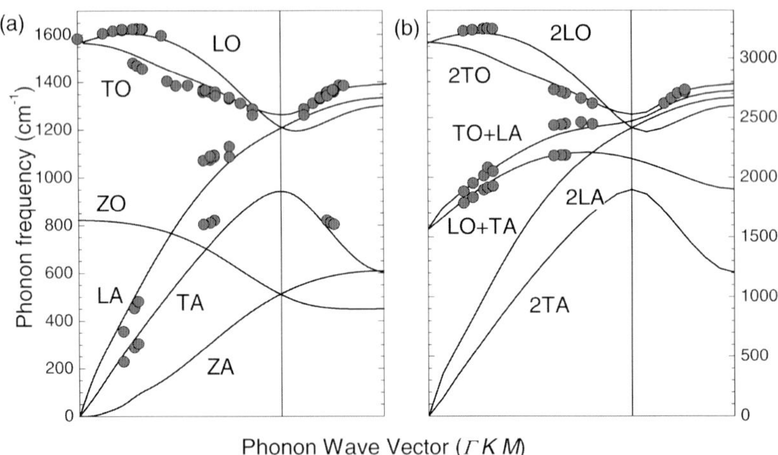

Fig. 68. (**a**) Phonon dispersion of graphite based on first-order disorder-induced double-resonant peaks in the Raman spectra. Data taken from various published experimental data in the literature, see [215, 216, 223, 296, 297] (**b**) Same as (**a**) but based on Raman peaks in the region of two-phonon spectra. *Solid lines* are dispersion curves from ab initio calculations [3] (**a**) or combinations and overtones with totally symmetric respresentations (**b**). From [125]

which leads to the density-of-states singularities, to curvature-induced effects on the phonon frequencies, and to the classification of nanotubes into families and branches with similar physical properties. Most recent progress for both physics and technology is the identification of the chirality by resolving resonance Raman spectra as a function of excitation energy *and* wave-number shift in the experimental Kataura plots. Individual or bundled metallic and semiconducting nanotubes can now easily be identified. We have given a prescription for reliably identifying the chirality of an individual single-walled carbon nanotube. Recently, excitonic energies were determined experimentally, further enhancing the understanding of this fascinating class of materials.

Carbon nanotubes are a model system for one-dimensional solids. Throughout this Chapter we demonstrated how the geometry and the large aspect ratio, as well as electron and phonon confinement, can be studied by Raman spectroscopy. The most prominent example of how geometry changes the Raman spectra is the depolarization effect. Because of the anisotropic polarizability, the totally symmetric phonons strongly dominate scattering in carbon nanotubes. An exciting manifestation of electron and phonon confinement and the interaction between the confined quasiparticles are the phonon–plasmon coupled modes, the Kohn anomalies of several phonon branches and the Peierls distortion of very small diameter nanotubes. Much of this awaits final experimental verification.

From a spectroscopic point-of-view, the introduction of Raman double resonance as a process abundant in graphite-based materials has contributed much to their understanding. It provides a basis for the systematic analysis of defects and permits determination of the phonon dispersion in at least parts of the Brillouin zone. Extension to other suitable materials is conceivable.

In this Chapter we have presented the first attempts towards a unified description of the Raman effect in single-walled carbon nanotubes. This includes excitons as intermediate states in the Raman process, electron–phonon and electron–photon coupling. We expect many further studies to appear in the near future, in particular, turning away from the band-to-band transition picture towards a model accounting for the large electron–hole interaction in carbon nanotubes and the presence of excitons at room temperature.

Acknowledgements

We are grateful to J. Maultzsch, M. Machón, J. Robertson, A. C. Ferrari, H. Telg, P. Rafailov, P. Ordejón, and M. Damnjanović for many useful discussions and collaborations on carbon nanotubes. We thank S. Piscanec and R. Krupke for preparing Figs. 16 and 50, respectively. We would like to thank Y. Miyauchi and S. Maruyama for sending us Fig. 33 prior to publication. S. R. was supported by the Oppenheimer Fund and Newnham College. Funding by the Deutsche Forschungsgemeinschaft through project Th 662/8-2 is gratefully acknowledged.

References

[1] S. Iijima: Helical microtubules of graphitic carbon, Nature **354**, 56 (1991)

[2] S. Iijima, T. Ichihasi: Single-shell carbon nanotubes of 1-nm diameter, Nature **363**, 603 (1993)

[3] J. Maultzsch, S. Reich, C. Thomsen, H. Requardt, P. Ordejón: Phys. Rev. Lett. **92**, 075501 (2004)

[4] C. Thomsen, S. Reich: Double-resonant Raman scattering in graphite, Phys. Rev. Lett. **85**, 5214 (2000)

[5] P. R. Wallace: The band theory of graphite, Phys. Rev. **71**, 622 (1947)

[6] S. Reich, J. Maultzsch, C. Thomsen, P. Ordejón: Tight-binding description of graphene, Phys. Rev. B **66**, 035412 (2002)

[7] V. N. Popov: New. J. Phys. **6**, 17 (2004)

[8] D. Sánchez-Portal, E. Artacho, J. M. Soler, A. Rubio, P. Ordejón: *Ab initio* structural, elastic, and vibrational properties of carbon nanotubes, Phys. Rev. B **59**, 12678 (1999)

[9] S. Reich, C. Thomsen, P. Ordejón: Electronic band structure of isolated and bundled nanotubes, Phys. Rev. B **65**, 155411 (2002)

[10] C. D. Spataru, S. Ismail-Beigi, L. X. Benedict, S. G. Louie: Phys. Rev. Lett. **92**, 077402 (2004)

[11] V. Perebeinos, J. Tersoff, P. Avouris: Phys. Rev. Lett. **94**, 027402 (2005)

[12] F. Wang, G. Dukovic, L. E. Brus, T. F. Heinz: The optical resonances in carbon nanotubes arise from excitons, Science **308**, 838 (2005)
[13] J. Maultzsch, C. Pomraenke, S. Reich, E. Chang, D. Prezzi, A. Ruini, E. Molinari, M. Strano, C. Thomsen, C. Lienau: Exciton binding energies from two-photon photoluminescence in single-walled carbon nanotubes, Phys. Rev. B **72**, 241402 (2005)
[14] Q. Zhao, H. D. Wagner: Raman spectroscopy of carbon-nanotube-based composites, in A. C. Ferrari, J. Robertson (Eds.): *Raman Spectroscopy in Carbon: from Nanotubes to Diamond*, vol. 362, Philosophical Transactions of the Royal Society A (The Royal Society, London 2004) p. 2407
[15] R. Krupke, F. Hennrich, H. v. Löhneysen, M. M. Kappes: Science **301**, 344 (2003)
[16] S. Reich, L. Li, J. Robertson: Control the chirality of carbon nanotubes by epitaxial growth, Chem. Phys. Lett. **421**, 469–472 (2006)
[17] N. N. Ilichev: SWNTs as a passive Q-switches for self mode-locking of solid state lasers, in H. Kuzmany, J. Fink, M. Mehring, S. Roth (Eds.): *Electronic Properties of Novel Materials*, AIP Conference Proceedings (AIP, College Park 2005)
[18] A. B. Dalton, S. Collins, E. Muñoz, J. M. Razal., V. H. Ebron, J. P. Ferraris, J. N. Coleman, B. G. Kim, R. H. Baughman: Super-tough carbon-nanotube fibres – these extraordinary composite fibres can be woven into electronic textiles, Nature **423**, 703 (2003)
[19] R. H. Baughman, C. Cui, A. Zakhidov, Z. Iqbal, J. N. Barisci, G. M. Spinks, G. G. Wallace, A. Mazzoldi, D. D. Rossi, A. G. Rinzler, O. Jaschinski, S. Roth, M. Kertesz: Carbon nanotube actuators, Science **284**, 1340 (1999)
[20] S. Reich, C. Thomsen, J. Maultzsch: *Carbon Nanotubes: Basic Concepts and Physical Properties* (Wiley-VCH, Berlin 2004)
[21] M. S. Dresselhaus, G. Dresselhaus, P. C. Eklund: *Science of Fullerenes and Carbon Nanotubes* (Academic, San Diego 1995)
[22] M. S. Dresselhaus, G. Dresselhaus, P. Avouris (Eds.): *Carbon Nanotubes: Synthesis, Structure, Properties, and Applications*, vol. 80, Topics in Applied Physics (Springer, Berlin, Heidelberg 2001)
[23] R. Saito, G. Dresselhaus, M. S. Dresselhaus: *Physical Properties of Carbon Nanotubes* (Imperial College Press, London 1998)
[24] P. J. F. Harris: *Carbon Nanotubes and Related Structures* (Cambridge University Press, Cambridge 1999)
[25] T. Ando: Theory of electronic states and transport in carbon nanotubes, J. Phys. Soc. Jpn. **74**, 777 (2005)
[26] M. S. Dresselhaus, G. Dresselhaus, R. Saito, A. Jorio: Raman spectroscopy of carbon nanotubes, Phys. Rep. **409**, 47 (2005)
[27] R. P. Raffaelle, B. J. Landi, J. D. Harris, S. G. Bailey, A. F. Hepp: Carbon nanotubes for power applications, Mater. Sci. Eng. B **116**, 233 (2005)
[28] J. Wang: Carbon-nanotube based electrochemical biosensors: A review, Electroanal. **17**, 7 (2005)
[29] U. D. Venkateswaran: Squeezing carbon nanotubes, phys. stat. sol. B **241**, 3345 (2004)
[30] C. Thomsen, H. Kataura (Eds.): *Focus on Carbon Nanotubes*, vol. 5, The New Journal of Physics (Institute of Physics and Deutsche Physikalische Gesellschaft, London 2003)

[31] A. C. Ferrari, J. Robertson (Eds.): *Raman Spectroscopy in Carbon: from Nanotubes to Diamond*, vol. 362, Philosophical Transactions of the Royal Society A (The Royal Society, London 2004)
[32] Y. Baskin, L. Meyer: Lattice constant of graphite at low temperatures, Phys. Rev. B **100**, 544 (1955)
[33] Y. X. Zhao, I. L. Spain: X-ray diffraction data for graphite to 20 GPa, Phys. Rev. B **40**, 993 (1989)
[34] M. Hanfland, H. Beister, K. Syassen: Graphite under pressure: Equation of state and first-order Raman modes, Phys. Rev. B **39**, 12 598 (1989)
[35] J. C. Boettger: All-electron full-potential calculation of the electronic band structure, elastic constants, and equation of state for graphite, Phys. Rev. B **55**, 11202 (1997)
[36] J. Kürti, V. Zólyomi, M. Kertesz, G. Sun: New J. Phys. **5**, 125 (2003)
[37] A. Burian, J. Koloczek, J. Dore, A. Hannon, J. Nagy, A. Fonseca: Radial distribution function analysis of spatial atomic correlations in carbon nanotubes, Diam. Relat. Mater. **13**, 1261 (2004)
[38] S. Reich, C. Thomsen, P. Ordejón: Eigenvectors of chiral nanotubes, Phys. Rev. B **64**, 195416 (2001)
[39] J.-C. Charlier, P. Lambin: Electronic structure of carbon nanotubes with chiral symmetry, Phys. Rev. B **57**, 15037 (1998)
[40] A. Thess, R. Lee, P. Nikolaev, H. Dai, P. Petit, J. Robert, C. Xu, Y. H. Lee, S. G. Kim, A. G. Rinzler, D. T. Colbert, G. E. Scuseria, D. Tománek, J. E. Fischer, R. E. Smalley: Crystalline ropes of metallic carbon nanotubes, Science **273**, 483 (1996)
[41] K. Hata, D. N. Futaba, K. Mizuno, T. Namai, M. Yumura, S. Iijima: Water-assisted highly efficient synthesis of impurity-free single-walled carbon nanotubes, Science **306**, 1362 (2004)
[42] L. X. Zheng, M. J. O'Connell, S. K. Doorn, X. Z. Liao, Y. H. Zhao, E. A. Akhadov, M. A. Hoffbauer, B. J. Roop, Q. X. Jia, R. C. Dye, D. E. Peterson, S. M. Huang, J. Liu, et al.: Ultralong single-wall carbon nanotubes, Nature Mater. **3**, 673 (2004)
[43] M. J. Bronikowski, P. A. Willis, D. T. Colbert, K. A. Smith, R. E. Smalley: J. Vac. Sci. Technol. A **19**, 1800 (2001)
[44] S. M. Bachilo, L. Balzano, J. E. Herrera, F. Pompeo, D. E. Resasco, R. B. Weisman: J. Am. Chem. Soc. **125**, 11186 (2003)
[45] Y. Miyauchi, S. Chiashi, Y. Murakami, Y. Hayashida, S. Maruyama: Chem. Phys. Lett. **387**, 198 (2004)
[46] N. Wang, Z. K. Tang, G. D. Li, J. S. Chen: Single-walled 4 Å carbon nanotube arrays, Nature (London) **408**, 50 (2000)
[47] R. R. Bacsa, A. Peigney, C. Laurent, P. Puech, W. S. Bacsa: Phys. Rev. B **65**, 161404 (2002)
[48] S. Bandow, M. Takizawa, K. Hirahara, M. Yudasaka, S. Iijima: Chem. Phys. Lett. **337**, 48 (2001)
[49] T. W. Ebbesen, P. M. Ajayan: Large-scale synthesis of carbon nanotubes, Nature **358**, 220 (1992)
[50] D. S. Bethune, C. H. Kiang, M. S. deVries, G. Gorman, R. Savoy, J. Vazques, R. Beyers: Cobalt-catalyzed growth of carbon nanotubes with single-atomic-layer, Nature **363**, 605 (1993)

[51] C. Journet, W. K. Maser, P. Bernier, A. Loiseau, M. L. de la Chapelle, S. Lefrant, P. Deniard, R. Lee, J. E. Fischer: Large-scale production of single-walled carbon nanotubes by the electric-arc technique, Nature **388**, 756 (1997)
[52] H. Dai: in M. S. Dresselhaus, G. Dresselhaus, P. Avouris (Eds.): *Carbon Nanotubes*, vol. 80, Topics in Applied Physics (Springer, Berlin, Heidelberg 2001) p. 29
[53] M. Terrones: Carbon nanotubes: synthesis and properties, electronic devices and other emerging applications, Int. Mater. Rev. **49**, 325 (2005)
[54] A. A. Puretzky, D. B. Geohegan, S. Jesse, I. N. Ivanov, G. Eres: In situ measurements and modeling of carbon nanotube array growth kinetics during chemical vapor deposition, Appl. Phys. A: Mater. Sci. Process. **81**, 223 (2005)
[55] M. C. Schabel, J. L. Martins: Energetics of interplanar binding in graphite, Phys. Rev. B **46**, 7185 (1992)
[56] L. A. Girifalco, M. Hodak: Van der Waals binding energies in graphitic structures, Phys. Rev. B **65**, 125404 (2002)
[57] R. Z. H. Ulbricht, T. Hertel: Interlayer cohesive energy of graphite from thermal desorption of polyaromatic hydrocarbons, Phys. Rev. B **69**, 155406 (2004)
[58] S. Reich, C. Thomsen, P. Ordejón: Phys. Rev. B **65**, 153407 (2002)
[59] M. J. O'Connell, S. M. Bachilo, C. B. Huffman, V. C. Moore, M. S. Strano, E. H. Haroz, K. L. Rialon, P. J. Boul, W. H. Noon, C. Kittrell, J. Ma, R. H. Hauge, R. B. Weisman, R. E. Smalley: Band gap fluorescence from individual single-walled carbon nanotubes, Science **297**, 593 (2002)
[60] S. M. Bachilo, M. S. Strano, C. Kittrell, R. H. Hauge, R. E. Smalley, R. B. Weisman: Structure-assigned optical spectra of single-walled carbon nanotubes, Science **298**, 2361 (2002)
[61] S. Lebedkin, F. Hennrich, T. Skipa, M. M. Kappes: Near-infrared photoluminescence of single-walled carbon nanotubes prepared by the laser vaporization method, J. Phys. Chem. B **107**, 1949 (2003)
[62] T. Hertel, A. Hagen, V. Talalaev, K. Arnold, F. Hennrich, M. Kappes, S. Rosenthal, J. McBride, H. Ulbricht, E. Flahaut: Spectroscopy of single- and double-wall carbon nanotubes in different environments, Nano Lett. **5**, 511 (2005)
[63] H. Telg, J. Maultzsch, S. Reich, F. Hennrich, C. Thomsen: Phys. Rev. Lett. **93**, 177401 (2004)
[64] C. Fantini, A. Jorio, M. Souza, M. S. Strano, M. S. Dresselhaus, M. A. Pimenta: Phys. Rev. Lett. **93**, 147406 (2004)
[65] M. S. Strano, S. K. Doorn, E. H. Haroz, C. Kittrell, R. H. Hauge, R. E. Smalley: Assignment of (n, m) Raman and optical features of metallic single-walled carbon nanotubes, Nano Lett. **3**, 1091 (2003)
[66] Z. F. Ren, Z. P. Huang, J. W. Xu, J. H. Wang, P. Bush, M. P. Siegal, P. N. Provencio: Synthesis of large arrays of well-aligned carbon nanotubes on glass, Science **282**, 1105 (1998)
[67] M. Chhowalla, K. B. K. Teo, C. Ducati, N. L. Rupesinghe, G. A. J. Amaratunga, A. C. Ferrari, D. Roy, J. Robertson, W. I. Milne: Growth process conditions of vertically aligned carbon nanotubes using plasma enhanced chemical vapor deposition, Appl. Phys. Lett. **90**, 5308 (2001)

[68] J. Lefebvre, Y. Homma, P. Finnie: Bright band gap photoluminescence from unprocessed single-walled carbon nanotubes, Phys. Rev. Lett. **90**, 217401 (2003)

[69] J. C. Meyer, M. Paillet, G. S. Duesberg, S. Roth: Electron diffraction analysis of individual single-walled carbon nanotubes, Ultramicroscopy **106**, 176 (2005)

[70] S. Hofmann, C. Ducati, B. Kleinsorge, J. Robertson: Direct growth of aligned carbon nanotube field emitter arrays onto plastic substrates, Appl. Phys. Lett. **83**, 4661 (2003)

[71] L. Guan, K. Suenaga, Z. Shi, Z. Gu, S. Iijima: Direct imaging of the alkali metal site in K-doped fullerene peapods, Phys. Rev. Lett. **94**, 045502 (2005)

[72] L. Henrard, A. Loiseau, C. Journet, P. Bernier: Synth. Met. **103**, 2533 (1999)

[73] L. Henrard, A. Loiseau, C. Journet, P. Bernier: Study of the symmetry of single-wall nanotubes by electron diffraction, Eur. Phys. B **13**, 661 (2000)

[74] S. Reich, C. Thomsen, J. Robertson: Exciton resonances quench the photoluminescence of zigzag carbon nanotubes, Phys. Rev. Lett. **95**, 077402 (2005)

[75] J. Jiang, R. Saito, A. Grüneis, S. G. Chou, G. G. Samsonidze, A. Jorio, G. Dresselhaus, M. S. Dresselhaus: Photoexcited electron relaxation processes in single-wall carbon nanotubes, Phys. Rev. B **71**, 045417 (2005)

[76] B. W. Smith, M. Monthioux, D. E. Luzzi: Encapsulated C_{60} in carbon nanotubes, Nature **396**, 323 (1998)

[77] H. Kataura, Y. Maniwa, T. Kodama, K. Kikuchi, K. Hirahara, K. Suenaga, S. Iijima, S. Suzuki, Y. Achiba, W. Kratschmer: Synth. Met. **121**, 1195 (2001)

[78] T. Pichler, H. Kuzmany, H. Kataura, Y. Achiba: Metallic polymers of C_{60} inside single-walled carbon nanotubes, Phys. Rev. Lett. **87**, 267401 (2001)

[79] J. Sloan, M. C. Novotny, S. R. Bailey, G. Brown, C. Xu, V. C. Williams, S. Friedrichs, E. Flahaut, R. L. Callender, A. P. E. York, K. S. Coleman, M. L. H. Green, R. E. Dunin-Borkowskib, J. L. Hutchison: Two layer 4:4 coordinated KI crystals grown within single walled carbon nanotubes, Chem. Phys. Lett. **329**, 61 (2000)

[80] J. Diefenbach, T. P. Martin: Model calculations of alkali halide clusters, J. Chem. Phys. **83**, 4585 (1985)

[81] M. Damnjanović, I. Milošević, T. Vuković, R. Sredanović: Full symmetry, optical activity, and potentials of single-wall and multiwall nanotubes, Phys. Rev. B **60**, 2728 (1999)

[82] M. Damnjanović, I. Milošević, T. Vuković, R. Sredanović: Symmetry and lattices of single-wall nanotubes, J. Phys. A **32**, 4097 (1999)

[83] J. W. Mintmire, B. I. Dunlap, C. T. White: Are fullerene tubules metallic?, Phys. Rev. Lett. **68**, 631 (1992)

[84] M. Damnjanović, T. Vuković, I. Milošević: Modified group projectors: Tight binding method, J. Phys. A: Math. Gen. **33**, 6561 (2000)

[85] J. Maultzsch, S. Reich, C. Thomsen, E. Dobardžić, I. Milošević, M. Damnjanović: Phonon dispersion of carbon nanotubes, Solid State Commun. **121**, 471 (2002)

[86] T. Vuković, I. Milošević, M. Damnjanović: Carbon nanotubes band assignation, topology, Bloch states, and selection rules, Phys. Rev. B **65**, 045418 (2002)

[87] I. Milošević, T. Vuković, S. Dmitrović, M. Damnjanović: Phys. Rev. B **67**, 165418 (2003)

[88] E. Chang, G. Bussi, A. Ruini, E. Molinari: Excitons in carbon nanotubes: An ab initio symmetry-based approach, Phys. Rev. Lett. **92**, 196401 (2004)

[89] T. Inui, Y. Tanabe, Y. Onodera: *Group Theory and its Application in Physics* (Springer Verlag, Berlin Heidelberg New York 1996)

[90] E. B. Wilson, J. C. Decius, P. C. Cross: *Molecular Vibrations* (Dover, New York 1980)

[91] M. S. Dresselhaus, G. Dresselhaus, R. Saito: Carbon fibers based on C_{60} and their symmetry, Phys. Rev. B **45**, 6234 (1992)

[92] R. A. Jishi, M. S. Dresselhaus, G. Dresselhaus: Symmetry properties of chiral carbon nanotubes, Phys. Rev. B **47**, 16671 (1993)

[93] M. S. Dresselhaus, P. C. Eklund: Adv. Phys. **49**, 705 (2000)

[94] A. Jorio, R. Saito, G. Dresselhaus, M. S. Dresselhaus: Determination of nanotube properties by Raman spectroscopy, in A. C. Ferrari, J. Robertson (Eds.): *Raman Spectroscopy in Carbons: From Nanotubes to Diamond* (Philosophical Transaction of the Royal Society 2004) p. 2311

[95] R. A. Jishi, L. Venkataraman, M. S. Dresselhaus, G. Dresselhaus: Phonon modes in carbon nanotubules, Chem. Phys. Lett. **209**, 77–82 (1993)

[96] D. L. Rousseau, R. P. Bauman, S. P. S. Porto: Normal mode determination in crystals, J. Raman Spectrosc. **10**, 253 (1981)

[97] A. Fainstein, P. Etchegoin, M. P. Chamberlain, M. Cardona, K. Tötemeyer, K. Eberl: Selection rules and dispersion of GaAs/AlAs multiple-quantum-well optical phonons studied by Raman scattering in right-angle, forward, and backscattering in-plane geometries, Phys. Rev. B **51**, 14 448 (1995)

[98] B. Jusserand, M. Cardona: Superlattices and other microstructures, in M. Cardona, G. Güntherodt (Eds.): *Light Scattering in Solids V*, vol. 66, Topics in Applied Physics (Springer, Berlin 1989) p. 49

[99] G. S. Duesberg, I. Loa, M. Burghard, K. Syassen, S. Roth: Polarized Raman spectroscopy of individual single-wall carbon nanotubes, Phys. Rev. Lett. **85**, 5436 (2000)

[100] A. Ugawa, A. G. Rinzler, D. B. Tanner: Far-infrared gaps in single-wall carbon nanotubes, Phys. Rev. B **60**, R11 305 (1999)

[101] S. Reich, C. Thomsen, G. S. Duesberg, S. Roth: Intensities of the Raman active modes in single and multiwall nanotubes, Phys. Rev. B **63**, R041401 (2001)

[102] A. Jorio, A. G. S. Filho, V. W. Brar, A. K. Swan, M. S. Ünlü, B. B. Goldberg, A. Righi, J. H. Hafner, C. M. Lieber, R. Saito, G. Dresselhaus, M. S. Dresselhaus: Phys. Rev. B **65**, 121402 (2002)

[103] E. Dobardžić, I. M. B. Nikolić, T. Vuković, M. Damnjanović: Single-wall carbon nanotubes phonon spectra: Symmetry-based calculations, Phys. Rev. B **68**, 045 408 (2003)

[104] M. Damnjanović, E. Dobardžić, I. Milošević: Chirality dependence of the radial breathing mode: a simple model, J. Phys.: Condens. Matter **16**, L505 (2004)

[105] G. D. Mahan: Phys. Rev. B **65**, 235402 (2002)

[106] R. Nicklow, N. Wakabayashi, H. G. Smith: Lattice dynamics of pyrolytic graphite, Phys. Rev. B **5**, 4951 (1972)

[107] O. L. Blakslee, D. G. Proctor, E. J. Seldin, G. B. Spence, T. Weng: Elastic constants of compression-annealed pyrolytic graphite, J. Appl. Phys. **41**, 3373 (1970)

[108] E. J. Seldin, C. W. Nezbeda: Elastic constants and electron-microscope observations of neutron-irradiated compression-annealed pyrolytic and single-crystal graphite, J. Appl. Phys. **41**, 3389 (1970)

[109] L. Wirtz, A. Rubio, R. A. de la Concha, A. Loiseau: Phys. Rev. B **68**, 45425 (2003)

[110] S. Niyogi, M. A. Hamon, H. Hu, B. Zhao, P. Bhowmik, R. Sen, M. E. Itkis, R. Haddon: Chemistry of single-walled nanotubes, Acc. Chem. Res. **35**, 1105 (2002)

[111] A. Jorio, C. Fantini, M. A. Pimenta, R. B. Capaz, G. G. Samsonidze, G. Dresselhaus, M. S. Dresselhaus, J. Jiang, N. Kobayashi, A. Grüneis, R. Saito: Resonance Raman spectroscopy (n, m)-dependent effects in small-diameter single-wall carbon nanotubes, Phys. Rev. B **71**, 075 401 (2005)

[112] V. N. Popov, L. Henrard: Phys. Rev. B **65**, 235415 (2002)

[113] E. Dobardžić, J. Maultzsch, I. Milošević, C. Thomsen, M. Damnjanović: phys. stat. sol. (b) **237**, R7 (2003)

[114] G. Wu, J. Zhou, J. Dong: Radial-breathing-like phonon modes of double-walled carbon nanotubes, Phys. Rev. B **72**, 115 418 (2005)

[115] C. Kramberger, R. Pfeiffer, H. Kuzmany, V. Zólyomi, J. Kürti: Assignment of chiral vectors in carbon nanotubes, Phys. Rev. B **68**, 235 404 (2003)

[116] R. Pfeiffer, H. Kuzmany, C. Kramberger, C. Schaman, T. Pichler, H. Kataura, Y. Achiba, J. Kürti, V. Zólyomi: Phys. Rev. Lett. **90**, 225501 (2003)

[117] A. Kasuya, Y. Sasaki, Y. Saito, K. Tohji, Y. Nishina: Evidence for size-dependent discrete dispersions in single-wall nanotubes, Phys. Rev. Lett. **78**, 4434 (1997)

[118] O. Dubay, G. Kresse: Accurate density functional calculations for the phonon dispersion relations of graphite layer and carbon nanotubes, Phys. Rev. B **67**, 035 401 (2003)

[119] C. Thomsen, S. Reich, P. Ordejón: *Ab initio* determination of the phonon deformation potentials of graphene, Phys. Rev. B **65**, 073403 (2002)

[120] S. A. Solin, A. K. Ramdas: Raman spectrum of diamond, Phys. Rev. B **1**, 1687 (1970)

[121] J. Kulda, H. Kainzmaier, D. Strauch, B. Dorner, M. Lorenzen, M. Krisch: Phys. Rev. B **66**, 241202(R) (2002)

[122] S. Reich, A. C. Ferrari, R. Arenal, A. Loiseau, I. Bello, J. Robertson: Resonant Raman scattering in cubic and hexagonal boron nitride, Phys. Rev. B **71**, 205201 (2005)

[123] G. Kern, G. Kress, J. Hafner: *Ab-initio* lattice dynamics and phase diagram of boron nitride, Phys. Rev. B **59**, 8851 (1999)

[124] S. Reich, H. Jantoljak, C. Thomsen: Shear strain in carbon nanotubes under hydrostatic pressure, Phys. Rev. B **61**, R13 389 (2000)

[125] S. Reich, C. Thomsen: Raman spectroscopy of graphite, in A. C. Ferrari, J. Robertson (Eds.): *Raman Spectroscopy in Carbons: From Nanotubes to Diamond* (Philosophical Transaction of the Royal Society 2004) p. 2271

[126] R. Saito, A. Jorio, J. H. Hafner, C. M. Lieber, M. Hunter, T. McClure, G. Dresselhaus, M. S. Dresselhaus: Chirality-dependent G-band Raman intensity of carbon nanotubes, Phys. Rev. B **64**, 085312 (2001)

[127] S. Piscanec, M. Lazzeri, F. Mauri, A. C. Ferrari, J. Robertson: Phys. Rev. Lett. **93**, 185503 (2004)

[128] M. Lazzeri, S. Piscanec, F. Mauri, A. C. Ferrari, J. Robertson: Electron transport and hot phonons in carbon nanotubes, Phys. Rev. Lett. **95**, 236802 (2005)

[129] J. Maultzsch, S. Reich, U. Schlecht, C. Thomsen: High-energy phonon branches of an individual metallic carbon nanotube, Phys. Rev. Lett. **91**, 087402 (2003)

[130] Z. M. Li, V. N. Popov, Z. K. Tang: A symmetry-adapted force-constant lattice-dynamical model for single-walled carbon nanotubes, Solid State Commun. **130**, 657 (2005)

[131] S. D. M. Brown, A. Jorio, P. Corio, M. S. Dresselhaus, G. Dresselhaus, R. Saito, K. Kneipp: Origin of the Breit-Wigner-Fano lineshape of the tangential G-band feature of metallic carbon nanotubes, Phys. Rev. B **63**, 155414 (2001)

[132] K. Kempa: Phys. Rev. B **66**, 195406 (2002)

[133] O. Dubay, G. Kresse, H. Kuzmany: Phys. Rev. Lett. **88**, 235506 (2002)

[134] P. Y. Yu, M. Cardona: *Fundamentals of Semiconductors* (Springer-Verlag, Berlin 1996)

[135] M. Damnjanović, T. Vuković, I. Milošević: Fermi level quantum numbers and secondary gap of conducting carbon nanotubes, Solid State Commun. **116**, 265 (2000) Table 1 in the reference contains a small error: For chiral tubes and the $\tilde{k}$ quantum numbers $\tilde{k}_F = 2q\pi/3na$ for $\mathcal{R} = 3$ tubes and $\tilde{k}_F = 0$ for $\mathcal{R} = 1$ nanotubes.

[136] X. Blase, L. X. Benedict, E. L. Shirley, S. G. Louie: Hybridization effects and metallicity in small radius carbon nanotubes, Phys. Rev. Lett. **72**, 1878 (1994)

[137] A. Kleiner, S. Eggert: Curvature, hybridization, and STM images of carbon nanotubes, Phys. Rev. B **64**, 113402 (2001)

[138] W. Kohn: Image of the Fermi Surface in the Vibration Spectrum of a Metal, Phys. Rev. Lett. **2**, 393 (1959)

[139] M. V. Klein: Raman studies of phonon anomalies in transition-metal compounds, in M. Cardona, G. Güntherodt (Eds.): *Light Scattering in Solids III*, vol. 51, Topics in Applied Physics (Springer, Berlin 1982) p. 121

[140] K.-P. Bohnen, R. Heid, H. J. Liu, C. T. Chan: Lattice dynamics and electron-phonon interaction in (3,3) carbon nanotubes, Phys. Rev. Lett. **93**, 245501 (2004)

[141] S. Piscanec, M. Lazzeri, A. C. Ferrari, F. Mauri, J. Robertson: private communication

[142] D. Connétable, G.-M. Rignanese, J.-C. Charlier, X. Blase: Room temperature Peierls distortion in small diameter nanotubes, Phys. Rev. Lett. **94**, 015503 (2005)

[143] J. Kürti, G. Kresse, H. Kuzmany: First-principles calculations of the radial breathing mode of single-wall carbon nanotubes, Phys. Rev. B **58**, 8869 (1998)

[144] M. Machón, S. Reich, H. Telg, J. Maultzsch, P. Ordejón, C. Thomsen: Phys. Rev. B **71**, 35416 (2005)

[145] C. Thomsen: Second-order Raman spectra of single and multi-walled carbon nanotubes, Phys. Rev. B **61**, 4542 (2000)

[146] A. C. Ferrari, J. Robertson: Raman spectroscopy of amorphous, nanostructured, diamond-like carbon, and nanodiamond, in (2004) p. 2477 of Ref. [31]

[147] H. H. Gommans, J. W. Alldredge, H. Tashiro, J. Park, J. Magnuson, A. G. Rinzler: Fibers of aligned single-walled carbon nanotubes: Polarized Raman spectroscopy, J. Appl. Phys. **88**, 2509 (2000)
[148] N. Hamada, S.-I. Sawada, A. Oshiyama: New one-dimensional conductors: Graphitic microtubules, Phys. Rev. Lett. **68**, 1579 (1992)
[149] R. Saito, M. Fujita, G. Dresselhaus, M. S. Dresselhaus: Electronic structure of graphene tubules based on C_{60}, Phys. Rev. B **46**, 1804 (1992)
[150] S. Reich, C. Thomsen: Chirality dependence of the density-of-states singularities in carbon nanotubes, Phys. Rev. B **62**, 4273 (2000)
[151] G. G. Samsonidze, R. Saito, A. Jorio, M. A. Pimenta, A. G. Souza, A. Gruneis, G. Dresselhaus, M. S. Dresselhaus: The concept of cutting lines in carbon nanotube science, J. Nanosci. Nanotech. **3**, 431 (2003)
[152] R. F. Willis, B. Feuerbacher, B. Fitton: Graphite conduction band states from secondary electron emission spectra, Phys. Lett. A **34**, 231 (1971)
[153] J. W. Mintmire, C. T. White: Universal density of states for carbon nanotubes, Phys. Rev. Lett. **81**, 2506 (1998)
[154] M. Machon, S. Reich, C. Thomsen, D. Sánchez-Portal, P. Ordejón: Phys. Rev. B **66**, 155410 (2002)
[155] Z. M. Li, Z. K. Tang, H. J. Liu, N. Wang, C. T. Chan, R. Saito, S. Okada, G. D. Li, J. S. Chen, N. Nagasawa, S. Tsuda: Polarized absorption spectra of single-walled 4 Å carbon nanotubes aligned in channels of an $AlPO_4$-5 single crystal, Phys. Rev. Lett. **87**, 127401 (2001)
[156] T. Miyake, S. Saito: Quasiparticle band structure of carbon nanotubes, Phys. Rev. B **68**, 155424 (2003)
[157] V. Barone, G. E. Scuseria: Theoretical study of the electronic properties of narrow single-walled carbon nanotubes: Beyond the local density approximation, J. Chem. Phys. **121**, 10376 (2004)
[158] V. N. Popov, L. Henrard: Comparative study of the optical properties of single-walled carbon nanotubes within orthogonal and nonorthogonal tight-binding models, Phys. Rev. B **70**, 115470 (2004)
[159] V. Zólyomi, J. Kürti: First-principles calculations for the electronic band structures of small diameter single-wall carbon nanotubes, Phys. Rev. B **70**, 085403 (2004)
[160] V. Perebeinos, J. Tersoff, P. Avouris: Phys. Rev. Lett. **92**, 257402 (2004)
[161] H. Zhao, S. Mazumdar: Electron-electron interaction effects on the optical excitations of semiconducting single-walled carbon nanotubes, Phys. Rev. Lett. **95**, 157 402 (2005)
[162] A. Grüneis, R. Saito, G. G. Samsonidze, T. Kimura, M. A. Pimenta, A. Jorio, A. G. S. Filho, G. Dresselhaus, M. S. Dresselhaus: Phys. Rev. B **67**, 165402 (2003)
[163] C. T. White, J. W. Mintmire: Density of states reflects diameter in nanotubes, Nature (London) **394**, 29 (1998)
[164] T. W. Odom, J. L. Huang, P. Kim, C. M. Lieber: Atomic structure and electronic properties of single-walled carbon nanotubes, Nature **391**, 62 (1998)
[165] J. W. G. Wildöer, L. C. Venema, A. G. Rinzler, R. E. Smalley, C. Dekker: Electronic structure of atomically resolved carbon nanotubes, Nature **391**, 59 (1998)
[166] H. Ajiki, T. Ando: Carbon nanotubes: Optical absorption in Aharonov-Bohm flux, Jpn. J. Appl. Phys. **Suppl. 34-1**, 107 (1994)

[167] H. Kataura, Y. Kumazawa, Y. Maniwa, I. Umezu, S. Suzuki, Y. Ohtsuka, Y. Achiba: Optical properties of single-wall carbon nanotubes, Synth. Met. **103**, 2555 (1999)

[168] P. M. Rafailov, H. Jantoljak, C. Thomsen: Electronic transitions in single-walled carbon nanotubes: A resonance Raman study, Phys. Rev. B **61**, 16 179 (2000)

[169] R. Saito, G. Dresselhaus, M. S. Dresselhaus: Trigonal warping effect of carbon nanotubes, Phys. Rev. B **61**, 2981 (2000)

[170] V. N. Popov, L. Henrard, P. Lambin: Nano Lett. **4**, 1795 (2004)

[171] R. B. Capaz, C. D. Spataru, P. Tagney, M. L. Cohen, S. G. Louie: Temperature dependence of the band gap of semiconducting carbon nanotubes, Phys. Rev. Lett. **94**, 036801 (2005)

[172] X.-F. He: Excitons in anisotropic solids: The model of fractional-dimensional space, Phys. Rev. B **43**, 2063 (1991)

[173] P. Lefebvre, P. Cristol, H. Mathieu: Unified formulation of excitonic spectra of semiconductor quantum wells, superlattices, and quantum wires, Phys. Rev. B **48**, 17308 (1993)

[174] C. L. Kane, E. J. Mele: Phys. Rev. Lett. **93**, 197402 (2004)

[175] S. G. Chou, F. Plentz, J. Jiang, R. Saito, D. Nezich, H. B. Ribeiro, A. Jorio, M. A. Pimenta, G. G. Samsonidze, A. P. Santos, M. Zheng, G. B. Onoa, E. D. Semke, G. Dresselhaus, M. S. Dresselhaus: Phonon-assisted excitonic recombination channels observed in DNA-wrapped carbon nanotubes using photoluminescence spectroscopy, Phys. Rev. Lett. **94**, 127402 (2005)

[176] S. Lebedkin, K. Arnold, F. Hennrich, R. Krupke, B. Renker, M. M. Kappes: FTIR-luminescence mapping of dispersed single-walled carbon nanotubes, New J. Phys. **5**, 140 (2003)

[177] X. H. Qiu, M. Freitag, V. Perebeinos, P. Avouris: Photoconductivity spectra of single-carbon nanotubes: Implications on the nature of their excited states, Nano Lett. **5**, 749 (2005)

[178] J.-S. Lauret, C. Voisin, G. Cassabois, C. Delalande, P. Roussignol, O. Jost, L. Capes: Phys. Rev. Lett. **90**, 57404 (2003)

[179] T. Hertel, G. Moos: Phys. Rev. Lett. **84**, 5002 (2000)

[180] G. N. Ostojic, S. Zaric, J. Kono, M. S. Strano, V. C. Moore, R. H. Hauge, R. E. Smalley: Phys. Rev. Lett. **92**, 117402 (2004)

[181] S. Reich, M. Dworzak, A. Hoffmann, C. Thomsen, M. S. Strano: Excited-state carrier lifetime in single-walled carbon nanotubes, Phys. Rev. B **71**, 033402 (2005)

[182] O. J. Korovyanko, C.-X. Sheng, Z. V. Vardeny, A. B. Dalton, R. H. Baughman: Phys. Rev. Lett. **92**, 17403 (2004)

[183] M. Cardona, G. Güntherodt (Eds.): *Light Scattering in Solids I–VIII* (Springer, Berlin, Heidelberg 1982–2000)

[184] M. Cardona: Resonance phenomena, in M. Cardona, G. Güntherodt (Eds.): *Light Scattering in Solids II*, vol. 50, Topics in Applied Physics (Springer, Berlin 1982) p. 19

[185] R. M. Martin, L. M. Falicov: Resonant Raman scattering, in M. Cardona (Ed.): *Light Scattering in Solids I: Introductory Concepts*, vol. 8, 2nd ed., Topics in Applied Physics (Springer-Verlag, Berlin Heidelberg New York 1983) p. 79

[186] J. Maultzsch: *Vibrational properties of carbon nanotubes and graphite*, Ph.D. thesis, Technische Universität Berlin, Berlin (2004)

[187] A. M. Rao, E. Richter, S. Bandow, B. Chase, P. C. Eklund, K. A. Williams, S. Fang, K. R. Subbaswamy, M. Menon, A. Thess, R. E. Smalley, G. Dresselhaus, M. S. Dresselhaus: Diameter-selective Raman scattering from vibrational modes in carbon nanotubes, Science **275**, 187 (1997)

[188] M. A. Pimenta, A. Marucci, S. A. Empedocles, M. G. Bawendi, E. B. Hanlon, A. M. Rao, P. C. Eklund, R. E. Smalley, G. Dresselhaus, M. S. Dresselhaus: Raman modes of metallic carbon nanotubes, Phys. Rev. B **58**, R16 016 (1998)

[189] M. Milnera, J. Kürti, M. Hulman, H. Kuzmany: Periodic resonance excitation and intertube interaction from quasicontinuous distributed helicities in single-wall carbon nanotubes, Phys. Rev. Lett. **84**, 1324 (2000)

[190] R. R. Schlittler, J. W. Seo, J. K. Gimzewski, C. Durkan, M. S. M. Saifullah, M. E. Welland: Science **292**, 1136 (2001)

[191] M. F. Chisholm, Y. H. Wang, A. R. Lupini, G. Eres, A. A. Puretzky, B. Brinson, A. V. Melechko, D. B. Geohegan, H. T. Cui, M. P. Johnson, S. J. Pennycook, D. H. Lowndes, S. Arepalli, C. Kittrell, S. Sivaram, M. Kim, G. Lavin, J. Kono, R. Hauge, R. E. Smalley: Science **300**, 5623 (2001)

[192] M. E. Welland, C. Durkan, M. S. M. Saifullah, J. W. Seo, R. R. Schlittler, J. K. Gimzewski: Science **300**, 1236 (2001)

[193] S. Hofmann, C. Ducati, J. Robertson: Low-temperature self-assembly of novel encapsulated compound nanowires, Adv. Mater. **14**, 1821 (2002)

[194] J. Maultzsch, S. Reich, C. Thomsen: Raman scattering in carbon nanotubes revisited, Phys. Rev. B **65**, 233402 (2002)

[195] Y. Miyauchi, S. Maruyama: Identification of excitonic phonon sideband by photoluminescence spectroscopy of single-walled carbon-13 nanotubes, Phys. Rev. Lett. (2005) submitted, cond-mat/0508232

[196] C. Fantini, A. Jorio, M. Souza, L. Ladeira, A. S. Filho, R. Saito, G. Samsonidze, G. Dresselhaus, M. S. Dresselhaus, M. A. Pimenta: One-dimensional character of combination modes in the resonance Raman scattering of carbon nanotubes, Phys. Rev. Lett. **93**, 087401 (2004)

[197] F. Simon, C. Kramberger, R. Pfeiffer, H. Kuzmany, V. Zólyomi, J. Kürti, P. M. Singer, H. Alloul: Isotope engineering of carbon nanotube systems, Phys. Rev. Lett. **95**, 017401 (2005)

[198] G. Bussi, J. Menéndez, J. Ren, M. Canonico, E. Molinari: Quantum interferences in the Raman cross section for the radial breathing mode in metallic carbon nanotubes, Phys. Rev. B **71**, 041404 (2005)

[199] C. Trallero-Giner, A. Cantarero, M. Cardona, M. Mora: Phys. Rev. B **45**, 6601 (1992)

[200] M. Canonico, G. B. Adams, C. Poweleit, J. Menéndez, J. B. Page, G. Harris, H. P. van der Meulen, J. M. Calleja, J. Rubio: Phys. Rev. B **65**, 201402(R) (2002)

[201] G. Bussi, J. Menéndez, J. Ren, M. Canonico, E. Molinari: Quantum interferences in the Raman cross section for the radial breathing mode in metallic carbon nanotubes, Phys. Rev. B **71**, 041404 (2005)

[202] J. Mathews, R. L. Walker: *Mathematical Methods of Physics*, 2nd ed. (Addison-Wesley, Redwood City 1970)

[203] A. Jorio, A. G. S. Filho, G. Dresselhaus, M. S. Dresselhaus, R. Saito, J. H. Hafner, C. M. Lieber, F. M. Matinaga, M. S. S. Dantas, M. A. Pimenta: Joint density of electronic states for one isolated single-wall carbon nanotube studied by resonant Raman scattering, Phys. Rev. B **63**, 245416 (2001)

[204] A. G. S. Filho, A. Jorio, J. H. Hafner, C. M. Lieber, R. Saito, M. A. Pimenta, G. Dresselhaus, M. S. Dresselhaus: Electronic transition energy E_{ii} for an isolated (n, m) single-wall carbon nanotube obtained by anti-Stokes/Stokes resonant Raman intensity ratio, Phys. Rev. B **63**, 241404 (2001)

[205] A. G. Souza Filho, S. G. Chou, G. G. Samsonidze, G. Dresselhaus, M. S. Dresselhaus, L. An, J. Liu, A. K. Swan, M. S. Ünlü, B. B. Goldberg, A. Jorio, A. Grüneis, R. Saito: Stokes and anti-Stokes Raman spectra of small-diameter isolated carbon nanotubes, Phys. Rev. B **69**, 115 428 (2004)

[206] J. Maultzsch, S. Reich, C. Thomsen: Chirality selective Raman scattering of the D-mode in carbon nanotubes, Phys. Rev. B **64**, 121407(R) (2001)

[207] J. Maultzsch, S. Reich, C. Thomsen: Double-resonant Raman scattering in graphite: Interference effects, selection rules, and phonon dispersion, Phys. Rev. B **70**, 155403 (2004)

[208] S. I. Gubarev, T. Ruf, M. Cardona: Doubly resonant Raman scattering in the semimagnetic semiconductor $Cd_{0.95}Mn_{0.05}Te$, Phys. Rev. B **43**, 1551 (1991)

[209] V. Sapega, M. Cardona, K. Ploog, E. Irchenko, D. Mirlin: Spin-flip Raman scattering in $GaAs/Al_xGa_{1-x}As$ multiple quantum wells, Phys. Rev. B **45**, 4320 (1992)

[210] R. P. Vidano, D. B. Fischbach, L. J. Willis, T. M. Loehr: Observation of Raman band shifting with excitation wavelength for carbons and graphites, Solid State Commun. **39**, 341 (1981)

[211] P.-H. Tan, Y.-M. Deng, Q. Zhao: Temperature-dependent Raman spectra and anomalous Raman phenomenon of highly oriented pyrolytic graphite, Phys. Rev. B **58**, 5435 (1998)

[212] V. A. Gaisler, I. G. Neizvestnyi, M. P. Singukov, A. B. Talochkin: JETP Lett. **45**, 441 (1987)

[213] D. Mowbray, H. Fuchs, D. Niles, M. Cardona, C. Thomsen, B. Friedl: Raman study of the Ge phonon side band, in E. Anastassakis, J. Joannopoulos (Eds.): *20th International Conference on the Physics of Semiconductors* (World Scientific, Singapore 1990) p. 2017

[214] M. Mohr, M. Machón, J. Maultzsch, C. Thomsen: Double-resonant Raman processes in germanium: Group theory and *ab initio* calculations, Phys. Rev. B **73**, 035217 (2006)

[215] I. Pócsik, M. Hundhausen, M. Koos, L. Ley: Origin of the D peak in the Raman spectrum of microcrystalline graphite, J. Non-Cryst. Solids **227–230B**, 1083 (1998)

[216] Y. Wang, D. C. Alsmeyer, R. L. McCreery: Raman spectroscopy of carbon materials: Structural basis of observed spectra, Chem. Mater. **2**, 557 (1990)

[217] M. J. Matthews, M. A. Pimenta, G. Dresselhaus, M. S. Dresselhaus, M. Endo: Origin of dispersive effects of the Raman D band in carbon materials, Phys. Rev. B **59**, R6585 (1999)

[218] F. Tuinstra, J. L. Koenig: Raman spectrum of graphite, J. Chem. Phys. **53**, 1126 (1970)

[219] V. Zólyomi, J. Kürti: Phys. Rev. B **66**, 73418 (2002)

[220] C. Mapelli, C. Castiglioni, G. Zerbi, K. Müllen: Phys. Rev. B **60**, 12710 (1999)

[221] C. Castiglioni, F. Negri, M. Rigolio, G. Zerbi: J. Chem. Phys. **115**, 3769 (2001)

[222] A. Ferrari, J. Robertson: Phys. Rev. B **64**, 75414 (2001)

[223] P. Tan, L. An, L. Liu, Z. Guo, R. Czerw, D. L. Carroll, P. M. Ajayan, N. Zhang, H. Guo: Phys. Rev. B **66**, 245410 (2002)

[224] L. D. Landau, E. M. Lifschitz: *Elektrodynamik der Kontinua* (Akademie-Verlag, Berlin 1967)

[225] L. X. Benedict, S. G. Louie, M. L. Cohen: Phys. Rev. B **52**, 8541 (1995)

[226] S. Tasaki, K. Maekawa, T. Yamabe: π -band contribution to the optical properties of carbon nanotubes: Effects of chirality, Phys. Rev. B **57**, 9301 (1998)

[227] C. Thomsen, S. Reich, P. M. Rafailov, H. Jantoljak: Symmetry of the high-energy modes in carbon nanotubes, phys. stat. sol. B **214**, R15 (1999)

[228] J. Hwang, H. H. Gommans, A. Ugawa, H. Tashiro, R. Haggenmueller, K. I. Winey, J. E. Fischer, D. B. Tanner, A. G. Rinzler: Polarized spectroscopy of aligned single-wall carbon nanotubes, Phys. Rev. B **62**, R13 310 (2000)

[229] A. Hartschuh, E. J. Sánchez, X. S. Xie, L. Novotny: Phys. Rev. Lett. **90**, 95503 (2003)

[230] C. Thomsen, S. Reich, J. Maultzsch: Resonant Raman spectroscopy of nanotubes, in A. C. Ferrari, J. Robertson (Eds.): *Raman Spectroscopy in Carbons: From Nanotubes to Diamond* (Philosophical Transaction of the Royal Society 2004) p. 2337

[231] H. Kuzmany, R. Pfeiffer, M. Hulman, C. Kramberger: Raman spectroscopy of fullerenes and fullerene-nanotube composites, in A. C. Ferrari, J. Robertson (Eds.): *Raman Spectroscopy in Carbons: From Nanotubes to Diamond* (Philosophical Transaction of the Royal Society 2004) p. 2375

[232] R. Pfeiffer, H. Kuzmany, T. Pichler, H. Kataura, Y. Achiba, M. Melle-Franco, F. Zerbetto: Electronic and mechanical coupling between guest and host in carbon peapods, Phys. Rev. B **69**, 035404 (2004)

[233] P. M. Rafailov, V. G. Hadjiev, H. Jantoljak, C. Thomsen: Raman depolarization ratio of vibrational modes in solid C_{60}, Solid State Commun. **112**, 517 (1999)

[234] W. Hayes, R. Loudon: *Scattering of Light by Crystals* (Wiley and Sons Inc., New York 1978)

[235] P. Rafailov, C. Thomsen, K. Gartsman, I. Kaplan-Ashiri, R. Tenne: Orientation dependence of the polarizability of an individual WS_2 nanotube by resonant Raman spectroscopy, Phys. Rev. B **72**, 205436 (2005)

[236] A. Jorio, R. Saito, J. H. Hafner, C. M. Lieber, M. Hunter, T. McClure, G. Dresselhaus, M. S. Dresselhaus: Structural (n, m) determination of isolated single-wall carbon nanotubes by resonant Raman scattering, Phys. Rev. Lett. **86**, 1118 (2001)

[237] H. Kuzmany, W. Plank, M. Hulman, C. Kramberger, A. Grüneis, T. Pichler, H. Peterlik, H. Kataura, Y. Achiba: Determination of SWCNT diameters from the Raman response of the radial breathing mode, Eur. Phys. J. B **22**, 307 (2001)

[238] C. L. Kane, E. J. Mele: Phys. Rev. Lett. **90**, 207401 (2003)

[239] G. G. Samsonidze, R. Saito, N. Kobayashi, A. Grüneis, J. Jiang, A. Jorio, S. G. Chou, G. Dresselhaus, M. S. Dresselhaus: Family behavior of the optical transition energies in single-wall carbon nanotubes of smaller diameters, Appl. Phys. Lett. **85**, 5703 (2005)
[240] C. Thomsen, H. Telg, J. Maultzsch, S. Reich: Chirality assignments in carbon nanotubes based on resonant Raman scattering, phys. stat. sol. **242** (2005)
[241] S. K. Doorn, D. A. Heller, P. W. Barone, M. L. Usrey, M. S. Strano: Resonant Raman excitation profiles of individually dispersed single walled carbon nanotubes in solution, Appl. Phys. A: Mater. Sci. Process. **78**, 1147 (2004)
[242] J. Maultzsch, H. Telg, S. Reich, C. Thomsen: Resonant Raman scattering in carbon nanotubes: Optical transition energies and chiral-index assignment, Phys. Rev. B **72**, 205438 (2005)
[243] I. Farkas, D. Helbing, T. Vicsek: Nature **419**, 131 (2002)
[244] C. Thomsen, S. Reich, A. R. Goñi, H. Jantoljak, P. Rafailov, I. Loa, K. Syassen, C. Journet, P. Bernier: Intramolecular interaction in carbon nanotube ropes, phys. stat. sol. B **215**, 435 (1999)
[245] M. J. O'Connell, S. Sivaram, S. K. Doorn: Near-infrared resonance Raman excitation profile studies of single-walled carbon nanotube intertube interactions: A direct comparison of bundled and individually dispersed hipco nanotubes, Phys. Rev. B **69**, 235 415 (2004)
[246] R. B. Weisman, S. M. Bachilo: Nano Lett. **3**, 1235 (2003)
[247] A. Hagen, G. Moos, V. Talalaev, T. Hertel: Electronic structure and dynamics of optically excited single-wall carbon nanotubes, Appl. Phys. A: Mater. Sci. Process. **78**, 1137 (2004)
[248] R. Krupke: (2005) private communication
[249] R. Saito, T. Takeya, T. Kimura, G. Dresselhaus, M. S. Dresselhaus: Raman intensity of single-wall carbon nanotubes, Phys. Rev. B **57**, 4145 (1998)
[250] M. Lazzeri, S. Piscanec, F. Mauri, A. C. Ferrari, J. Robertson: Phonon linewidths and electron phonon coupling in nanotubes, Phys. Rev. B **73**, 155426 (2006)
[251] H. Telg, J. Maultzsch, S. Reich, C. Thomsen: Chirality dependence of the high-energy Raman modes in carbon nanotubes, in H. Kuzmany, J. Fink, M. Mehring, S. Roth (Eds.): *Electronic Properties of Novel Materials*, IWEPNM Kirchberg (2005) p. 162
[252] M. Paillet, P. Poncharal, A. Zahab, J.-L. Sauvajol, J. C. Meyer, S. Roth: Vanishing of the Breit-Wigner-Fano component in individual single-wall carbon nanotubes, Phys. Rev. Lett. **94**, 237 401 (2005)
[253] C. Thomsen, J. Maultzsch, H. Telg, S. Reich: private communication
[254] A. J. Shields, M. P. Chamberlain, M. Cardona, K. Eberl: Raman scattering due to interface optical phonons in GaAs/AlAs multiple quantum wells, Phys. Rev. B **51**, 17 728 (1995)
[255] C. Liu, A. J. Bard, F. Wudl, I. Weitz, J. R. Heath: Electrochem. Solid-State Lett. **2**, 577 (1999)
[256] A. S. Claye, J. E. Fisher, C. B. Huffman, A. G. Rinzler, R. E. Smalley: J. Electrochem. Soc. **147**, 2845 (2000)
[257] S. J. Tans, A. Verschueren, C. Dekker: Room-temperature transistor based on a single carbon nanotube, Nature (London) **393**, 6680 (1998)
[258] A. Modi, N. Koratkar, E. Lass, B. Wei, P. M. Ajayan: Nature (London) **424**, 171 (2003)

[259] A. M. Rao, P. C. Eklund, S. Bandow, A. Thess, R. E. Smalley: Evidence for charge transfer in doped carbon nanotube bundles from Raman scattering, Nature **388**, 257 (1997)

[260] S. Kazaoui, N. Minami, R. Jacquemin: Phys. Rev. B **60**, 13339 (1999)

[261] L. Grigorian, K. A. Williams, S. Fang, G. U. Sumanasekera, A. L. Loper, E. C. Dickey, S. J. Pennycook, P. C. Eklund: Phys. Rev. Lett. **80**, 5560 (1998)

[262] N. Bendiab, L. Spina, A. Zahab, P. Poncharal, C. Marliere, J. L. Bantignies, E. Anglaret, J. L. Sauvajol: Phys. Rev. B **63**, 153407 (2001)

[263] N. Bendiab, E. Anglaret, J. L. Bantignies, A. Zahab, J. L. Sauvajol, P. Petit, C. Mathis, S. Lefrant: Phys. Rev. B **64**, 245424 (2001)

[264] T. Pichler, A. Kukovecz, H. Kuzmany, H. Kataura, Y. Achiba: Phys. Rev. B **67**, 125416 (2003)

[265] L. Kavan, P. Rapta, L. Dunsch: Raman and Vis-NIR spectroelectrochemistry at single-walled carbon nanotubes, Chem. Phys. Lett. **328**, 363 (2000)

[266] L. Kavan, L. Dunsch, H. Kataura: Carbon **42**, 1011 (2004)

[267] P. Rafailov, J. Maultzsch, C. Thomsen, H. Kataura: Electrochemical switching of the Peierls-like transition in metallic single-walled carbon nanotubes, Phys. Rev. B **72**, 045411 (2005)

[268] P. Corio, A. Jorio, N. Demir, et al.: Spectro-electrochemical studies of single wall carbon nanotubes films, Chem. Phys. Lett. **392**, 396 (2004)

[269] L. Kavan, P. Rapta, L. Dunsch, M. J. Bronikowski, P. Willis, R. Smalley: J. Phys. Chem. B **105**, 10764 (2001)

[270] J. N. Barisci, G. G. Wallace, R. H. Baughman, L. Dunsch: J. Electroanalyt. Chem. **488**, 92 (2000)

[271] M. Stoll, P. M. Rafailov, W. Frenzel, C. Thomsen: Chem. Phys. Lett. **375**, 625 (2003)

[272] S. Gupta, M. Hughes, A. H. Windle, J. Robertson: Charge transfer in carbon nanotube actuators investigated using in situ Raman spectroscopy, J. Appl. Phys. **95**, 2038 (2004)

[273] P. Rafailov, M. Stoll, C. Thomsen: J. Phys. Chem. Solids **108**, 19241 (2004)

[274] P. Corio, P. S. Santos, V. W. Brar, G. G. Samsonidze, S. G. Chou, M. S. Dresselhaus: Chem. Phys. Lett. **370**, 675 (2003)

[275] L. Kavan, L. Dunsch, H. Kataura: Chem. Phys. Lett. **361**, 79 (2002)

[276] N. W. Ashcroft, N. D. Mermin: *Solid State Physics* (Saunders College, Philadelphia 1976)

[277] J. B. Renucci, R. N. Tyte, M. Cardona: Resonant Raman scattering in silicon, Phys. Rev. B **11**, 3885 (1975)

[278] R. Trommer, M. Cardona: Phys. Rev. B **17**, 1865 (1973)

[279] J. M. Calleja, J. Kuhl, M. Cardona: Phys. Rev. B **17**, 876 (1978)

[280] S. V. Goupalov: Chirality dependence in the Raman cross section, Phys. Rev. B **71**, 153404 (2005)

[281] F. S. Khan, P. B. Allen: Deformation potentials and electron-phonon scattering: Two new theorems, Phys. Rev. B **29**, 3341 (1984)

[282] J. Jiang, R. Saito, A. Grüneis, S. G. Chou, G. G. Samsonidze, A. Jorio, G. Dresselhaus, M. S. Dresselhaus: Intensity of the resonance Raman excitation spectra of single-wall carbon nanotubes, Phys. Rev. B **71**, 205420 (2005)

[283] R. E. Peierls: *Quantum Theory of Solids* (Oxford University Press, New York 1955)

[284] A. Jorio, A. G. Souza Filho, G. Dresselhaus, M. S. Dresselhaus, A. K. Swan, M. S. Ünlü, B. B. Goldberg, M. A. Pimenta, J. H. Hafner, C. M. Lieber, R. Saito: Phys. Rev. B **65**, 155412 (2002)
[285] J. Kürti, V. Zólyomi, A. Grüneis, H. Kuzmany: Phys. Rev. B **65**, 165433 (2002)
[286] C. Jiang, J. Zhao, H. A. Therese, M. Friedrich, A. Mews: J. Phys. Chem. B **107**, 8742 (2003)
[287] L. Cançado, M. A. Pimenta, B. R. A. Nevesand, M. S. S. Dantas, A. Jorio: Influence of the atomic structure on the Raman spectra of graphite edges, Phys. Rev. Lett. **93**, 247401 (2004)
[288] M. Machón, S. Reich, J. Maultzsch, H. Okudera, A. Simon, R. Herges, C. Thomsen: Structural, electronic and vibrational properties of (4,4) picotube crystals, Phys. Rev. B **72**, 155 402 (2005)
[289] S. Kammermeier, P. G. Jones, R. Herges: Angew. Chem. Int. Ed. Engl. **35**, 2669 (1996)
[290] S. Kammermeier, P. G. Jones, R. Herges: Angew. Chem. **108**, 2834 (1996)
[291] M. Souza, A. J. C. Fantini, B. R. A. Neves, M. A. Pimenta, R. Saito, A. Ismach, E. Joselevich, V. W. Brar, G. G. Samsonidze, G. Dresselhaus, M. S. Dresselhaus: Single- and double-resonance Raman G-band processes in carbon nanotubes, Phys. Rev. B **69**, 241403 (2004)
[292] J. Maultzsch, S. Reich, C. Thomsen, S. Webster, R. Czerw, D. L. Carroll, S. M. C. Vieira, P. R. Birkett, C. A. Rego: Raman characterization of boron-doped multiwalled carbon nanotubes, Appl. Phys. Lett. **81**, 2647 (2002)
[293] R. J. Nemanich, S. A. Solin: First- and second-order Raman scattering from finite-size crystals of graphite, Phys. Rev. B **20**, 392 (1979)
[294] M. Reedyk, C. Thomsen, M. Cardona, J. S. Xue, J. E. Greedan: Observation of the effects of phonon dispersion on the Fröhlich-interaction-induced second-order Raman scattering in $Pb_2Sr_2PrCu_3O_8$, Phys. Rev. B **50**, 13762 (1994)
[295] R. Saito, A. Jorio, A. G. Souza-Filho, G. Dresselhaus, M. S. Dresselhaus, M. A. Pimenta: Phys. Rev. Lett. **88**, 27401 (2002)
[296] Y. Kawashima, G. Katagiri: Phys. Rev. B **52**, 10053 (1995)
[297] P. Tan, C. Hu, J. Dong, W. Shen, B. Zhang: Phys. Rev. B **64**, 214301 (2001)

Index

2D hydrogen model, 158

acoustic
  (ZA) branch, 141
    of graphene, 141
anisotropic polarizability, 153, 174
arc-discharge, 122
armchair
  nanotubes, 140
  tubes, 133
atomic-microscope image
  carbon nanotube, 209

band-to-band transition, 151, 167
Bethe–Salpeter equation, 159
buckypaper, 194

C_{60} molecules, 125
C_{60} peapod, 125
Carbon Nanotubes, 115
carbon nanotubes, 140
  metallic, 168
  optical selection rules, 153
  single-wall, 140

character table, 129
  D_{qh}, 128
chemical vapor deposition, 122, 123
chirality assignment, 177
curvature, 185
CVD, 122
cyclic voltammetry, 194

D mode, 136, 163, 207, 208, 213
D^* mode, 213
defects, 214
depolarization
  effect, 174
  ratio, 175
dielectric
  constant, 158
  function, 152
dipole
  approximation, 152
double-resonance
  conditions, 171
  process, 192
double-resonant
  Raman scattering, 169
  scattering, 215
double-walled nanotubes, 134

eigenmodes
  of nanotubes, 134
eigenvector
  of nanotube vibrations, 136
  Raman-active, 173
electrochemical doping, 194
electron–phonon coupling, 191, 196, 201
electronic band structure, 148
exciton, 151, 158
  binding energy, 158–160
  exciton–phonon excitations, 161
  phonon side bands, 161
excitonic transitions, 168
extended tight-binding model, 148

Fano lineshape, 190
Fermi level, 195
frozen-phonon approximation, 197, 199

G line, 136
G mode, 209, 213

Ge, 207
graphene, 117, 118, 126, 199
  cylinder limit, 206
  electronic band structure, 146
  empirical formula, 186
  Kohn anomalies, 143
  semimetal, 145
  trigonal distortion, 155
graphene LO mode
  overbending, 140
graphite, 116, 137, 140, 168
  D mode, 116, 172
  optical modes, 139
  phonon dispersion, 137
  Raman spectrum, 212
  sound velocities, 131
  tight-binding parameters, 116

helical quantum numbers, 143
high-energy mode (HEM), 188, 191
  single-walled tubes, 163
high-energy phonon, 130

Iijima, 115
inelastic X-ray data
  on graphite, 215
inorganic molecules, 125
interferences, 168
interlayer coupling, 134
intermediate states, 165
isogonal point groups, 127
isotope shift, 164

joint density of electronic states, 152

Kataura plot, 154, 155, 160, 178, 180, 187, 204
Kohn anomaly, 139, 140, 190, 196, 204, 206
  graphene, 143

laser
  ablation, 122
line groups, 126

matrix element, 203
  electron–phonon, 197, 207
metallic tubes, 155
  condition for, 150

nanotube, 118
  achiral, 121
  armchair, 121, 130
  axial vibrations, 130
  branches and families, 182
  bundled, 191
  bundles, 123
  charge transfer, 193
  chiral, 121
    angle, 118
    index, 118
  chiralities, 123
  circumferential vibrations, 130
  classifications, 157
  curvature, 132, 150
  diameter, 118
  doping, 193
  double-walled, 135
  families, 157
    and branches, 155
  forest, 124
  interference effects, 202
  light-absorption, 174
  luminescence, 177
  metallic, 121, 149, 187, 193
  multiwalled, 174
  neighbor rule, 157
  phonon branch, 192
  plasmons, 190
  point group, 129
  Raman active, 130
  semiconducting, 121, 149
  single-walled, 127
  structural parameters, 119
  symmetry, 126
    operations, 127
  zigzag, 121, 130

optical absorption, 152
optical matrix elements, 152
optical selection rules
  carbon nanotubes, 153
overbending, 138
  graphene LO branches, 138
  graphene LO mode, 140
overtone, 172, 173, 212
  Raman spectrum, 163

peapods, 125
Peierls distortion, 204
Peierls-like
  mechanism, 190
  softening, 196
phonon
  dispersion
    graphite, 215
  dispersion curves, 142, 171
    armachair tube, 142
  eigenvectors, 129
  plasmon coupled modes, 188, 190, 194
  symmetries, 129
    carbon nanotubes, 129
π wavefunctions, 145
picotubes, 211
pyramidalization angle, 132

quantization condition
  electronic band structure, 144
quasimetallic tubes, 140

radial-breathing mode (RBM), 130, 162, 177, 196
radial-breathing vibration, 130
Raman
  cross section, 165
  double resonance, 211
    in Ge, 172
  effect
    carbon nanotubes, 162
  efficiency, 167
    Stokes scattering, 165
  matrix element, 165, 166, 170, 188
  scattering
    double-resonant, 163
  spectra
    single-walled nanotubes, 162
Raman-active vibrations, 136
ratio problem, 179
resonance
  incoming, 165
  outgoing, 165
  profiles
    of nanotubes, 180
  Raman scattering, 162
resonant
  Raman effect, 150

scissors operator, 182

secondary gap, 140
selection
  rule, 148, 152
    $\mathcal{R} = 3$, 210
semiconducting tubes, 155
  condition for, 150
semimetal
  graphene, 145
shape anisotropy, 174
shape depolarization, 212
single resonance, 164
sodium dodecyl sulfate, 177
space-resolved imaging, 209
spectroelectrochemistry, 193
Stokes and anti-Stokes frequencies, 172

tangential modes, 136
third-neighbor interaction, 145
tight-binding
  approximation, 197
  energies, 185
  model, 145, 159
transport properties, 189
two-photon absorption, 160

valence force field model (VFF), 134
van-der-Waals
  forces, 123
  interaction, 185
van-Hove singularity, 154

wavevector
  scan, 171
WS_2, 176

zeolite, 174
zigzag tubes, 133
zone folding, 137, 138, 144, 155, 185

Resonant Raman Scattering by Acoustic Phonons in Quantum Dots

Adnen Mlayah and Jesse Groenen

Laboratoire de Physique des Solides, Université Paul Sabatier, CNRS, 118 route de Narbonne, 31062, Cedex 4 Toulouse, France
{Jesse.Groenen,Adnen.Mlayah}@lpst.ups-tlse.fr

Abstract. In this Chapter we discuss the use of resonant Raman scattering by acoustic phonons and its applications to the study of the spatial localization and spatial correlation of electronic wavefunctions in low-dimensional semiconductor and metallic systems. Special attention is paid to the simulation of the Raman spectra and their direct comparison with experimental data. We review the basic concepts of this field and results reported in the literature.

1 Introduction

The optical properties of metal and semiconductor quantum dots exhibit strong differences with respect to the bulk materials, and have therefore attracted much interest from the viewpoint of the fundamental physics and of the potential applications. These differences lie in the size quantization of the electron energy levels. Quantum dots are often named artificial atoms to underline the three-dimensional confinement of the electronic states. Like atoms, quantum dots exhibit many-electron effects such as Pauli blockade [1] and Coulomb blockade [2]. However, the important difference from real atoms is that quantum-dot electrons are embedded in a phonon bath. This has a significant impact on their optical properties since electron–phonon interaction determines the broadening of energy levels and relaxation rates of electronic states. For instance, the phonon bottleneck effect [3, 4] is responsible for the suppression of interlevel electron-relaxation channels. It is therefore a limiting factor for radiative excitonic recombinations and hence for the optical emission of QD-based lasers. Moreover, in certain situations, the simple picture of the artificial atom cannot capture the physics of quantum dots any longer. Indeed, at low-temperature, electronic and vibronic states of quantum dots can strongly mix together and form polaronic states [5–7]. In that case, the energy of the infrared optical absorption bands cannot be explained only on the basis of size-induced quantization of electron levels. Anticrossing of confined electronic transitions and optical phonon energies is needed to account for the experimental data [5, 7].

Electron–phonon interaction in quantum dots, investigated by resonant Raman scattering, is the central topic of this Chapter. We purposely focused on acoustic phonons and their interactions with QD electronic states. Optical

M. Cardona, R. Merlin (Eds.): Light Scattering in Solid IX,
Topics Appl. Physics **108**, 237–316 (2007)

phonons have been extensively investigated in connection with size-induced quantization and strain effects. However, optical phonons exhibit only small changes with QD size because their frequency dispersion around the Brillouin zone center is rather small (tens of cm^{-1}). In addition, these changes are weaker than the frequency shifts induced by the strain usually present in semiconductor QD, particularly in those grown in the Stranski–Krastanov regime. Therefore, it is difficult to distinguish between strain and confinement effects on optical vibrations. On the contrary, acoustic phonons exhibit a strong (hundreds of cm^{-1}) and linear dispersion over a wide range of the Brillouin zone. Moreover, their wavelengths (in the THz frequency range) is of the order of a few nanometers and are thus comparable to the size of QDs. Another important advantage of using acoustic phonons for studying quantum dots is the delocalized nature of the vibration displacement field. Indeed, optical phonons are usually confined within the QD because of the vibration-frequency gap between the QD and the barrier materials. Hence, when looking at optical phonons, each QD can be considered independently in the sense that its vibrational and light-scattering properties can be understood in terms of quantum confinement of electronic and vibronic states. Correlations in the spatial distribution of QD are disregarded. On the other hand, acoustic vibrations could be delocalized over distances much larger than the QD average separation. As a result, collective behavior related to the spatial arrangement of the QD appears in the electron–phonon interaction and in the light-scattering efficiency.

The main concepts to be developed in this Chapter are centered around the spatial localization and spatial correlation of electrons and acoustic phonons in QD nanostructures. The electronic excitations are either excitons and free electron–hole pairs or plasmons, depending on the semiconducting or metallic nature of the QD. First, a brief introduction to the effect of wave-function localization on the electron–phonon interaction and on the resonant Raman scattering is presented for quantum-well structures (one-dimensional confinement). These basic ideas, initially developed by *Kop'ev* et al. [8], *Sapega* et al. [9] and *Mirlin* et al. [10], have been recently extended to semiconductor quantum dots by *Huntzinger* et al. [11], *Cazayous* et al. [12–14], and *Milekhin* et al. [15–17]. In a first step, only QD structures with a rather weak acoustic impedance mismatch between the QD and the barrier are considered in order to separate electron- and phonon-confinement effects. Raman interferences in QD multilayers are investigated in connection with three-dimensional electron confinement and spatial ordering along the growth direction as well as in the plane of the layers. Acoustic mirror effects are discussed for QD layers located in the vicinity of a surface. We show how the surface can modulate in real space and in the frequency domain the acoustic-phonon-displacement fields and hence the electron–phonon interaction. This has an important impact on the electron-relaxation rates and dephasing processes.

Systems exhibiting strong phonon confinement are considered in the last section of the Chapter. This concerns semiconductor and metal particles embedded in glass and polymers. Confinement of acoustic phonons is discussed in the framework of the widely used *Lamb's* model [18, 19] describing the vibrations of a homogeneous elastic free sphere. Matrix effects are discussed according to a core–shell model recently reported by *Murray* and *Saviot* [20]. This model describes not only the frequency shift and damping of acoustic vibrations but also the localization/delocalization of the associated displacement field. It can be considered as the ultimate development of the continuous elastic-medium approach. Microscopic approaches [21, 22] based on valence force field interatomic interactions are examined and their results compared to Lamb's model. This allows us to address the validity of Lamb's model for very small clusters in which the number of surface atoms is much larger than that of inner atoms. A review of the Raman scattering and time-resolved optical absorption experiments involving confined acoustic phonons is then presented. Both techniques are rather complementary, since they imply different vibration modes. The optical activity of confined phonons is discussed in terms of coupling mechanisms between vibrations and electronic excitations for both semiconductor and metallic particles.

2 Introduction to Wavefunction Localization and Electron–Acoustic Phonon Interaction

An important aspect of the resonant Raman scattering by acoustic phonons, in quantum-dot nanostructures, is the spatial localization and spatial correlation of the electronic states involved in the light-scattering process. The first observations of acoustic-phonon Raman scattering due to electron wavefunction localization were reported by *Kop'ev* et al. [8], *Sapega* et al. [9] and *Mirlin* et al. [10] in GaAs/AlAs superlattices and multiple quantum well structures. In addition to the scattering by folded acoustic phonons [23], a continuous emission background centered around the Rayleigh line was observed in the low-frequency range of the Raman spectra excited close to resonance with confined electronic transitions [8, 9, 24, 25]. The strong resonant character of this signal was demonstrated by magneto-Raman measurements [26–28]. Its origin was attributed to the activation of the acoustic-phonon density of states due to selection of individual quantum wells [8–10, 26]. The selection occurs because of inhomogeneous broadening of the confined electronic transitions and resonance with the probe laser line. The important characteristic of Raman scattering from individual quantum wells is the absence of wavevector conservation, in the direction perpendicular to the QW plane, for both the electron–photon and electron–phonon interaction steps of the scattering process [8–10, 24, 26–28]. This breakdown of the wavevector-conservation law is the consequence of the spatial localization of the electronic states. Raman measurements from single GaAs/AlAs quantum wells confirmed this effect and suggested that acoustic-phonon Raman scattering could be used for studying the spatial localization of electronic states [29] themselves. In ideal two-dimensional structures the wavefunctions are localized in one direction due to quantum confinement along the growth direction, whereas in the plane of the quantum wells translational symmetry still holds and hence the electronic states are delocalized. However, it is well known that interface roughness and interdiffusion lead to inplane localization of the electronic states. Depending on temperature and on the disorder (magnitude and spatial correlation) the quantum wells can therefore exhibit a QD-like behavior. This has been pointed out by *Ruf* et al. [30, 31] and *Belitsky* et al. [32] who extracted the length scale of interface roughness using a detailed spectral analysis of the disorder-induced acoustic-phonon Raman scattering. In order to illustrate the concepts introduced above we first present the studies performed on two-dimensional systems consisting of very few GaAs/GaP quantum wells. In these structures the acoustic vibrations can be approximated by plane waves since the mechanical properties of the QW and the barrier layers are very similar and the number of interfaces limited. Thus, the acoustic vibrations are extended phonon states whereas the electron and the hole states are spatially localized in one direction. The aim is to show that, independently of any confinement effect of acoustic vibrations, the low-frequency Raman scattering can exhibit broad bands or peaks depending only on the spatial distribution of the excited electronic density.

2.1 Low-Frequency Resonant Raman Scattering in Quantum Wells

Figure 1 presents low-frequency Raman spectra of GaAs/GaP quantum wells [33–35]. The thickness of each QW is two monoatomic (ml) layers only; they are separated by 38 nm GaP barriers. The structure contains three quantum wells. The band alignement of GaAs strained to match GaP and the electronic structure of the GaAs QW have been studied by *Prieto* et al. [36]: the lowest-energy electron states of GaAs fully matched to GaP arise from the X-edge of the Brillouin zone (type-II band alignment) [36]. Therefore, electrons photoexcited in the zone-center Γ states of the confined e_1 subband can rapidly thermalize to the zone-edge X states, and then recombine with the heavy holes at the Γ point of the confined hh_1 valence subband. This luminescence process is assisted by zone-edge acoustic and optical phonons. For the 2 ml GaAs/GaP QW the luminescence peak was observed at 2 eV, whereas the $e_1 - hh_1$ absorption edge was found at 2.54 eV from low-temperature electroreflectance spectroscopy [36]. This large Stokes shift allows resonant excitation and detection of Raman scattering with no overlapping with the luminescence signal. The spectra shown in Fig. 1 were excited close to resonance with the confined $e_1 - hh_1$ transition and detected in the frequency range of acoustic phonons (Scattering by GaAs and GaP optical phonons has

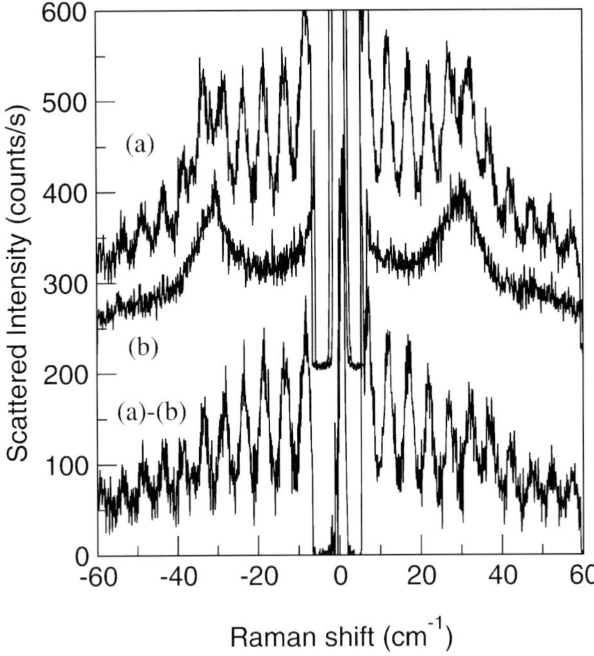

Fig. 1. Room-temperature low-frequency Raman scattering from 2 ml thick GaAs/GaP quantum wells (**a**) and from a bulk GaP (**b**) reference layer. The excitation energy is 2.47 eV, close to resonance with the $e_1 - hh_1$ transition of the QWs. The band around 30 cm^{-1} in spectrum (**b**) is due to second-order scattering by a subtractive combination LO–TO of longitudinal and transverse optical phonons of GaP [33]

been studied by *Castrillo* et al. [37] in connection with strain and quantum-confinement effects). A series of peaks, superimposed on the second-order scattering by longitudinal (LO) and transverse (TO) optical phonons of GaP are observed in Fig. 1. Spectrum (a)-(b) in this figure shows the contribution of the GaAs quantum wells after subtraction of spectrum (b) recorded from a GaP reference layer.

The dependence upon excitation energy is presented in Fig. 2 (from [33, 35]). One can see the resonant character of the scattering: the energy of the $e_1 - hh_1$ transition is around 2.44 eV (at 300 K) and the Raman peaks disappear when the detuning between the laser energy and the $e_1 - hh_1$ transition energy is of the order of ±30 meV. This value is comparable to the broadening of the $e_1 - hh_1$ transition deduced from a lineshape analysis of the electroreflectance spectra [36]. The spectra of Figs. 1 and 2 were obtained in the $z(x'x')\bar{z}$ configuration, for which scattering by LO (and LA) phonons is allowed. This is why second-order scattering due to subtractive combinations of LO and TO phonons appears in the low-frequency range. In order to get rid

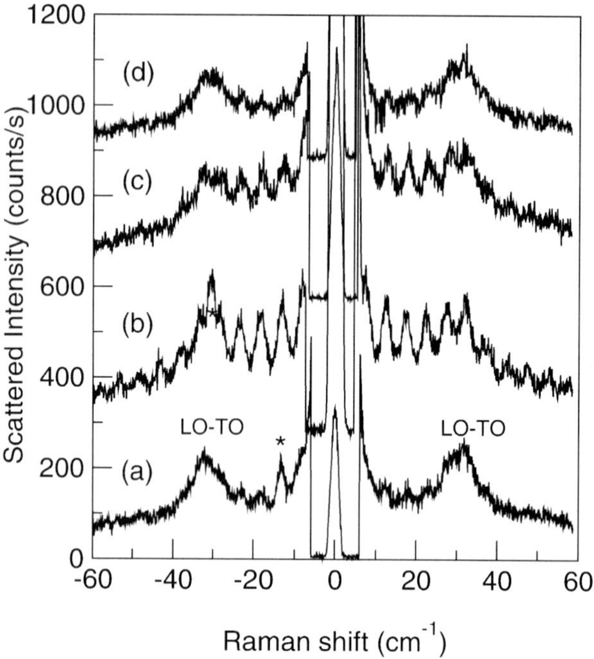

Fig. 2. Room-temperature Raman spectra of the QW structure (same as in Fig. 1), showing the effect of resonance around the $e_1 - hh_1$ transition. Spectra (**a**), (**b**), (**c**) and (**d**) were excited at 2.41, 2.47, 2.50 and 2.54 eV, respectively. The stars indicate laser plasma lines

of this scattering one should use the $z(xx)\bar{z}$ configuration preferably, as shown in Fig. 3. Indeed, according to the Raman and Brillouin selection rules [24] for deformation potential electron–phonon interaction, LO phonon scattering is forbidden, whereas LA phonon scattering is allowed in this configuration.

2.2 Diffraction and Interference

The peaks observed in Figs. 1–3 are in fact oscillations of the scattered intensity. They are due to interference between the contributions of each localized state to the Raman scattering by acoustic phonons. The mechanical properties of GaAs and GaP being very similar, acoustic waves can propagate as plane waves. Their reflection or scattering at the interfaces can be neglected in a first step. Moreover, their coherence length is much longer than the QW separation. Therefore, a given acoustic mode can interact with electrons and holes confined in each of the quantum wells. Hence, its emission or absorption is spatially coherent. It can be either enhanced or inhibited depending on whether the vibration wavelength is an integer or half-integer multiple of the separation d between quantum wells (Fig. 4). This effect is very similar

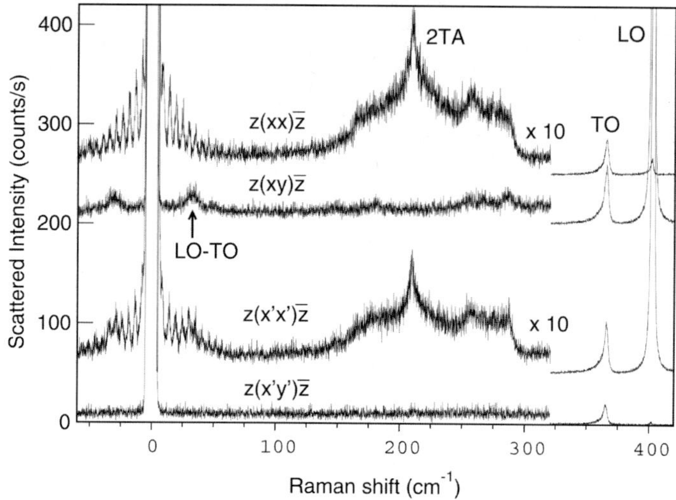

Fig. 3. Raman selection rules showing acoustic and optical phonon scattering in different scattering configurations; x, x', y, y', z correspond to the [100], [110], [010], $[1\bar{1}0]$ and [001] crystallographic axes, respectively. For (100) surface orientation, TO-phonon scattering is normally forbidden in all scattering configurations. However, it can be activated because of deviations from the true backscattering configuration. Scattering by GaP LO phonons is forbidden in the $z(xx)\bar{z}$ configuration and hence second-order scattering by LO–TO phonons is absent

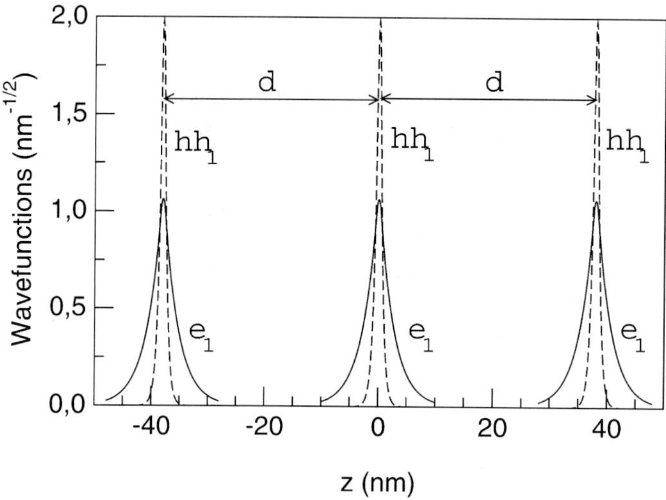

Fig. 4. 7×7 $k.p$ electron (e_1) and heavy hole (hh_1) wavefunctions of 2 ml GaAs QW in GaP [36]. The wavefunctions penetrate deeply into the barrier layers, especially the e_1 wavefunctions. The distance (38 nm) between QWs is much larger than the penetration depth of the wavefunctions. Tunneling between QWs is therefore negligible

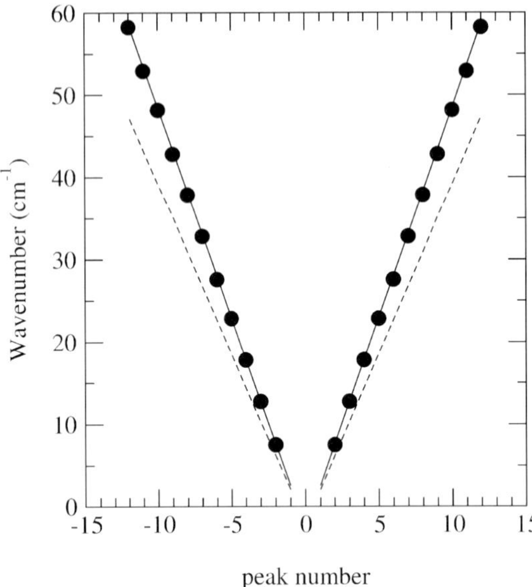

Fig. 5. *Dots*: peak frequencies measured from spectrum (**a**) of Fig. 1 versus peak index (ℓ in 1). The lowest-frequency peak was numbered 2. The first peak is not observed because of the intense Rayleigh wings. The *continuous* and *dotted lines* show the linear dispersion of LA phonons in GaP and GaAs, respectively. From [29, 35]

to the well-known interference picture one can generate by shining light on a pair of Young's slits. By analogy with optics, one can write the constructive interference condition

$$q_z = \frac{2\ell\pi}{d} + \Delta k_z \,, \tag{1}$$

where ℓ is an integer and $\Delta k_z = k_{iz} - k_{sz}$ the exchanged wavevector. In backscattering, the wavevectors are oriented along the growth axis (labeled z). The frequencies of the scattered intensity maxima can be deduced from (1) giving the separation d between the quantum wells and the dispersion relation $\omega(q_z)$ of LA phonons. In the frequency range under investigation, $\omega(q_z)$ is quasilinear both for GaAs and GaP [38]. Moreover, according to the $k.p$ model [36], the electronic wavefunctions strongly penetrate into the GaP barriers (Fig. 3). As a matter of fact, for a 2 ml GaAs QW, 80 % of the electron density distribution is located outside the QW (Fig. 4). Hence, the electrons interact more strongly with the acoustic vibrations of GaP than of GaAs type. So, using (1) and the sound velocity of LA phonons in GaP ($v_l = 5830\,\mathrm{m \cdot s^{-1}}$ [39]) one obtains the frequencies of the intensity maxima. As shown in Fig. 5 the calculated frequencies are in agreement with the measured ones.

2.3 Scattering Efficiency

To go deeper into the analysis of the Raman-scattering process, one may calculate the spectral dependence of the scattered-light intensity. For resonant excitation, the scattering efficiency is governed by the three-step light-scattering process. It is proportional to [9, 24, 40]

$$\Gamma = \left| \sum_{e_1, e_1'} \frac{\langle hh_1 | H_{\mathrm{e-pht}} | e_1' \rangle \langle e_1' | H_{\mathrm{e-phn}} | e_1 \rangle \langle e_1 | H_{\mathrm{e-pht}} | hh_1 \rangle}{\left(E_{e_1'-hh_1} + \mathrm{i}\gamma_{e_1'-hh_1} - E_{\mathrm{s}}\right)\left(E_{\mathrm{i}} - E_{e_1-hh_1} - \mathrm{i}\gamma_{e_1-hh_1}\right)} \right|^2 , \qquad (2)$$

where, e_1 and hh_1 are electron and heavy hole states forming the two-dimensional confined subbands. According to the conduction- and valence-band offsets of GaAs pseudomorphically strained by GaP [36], and to the thickness of the GaAs QW, there is only one electron, one heavy hole and one light hole confined subbands. $E_{e_1-hh_1}$ and $\gamma_{e_1-hh_1}$ are the energy and the homogeneous linewidth of the $e_1 - hh_1$ transition and E_{i} (E_{s}) is the energy of the incident (scattered) photons. $H_{\mathrm{e-pht}}$ and $H_{\mathrm{e-phn}}$ are the electron–photon and electron–phonon interaction Hamiltonians. Either conduction or valence states can be the intermediate states in the light-scattering process. Their contributions to the scattering efficiency must be summed coherently, i.e., before squaring them (see (2)) because it is the electron–hole pair created at the photon-absorption step that recombines after phonon emission (or absorption) and gives rise to the emitted photon. In order to emphasize the fact that phase coherence between the photocreated electron and hole is preserved in the light-scattering process, radiative recombination of the coherent electron–hole pairs has been termed geminate recombination [8]. Using the envelope wavefunction approximation

$$\psi_{e_1}(z, \rho) = \phi_{e_1}(z) \mathrm{e}^{\mathrm{i}k_{//}\rho} , \qquad (3)$$

the matrix element for deformation potential electron–phonon interaction reads (for a Stokes process) [41]

$$\langle e_1' | H_{\mathrm{e-phn}} | e_1 \rangle = \mathrm{i}D_{\mathrm{e}} \left(n_q + 1\right)^{1/2} q u_0 \delta\left(\mathrm{k}_{||}' - \mathrm{k}_{||} + \mathrm{q}_{||}\right) \int \left|\phi_{e_1}(z)\right|^2 \mathrm{e}^{\mathrm{i}q_z z} \mathrm{d}z , \qquad (4)$$

where q, u_0 and n_q are the phonon wavevector, amplitude and population factor, respectively; D_{e} is the deformation potential energy [24, 42]. $\phi_{e_1}(z)$ is the electron wavefunction of the QW structure along the direction of confinement. The emission of acoustic phonons involving hole states can be expressed similarly.

The wavevector conservation, in the plane of the QW (delta in (4)), arises from the 2D character of the electron wavefunction, i.e., from the uniform

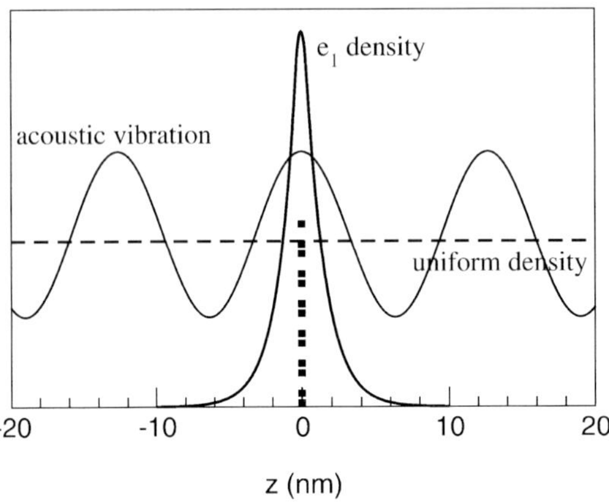

Fig. 6. Electron–acoustic phonon interaction induced by the spatial localization of the electronic states. In the case of delocalized states (uniform probability density distribution), the average value of the electron–phonon interaction energy (diagonal matrix element) is zero because negative and positive values cancel exactly. In the case of localized electronic states (shown is the 7×7 $k.p$ electron density e_1 of a 2 ml GaAs/GaP QW), the average value of the probability density multiplied by the displacement field is not zero and depends on the spatial extent of the electronic state. The *dotted vertical line* shows the thickness of the QW

inplane distribution of the electronic density. It also applies in (2) to the electron–photon interaction at the absorption and emission steps of the incident and scattered photons. It is interesting to note that in the case of bulk crystals there are no diagonal matrix elements for the electron–phonon interaction. Indeed, when the intermediate electronic states are spatially extended states, the wavevector conservation law implies that $\langle e\,|H_{e-\mathrm{phn}}|\,e\rangle = 0$ and $\langle hh\,|H_{e-\mathrm{phn}}|\,hh\rangle = 0$ (e and hh being bulk electron and hole states). In low-dimensional systems, the electronic states are repeatedly localized in real space (at least in one direction), and therefore diagonal matrix elements for the electron–phonon interaction become allowed ($k'_{//} = k_{//}$ and $q_{//} = 0$ in (4)). They give rise to the major contribution in the light-scattering efficiency. As illustrated schematically in Fig. 6, this localization-induced Raman scattering, is the main difference between bulk crystals and low-dimensional systems. It is similar to the disorder-induced Raman scattering in the sense that disorder also leads to wavevector nonconserving Raman scattering, although the localization does not exhibit the superlattice periodicity. *Takagahara* [41] and *Zimmermann* [43] showed that such an effect determines the dephasing rate of electronic states in quantum wells and quantum dots and hence their luminescence properties. These authors introduced two types of dephasing, namely a real and a pure dephasing [44, 45] corresponding to offdiagonal (final-state interaction) and on-diagonal electron–phonon matrix elements, respectively. On the other hand,

Zimmermann and *Runge* [44] and *Belistky* et al. [46] studied theoretically the effect of electron wavefunction localization on Rayleigh scattering [47] and pointed out the importance of pure dephasing processes. As illustrated in Fig. 4, tunneling between quantum wells is negligible. Therefore, assuming independent and identical electron wavefunctions confined in each of the three QW the diagonal terms of (4) read

$$\langle e_1 | H_{e-\mathrm{phn}} | e_1 \rangle = \mathrm{i} D_e (n_q + 1)^{1/2} q u_0 M(q_z) S(q_z) \,, \tag{5}$$

in which $M(q_z)$ represents the integral in (4), i.e., the Fourier transform of the electronic density localized around a single quantum well (SQW); $M(q_z)$ can be viewed as a form factor, whereas $S(q_z)$ is a structure factor

$$S(q_z) = 1 + \mathrm{e}^{\mathrm{i}(q_z - \Delta k_z)d} + \mathrm{e}^{\mathrm{i}(q_z - \Delta k_z)2d} \,, \tag{6}$$

which expresses the layering of the electronic density along the z-direction. More generally, $M(q_z)$ and $S(q_z)$ are, respectively, associated with the spatial localization and spatial correlation of the electronic density. Equation (5) is very similar to that representing the elastic scattering amplitude of X-rays, except that the wavevector of the probing radiation is replaced by the wavevector of sound waves. Replacing (5) into (2), one can see that the Raman efficiency is proportional to $|S(q_z)|^2$, which is an oscillatory function of q_z and reflects the interference effects discussed above. $|M(q_z)|^2$ is the spectral envelope of the intensity oscillations. Owing to the linear dispersion of acoustic phonons $\omega(q_z)$, the wave-number axis of the Raman spectra (Figs. 1–5) can be directly converted into the Fourier components of the electronic density involved in the resonant light-scattering process.

2.4 Intermediate-State Selection

Figure 7 presents calculated and measured Raman spectra for resonant excitation with both the $e_1 - hh_1$ and $e_1 - lh_1$ transitions. The confined wavefunctions, used in the calculations, were obtained from a 7×7 $k.p$ model [36]. Contributions of both electron and hole states to the Raman scattering were taken into account. Their relative importance depends on the values of the deformation potential energy and on the density of states associated with each subband. As mentioned above, the spectral envelope of the oscillations is the Fourier transform of the electronic density distribution selected by the optical excitation. Since heavy holes are more localized in real space than light holes, they are more delocalized in reciprocal space and their contribution to the Raman scattering extends to higher frequencies. This is clearly observed in Fig. 7: when tuning the optical excitation to the electron-light hole transition the spectral envelope of the intensity oscillations becomes narrower because light holes are less localized in real space than heavy holes. This underlines the sensitivity of the low-frequency Raman scattering to the electronic wavefunction localization [8–10, 24].

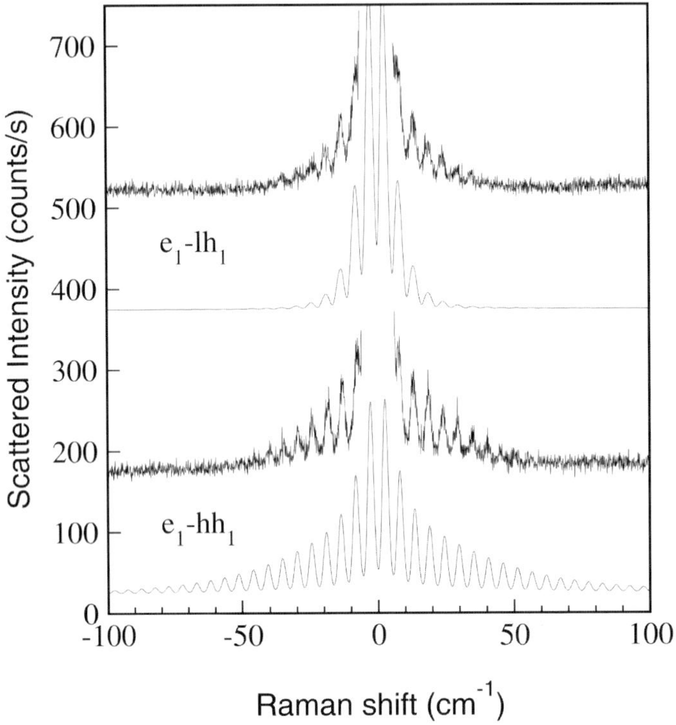

Fig. 7. Calculated and measured room-temperature Raman spectra of GaAs quantum wells in GaP. The excitation energy is resonant with either the $e_1 - hh_1$ transition (2.602 eV) or with the $e_1 - lh_1$ transition (2.707 eV). From [35]

The basic concepts associated with resonant Raman scattering by acoustic phonons due to wavefunction localization of the intermediate electronic states have been introduced. Only one-dimensional localization due to quantum confinement has been considered. These concepts will be now extended to quantum dots.

3 Raman Interferences in Self-Assembled Quantum Dots

Quite a number of studies on Raman scattering by optical phonons have been reported; they include effects of strain, alloying, confinement, interfaces and resonance; a nonexhaustive list can be found in [15–17, 48–57]. Less attention has been devoted to Raman scattering by acoustic phonons from self-assembled QDs [11–17, 57–62]. Within the context of resonant RS by acoustic phonons in QDs, self-assembled QDs are of particular interest as they display efficient confinement for electrons and holes but not for acoustic phonons. Indeed, the acoustic mismatch between QD and barrier materials to be discussed below is weak. We may therefore discuss Raman scattering using a rather simple description for the acoustic phonons: we will consider either bulk-like phonons or acoustic modes in 2D layered systems (a description in which alternating barriers and effective layers corresponding to the QDs are considered) extending over the whole QD multilayer. More refined modeling of acoustic properties of QDs (including size, shape and matrix effects) will be discussed in Sect. 4: they are, however, limited to single QDs. As we are interested in collective effects, i.e., interference effects, it is crucial to adopt a tractable, and thus simple, description. Raman scattering by acoustic phonons will be discussed considering deformation potential interaction between spatially distributed confined electronic states and longitudinal acoustic modes extending over the whole structure.

Raman-scattering interferences, similar to those discussed above for GaAs/GaP QW structures, have been reported for self-assembled QDs multilayers: the first results were presented by *Cazayous* et al. [12] following a very basic experiment performed on Ge/Si QD bilayers. Later, studies on multilayers were reported [13, 14]. This section is devoted to Raman-scattering interferences in self-assembled QDs. It is organized as follows. After a brief presentation of QD self-assembly, we will discuss the spatial localization (form factor) and spatial correlation (structure factor) of the confined electronic states. Finally, surface effects will be addressed.

3.1 Self-Assembly and -Ordering of Quantum Dots

Quantum dots *self-assemble* according to the Stranski–Krastanow growth mode during growth of lattice-mismatched semiconductors (for instance, Ge on Si or InAs on GaAs). Below a critical thickness (typically a few monolayers) growth proceeds in a two-dimensional mode. On top of this so-called wetting layer (WL), three-dimensional islands form, providing efficient strain relief. Once capped, effective three-dimensional electronic confinement is obtained. These QDs are coherently strained and exhibit narrow size distributions. Various shapes (from faceted pyramids to rounded domes) have been reported; one should note that the height/width or diameter ratio is often typically one to ten. Self-assembled QDs are rather complex systems. Physical properties are determined by the interplay between shape, size and strain effects (see, for instance, [63]). Moreover, material interdiffusion and mixing often occur during QD growth and capping (see, for instance, [64]). The *self-assembly* may provide spectacular *self-ordering.* When QD layers are stacked, the burried dots influence the nucleation in the subsequent layers. This interaction occurs via elastic strain fields and may induce, for instance, vertical QD alignment [66–70]. Figure 8 shows a transmission electron microscopy (TEM) image by *Kienzle* et al. [65] that demonstrates the vertical alignment of Ge/Si QDs. Self-organisation also includes inplane ordering and three-dimensional QD superlattice formation [71, 72] and it was shown to improve significantly

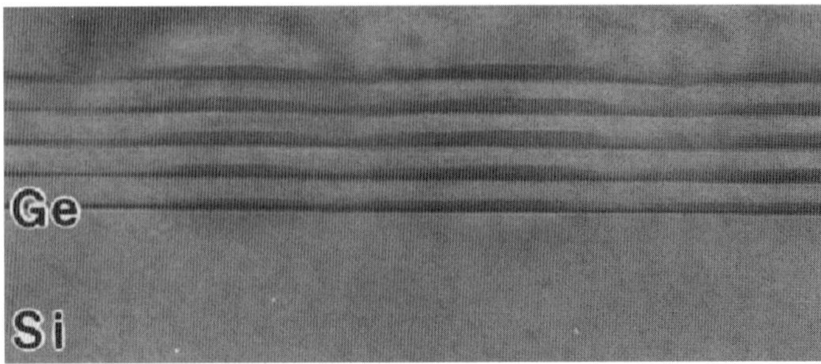

Fig. 8. Tranmission electron microscopy image of a five-fold Ge/Si QDs multilayer. From [65]

size homogeneity [68,69,71]. A detailed review of structural properties of self-organized QDs was presented recently by *Stangl* et al. [73]. Coupling between spatially correlated QDs provides a means of tuning electronic properties [74].

3.2 Spatial Localization

Interaction between a bulk-like acoustic wave (defined by the wavevector components q_z and $q_{\|}$) and electronic states confined within a QD provides the following form factor

$$\int \varphi_n^*(z)\varphi_{n'}(z)\,\mathrm{e}^{-\mathrm{i}q_z z}\,\mathrm{d}z \int \psi_m^*(\boldsymbol{r})\psi_{m'}(\boldsymbol{r})\,\mathrm{e}^{-\mathrm{i}\boldsymbol{q}_{\|}\cdot\boldsymbol{r}}\,\boldsymbol{d^2r}\,, \quad (7)$$

where $\varphi_{n\,\mathrm{or}\,n'}(z)$ and $\psi_{m\,\mathrm{or}\,m'}(\boldsymbol{r})$ are electronic wavefunctions that account for the confinement of the states (n,m) and (n',m') along the growth direction (z) and in plane $(\boldsymbol{r})$, respectively. Due to the lack of translation invariance the wavevector-conservation law breaks down. As a result of the three-dimensional (3D) confinement neither q_z nor $q_{\|}$ are conserved. The form factor is thus given by the Fourier transform of electronic density along the growth direction and in plane.

Selection of electronic states occurs via resonance. Resonant excitation is a critical issue for Raman scattering by acoustic and optical vibrations in self-assembled QDs. Indeed, the material involved in a QD layer is often as little as a few atomic layers. For instance, Ge LO phonons in Ge/Si QDs multilayers were shown by Milekhin et al. to display strong resonances (Fig. 9). The resonance peak at about 2.34 eV is attributed to the E_1 transitions of the Ge QDs [17,53]. The downshift of the LO frequency (from 2.5 to 2.7 eV) results from the resonant contribution of small Ge QDs [17]. The E_1 transitions provide strong but broad resonances, [12,17,53] analogous to those in bulk material [75,76]. As a matter of fact, studies of resonant Raman scattering by acoustic phonons in self-assembled QDs deal usually with E_1 resonances.

One should note that for a given QD, due to the large effective masses around the L point, the E_1 confinement-induced energy splittings and shifts

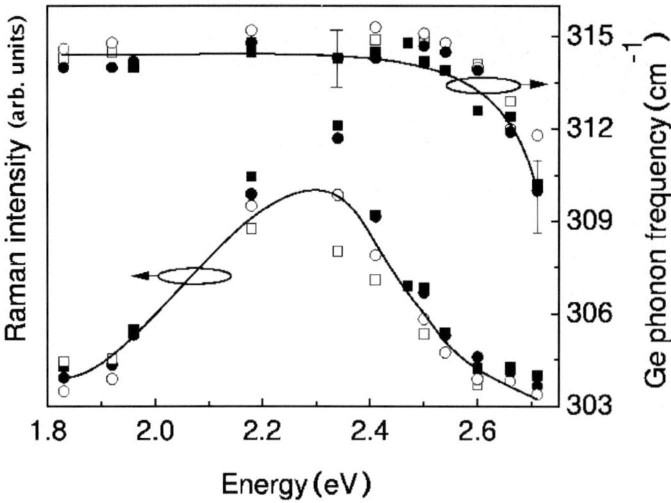

Fig. 9. Dependence on excitation energy of the Ge LO-phonon Raman efficiency and frequency for a Ge/Si QD multilayer. From [17]

are very small [77]. It is thus appropriate to consider a regime with localization at nanoscale, rather than a quantum regime. This implies that: 1. no particular quantum states are selected via resonance. 2. owing to the small acoustic phonon energies, both incoming and outgoing resonance conditions (double resonance) are fullfilled. Owing to the very broad E_1 resonances one may often assume that all QDs have the same resonance factor. In the size-distribution tails, however, precautions should be taken (see, for instance, Fig. 9). Size-selective resonance can be included easily in the Raman intensity calculations (see, for instance, *Sirenko* et al. [25]).

Figure 10 shows Raman scattering from InAs/InP QDs excited in the vicinity of the InAs E_1 transitions and detected in both optical- and acoustic-phonon frequency regions. Two resonances – one corresponding to the 2.41–2.54 eV range and another to the 2.54–2.73 eV range – can be identified in Fig. 10; they were assigned to the wetting layer and QDs, respectively. In the spectral range of the InAs optical phonons, the spectra display features related to confined and interface modes. The latter have their counterpart in the InP frequency range; the InP LO peak is asymmetric towards its low-frequency side, due to scattering by InP-like interface phonons. The intensity oscillations observed in the acoustic-phonon spectral range are related to interference effects to be discussed in Sect. 3.3. Strikingly, the interference envelopes differ significantly for the two resonances. According to (7), the localization of the electronic states differs. The QD-related electronic states are well confined within the QDs, whereas the wetting-layer-related states spread with an exponential decay into the InP barriers (the wetting layer is about 2 monolayers thick, similar to the GaAs/GaP case discussed previously

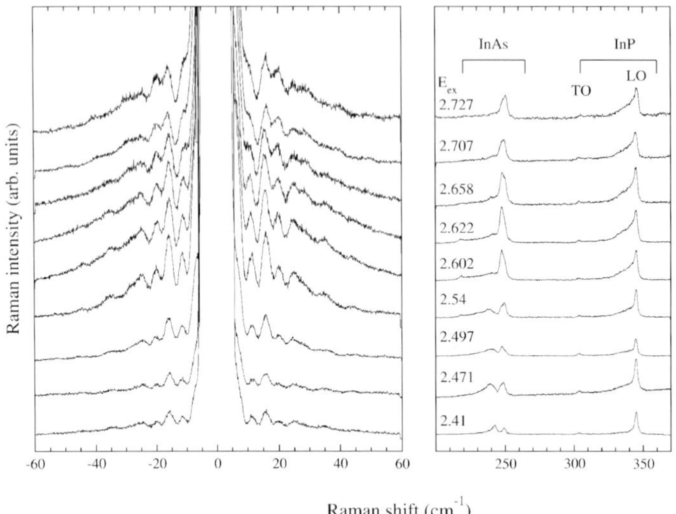

Fig. 10. Room-temperature Raman spectra of InAs/InP QDs (h = 3 nm and ϕ = 20 nm) recorded in the acoustic- and optical-phonon frequency regions (*left* and *right panel*, respectively) with excitation energies ranging from 2.41 to 2.72 eV. From [11]

in connection with Figs. 4 and 7) [78]. Noteworthy, samples with uncapped QDs solely display the QD-related resonance; this is likely due to oxidation of the initial wetting layer [11].

Confinement determines the acoustic phonons that may contribute to the Raman signal. The more the electronic state is localized along a given direction, the larger the wavevector component along this direction that contributes to the scattering. For most of the self-assembled QDs, the diameter ϕ is about one order of magnitude larger than the height h. Hence, the phonons giving a significant contribution do have small inplane wavevector components $q_{||}$. Their wavevector orientation is thus close to the growth axis. One may therefore assume pure longitudinal acoustic modes with isotropic dispersion. Under such circumstances, the interference envelope is more sensitive to h than ϕ, as shown in Fig. 11. One should note that, unlike the case of perfect quantum wells, many acoustic modes contribute to the scattered intensity at a given wave number $I(\omega)$; i.e., all those satisfying $\omega = v\sqrt{q_z^2 + q_{||}^2}$ (where v is the speed of sound), their contributions being weighted by (7).

Figure 12 shows experimental and calculated Raman spectra from Ge/Si QDs with various QDs sizes. According to the form factor (7), the smaller the QDs, the farther the envelope extends in the spectrum. One should note that this extent is determined not only by the QDs size but also by the speed of sound. The relevant speed is the one corresponding to the material in the QD. Intermixing effects that occur during the QD self-assembly and capping

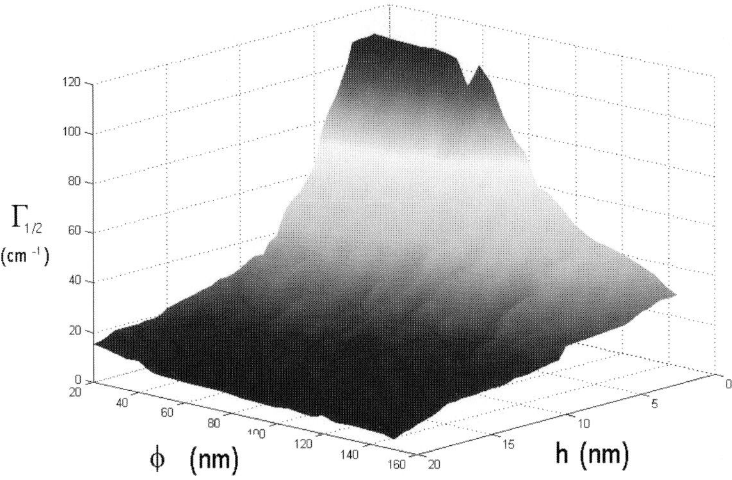

Fig. 11. Width at half-maximum of the interference envelope (Fourier transform of the electronic density) calculated for Ge/Si QDs as a function of QD height h and diameter ϕ [78]

have to be taken into account; for instance, the actual Ge content in nominal Ge/Si QDs might be about 75 % [12, 14].

The actual composition can be derived rather accurately from the interference effects discussed in Sect. 3.3. Raman scattering by optical phonons has been shown to provide an additional and reliable means to measure the chemical composition in self-assembled QDs [49, 81]. Note that the interference envelope depends also on the inplane extension of the electronic state. The latter is systematically found to be smaller than the island diameter (16, 20, 60 and 70 nm for A, B, C and D, respectively). This is likely due to the actual island shape and strain fields. Unlike the QW spectra presented in Fig. 7, the spectra calculated for the QDs vanish at zero wave number. This behavior is a direct signature of the three-dimensional confinement of the electronic states. Indeed, close to zero the intensity scales like q^2, according to the activation of a 3D acoustic phonon DOS (scales like q^2), the Bose-Einstein population factor ($1 = q$ for small ω), the deformation potential interaction (q^2), and the normalisation of the displacement elds $(1/\sqrt{q})^2$. The strong Rayleigh scattering hinders its observation; see, however, A in Fig. 12.

The interference envelope is quite sensitive to small size changes. Figure 13 shows the average QD heights deduced from the interference envelope in fivefold Ge/Si QDs multilayers versus the interlayer spacing. The elastic interaction between layers modifies the QD heights during the stacking process. For thin spacing this interaction is strong: it yields preferential QD nucleation sites with lowered misfit. Within this strong-interaction regime, the QD height thus increases in the first two layers and stabilizes in the next

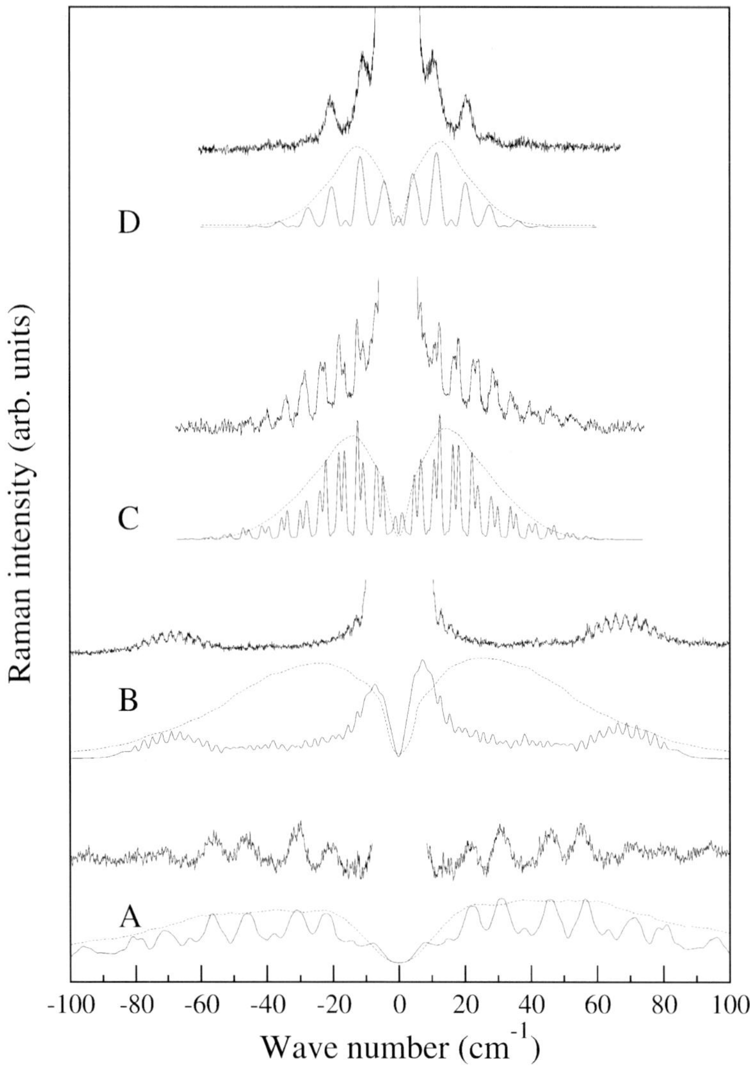

Fig. 12. Experimental (noisy curves) and calculated Raman spectra from Ge/Si QDs with different QD sizes h and ϕ. Calculated interference envelopes are also plotted. A: small islands ($h = 2\,\mathrm{nm}$ and $\phi = 20\,\mathrm{nm}$) [79]; B: hut clusters ($h = 3\,\mathrm{nm}$ and $\phi = 25\,\mathrm{nm}$) [80] C and D: planoconvex islands similar to those shown in Fig. 8 ($h = 6\,\mathrm{nm}$ and $8\,\mathrm{nm}$, $\phi = 80\,\mathrm{nm}$ and $100\,\mathrm{nm}$, respectively)

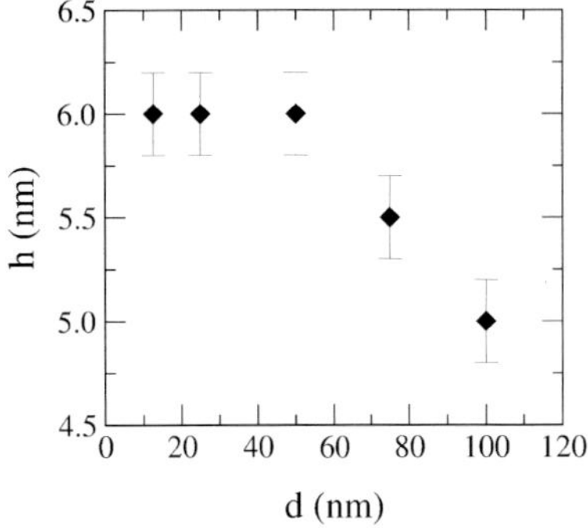

Fig. 13. Average QD height h deduced from the interference envelope in fivefold Ge/Si QDs multilayers versus interlayer spacing d. From [14, 78]

ones [64]. The mean height in the QD layer stack is therefore larger than that observed for a single layer. This increase does not occur for thick spacings (low elastic interaction), resulting in a lower mean height. The height changes reported in Fig. 13 are consistent with these growth regimes.

3.3 Spatial Correlation

In this section we discuss the effects of many QDs by considering the interaction between an ensemble of localized electronic states and delocalized acoustic modes. In particular, we shall address how Raman scattering depends on the QDs spatial distribution within multilayers, i.e., the stacking of QDs layers along the growth direction and the inplane distributions of QDs within the layers.

3.3.1 Layer Stacking

In order to discuss finite stacking of QDs layers, we focus on the part of the structure factor related to the growth axis z. A simplified, but useful, structure factor can be derived if one assumes that the localized electronic states are distributed within a single acoustic medium (no acoustic mismatch between QDs and barriers). This medium is semi-infinite: the QD layer stack is enclosed between an infinite substrate and a finite cap layer. Each acoustic mode is given by the superposition of two counterpropagating plane-wave components, labelled $+q_z$ and $-q_z$ for the component propagating away and towards the sample surface, respectively. Stress-free boundary conditions are applied at the surface. The light is assumed to propagate in a homogeneous medium as well. QDs with vertically correlated distributions (the inplane distribution of the QDs is the same in each layer) and indentical form and resonance factors are considered. Performing the coherent sum over the N

QD layers, one finds that the contribution of a given mode to the Raman-scattering intensity is proportional to [14]:

$$H^2(q_z + \Delta k_z) + H^2(q_z - \Delta k_z) - 2H(q_z + \Delta k_z) \times H(q_z - \Delta k_z) \cos\left[2q_z\left(t + (N-1)\,d/2\right)\right] , \quad (8)$$

with

$$H(Q) = \frac{\sin(\frac{Nd}{2}Q)}{\sin(\frac{d}{2}Q)} , \quad (9)$$

where d is the spacing between QD layers and H the interference function one usually finds in optics textbooks [82]. The first term in (8) is due to the $+q_z$ component and the second one to $-q_z$. The third term in (8) includes contributions of both components and depends on the location of QDs layers stack with respect to the sample surface. $t + (N-1)d/2$ is the distance between the middle of the QD layer stack and the sample surface (t gives the location of the first QD layer with respect to the sample surface). The cosine function in the third term corresponds to the surface effects to be discussed in Sect. 3.4. For thick cap layers (for instance $t \approx 100\,\mathrm{nm}$), the corresponding oscillations are not resolved experimentally. In the current section we disregard this modulation. Equation (8) is of particular interest as it allows one to easily understand how the main features observed in the experimental and calculated spectra depend on the number of QDs layers N and the interlayer spacing d. The well-known interference function H yields intensity oscillations. Intensity maxima (minima) correspond to constructive (destructive) interferences comparable to the bright and dark fringes in wave optics. The phase conditions yielding maxima differ for the $+q_z$ and $-q_z$ components of a given vibrational mode. Indeed, intensity maxima are obtained for

$$(q_z \pm \Delta k_z)\frac{d}{2} = \pi n , \quad (10)$$

n being an integer. For a given wave number in the spectrum, the contributions $H(\Delta k_z + q_z)$ and $H(\Delta k_z - q_z)$ are thus different. Consequently, if one considers the whole spectrum, i.e., different wave numbers, series of doublets appear. As displayed in Fig. 14, the interference period varies as $1/d$ (10). The doublet splitting is given by the shift between $H^2(\Delta k_z + q_z)$ and $H^2(\Delta k_z - q_z)$ and therefore displays beats vs $\Delta k_z d$. Δk_z is the exchanged wavevector (determined by the scattering geometry and excitation wavelength). Zero splittings in Fig. 14 are obtained for $\Delta k_z d = \frac{\pi}{2}m$, m being an integer. Interestingly, in the spectra reported in Fig. 1 the $+q_z$ and $-q_z$ contributions are superposed (i.e., they yield equivalent interference conditions). Simultaneous contribution of the $+q_z$ and $-q_z$ components requires overlapping between $H(q_z + \Delta k_z)$ and $H(q_z - \Delta k_z)$ (8).

As an example, calculated and measured resonant Raman spectra of Ge/Si QDs multilayers with $N = 2, 5$ and 20 are reported in Fig. 15. Peaks corresponding to constructive interferences are observed. Their frequencies depend

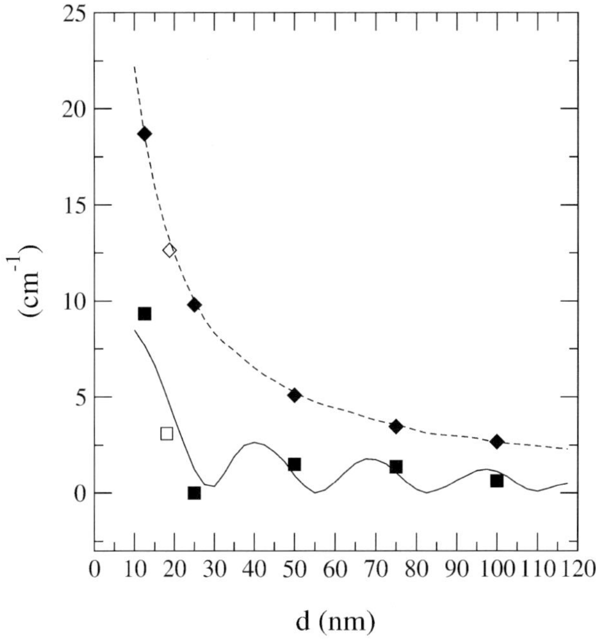

Fig. 14. Interference period (*dashed line*) and doublet splitting (*solid line*) calculated for Ge/Si QDs multilayers with $N = 5$ as a function of spacing d. Symbols are experimental data. From [14]

on Δk_z, whereas their interference period does not (10). Sound velocities of the QDs and barrier materials and their acoustic mismatch (i.e., some of the shortcomings from the above calculation; see (8)) are included in the Raman-scattering simulations presented below. The counterpropagating components of each acoustic mode are obtained by considering displacement and stress-field continuity at layer interfaces and a free sample surface. Light absorption is included as well.

Figure 15 clearly shows that the observation of doublet features depends greatly on the number of QDs layers. One has to compare the width of the constructive interference peaks and the doublet splitting. For large N values, one can distinguish easily between the $H^2(\Delta k_z + q_z)$ and $H^2(\Delta k_z - q_z)$ contributions: doublets are indeed observed (see $N = 20$ in Fig. 15). For small N, the peaks of $H^2(\Delta k_z \pm q_z)$ are broad; the overlap between those two contributions hinders the observation of the doublet features. It is noteworthy that the $N - 2$ weak secondary intensity maxima one could expect from the interference function H are not resolved.

The calculations account rather well for the peak frequencies, doublets splitting and relative intensities within each doublet and between doublets. These relative intensities are determined by the interference envelope (i.e., the electronic confinement) and the acoustic impedance mismatch. In or-

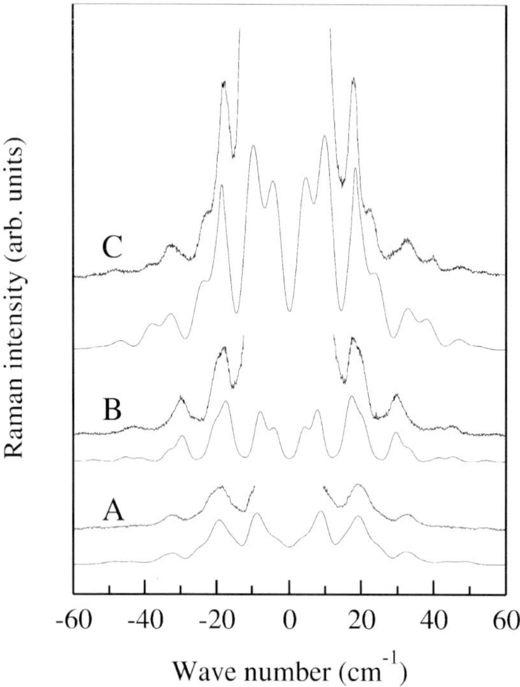

Fig. 15. Experimental and calculated resonant Raman spectra of structures consisting of $N = 2$, 5 and 20 Ge/Si QD layers (samples A, B and C). Note that the Si interlayer spacing equals 18.5 nm for samples A and B, and 20 nm for sample C; these thin Si spacers ensure vertical correlation. Calculated spectra include a 1 cm^{-1} experimental resolution. From [14]

der to derive analytically the structure factor (8) the acoustic impedance mismatch was not considered. It is included in the simulations (here $\eta = \rho_{\mathrm{Si}} v_{\mathrm{Si}} / \rho_{\mathrm{Ge}} v_{\mathrm{Ge}} = 0.78$).

We emphasize that neither the peaks nor the doublets are due to Brillouin-zone folding, phonon energy gaps or accumulations in the phonon density of states [83, 84]. The experiments and calculations presented here deal with finite multilayers, having few QDs layers. Consequently, Brillouin-zone folding, which requires periodicity of the vibrational properties, i.e., superlattices (SL), is not relevant. Phonon energy-gap opening requires multiple constructive wave reflections, i.e., many layers [83].

At this stage, a somewhat puzzling fact – at first sight – has to be discussed. *Rytov's* model [85], designed for ideal periodic SL, accounts well for the peak positions in the spectra of finite-size QDs multilayers [14–16, 57, 58]. Figure 16 shows results obtained by *Milekhin* et al. on a 10-fold Ge/Si QD layer stack [57]. Up to 6 doublet features are observed; their frequencies fit well with the dispersion calculated assuming an infinite SL (Fig. 16).

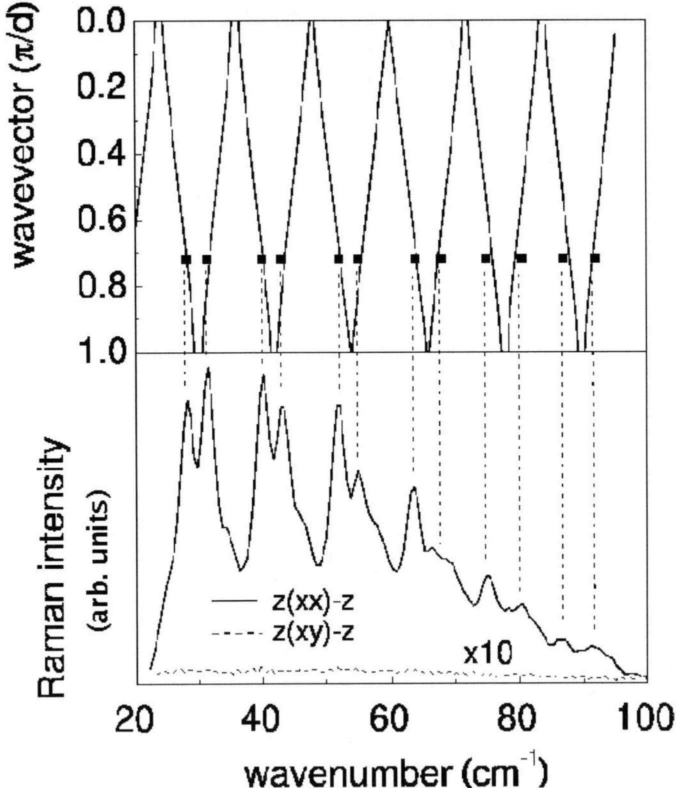

Fig. 16. *Lower panel*: spectrum of a 10-fold Ge/Si QDs layer stack with 20 nm spacing layers. *Top panel*: calculated SL dispersion. Note the small phonon energy gaps at the minizone boundaries. From [57]

In superlattices (i.e., periodic multilayers extending both sides to infinity) the Brillouin zone is folded into a minizone (delimited by $q = 0$ and $q_m = \pm\pi/d$) and phonon energy gaps are opened at the minizone boundaries [86]. *Rytov's* model is usually used to calculate the acoustic-phonon dispersion within the SL Brillouin minizone $\omega(q_{\mathrm{SL}})$ [85, 87]. Assuming no acoustic mismatch, the SL dispersion relation derived from Rytov's model can be written in the following simplified form:

$$\omega = v\left(\pm q_{\mathrm{SL}} + 2\pi\frac{m}{d}\right) , \tag{11}$$

where v is the mean sound velocity and m the folding index [88,89]. Owing to the SL periodicity, only the phonons satisfying the wavevector-conservation rule contribute to the scattering: $q_{\mathrm{SL}} = \Delta k_z$. One can thus conclude that deriving the positions of intensity maxima from (10) (with $\omega = vq_z$) or from (11) (with $q_{\mathrm{SL}} = \Delta k_z$) is equivalent.

One should, however, note that, when the sound velocities in the QD and barriers are considered in the simulations (instead of a mean velocity), the frequencies predicted by the interference scheme differ slightly from those predicted by Rytov's model for small N values. However, the former tend towards the latter rather rapidly (typically for $N \geq 5$) [14].

May we infer that the zone-folding scheme and the interference one are equivalent? Certainly, not. Significant differences have to be pointed out. Because of the periodicity of an infinite SL, crystal momentum is conserved in the scattering process and sharp peaks (doublets) are observed. In the interference scheme no periodicity is required. Wavevector is not conserved, all modes are likely to contribute, and peaks appear because of constructive interferences (their width depends on the number of layers N (Fig. 15)). In Rytov's model zone folding and gap opening originate in the difference between the acoustic properties of the two materials. As Raman-scattering interferences are due to the spatial localization and distribution of the electronic states, they do not require acoustic mismatch: localized electronic states distributed within a single acoustic medium are likely to lead to interferences.

In the interference scheme no periodicity has to be assumed; it can therefore be applied whatever the number of QDs layers, i.e., from a few layers to SL [14]. In contrast, the zone-folding scheme is valid only for ideal periodic SL. The low-frequency Raman scattering in finite QD multilayers cannot be assigned to zone folding and subsequent wavevector conservation. The zone-folding scheme has a solely practical interest as it allows one to easily predict how peaks or doublets shift and merge together when varying the interlayer spacing. Changing d is equivalent to exploring minizone dispersion branches.

There have been some reports on Raman scattering in QD multilayers invoking folded acoustic modes contributions (despite the often small number of QD layers involved) [15, 16, 57, 58]. Assignments were made by considering only the peak frequencies. To determine the scattering mechanism one very often has to go beyond this simple procedure and perform Raman-scattering intensity simulations.

Milekhin et al. [15, 16] reported the simultaneous observation of oscillations and a broad continuous emission. They suggested that the oscillations are superimposed on the continuous emission, and invoked different scattering mechanisms: oscillations were assigned to folded acoustic phonons and continuous emission to the breakdown of the crystal momentum conservation. Within the interference scheme discussed above, it is possible to assign the simultaneous observation of oscillations and continuous emission to a single mechanism: it may result from interferences with a limited or reduced interference contrast (to be discussed below; Sect. 3.3).

Low-frequency Raman spectra of GaAs and AlAs QDs embedded in InAs were reported by *Tenne* et al.; fivefold stacks were shown to display doublet peaks that were assigned to folded acoustic phonons [58]. There is a noticeable difference from the results discussed above. The measurements were performed in resonance with the E_1 transition of the InAs barriers (GaAs and AlAs dots embedded in InAs actually form antidots). Identifying the electronic states involved in the scattering is thus a crucial issue.

Resonance is, however, not a prerequisite for observing Raman-scattering interferences. Indeed, *Giehler* et al. [90] reported interference effects in acoustic Raman scattering from GaAs/AlAs finite-size mirror-plane superlattices. Interferences were shown to arise from phase shifts between the contributions of different layers to the scattering intensity. Experimental results were compared to calculations performed assuming photoelastic scattering. The interferences depend sensitively on the stacking sequence and the modulation of the photoelastic constants [90]. The photoelastic model is useful far from electronic resonances [86] and has been widely used to calculate low-frequency Raman spectra of planar superlattices [83, 87, 91]. The scattering efficiency within each layer is included by means of photoelastic constants (instead of the electron–phonon interaction in (7)) and the contributions of the different layers are summed coherently. Obviously, similarities with the interferences arising from resonant excitation of localized electronic states exist. The modulation of the photoelastic constants and the localization of the electronic states play equivalent roles; they determine the form factor and the envelope of the signal, and thus the relative intensity between and within doublets. The intensities reflect the Fourier components of the profiles [86, 92]. We emphasize that a very strong modulation of the photoelastic constants in the photoelastic model is equivalent to considering strongly localized electronic states in the resonant Raman-interference model. For instance, the modulation $P_{\mathrm{Ge}} = 10$ and $P_{\mathrm{Si}} = 1$ considered in [84] can be transposed into the corresponding step profile of the photoexcited electronic states. One should keep in mind that photoelastic constants may be complex numbers and undergo noticeable changes versus photon energy [89, 93, 94]. Most measurements are performed in energy regions where at least one of the constituents absorbs light and hence resonance effects cannot be ruled out. In particular, the Ge/Si and InAs E_1 resonances were shown to be very broad [75, 76].

Within the photoelastic model, the scattering intensity is written as the square modulus of the Fourier transform of the polarization field [95]. Unlike for resonant scattering from localized electronic states, one performs a single integral, including the phonon- and photon-related terms. The contributions of the photons therefore differ in those two descriptions. It is noteworthy that in the calculation of photoelastic scattering in superlattices (including elastic, photoelastic and complex refractive-index modulations) presented by *He* et al. [95], expressions given for backward and forward scattering include the interference function defined above (9).

The interference effects discussed in [90] and in [12–14] are basically similar: they originate in the coherent sum from different parts of the structures and can be discussed in terms of spatial distributions (structure factors). They allow the investigation of finite-size effects otherwise hardly accessible [14, 90].

It is worth mentioning that interferences may also arise when different electron–phonon coupling mechanisms have to be considered simultaneously [96, 97].

3.3.2 Vertical Ordering

The Raman-scattering interferences depend on electronic confinement within the QD (form factor) and on relative QD positions (structure factor). It is well known that the structure factor determines not only the interference period but also the interference constrast. Unlike the interference envelope and interference period, the interference contrast has not been discussed yet. In order to illustrate how the Raman-scattering interferences depend on the spatial distributions of the localized electronic states, we shall now consider explicitly three-dimensional distributions of QDs in multilayers.

Different spatial-correlation regimes can be explored by varying the interlayer spacing in QD multilayers [65]. The results presented below deal with fivefold Ge/Si QDs stacks belonging to the series presented in [65, 98]. The vertical alignment of the QDs was investigated by means of TEM [65] and X-ray diffraction [98].

Figure 17 presents both experimental and simulated spectra of Ge/Si QD multilayers [13]. Together with each experimental spectrum, two calculated spectra are shown; they were calculated considering perfect vertically correlated island distributions (QDs have identical positions within the five layers, no inplane ordering) and random QD distributions in all layers (finite sampling effects were avoided by summing spectra calculated for many QD distributions).

Prior to focusing on spatial-correlation effects, it is worth commenting on the stacking effects discussed in the previous section. The experimental data reported in Fig. 14 were actually taken from Fig. 17. Unlike for spectra C, D, and E, identifying the interference period and doublet splitting is not straightforward for spectra A and B (Fig. 17). Indeed, the doublet splitting vanishes for spectrum B and is about half the interference period for spectrum A (Fig. 14). Once the contributions of the $+q_z$ and $-q_z$ components are identified, one can measure the interference period; it decreases when d increases. For a given number of QD layers (here $N = 5$), the width of the constructive-interference peaks decreases rapidly when d increases, as expected from (9).

Let us now discuss the interference contrast. Concerning the experimental data, Fig. 17 clearly shows that the interference contrast decreases with increasing interlayer spacing d. For a given interlayer spacing, simulations show that the interference contrast is strong for vertically correlated QDs and weak for random QD distributions. On the one hand, the experimental spectra A, B and C are similar to those calculated with vertically correlated QDs. On the other hand, spectra D and E are rather similar to those calculated with random QD distributions.

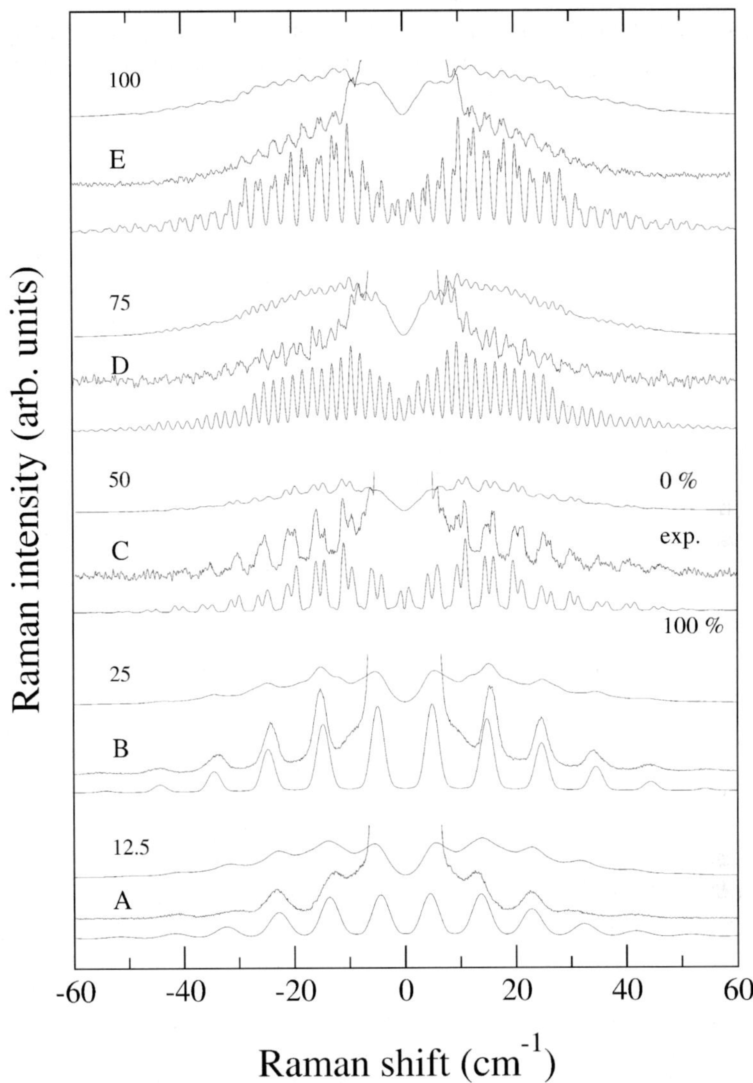

Fig. 17. Experimental (noisy curves labelled exp.) and simulated (0 % and 100 % stand for random and vertically aligned QDs, respectively) spectra of Ge/Si QDs multilayers ($N = 5$). d ranges from 12.5 to 100 nm. The QDs have the shape of planoconvex lenses with a mean height $h \simeq 6$ nm and base diameter $\phi \simeq 85$ nm. Spectra A and B were obtained with an experimental resolution of $2\,\mathrm{cm}^{-1}$; spectra C, D and E were recorded with $1.2\,\mathrm{cm}^{-1}$ resolution in order to resolve fine structures. From [13]

In order to illustrate how the interferences depend on the QD's spatial distributions, let us consider localized electronic states that are distributed within an infinite homogeneous acoustic medium. These electronic states interact with an acoustic mode, i.e., a plane wave. Light is assumed to propagate (along the growth axis) in a homogeneous medium as well. The corresponding Raman intensity $I(q_z, q_{||})$ is proportional to [13]:

$$\sum_{\mathrm{p}} S_{pp}(\boldsymbol{q}_{||}) + 2\Re e \left(\sum_{p \neq p'} \mathrm{e}^{\mathrm{i}(\Delta k_z + q_z)(z_p - z_{p'})} S_{pp'}(\boldsymbol{q}_{||}) \right) , \quad (12)$$

where

$$S_{pp'}(\boldsymbol{q}_{||}) = \sum_{l_p, l'_{p'}} \mathrm{e}^{\mathrm{i}\boldsymbol{q}_{||} \cdot (\boldsymbol{r}_{l_p} - \boldsymbol{r}_{l'_{p'}})} , \quad (13)$$

where $S_{pp}(\boldsymbol{q}_{||})$ and $S_{pp'}(\boldsymbol{q}_{||})$ are spatial correlation factors between electronic states within layer p and between layers p and p', respectively. The electronic state location (QD position) is given by z_p and $\boldsymbol{r}_{l_p}$ along the growth direction and inplane, respectively. The electronic states under consideration are those that are excited optically and play the role of intermediate states in the resonant light-scattering process.

If one considers identical QDs inplane distribution in each layer, $S_{pp} = S_{pp'}$ with p and p' arbitrary; the Raman intensity is proportional to $S_{pp}(\boldsymbol{q}_{||})H^2(q_z + \Delta k_z)$ (8). For the sake of simplicity, a single acoustic-wave component ($+q_z$) was considered to derive (12). By including the $-q_z$ component, the second interference term ($H^2(q_z - \Delta k_z)$) and the mixed term ($H(q_z + \Delta k_z)H(q_z - \Delta k_z)$) would appear; see (8). Whereas interlayer inplane distributions $S_{pp'}$ might be correlated, within each layer the QD distribution is random by nature. No constructive interferences are therefore expected from the S_{pp} structure factors.

Equation (8) refers to a given acoustic mode (q_z and $q_{||}$). At a given energy or wave number, the Raman-scattering intensity $I(\omega)$ includes the contributions of many acoustic modes; these modes have the same energy but different q_z and $q_{||}$ components ($\omega = v\sqrt{q_z^2 + q_{||}^2}$). One has to keep in mind that the weight of these contributions is given by the Fourier transform of the probability density distribution of the photoexcited electronic states (7).

Electronic confinement in the QDs determines the phonons that may contribute to the Raman signal. The more the electronic state is localized along a given direction, the more the wavevector components along this direction contribute. Here, the QD diameter is about one order of magnitude larger than its height. Hence, the phonons giving a significant contribution do have small inplane wavevector components $q_{||}$. Their wavevector orientation is thus close to the growth axis. One may therefore consider pure longitudinal acoustic modes with isotropic dispersion.

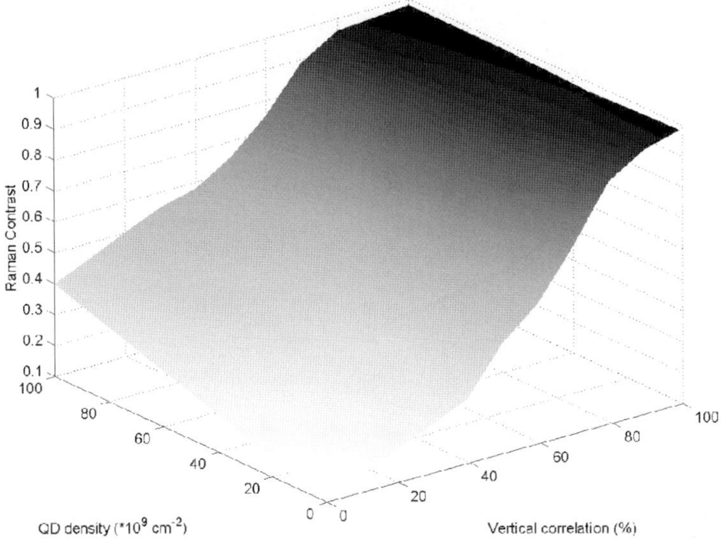

Fig. 18. Raman-interference contrast as a function of QD density and vertical ordering degree. From [99]

The interference contrast thus depends on both the localization of the electronic states excited within the QDs and the spatial correlation factors $S_{pp'}$ between QD layers. To probe interlayer correlations, the contributions of $q_{||}$ components are required; they are provided by the lateral localization of the electronic states. Strong lateral localization implies activation of large $q_{||}$ components. On the contrary, for very large QD diameters, only $q_{||}$ components close to zero are activated: this is similar to the two-dimensional quantum-well case. As far as spatial correlations are concerned, perfectly correlated QD multilayers behave similarly to quantum-well multilayers.

The interference contrast may also strongly depend on the QD density within the layers [99]. The higher the density, the more terms have to be summed in the $S_{pp'}$ correlation factors. Figure 18 shows how the interference contrast depends on both the ordering and the QD density. For a given QD density, the contrast increases progressively with the degree of vertical correlation. For perfect vertical alignment, the contrast is independent of the QD density. For random distributions a residual contrast is observed. Unlike for perfect vertical correlation (100 %), it much depends on the density. This residual constrast is a signature of a well-defined spatial correlation that remains whatever the inplane distributions and corresponds to the stacking along the growth axis. The ordering can be probed via the $S_{pp'}$ correlation factors. The way a given spatial distribution (i.e., a specific density and vertical alignment degree) is probed depends on the $q_{||}$ components that are activated (i.e., the lateral electronic confinement). For a limited degree of

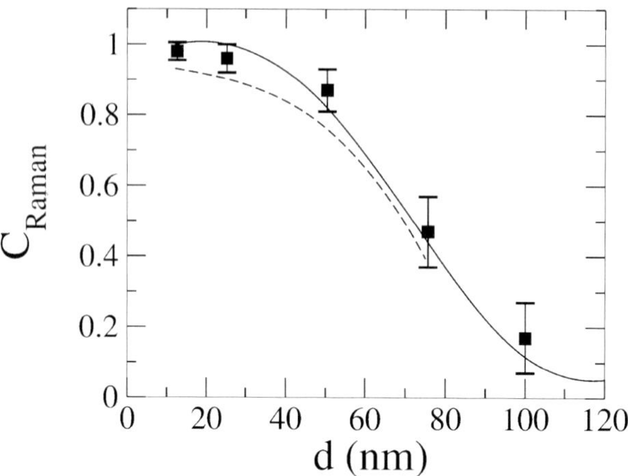

Fig. 19. Raman-interference contrast C_{Raman} vs. interlayer spacing d (*symbols*) [13, 100]. Comparison with the degree of vertical ordering measured by TEM (*solid line*) [65] and X-ray diffraction (dashed line) [98]. *Error bars* correspond to 15 % heigth fluctuations [13, 100]

correlation, one expects the contrast to decrease when the lateral localization gets stronger.

It is worth noting that light absorption may also limit the interference constrast. It determines, via the photon–electron interaction (for both the excitation and emission steps), the contribution of the QDs belonging to each layer; light has to reach the QD layer and to escape from the multilayer. Optical absorption must thus be included in the simulations.

In order to derive quantitative information on the degree of order from the experimental spectra (Fig. 17), one can define the following normalized Raman-interference contrast: $C_{\mathrm{Raman}} = (C^{\mathrm{exp}} - C^{\mathrm{ran}})/(C^{\mathrm{cor}} - C^{\mathrm{ran}})$, where C^{exp} is the experimental contrast, C^{ran} the contrast calculated with random QD distribution and C^{cor} the contrast calculated with vertically correlated QD's [13]. Perfect vertical correlation yields $C_{\mathrm{Raman}} = 1$, whereas a random distribution gives $C_{\mathrm{Raman}} = 0$. C_{Raman} thus provides a means of measuring spatial correlations. C_{Raman} is reported as a function of the interlayer spacing in Fig. 19. Clearly $C_{\mathrm{Raman}} \to 1$ for short spacings and $C_{\mathrm{Raman}} \to 0$ for larger ones; C_{Raman} undergoes a rather steep transition for $d \approx 70\,\mathrm{nm}$. Strikingly, this behavior is similar to that reported for the QD alignment degree P measured by transmission electron microscopy (TEM) and by X-ray diffraction [65, 98, 100].

Raman interference provides good statistical averaging over very large QD ensembles, similarly to X-ray diffraction. In addition to the spatial correlations discussed above, any structural fluctuation and inhomogeneity is likely to reduce the interference constrast. For instance, QD-height fluctuations (obviously present in real samples) reduce the interference constrast.

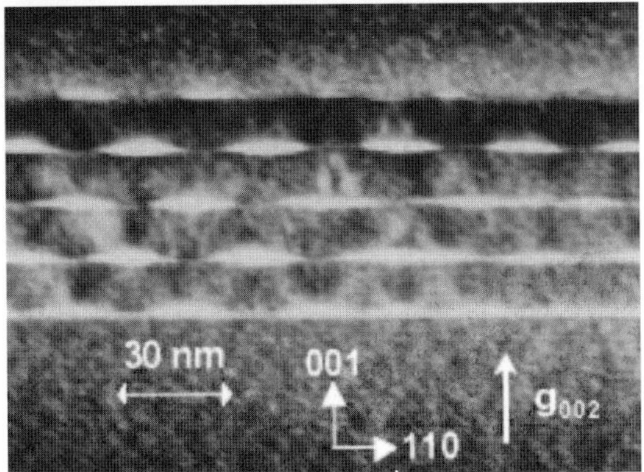

Fig. 20. Cross-sectional TEM image of a InAs/InAlAs multilayer. From [101]

Note that these fluctuations affect both the form factor (the individual QD response or envelope is changed) and the structure factor (the projection along the growth axis of the distance between QDs belonging to different layers fluctuates); they can be included easily in the simulations (Fig. 19).

The example provided above demonstrates that relevant structural information can be derived from Raman-scattering interferences. Here one takes advantage of the broad E_1 resonance of the QDs, i.e., resonance and form-factor effects are weak. Quantitative analysis requires, however, to compare the experimental data to simulations, like in X-ray diffraction experiments.

3.3.3 In-plane Ordering

The Ge/Si QD multilayers discussed above display vertical ordering but random inplane distributions. We shall now discuss structures displaying both vertical and inplane ordering.

Self-assembled InAs/InP QD multilayers display vertical ordering (similar to Ge/Si and InAs/GaAs QD multilayers). When InAlAs spacers (lattice matched to InP) are grown instead of InP ones, remarkably staggered arrangements are obtained [101, 102]. InAs actually forms wires elongated along $[1\bar{1}0]$. The cross-sectional TEM image shown in Fig. 20 evidences the staggered arrangement.

This arrangement results mainly from a phase separation taking place in the InAlAs spacer layers. In-rich V-like arms originate from each InAs wire (see the contrast changes in Fig. 20). The crossing points of two adjacent arms is a preferential nucleation site for a new InAs wire in the subsequent layer. With respect to its nearest neighbors in the layer underneath, its position is shifted by half the inplane separation between wires. One may wonder whether one can find some signature of this staggered arrangement (or vertically anticorrelated distribution) in the Raman-scattering spectra.

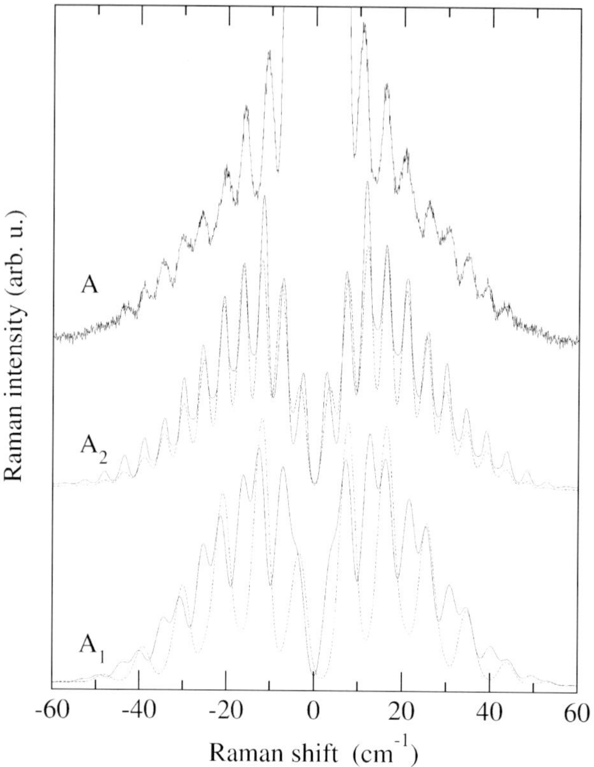

Fig. 21. Spectra of a InAs/AlInAs quantum wire multilayer with 15 nm spacer layers. Spectra were simulated considering vertically correlated and staggered distributions (A_1 and A_2, respectively). *Dashed lines* correspond to spectra calculated considering separately the $+q_z$ and $-q_z$ acoustic-mode components. From [78, 99]

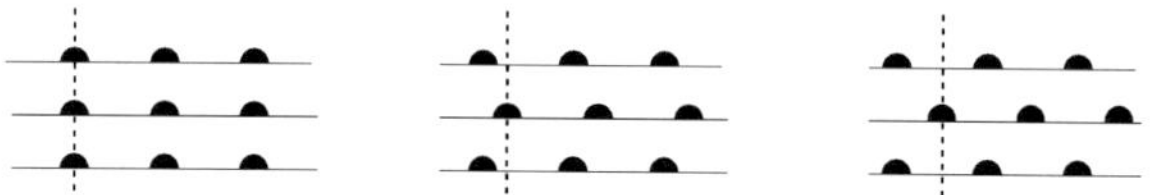

Fig. 22. Schematic plots of quantum wire multilayers (cross section). All plots display inplane ordering: within each layer wires are regularly spaced. The *left plot* shows vertical alignment. The *middle* and *right plots* are obtained from the left one by rigidly shifting the second layer (see the shift of the *dashed line*). When this lateral shift equals half the inplane spacing between wires, a staggered arrangement is obtained (*right*). An intermediate case corresponding to a smaller shift is also shown (*middle*)

Figure 21 shows the experimental spectrum (A) of a fivefold InAs/InAlAs multilayer [99]. In order to evidence how the Raman scattering depends on the specific arrangement simulated spectra are reported as well. They were calculated considering staggered (A_1) and vertically correlated (A_2) arrangements. Schematic plots of these arrangements are shown in Fig. 22. By comparing, for a given interlayer spacing, the staggered and vertically correlated distributions, one expects the interference period (related to the layer stacking) to be half that for the staggered arrangement. One should note that, owing to the 15 nm InAlAs interlayer spacing and laser excitation used, the $+q_z$ and $-q_z$ contributions in spectrum A_1 are superimposed, whereas they are shifted in spectrum A_2. This can be evidenced easily by performing simulations considering separately the $+q_z$ and $-q_z$ components (Fig. 21) [78]. Therefore, spectrum A_1 displays regularly spaced peaks, whereas spectrum A_2 displays doublet features. The interference period in spectrum A_1 is half that in spectrum A_2, as expected. A clear signature of the staggered arrangement can thus be found in spectrum A, i.e., in reciprocal space.

Notice that since a lateral superlattice is formed, wavevector conservation applies for the inplane component $q_{||} : q_{||} = m2\pi/d + (\Delta k_{||})$. Owing to the large lateral dimensions of the wires (Fig. 20), large $q_{||}$ do not contribute much.

InAlAs spacers in the 10–15 nm range allow adjacent In-rich V-like arms to join and provide well-ordered stacking sequences and improved wire-size homogeneity [101]. For thinner spacers, these arms do not join: during growth they reach separately the spacer-layer surface. Instead of a single nucleation site, one gets two potential nucleation sites. Rather complex stackings are observed [102]. One can, however, identify domains for which nucleation has occurred according to the same nucleation-site type (one of the two). Within these domains, the wire positions between subsequent layers are rigidly shifted, by less than half (50 %) of the inplane separation between wires (see middle panel in Fig. 22).

Simulations performed considering this simple spatial-distribution scheme compare actually rather well with experimental data. Figure 23 shows the experimental spectrum of a fivefold InAs/InAlAs multilayer with 5 nm InAlAs spacers and simulated spectra. The lateral positions of the wires between subsequent layers were progressively shifted in the simulations from 0 % to 50 % of the inplane wire spacing. The experimental spectrum displays some of the characteristic features (see symbols in Fig. 23) that are obviously present in the spectra with 30 and 40 % lateral-position shifts. From simple geometrical considerations, one can show that a 35 % shift is consistent with the spacer-layer thickness and the orientation of the In-rich V-like arms. Despite the rather simple structural picture considered in the simulations, characteristic features in the spectra are rather well accounted for.

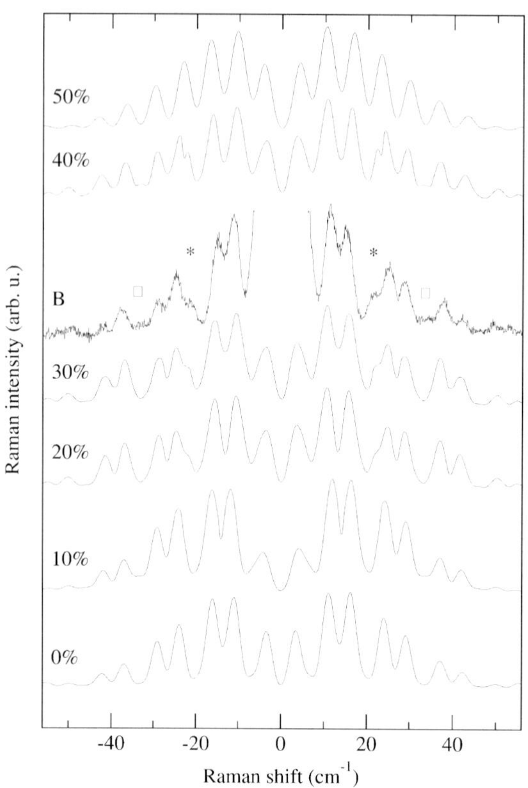

Fig. 23. Experimental spectrum (B) of a fivefold InAs/InAlAs multilayer with 5 nm InAlAs spacers and simulations. Lateral positions of the wires were progressively shifted from vertical ordering (0 %) to staggered ordering (50 %). The symbols (□, ∗) indicate some characteristic features [78]

3.3.4 Impurity Distributions

The substitution of As by N in GaAs leads to the formation of electronic states strongly localized around the nitrogen impurity. Hence, at low concentration the nitrogen impurities can be viewed as isolated quantum dots randomly distributed in a host matrix. This characteristic feature of dilute nitride systems is responsible for their special electronic and optical properties: giant bandgap bowing [103–105], large effective masses [106, 107] and unusual pressure dependence of the bandgaps [108, 109]. Calculations of the electronic structure of GaAsN and GaPN, based on a pseudopotential supercell technique, were developed by *Bellaiche* et al. [110], *Mattila* et al. [111] and *Kent* and *Zunger* [112, 113]. The composition and pressure dependence of the lowest-energy conduction states were calculated and compared with experiments [110–113]. It was shown that the optically active transitions involve conduction electron states strongly localized around the nitrogen im-

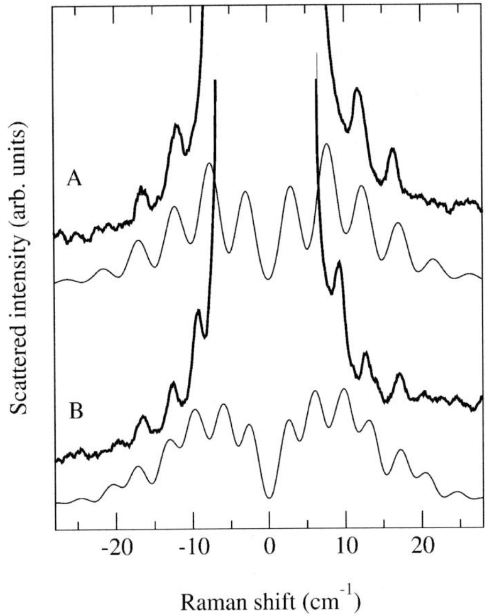

Fig. 24. Calculated (*thin lines*) and measured (*bold lines*) Raman spectra of the GaAsN/GaAs quantum-well structures. Optical excitation has been performed at 1.91 eV, close to resonance with the E_+ transition. The thickness of the GaAs spacers is 12 nm and 24 nm for samples A and B, respectively. From [115]

purity and mixed with host-crystal states from the Γ, L and X edges of the Brillouin zone. Mixing and localization of the electronic states were estimated [111–113]. According to *Kent* and *Zunger* [112] and *Wang* [114] nearly 80 % of the electronic density localized around an isolated N impurity is inside a sphere of 2 nm radius. It is interesting to use the Raman-interference effects discussed above in order to investigate the electronic density distribution in GaAs:N and to compare with theoretical predictions.

Figure 24 presents [115] Raman spectra measured on GaAsN/GaAs structures consisting of five GaAsN layers (9 nm thick) separated by 12 nm and 24 nm GaAs spacers. The nitrogen concentration in each layer is 0.8 %. The scattering was excited in resonance with the highest-energy transition, known in the literature as the E_+ transition [105, 108, 116, 117]. Oscillations of the scattered intensity are observed. Their period increases with decreasing separation between the GaAsN layers. They were interpreted as due to confinement of the resonantly excited state of GaAsN layers and to interference between the contributions of each layer to the Raman-scattering process [115].

Following *Mattila* et al. [111] and *Kent* and *Zunger* [112, 113], the electronic states involved in the E_+ transition consist of the nitrogen impurity state hybridized with conduction host states from the Γ and L points of the Brillouin zone. The localization and degree of mixing of these so-called perturbed host states (PHS) depends on the impurity concentration [111].

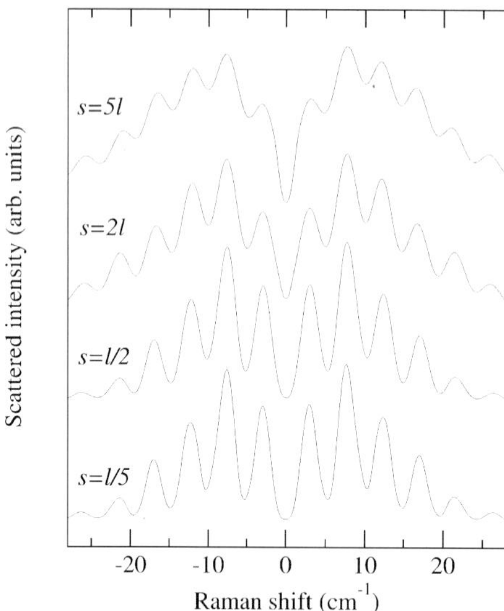

Fig. 25. Calculated Raman spectra for fixed localization length $\ell = 1.25\,\mathrm{nm}$ and for various average separations s between nearest-neighbor impurities. The calculations were performed assuming random distributions of the nitrogen in each GaAsN layer. From [115].

The spatial extension of the PHS wavefunctions can be comparable with the average separation between impurities [112–114]. Hence, overlapping of the wavefunctions, which should lead to the formation of an impurity PHS band [114], is likely to occur.

The calculated Raman spectra of Figs. 24 and 25 were generated by assuming the model presented previously for quantum wells and quantum dots: resonant excitation of intermediate electronic states (conduction states involved in the E_+ transition) and interaction with longitudinal acoustic phonons via deformation-potential mechanism. The electronic density localized around an isolated nitrogen impurity was described [115] by a Gaussian with half-width at half-maximum ℓ. Actually, the wavefunctions are more complex because of mixing with the host states [112–114]. The impurities were randomly distributed in each GaAsN layer (in the plane and along the growth direction). Their spatial distribution is characterized by the average separation between impurities s. Thus, depending on the localization length ℓ and on the average separation s, the inplane distribution of the electronic density can be either uniform (impurity-band formation and delocalization of the PHS) or strongly fluctuating (impurity-like states). Figure 25 plots the Raman spectra calculated for different (s/ℓ) ratios (ℓ is fixed). Since the spatial distribution of nitrogen atoms is random, the dependence of the interference Raman contrast on d/ℓ is the same as that already discussed in

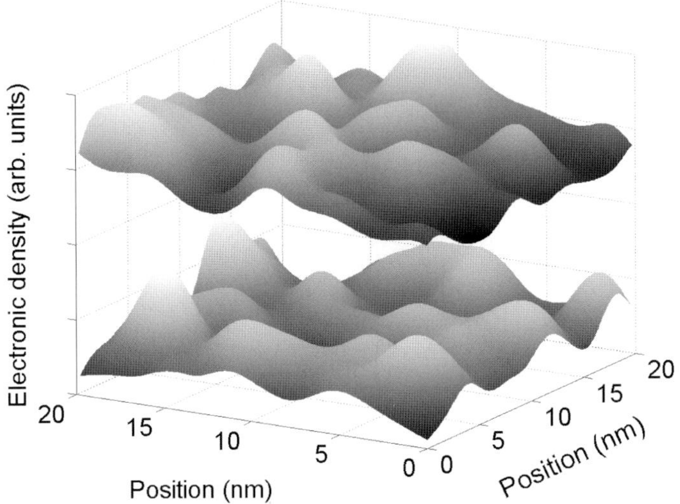

Fig. 26. Inplane electron density distribution (cross section) used for the calculations of the resonant Raman spectra (shown only for two GaAsN/GaAs quantum wells) and generated assuming coherent superposition of randomly distributed localized states centered on the nitrogen impurities. From [115].

Fig. 18 for quantum-dot multilayers with a larger interlayer spacing (i.e., with no vertical alignment of the QD).

The maximum interference contrast is obtained for a probability density distribution of the excited electronic states that is quasiuniform in the plane of the quantum wells ($s = \ell/5$). On the other hand, localized and isolated states ($s = 5\ell$) lead to a vanishing contrast. For a nitrogen concentration of 0.8 %, s is around 1.6 nm. The calculated spectra in Fig. 24 were generated with $\ell = 1.25$ nm and $s = 1.6$ nm. The agreement with the experimental spectra in terms of interference contrast is satisfactory given the simplicity of the electronic wavefunctions used for the calculations. The strong interference contrast in the measured spectra is due to overlapping of the wavefunctions located at neighboring nitrogen impurities (impurity-band formation). The electronic density distributions used for the calculation are shown in Fig. 26.

3.4 Surface Effects

3.4.1 Acoustic Mirror

We shall now discuss how the surface may play a role in the Raman-scattering interferences. Acoustic modes consisting of two counterpropagating wave components in each layer were considered when deriving the structure factor given in (8). $+q_z$ and $-q_z$ refer to the components propagating away and towards the sample surface, respectively. These components are linked together

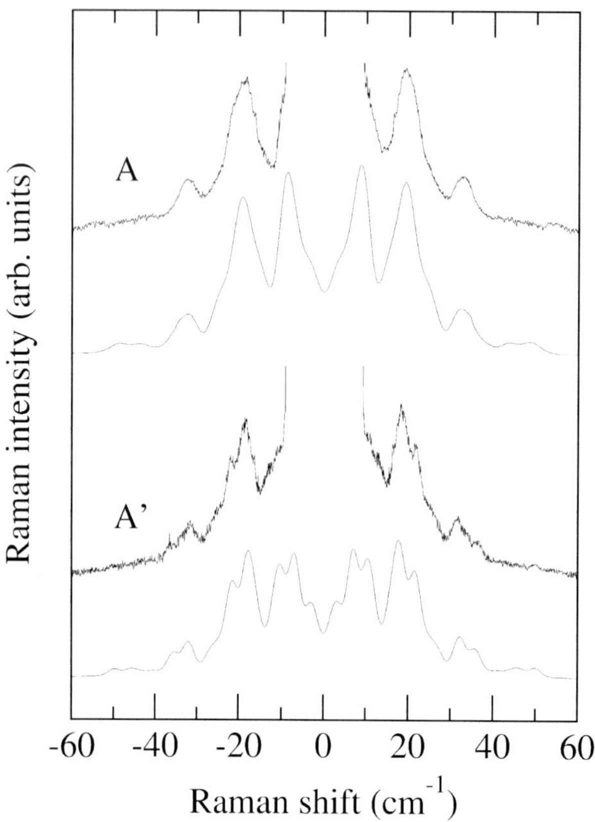

Fig. 27. Experimental and simulated Raman spectra for Ge/Si QDs bilayers ($N = 2$). The cap-layer thickness t is 65 and 35 nm for A and A′, respectively. From [14]

by the reflection at the sample surface: one thus deals with standing acoustic waves. Unlike the two first terms in the structure factor, $H^2(q_z + \Delta k_z)$ and $H^2(q_z - \Delta k_z)$ we have focused our attention on so far, the third term depends on the location of the QD layers stack with respect to the sample surface. It includes simultaneous contributions of the two wave components:

$$2H(q_z + \Delta k_z)H(q_z - \Delta k_z)\cos\left[2q_z\left(t + (N-1)\,d/2\right)\right] , \tag{14}$$

where $t+(N-1)d/2$ is the distance between the middle of the QDs layer stack and the sample surface and t gives the location of the first QD layer with respect to the sample surface. This term corresponds to a periodic surface-related modulation of the Raman intensity. Its period varies inversely with $t + (N-1)d/2$. Its spectral envelope is determined by the overlap between $H(q_z + \Delta k_z)$ and $H(q_z - \Delta k_z)$.

Experimental evidence for the surface-related intensity modulation has been found for Ge/Si QD bilayers with different Si final cap layers [14]. Both

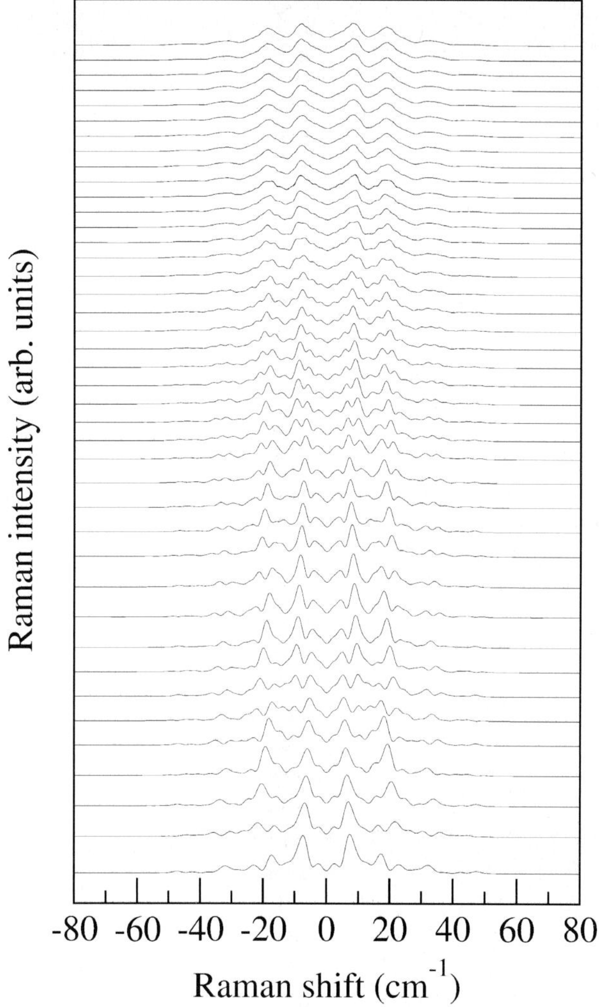

Fig. 28. Simulated Raman spectra of a double ($N = 2$) Ge/Si QDs layer stack for the interlayer spacing $d = 18.5$ nm and with the distance between the first QD layer and sample surface, t, ranging from 11 to 72.5 nm (*from bottom to top*). From [14]

experimental and simulated spectra are reported in Fig. 27. Interferences similar to those discussed above are observed for the thick Si cap layer (65 nm). An additional modulation could be resolved for the thin Si cap layer (35 nm). The rather good agreement between simulations and experiments allows us to assign this modulation to the surface-related term in the structure factor.

Simulations reported in Fig. 28 show that the scattered intensity depends considerably on the final cap-layer thickness. The actual spectra are determined by the interplay between the interlayer interferences and the surface-related modulation; the interlayer interference maxima are given by $q_z \pm \Delta k_z = \frac{2\pi}{d} n$ and the maxima of the surface-related modulation by $q_z = \frac{\pi}{t+d/2} m$ (see (10) and (14)), respectively. The surface-related modulation is clearly observed for thin cap layers (lower part of Fig. 28). It is worth noting that although the interlayer spacing is kept constant, the apparent maxima shift when the cap layer thickness changes. For thick cap layers (upper part of Fig. 28) the surface-related modulation is not resolved, owing to the limited experimental resolution included in the calculations.

To account for Raman-intensity oscillations in Figs. 27 and 28 one can propose an equivalent scheme. A standing acoustic wave (built up of two counterpropagating plane waves) interacting with a QD bilayer can be viewed as a single propagating plane wave interacting with two QD bilayers. One QD bilayer is real and located at the distance z_1 below the sample surface and the second one is virtual and located at the distance z_1 above the sample surface (Fig. 29). This picture involves a virtual QD bilayer that is the acoustic image of the real QD bilayer with respect to the sample surface. The surface acts as a mirror for the sound waves (Fig. 29). The interaction (via a single acoustic-wave component) between the QD bilayer and its acoustic mirror image yields an oscillation period that scales inversely with the distance between the double QD bilayer and its acoustic image (from center to center), i.e., twice the distance between the QD layer and the surface (Fig. 29). This simple picture allows one to recover the interference scheme discussed previously for multiple QD layers: one observes interferences within the QD bilayer and between the QD bilayer and its acoustic image. As the structure has two characteristic distances, two interference periods are expected. The interference contrast was shown to depend on the spatial correlations between QD layers; the acoustic mirror yields a perfect spatial correlation, as far as scattering of the acoustic waves by the surface inhomogeneities is negligible.

Let us now discuss structures containing many QD layers ($N > 2$). The distance with respect to the sample surface increases obviously with the number of QD layers N (for a given interlayer spacing): hence the period of the surface-related modulation decreases (even when the first QD layer is close to the sample surface). Moreover, the number of QD layers determines also the amplitude of the surface-related modulation. This amplitude is given by the overlap between $H(q_z + \Delta k_z)$ and $H(q_z - \Delta k_z)$ (14), which depends on the splitting between $H(q_z + \Delta k_z)$ and $H(q_z - \Delta k_z)$ and the spectral linewidths of the interference function $H(q_z \pm \Delta k_z)$. The splitting beats as a function of $+\Delta k_z d$ (Fig. 14) and the linewidths vary inversely with N (similarly to optical gratings, the total length determines the spectral linewidths). The amplitude of the surface-related modulation therefore rapidly decreases when N increases. For $N = 2$, the observation of the surface-related modulation is favored, because the overlap between $H(q_z + \Delta k_z)$ and $H(q_z - \Delta k_z)$ is significant (the corresponding splittings in the interlayer interferences are barely observed).

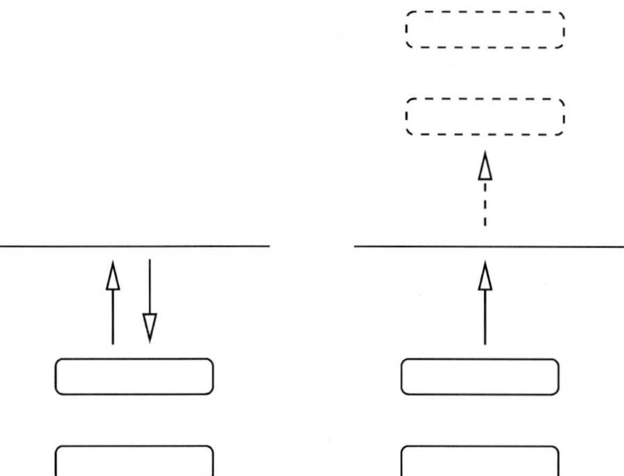

Fig. 29. (*left*) Schematic view of a QD bilayer interacting with a standing acoustic wave. (*right*) Equivalent picture where the surface acts as an acoustic mirror: the real QD bilayer and its virtual acoustic image interact with a propagating plane wave

3.4.2 Acoustic Cavity

One may wonder what happens in the Raman scattering of single QD layers. Following the structure factor (8) or the interference scheme (Fig. 29), one can expect surface-related modulation to occur in a single QD layer located close to surface. This surface-related modulation was reported for single layers of InAs/InP QDs by *Huntzinger* et al. [11]. *Milekhin* et al. [17] reported also periodic oscillations for a single layer of Ge/Si QDs (they were attributed to LA modes localized in the cap layer). The spectra reported in Fig. 10 do actually correspond to a structure containing a single layer of InAs/InP QDs [11]. The spectra display well-resolved oscillations in the acoustic-phonon frequency range. Oscillations are observed for excitation in resonance with the QDs (upper part of Fig. 10) and in resonance with the wetting layer (lower part of Fig. 10).

Unlike for QDs multilayers that may exhibit complex intensity modulations due to the interplay between stacking and surface-related modulation mechanisms (Figs. 27 and 28), one expects a rather simple behavior for a single layer of QDs or a single QW. The presence of the surface solely generates oscillations in the Raman-scattering intensity. Indeed, the structure factor (8) simply reads [118]:

$$(1 - \cos(2q_z t)) \;, \tag{15}$$

where t is the distance between the center of the QD layer and the sample surface. Their period Δq_z equals π/t; it decreases when increasing the final cap-layer thickness.

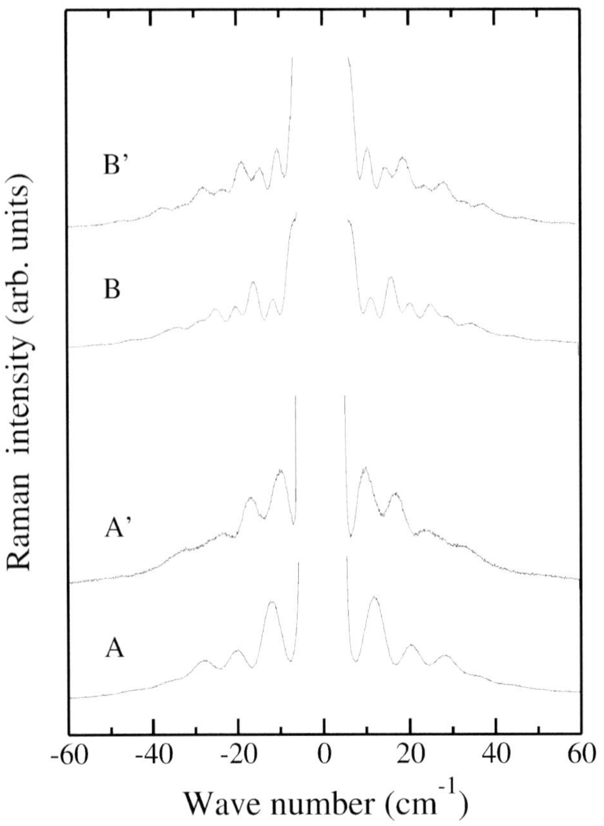

Fig. 30. Experimental Raman spectra of a single InAs/InP QD layer before oxidation (A and B) and after oxidation (A' and B') excited with the 488 nm Ar laser line (i.e., close to resonance with the E_1 gap of InAs). A and B have 9 and 15 nm nominal InP cap layers, respectively. From [118]

Comparing (15) with experimental data, like those reported in Fig. 10, is, however, somewhat disappointing. Whereas (15) accounts reasonably for the frequencies of the intensity maxima, it fails to reproduce their intensities. Sequences of strong and weak peaks can be identified easily, suggesting the presence of an additional and slower modulation superimposed on the surface-related oscillations discussed above.

The origin of this additional modulation was elucidated by varying systematically the thickness of the surface oxide layer [118]. The samples have a thin native oxide layer. They were reoxidized using a UV-ozone treatment in air. Figure 30 shows spectra of a single layer of InAs/InP QDs before and after UV-ozone treatment [118]. Both spectra display sequences of strong and weak peaks (i.e., the additional modulation just discussed). Obviously, oxida-

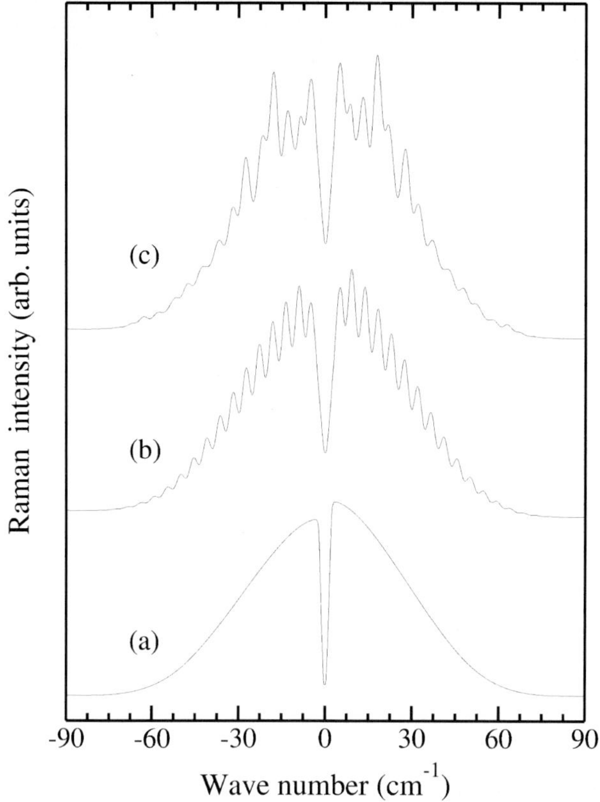

Fig. 31. Simulated Raman spectra of: (**a**) a single InAs QD layer embedded in an infinite InP medium and (**b**) a single InAs QD layer embedded in InP and located 15 nm beneath the sample surface. (**c**) same as in (**b**) plus a 1 nm oxide layer at the surface. The height and diameter of the InAs QDs are 3 nm and 25 nm, respectively. From [118]

tion yields pronounced intensity-modulation changes: intensity maxima shift and that their relative intensities (from one maximum to the next) change.

Raman-scattering simulations provide an easy means of identifying, successively, the roles of the surface and the oxide layer. Spectrum (a) in Fig. 31 was calculated considering a single InAs QD layer embedded in an infinite InP matrix (i.e., no surface). Its spectral shape is determined by the Fourier transform of the electronic density distribution. The spectrum vanishes at zero wave number according to the activation of the acoustic phonon density of states by the three-dimensional electronic confinement.

Except close to zero wave number, one observes a rather smoothly varying spectrum. Spectrum (b) in Fig. 31 was calculated by considering the previous QD layer located 15 nm underneath the sample surface (i.e., embedded between an infinite InP barrier and a 15 nm InP cap). It displays periodic oscillations, according to (15). They originate from the interaction of the QD electronic states with standing acoustic waves. When the wavelength of the acoustic mode varies (and thus the wave number in the spectrum), the electron centered on the QD experiences alternatively nodes and antinodes of the standing wave: as a result the Raman intensity oscillates.

In spectrum (b), the amplitude of the oscillations changes smoothly if one compares one intensity maximum to the next, since the envelope of the surface-induced modulation is given by the Fourier transform of the electronic density, i.e., spectrum (a). This contrasts with the behavior pointed out for the experimental data reported in Fig. 30 (see, for instance, spectra B and B'). Simulations considering simply the acoustic-wave reflection at the sample surface are obviously not able to account for the observed intensity modulation. The intensity modulations in experimental spectra depend considerably on the thickness of the oxide layer (Fig. 30). Spectrum (c) in Fig. 31 was calculated including an oxide layer at the surface. It displays a rather complex intensity modulation, with respect to the oxide-free case. Unlike in spectrum (b), the peaks are not regularly spaced. Below 30 cm^{-1}, one observes an irregular sequence of strong and weak maxima. Above 30 cm^{-1}, the oscillations are strongly damped.

The mismatch between the oxide layer and InP is significant (unlike the acoustic mismatch between InAs/InP, which is less than 2 %); it was estimated to be about 35 %. The oxide layer behaves as an additional acoustic cavity. With respect to the oxide-free structure, the acoustic-wave characteristics are changed. Consequently, the phase conditions yielding intensity maxima or minima in the interaction between the standing waves and the localized electronic states are modified. The intensity modulation therefore depends on the oxide layer thickness and on the acoustic mismatch between the oxide layer and the structure beneath.

Figure 32 presents a comparison between simulations and experiments. Striking changes are visible if one compares the experimental spectra B and B' (before and after oxidation, respectively). Whereas systematic doublet features are observed before oxidation (B), a less-regular intensity modulation is observed after oxidation (B'). In particular, one may identify in the 10 to 25 cm^{-1} range a characteristic triplet feature (a strong maximum surrounded by two weaker ones) for B'. Simulations are also displayed. Good agreement with experiments is obtained for both samples B and B' (regular doublets are obtained for B and the triplet feature is present for B'). The oxide-layer obviously plays a crucial role in the scattering process. The cap-layer and oxide-layer thicknesses considered in the simulations are consistent with the oxidation sequence. The oxide layer grows, whereas the InP cap layer shrinks, as expected. The cap-layer and oxide-layer thicknesses are 13.8 nm and 2.3 nm for sample B and 12.8 nm and 3.2 nm for sample B'. One should note that, owing to the very high phase sensitivity of the electron–acoustic-phonon interaction (i.e., the interference-like nature of the phenomenon), very small layer thickness changes can be detected.

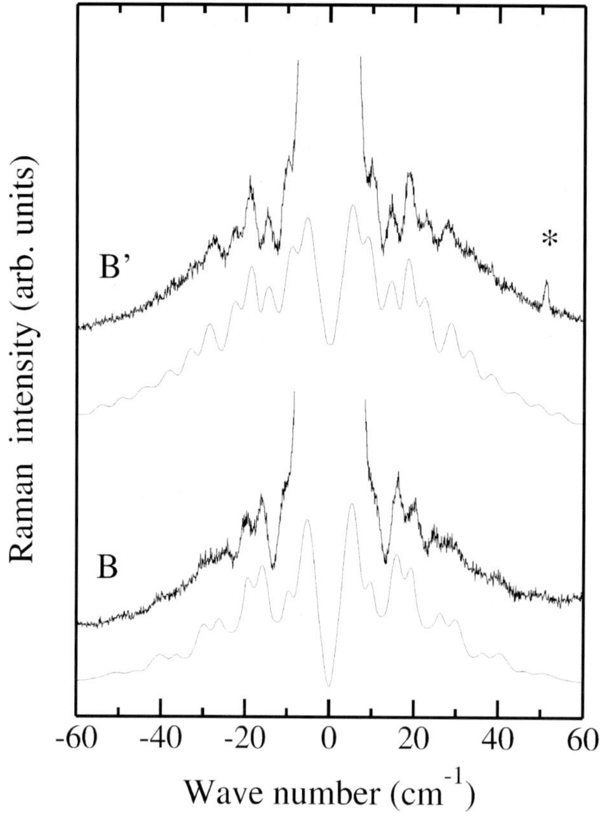

Fig. 32. Experimental (*upper trace*) and simulated (*lower trace*) Raman spectra of a single InAs/InP QD layer (15 nm InP cap) before and after oxidation (B and B', respectively). The *star* denotes a plasma line. From [118]

No attempt to add more details in the model was made since it would have obscured the fact that good agreement with experiment is already reached using the current simple and verifiable model. Including QD size or thickness fluctuations (which are obviously present in real samples) could be relevant, however. Fluctuations or inhomogeneities reduce the oscillation contrast. At zero wave number, the surface-induced modulation (15) has the same phase whatever the cap-layer thickness (unlike the interlayer interferences). In a single QD layer with QD size or cap-layer-thickness fluctuations, one expects therefore the contrast to be reduced more and more when the wave number increases. This is likely to explain why the oscillation contrast above $\approx 25\,\mathrm{cm}^{-1}$ is lower in the experimental spectra than in the simulated ones (Fig. 32).

For a given acoustic mode, its interaction with electronic states in QDs (or in a QW) depends on the exact location with respect to the sample surface. Spectra excited in resonance with the QDs therefore do not display the same modulation period as those excited in resonance with the wetting layer (Fig. 10); the "acoustic paths" indeed differ. Choosing an appropriate excitation energy allows us to select – by means of resonance – either the QDs or the WL (upper and lower part in Fig. 10, respectively); see [11, 119].

The acoustic mirror and cavity effects are also relevant for the Raman scattering in uncapped QDs. In this case, the spectral modulation of the scattered intensity depends on the height of the QDs and on the thickness of the oxide layer. Mean QD heights can be derived by comparing experiments and simulations. As one deals with very short "acoustic paths", even the height fluctuations of the uncapped QDs can be estimated with a good sensivity.

Phase shifts are a key issue in the acoustic mirror and cavity effects discussed here. *Giehler* et al. evidenced interference effects in acoustic-phonon Raman scattering from GaAs/AlAs mirror-plane superlattices [90]. The mirror plane symetry allows the introduction of controlled phase shifts. Additional phase shifts were obtained by introducing buffer layers in the center of the mirror-plane superlattices. Raman scattering is very sensitive to these phase changes. Rather complex stacking sequences were considered in [90]. The results discussed above deal with simple surface-related phase shifts.

4 Semiconductor and Metal Particles in the Strong Acoustic Confinement Regime

In the previous sections we focused on the quantum confinement of electrons and holes in QD and its consequences on the electron–acoustic-phonon interaction. As underlined above, acoustic phonons are only weakly confined in Ge/Si or InAs/InP QD because the mismatch between the acoustic impedances of the dot and the barrier is rather small. Therefore, acoustic vibrations could extend over distances much larger than the QD size and even than the average separation between quantum dots. This is at the origin of the spatially coherent nature of the light-scattering process. In contrast, confinement of acoustic phonons [120, 121] is a characteristic feature of semiconductor CdS [122–124], Si [125, 126], Ge [127] and noble metal particles Ag [128–133], Au [132], embedded in glass and polymers. In these systems, the coherence length of the particle vibrations can be equated to the quantum-dot size since penetration of the displacement field into the surrounding medium is negligible. An important consequence is that the light-scattering process loses its coherence in the sense that the overall scattering cross section from the collection of quantum dots is simply the sum of the contributions of each quantum dot (the scattering amplitudes must be squared and then summed). Thus, the spatial arrangement of the QD inside the matrix will have no signature in the low-frequency Raman spectra. This is why, till now, all experimental data reporting the observation of acoustic phonons in matrix-embedded

QD were interpreted in terms of single QD-effects. Collective effects are considered only in connection with inhomogeneous broadening of spectral lines due to QD-size distribution. One can argue that collective effects, due to interactions between QDs, should appear with increasing QD density, i.e., with decreasing average separation. This situation occurs for silver particles that were synthesized in reverse micelles and then deposited on a cleaved graphite substrate [134,135]. The nanoparticles spontaneously self-organize and form a monolayer of a perfectly ordered hexagonal network. The particles are spherical (5 nm average diameter) and coated with dodecanethiol. The average distance between particles (surface to surface) is 1.8 nm, which is comparable to the length (2 nm) of a dodecanethiol chain [134,135]. Hence, the density of particles is very large. Reflectivity and absorption measurements showed clear evidence for interparticle interaction reflecting the delocalization of the electromagnetic field associated with the surface-plasmon oscillations [136,137]. In contrast, the vibrational properties of these high-density and ordered particles can still be interpreted in terms of acoustic-phonon confinement in a single particle [133]. Collective effects due to spatial ordering are absent because the dodecanethiol chains act as acoustic insulators. This underlines the importance of the surrounding medium in determining not only the vibrational properties of the quantum dots but also their optical properties.

4.1 Eigenfrequencies and Eigenmodes

4.1.1 Lamb's Model

The vibrational dynamics of a continuous elastic and isotropic free sphere has been first described by *Lamb* [18, 19]. The displacement field $\boldsymbol{u}$ is obtained from Navier's equation

$$\rho\,\partial^2\boldsymbol{u}/\partial t^2 = (\lambda+\mu)\nabla(\nabla\boldsymbol{u}) + \mu\nabla^2\boldsymbol{u}\,, \tag{16}$$

where ρ is the mass density and λ and μ are the Lamé coefficients. Equation (16) is solved by introducing a scalar potential and a vector potential involving spherical harmonics labelled (l,m) and spherical Bessel functions of the first kind. Continuity of the displacement and stress fields at the particle surface are imposed [18,19]. These boundary conditions lead to secular equations that are solved numerically. The eigenmodes are labeled by the integer $n \geq 1$ in increasing order of frequencies. They can be classified into spheroidal (breathing and twisting motion) and torsional (only twisting) modes. Each frequency depends on the branch index n and angular momentum l, and is $2l+1$-fold degenerate. The quantized frequencies of confined acoustic vibrations scale inversely with the particle size [138].

$$\omega_{n,l} = S_{n,l}\frac{v_{\mathrm{T}}}{cR_{\mathrm{p}}}\,, \tag{17}$$

where v_T is the transverse sound velocity, c the light velocity and R_p the particle radius; $S_{n,l}$ is a factor depending on the mode (n, l) under consideration. The frequencies of torsional modes depend only on the transverse sound velocity, whereas those of spheroidal modes depend on the ratio of the longitudinal and transverse sound velocities. For embedded particles $S_{n,l}$ is also affected by the acoustic-impedance mismatch between the particle and the matrix [138–140].

4.1.2 Anisotropy in Lamb's Model

The frequencies of fundamental ($n = 1$) spheroidal and torsional modes calculated for a stress-free silicon particle [140] are quoted in Table 1. In the first line of this table, the silicon nanocrystal is described as a continuous isotropic sphere by uniformly averaging the longitudinal and transverse sound velocities over all crystallographic directions. The second line was obtained from molecular dynamics simulations [141] recently reported by *Saviot* et al. [140]. These simulations give an estimation of the effect of anisotropy of the longitudinal and transverse sound velocities on the frequencies of the confined acoustic-phonon modes. Table 1 shows that using averaged sound velocities in Lamb's model gives a reasonable agreement with the exact anisotropic calculations based on molecular dynamics.

Table 1. Frequencies in cm^{-1} of the fundamental spheroidal and torsional vibration modes of a free silicon nanoparticle. The particle radius is $R_p = 3.4\,\text{nm}$. In Lamb's model, sound velocities uniformly averaged over all directions were used: $v_L = (9017 \pm 233)\,\text{m s}^{-1}$ and $v_T = (5372 \pm 385)\,\text{m s}^{-1}$, where deviations due to anisotropy are indicated. The frequencies calculated using molecular dynamics (MD) for a sphere with cubic crystal elasticity are also shown. From [140]

Calculation method	Lamb's model	Anisotropic MD
Spheroidal		
$l = 1$ $l = 2$ $l = 3$	35.4 28.2 22.0	34.9 27.0 19.4
Torsional		
$l = 1$ $l = 2$	48.3 21.0	50.4 19.4

4.1.3 Limitations of Lamb's Model and Microscopic Approaches

With decreasing particle size, the validity of Lamb's model becomes questionable. In particular the surface/volume ratio increases and the vibrational dynamics of the particle becomes governed by the surface atoms rather than by the inner atoms. *Cheng* et al. [21] used a microscopic valence force field (VFF) model to investigate the phonon properties of Ge particles with sizes between 47 and 7289 atoms. Two bond-stretching and bond-bending parameters were used for the calculation of the total energy change due to crystal

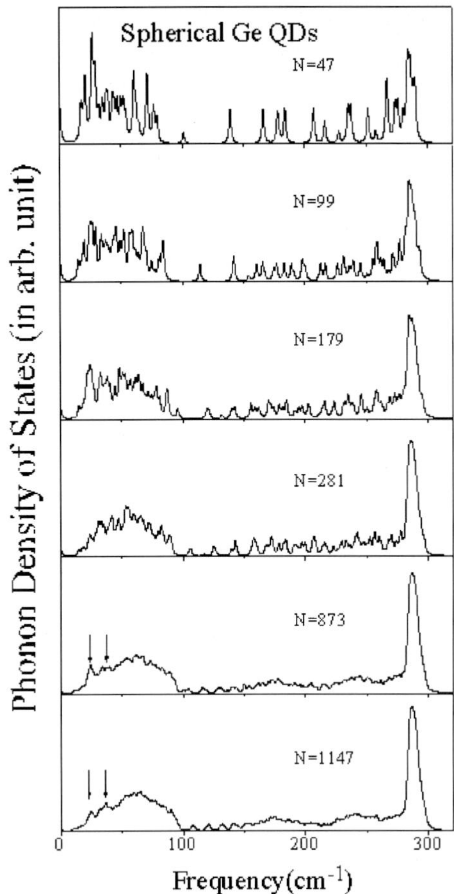

Fig. 33. Phonon density of states of spherical Ge quantum dots. The largest nanocrystal with $N = 1147$ has a radius $R_{\mathrm{p}} = 1.85\,\mathrm{nm}$. The *arrows* indicate the frequencies of surface vibrational modes that arise with decreasing QD size. From [21]

vibrations. Figure 33 shows the phonon density of states (PDOS) of the Ge quantum dots. Two peaks, indicated by arrows are visible at 25 and $37\,\mathrm{cm}^{-1}$ in the acoustic PDOS (0–$100\,\mathrm{cm}^{-1}$). They emerge as sharp lines with decreasing quantum-dot size. Their frequencies are independent of the dot size. In order to identify the origin of these peaks, *Cheng* et al. [21] calculated the vibration squared-amplitude (VSA) of each atom and then selected the atoms with the maximum VSA (MVSA). This allowed the atoms that participate in a given displacement mode to be pinpointed. Thus, the vibration modes could be classified as surface or volume modes with respect to the location of the atoms having the maximum VSA. It is clear from Fig. 34, that the modes below $50\,\mathrm{cm}^{-1}$ are basically surface modes. In particular, the surface charac-

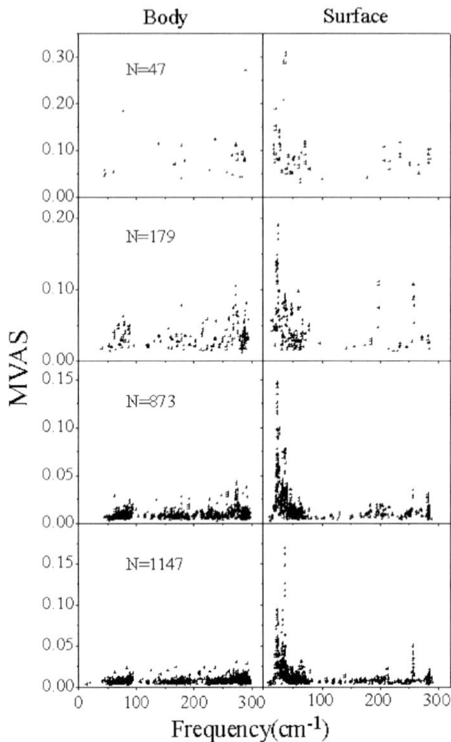

Fig. 34. Frequency dependence of the maximum value of squared amplitude (MVSA). The VFF eigenmodes are classified into surface-like and body-like modes according to the location of atoms with the maximum VSA. Plots of MVSA are shown for different QD sizes; N indicates the number of atoms of the QD. From [21]

ter of the two low-frequency lines observed in the PDOS (Fig. 33) is strong. What is the connection between the VFF eigenmodes of Figs. 33 and 34 and the confined vibration modes of Lamb's model? To connect the microscopic and the continuous-medium approaches *Cheng* et al. [21] used a projection of the atomic displacements obtained from the lattice-dynamical calculation, along the displacement pattern calculated for the spheroidal ($n = 1$, $l = 0$) and torsional ($n = 1$, $l = 3$) Lamb modes. The displacement-amplitude projections are then summed. The square of this sum is proportional to the total vibration energy in the Lamb mode. It is therefore a measure of the contribution of the particle eigenmodes to the Lamb mode onto which they are projected [21].

For N larger than 800 (i.e., dot diameter $\approx 3\,\mathrm{nm}$), the spectra of the displacement-amplitude projections exhibit peaks at the frequencies predicted by Lamb's model. These frequencies scale inversely with the dot diameter, in agreement with (17).

However, for N less than or about 800 the projections of the VFF eigenmodes produce no peaks at the frequencies of the spheroidal ($n = 1, l = 0$) and torsional ($n = 1, l = 3$) Lamb modes. This is attributed by *Cheng* et al. [21] to a breakdown of the Lamb model. Instead of confined Lamb's modes, surface lattice modes dominate the acoustic PDOS states and become prominent for particles with $N < 179$ (Figs. 33 and 34). As a matter of fact, for $N \approx 1700$ nearly half of the atoms are located at the particle surface and are therefore subjected to forces different from those experienced by the inner atoms. Hence, the vibrational properties of very small clusters consisting of a few hundreds of atoms can no longer be related to the bulk-material parameters such as mass density and sound velocities. The concept of phonon confinement itself becomes irrelevant.

The work of *Fu* et al. [22] on the vibrational properties of spherical GaP quantum dots is also worth discussing. These authors investigated theoretically both acoustic and optical vibrations using an atomic force field model developed for bulk GaP. Long-range Coulomb interaction was simulated by points having the anion and cation effective charges and located at the ionic positions. Short-range interactions were described by a VFF model [142, 143] taking into account coupling between bond stretching and bond bending and between bond stretching of two nearest bonds. *Fu* et al. [22] used a dual technique (in real and reciprocal spaces) to analyze the vibrational eigenmodes in terms of mode localization and mode mixing. This technique has been also used by *Qin* et al. [144] to investigate the phonon modes of GaAs/AlAs core-shell quantum dots.

In real space, the localization radius R_λ indicates the region of the QD in which the vibration mode is localized (similarly to the MVAS parameter of *Cheng* et al. [21])

$$R_\lambda^2 = \sum_i \sum_\rho |Q_\lambda^\rho(i)|^2 |r_i - r_\mathrm{c}|^2 , \tag{18}$$

where r_c is the dot center and $Q_\lambda^\rho(i)$ the ρ component of the phonon eigenvector of atom i vibrating in mode λ. Surface-like modes involve atoms at the dot periphery $R_\lambda \approx R_\mathrm{p}$, whereas confined modes will be localized inside the dot $R_\lambda < R_\mathrm{p}$.

Moreover, instead of projecting the VFF vibration modes on Lamb's modes, as *Cheng* et al. [21] did, *Fu* et al. [22] used a projection on the bulk vibrational states. The dot vibrations are analyzed in terms of spreading into the Brillouin zone and branch mixing. The projection on Lamb modes allows one to test the validity of the phonon-confinement model, based on the continuous elastic-medium approximation, whereas the spread on bulk states provides a valuable means of studying how the vibrations are modified by surface and finite-size effects with respect to the infinite parent crystal. In that sense, the approaches of *Fu* et al. [22] and *Cheng* et al. [21] are complementary.

Following *Fu* et al. [22] the displacement vector $\boldsymbol{u}_\lambda$ is expanded in terms of the bulk vibration states $\boldsymbol{u}_{n,\boldsymbol{k}}^{\text{bulk}}$ belonging to branch n and with wavevector $\boldsymbol{k}$

$$\boldsymbol{u}_\lambda(i) = \sum_n \int \mathrm{d}\boldsymbol{k}\, C_{n,\boldsymbol{k}}^\lambda\, \boldsymbol{u}_{n,\boldsymbol{k}}^{bulk}(i)\,. \tag{19}$$

The coefficients $C_{n,\boldsymbol{k}}$ are calculated using the orthogonality of $\boldsymbol{u}_{n,\boldsymbol{k}}^{\text{bulk}}$; the extension of vibration mode λ into the Brillouin zone [22] is then defined as

$$P_\lambda(k) = \int \mathrm{d}\boldsymbol{k}\delta\left(|\boldsymbol{k}| - k\right) \sum_n |C_{n,\boldsymbol{k}}^\lambda|^2\,. \tag{20}$$

$P_\lambda(k)$ is a *Brillouin-zone parentage coefficient* measuring the contribution of bulk vibration states with wavevector length k to the dot vibration mode λ.

Figure 35 shows the acoustic and optic phonon density of states of GaP quantum dots with diameter $2R_\mathrm{p}$ ranging from 2.22 nm to 4.34 nm. The PDOS of bulk GaP is also shown. With decreasing dot radius, new modes emerge in the frequency gaps between acoustic and optical phonons, and between transverse (TO) and longitudinal (LO) optical phonons. They can be identified owing to their localization radius R_λ. Indeed, Fig. 35 presents the frequency dependence of R_λ calculated using (18) and for a dot radius $R_\mathrm{p} = 1.62\,\mathrm{nm}$. The localization radii of gap modes are comparable to the dot radius and are therefore classified as surface-like modes. This is particularly verified for the modes located in the acoustic–optical phonon gap for which $R_\lambda = R_\mathrm{p}$. In the acoustic-phonon frequency range ($\nu^2 = 0 - 50\,\mathrm{THz}^2$) the surface character of the modes is rather strong: except for the modes around $9\,\mathrm{THz}^2$ and $40\,\mathrm{THz}^2$, the localization radii are larger than 1.1 nm, which is around $0.7\,R_\mathrm{p}$. The modes at 9, 40, 95, 113 and $138\,\mathrm{THz}^2$, showing localization inside the dot ($R_\lambda \approx 0.8\,\mathrm{nm}$) correspond to the sharp peaks in bulk PDOS. The *Brillouin-zone parentage* $P_\lambda(k)$ (20) indicate that these modes have a strong contribution from zone-edge bulk states.

4.1.4 Matrix Effects

The investigated semiconductor and metal particles are far from being free: They are embedded in a host matrix. However, the acoustic impedance of the matrix is often either much smaller or much larger than that of the particle and hence the particle surface can move freely or be rigidly fixed. In these two limiting cases Lamb's theory is still applicable and gives discrete sets of vibrational eigenmodes [145]. Their frequencies depend on the size and on the material parameters (density and sound velocities) of the particle only. The eigenmodes are not broadened.

Dubrovski et al. [138] investigated theoretically the vibrations of an elastic particle embedded in a homogeneous and infinite elastic matrix. Spherical Hankel functions of the second kind were used for the displacement field inside

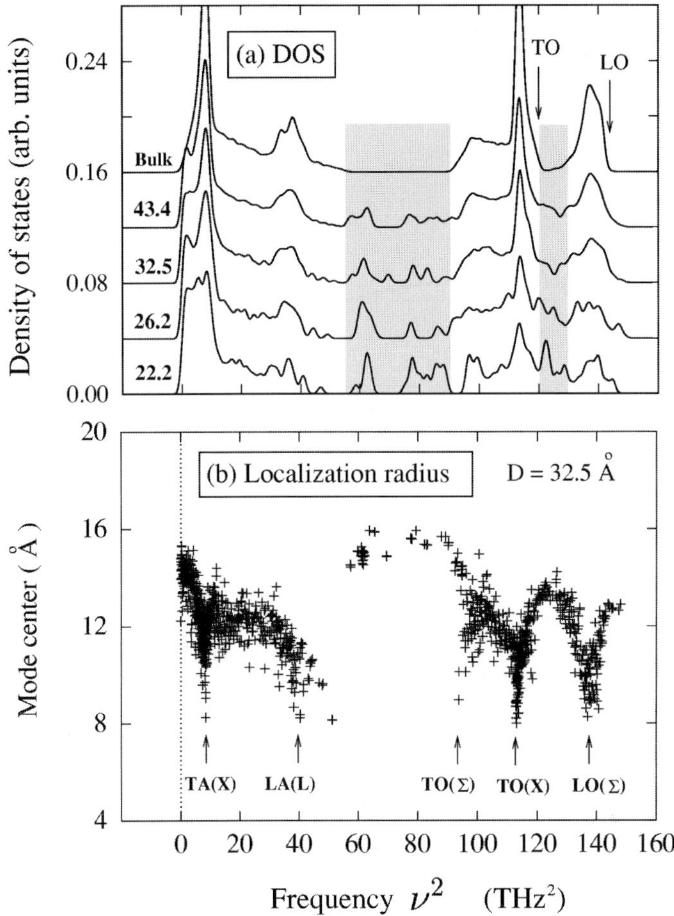

Fig. 35. (**a**) Phonon density of states of GaP quantum dots and of bulk GaP as a function of the square of the vibration frequency ($1\,\text{THz} = 33\,\text{cm}^{-1}$). The dot diameter $2R_\text{p}$ ranges from 2.22 nm to 4.34 nm (given in Å in the figure). The *arrows* indicate the bulk TO and LOphonons close to the Brillouin zone center. The *shaded areas* show the gaps between acoustic and optic phonons and between the TO and LO phonons of bulk GaP. (**b**) Localization radii from (18). The dot diameter is $2R_\text{p} = 3.25\,\text{nm}$. The *arrows* indicate the bulk states that contribute to the modes with localization radii inside the dot interior. X, Σ and L refer to the Brillouin zone edges in the [100], [110] and [111] directions, respectively. From [22]

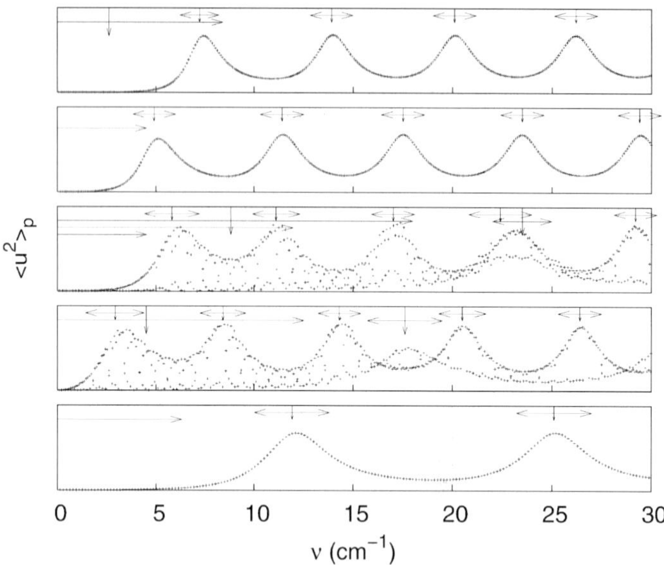

Fig. 36. Normalized mean-square displacements evaluated inside a silver particle ($R_p = 4.9\,\text{nm}$) embedded in BaO$-$P$_2$O$_5$ glass ; $R_m/R_p = 100$. Plots are shown from *bottom to top* for spheroidal modes $l = 0, 1, 2$ and for torsional modes $l = 1, 2$. *Vertical* and *horizontal arrows* indicate the frequencies and damping calculated using the complex frequency model. The modes corresponding to *horizontal arrows* that start at zero wave number are overdamped. From [20]

the matrix because it corresponds to an outgoing propagating wave. Therefore, radiation of the vibrational energy from the particle to the surrounding medium becomes possible and broadening of the vibrational eigenmodes takes place, leading to complex-valued eigenfrequencies [138–140]. The radiated vibrations can be viewed as arising from the elastic scattering of the acoustic plane waves, propagating inside the matrix, by the embedded particle [146]. In this complex frequency model (CFM) the eigenfrequencies are deduced from the roots of the secular equations (continuity of the displacement and stress fields) that can be solved numerically [140]. The CFM provides a good description of the matrix influence on the nanoparticle vibrations in terms of frequency shift and broadening of eigenmodes. However, othonormalized displacement fields could not be obtained from the CFM, because their complex amplitudes diverge with increasing particle size. This is due to the fact that in the CFM the particle is embedded in an infinite matrix. Properly normalized displacement fields are necessary not only for a complete understanding of the matrix influence but also for the modeling of the interaction between electronic excitations and acoustic vibrations [25].

Extending the work of *Montagna* and *Dusi* [147], *Murray* and *Saviot* [20] and *Saviot* et al. [140], have recently proposed a model that allows one to obtain orthonormalized displacement fields for the particle vibrations. In this core shell model (CSM) the spherical particle with radius R_p occupies the center of a spherical cavity with radius R_m representing the matrix [148]. R_m is much larger than R_p but finite. The particle and the matrix are homogeneous and isotropic as in the CFM. The equation of motion is solved for R_m and R_p; continuity of the displacement and stress fields are fullfilled at both R_m and R_p. In the macroscopic limit $R_\mathrm{m}/R_\mathrm{p} \gg 1$, the boundary conditions at R_m (rigid or stress-free surface) have no significant effect on the calculated displacement fields. The displacement vectors are orthonormalized according to total-energy conservation (time-independent energy in the absence of dissipation) [20]. In order to evaluate the internal motion of the particle vibrating in one of its eigenmodes, the displacement field is squared and its mean value inside the particle calculated. Figure 36 shows plots of the mean-square displacement as a function of frequency for spheroidal and torsional modes of a silver particle ($R_\mathrm{p} = 4.9\,\mathrm{nm}$) embedded in $\mathrm{BaO} - \mathrm{P_2O_5}$ glass. Vibration eigenmodes are obtained at all frequencies. They form a continuum, in contrast to the quantized eigenfrequencies of the free (Table 1) and rigid-surface models. This is due to the fact that the eigenmodes are no longer totally confined inside the particle. They are composed of particle and matrix vibrations. For a given type of particle motion (breathing and twisting or only twisting) and fixed value of the angular momentum l, the amplitude of the displacement field inside the particle depends on frequency. The first maximum of the mean-square displacement corresponds to the fundamental mode $n = 1$ and the maximum at higher frequencies to harmonics ($n > 1$). The spheroidal mode $l = 0$ and all torsional modes are pure breathing and twisting motions of the particle. For such modes, the mean-square displacement changes rather smoothly with frequency. For the dipolar $l = 1$ and quadrupolar $l = 2$ spheroidal modes the plots are more complex because these types of motion are composed of longitudinal and transverse vibrations mixed together by the boundary conditions at R_p and R_m. The arrows in Fig. 36 indicate the real and imaginary values of the complex frequencies, calculated according to the CFM model. The CFM accounts well for the frequency and broadening of the mean-square displacement from the CSM. Some modes are predicted by the CFM but do not show up in the CSM. In fact, the mean-square displacements displayed in Fig. 36 are representative of the particle motion only. Matrix modes, predicted by the CFM, do not appear in these plots.

As pointed out by *Murray* and *Saviot* [20] the matrix introduces new modes [20, 140]. For instance, in the plot of spheroidal vibrations $l = 1$ (Fig. 36), both CSM and CFM predict modes around $3\,\mathrm{cm}^{-1}$, which are identified as rattling modes (translational oscillations of the particles). They originate from the restoring force imposed by the matrix on the particles. Similarly, librational modes, corresponding to rotational oscillations of a nearly rigid particle, were also found. An important issue of the core–shell model is that it allows one to investigate the localization/delocalization of the vibra-

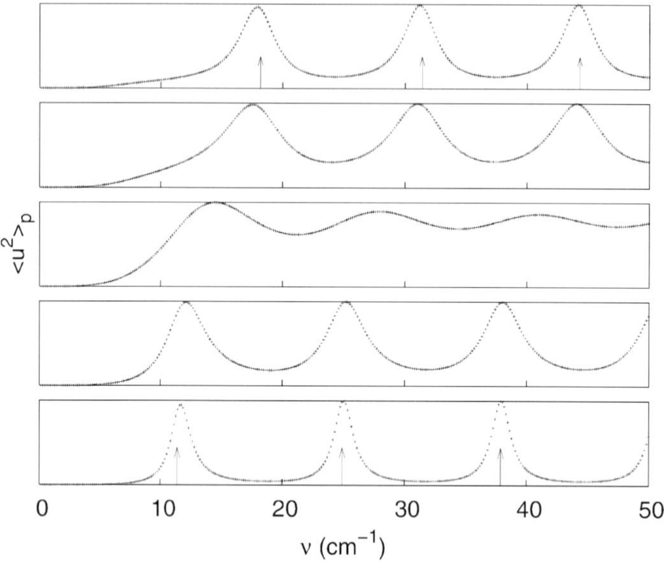

Fig. 37. Normalized mean square displacement inside a silver particle ($R_\mathrm{p} = 4.9\,\mathrm{nm}$) embedded in BaO–$P_2O_5$ glass; all plots are for spheroidal modes $l = 0$. The sound speeds of the matrix are for BaO–P_2O_5. The mass density of the matrix is scaled from BaO–P_2O_5 by factors 0.5, 1, 2.24, 4 and 6 (*bottom to top*). From [20]

tion modes by directly examining the displacement field. Figure 37 presents plots of the mean-square displacement of the spheroidal mode $l = 0$ (same as in Fig. 36) for various mass densities of the matrix [20]. The longitudinal and transverse sound velocities are taken to be those of BaO–P_2O_5. Because, for $\rho_\mathrm{m} = 0.5\rho_{\mathrm{BaO-P_2O_5}}$ and $\rho_\mathrm{m} = 6\rho_{\mathrm{BaO-P_2O_5}}$, the acoustic-impedance mismatch $\rho_\mathrm{m}v_\mathrm{m}/\rho_\mathrm{p}v_\mathrm{p}$ between the particle and the matrix is large, the maxima of the CSM agree well with the frequencies calculated for free and fixed particle surfaces (arrows in Fig. 37). In these two situations, the oscillations of the mean-square displacement are well defined, reflecting the strong confinement of the vibrations inside the particle. For $\rho_\mathrm{m} = 2.24\rho_{\mathrm{BaO-P_2O_5}}$ the acoustic impedance of the particle matches that of the matrix for longitudinal waves. The oscillations become weaker because the eigenmodes now also imply vibrations of the matrix. They can no longer be considered as solely modes of the particle but rather as those of the particle/matrix system. In other words, delocalization of the acoustic displacement field takes place when the mechanical properties (mass density and sound speed) of the particle and the matrix are very similar.

It is worthwhile mentioning that the concept of acoustic-impedance mismatch $\rho_\mathrm{m}v_\mathrm{m}/\rho_\mathrm{p}v_\mathrm{p}$ is defined for longitudinal (or transverse) plane waves suitable for planar geometries (two-dimensional layering of the mechanical properties) and must therefore be used with caution in the case of embedded quantum dots. Indeed, in spherical geometry the acoustic impedance depends on both the longitudinal and transverse sound velocities. As a result, matching of the boundary conditions at the matrix/particle interface could lead to strong confinement of acoustic vibrations even when the longitudinal acoustic impedances of the matrix and the particle are very similar. This has been recently pointed out by *Saviot* and *Murray* [149] for the case of gold particles in diamond.

The CSM has been used [20] to study the vibrational properties of Na in $BaO–P_2O_5$ glass, Si in glassy SiO_2, CdS in SiO_2 and in glassy GeO_2. This model provides the acoustic-phonon displacement fields of a spherical elastic particle embedded in an elastic matrix. It is an important step towards a realistic description of the interaction between acoustic vibrations and electronic excitations in quantum dots. Such a description is needed particularly for a clear interpretations of the resonant Raman scattering and time-resolved pump-probe experiments.

4.2 Optical Studies of Embedded Particles

The vibrations of matrix-embedded nanoparticles have been extensively studied using low-frequency Raman scattering [20, 120–133, 139, 140, 145, 148, 150–152] and time-resolved pump-probe experiments [153–159]. Persistent spectral hole burning was also used to investigate acoustic-phonon confinement in NaCl, CuCl and KCl QD embedded in silicate [160–162]. Moreover, acoustic phonons were observed recently as side bands in the tails of sharp luminescence lines emitted by single CdTe/ZnTe [163] and InAs/GaAs [164] quantum dots. In these systems the acoustic vibrations are bulk-like. We shall focus on systems exhibiting strong localization of acoustic vibrations. Therefore, we will not discuss the photoluminescence measurements performed on single quantum dots even though they are very important for the understanding of exciton dephasing processes and electron–phonon interaction in these systems [41, 163, 164].

Figure 38 shows low-frequency Raman spectra of silver nanoparticles embedded in alumina [130]; The particles are roughly spherical and their diameter ranges from 2.5 nm to 4.5 nm. The spectra were excited at 2.707 eV close to resonance with the surface plasmon absorption band of silver QDs (around 3 eV). A peak is observed around $10\,\mathrm{cm}^{-1}$ for $2R_\mathrm{p} = 4.5\,\mathrm{nm}$; its frequency increases with decreasing particle size.

Figure 39 presents the time-resolved transmission of silver particles ($R_\mathrm{p} = 13\,\mathrm{nm}$) embedded in a $BaO - P_2O_5$ matrix [158]. Measurements were performed using a two-color femtosecond pump-probe technique: a 80 fs near-infrared pump first excites intraband transitions of the confined electron gas. Then a 100 fs blue pulse probes the transmission change ΔT induced by the pump around the plasmon absorption band. In these samples the reflectivity modulation is very weak and hence $\Delta T/T$ reflects mainly the change of absorption. $\Delta T/T$ exhibits oscillations as a function of the pump-probe

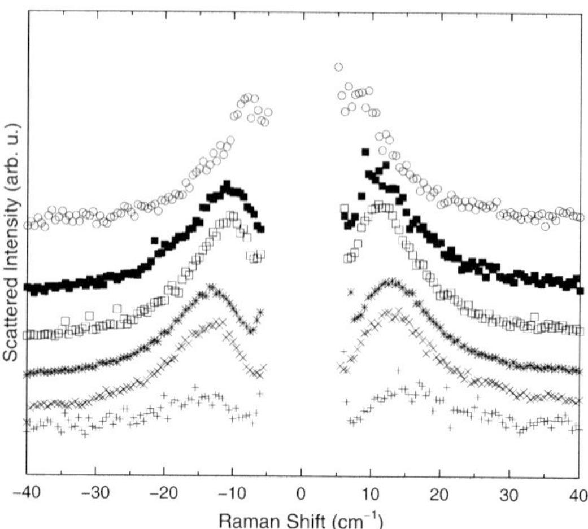

Fig. 38. Room-temperature Raman spectra of silver nanoparticles embedded in alumina. The average diameter $2R_{\rm p}$ of the particles is, from *bottom to top*, 2.5, 2.7, 3, 3.5, 3.8 and 4.5 nm. The excitation energy 2.707 eV is close to resonance with the plasmon absorption band. From [130]

time delay (Fig. 39). The oscillations period and damping were extracted by fitting an exponentially decaying cosine function to the experimental data. For $R_{\rm p} = 13\,{\rm nm}$, an oscillation period $T_{\rm osc} = 8.05\,{\rm ps}$ and a damping time $\tau = 7.7\,{\rm ps}$ were deduced [158]. *Del Fatti* et al. [158] showed that the oscillations period increases with the particle size.

Figure 40 displays the frequencies of the Raman peak (Fig. 38) and of the transmission modulation (Fig. 39) versus $1/R_{\rm p}$. Both frequencies scale inversely with the average radius of the particles. The slopes are, however, different. The vibration frequencies calculated using (17) and assuming a free particle are shown in Fig. 40. The sound velocities of bulk silver were used [158, 165]. The calculated frequencies of the fundamental quadrupolar mode ($n = 1, l = 2$) agree rather well with the Raman measurements, whereas those of the fundamental breathing mode ($n = 1, l = 0$) account well for the time-resolved pump-probe experiments.

In Raman scattering, as well as in time-resolved pump-probe experiments, the vibrational modes are revealed owing to their interaction with excitons or plasmons depending on the semiconducting or metallic nature of the particle. Therefore, only a few modes of the phonon density of states are optically active: only those that couple efficiently to the electronic states are able to scatter the light and modulate the optical absorption.

The Raman selection rules for a vibrating free sphere were discussed by *Duval* [166]. According to the group symmetry of the proper and improper rotations and to the symmetry of the polarizability tensor, only breathing ($n, l = 0$) and quadrupolar spheroidal modes ($n, l = 2$) are Raman active.

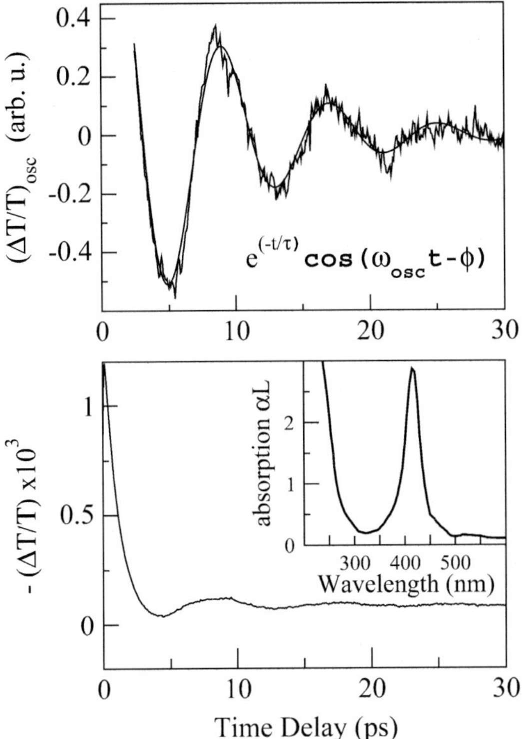

Fig. 39. *Lower panel*: time-resolved transmission of silver nanoparticles R_p = 13 nm embedded in BaO – P_2O_5. The *insert* shows the plasmon absorption peak around 400 nm. *Upper panel*: oscillatory part of the transmission. The frequency $\omega_{osc} = 0.78\,\text{ps}^{-1}$ and damping of the oscillations $\tau = 7.7\,\text{ps}$ were determined using a fit of an exponentially decaying cosine function to the experimental data. From [158]

Spheroidal modes with odd angular momentum $(n, l = 1, 3, 5)$ and torsional modes are forbidden. Moreover, *Montagna* and *Dusi* [147] predicted that for cubic Bravais lattices and for scattering via the dipole–induced-dipole mechanism only spheroidal modes $(n, l = 2)$ are Raman active. These selection rules hold for spherical free particles and nonresonant excitation of the Raman scattering. Deviations are expected in the case of nonspherical particles, noncubic lattices and for resonant excitation of the light scattering. For instance, torsional modes become Raman active for ellipsoidal free particles [166] and matrix-embedded particles [140]. Moreover, for resonant excitation the details of the coupling mechanisms between the confined phonons and the electronic excitations should be taken into account in order to predict the Raman activity [147, 167].

The agreement between calculated and measured frequencies in Fig. 40 allows one to assign the Raman scattering of Fig. 38 to the fundamen-

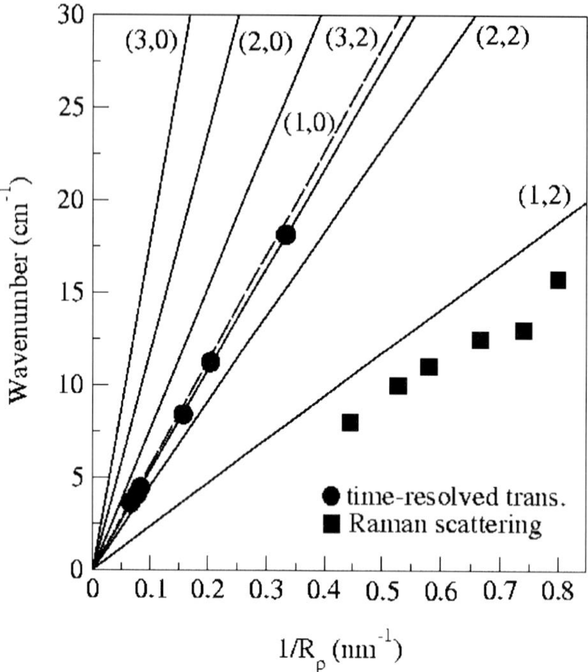

Fig. 40. Frequencies of the vibration modes observed by resonant Raman scattering [130] and time-resolved pump-probe experiments [158] versus the average radius of the particles. The *solid lines* show the frequencies of the (n, l) vibration modes calculated using Lamb's model ((17) and assuming stress-free boundary conditions at the particle surface). The *dashed line* (mode (1,0)) was calculated for a silver particle embedded in $BaO - P_2O_5$ glass. From [130, 158]

tal quadrupolar vibrational eigenmode ($n = 1, l = 2$) [129, 130, 166]. Although allowed, Raman scattering by breathing eigenmodes ($n, l = 0$) is very weak; it was reported only recently for silver particles with narrow size distributions [132, 133]. All published measurements [120, 128, 129, 131–133, 150] confirm that Raman scattering in metal nanoparticles is dominated by quadrupolar vibration eigenmodes. On the other hand, Fig. 40 shows that the fundamental breathing mode ($n = 1, l = 0$) is responsible for the time modulation of the optical transmission [158]. Quadrupolar vibration modes ($n, l = 2$) of metal particles are not observed in time-resolved pump-probe experiments. The vibration frequencies calculated using Lamb's model agree, for a stress-free particle, with both Raman and time-resolved measurements (Fig. 40). This is remarkable since the particles (studied in Fig. 40) are not free but matrix embedded. In fact, according to the CSM model of *Murray* and *Saviot* [20], the $BaO–P_2O_5$ matrix has only a weak influence on the vibration frequencies (see Fig. 37). As a matter of fact, the frequencies of mode $(1, 0)$ estimated assuming perfect particle/matrix contact (dashed line in Fig. 40) are only $\sim 5\,\%$ larger than those calculated for a stress-free particle [158]. In Fig. 40 one can notice that the calculated frequencies of mode $(n = 1, l = 2)$ are systematically somewhat overestimated with respect to the measured Raman frequencies. As discussed by *Palpant* et al. [130] this is likely due to resonance with the surface plasmon excitations. A matrix effect would have shifted the vibration frequencies to higher values with respect to the case of a free particle (dashed line in Fig. 37).

It is clear from Fig. 40 that resonant Raman scattering and time-resolved optical measurements give complementary information on the vibrational properties of QD since they measure different eigenmodes. This has been noticed not only for metal particles but also for the semiconductor PbSe [168] QD. Following the work of *Montagna* and *Dusi* [147], *Ikezawa* et al. [168] suggested that, as for silver particles, the absence of spheroidal modes $(n, l = 0)$ in the Raman spectra of PbSe QD is due to the cubic-type crystal structure. As a matter of fact, *Verma* et al. [139], *Tanaka* et al. [122], *Saviot* et al. [124] and *Ivanda* et al. [169] observed Raman scattering involving quadrupolar $(n = 1, l = 2)$ as well as pure radial $(n = 1, 2, 3, l = 0)$ vibration modes in CdS and CdSe nanoparticles that have a wurtzite-type crystal structure. From the preceding discussion, some questions, which continue to stimulate experimental and theoretical studies of the vibrational and optical properties of QDs, can be raised: 1. Why, in the metal particles, do quadrupolar vibrations $(n = 1, l = 2)$ dominate the Raman scattering, and radial vibrations $(n = 1, l = 0)$ the time dependence of the optical absorption? 2. What determines the Raman-intensity ratios in the semiconductor quantum dots where both quadrupolar and pure radial vibrations are observed [122, 124, 139]? 3. What are the main coupling mechanisms between confined phonons and electronic excitations in metal and semiconducor quantum dots?

4.3 Interactions between Confined Phonons and Electronic Excitations

It is interesting to note that the Raman experiments reported in semiconductor and metal particles have been performed in resonance, or close to resonance, with excitonic or plasmonic transitions. For offresonance excitation the particle signal becomes rarely detectable since scattering by acoustic vibrations of the matrix dominates [124]. Therefore, selection rules, optical activity of vibration modes and spectral lineshapes predicted for nonresonant excitation of the Raman scattering [147, 166] should be reconsidered in the case of resonant excitation. In particular, the electronic states involved as intermediate states in the resonant light scattering process and their coupling mechanisms to the vibrations, play important roles.

4.3.1 Exciton–Phonon Interaction in Semiconductor Particles

Sirenko et al. [25] investigated spin-flip and acoustic-phonon resonant Raman scattering in CdS QD embedded in silica glass. The acoustic impedances of the particle and the silica differ little, and hence phonon size-quantization was not observed although it does exist [20]. Consequently, as for the Ge/Si and InAs/InP QD, the scattering in CdS/silica glass is dominated by electronic confinement: the spectral shape of the Raman-scattering reflects the probability density distribution of the electronic states excited inside the quantum dots [25]. However, contrary to the Ge/Si and InAs/InP QD structures in which the electronic density is layered along (at least) one direction, the CdS particles are randomly distributed everywhere in the matrix. Therefore, Raman-interference effects (as those discussed for Ge/Si multilayers) were not observed [25]. *Sirenko* et al. [25] showed that both spin-flip and acoustic-phonon Raman scattering are well explained by the interaction of confined excitons with bulk-like acoustic phonons via the deformation-potential (DP) mechanism. *Gupalov* and *Merkulov* [167] extended the calculations of *Sirenko* et al. [25] by taking into account spin-orbit interaction in the CdS valence band and acoustic-phonon confinement (Lamb's model). In particular, they obtained the Raman selection rules for deformation potential exciton–phonon interaction: these selection rules are the same as those deduced from the symmetry of the polarizability tensor [147, 166], i.e., only spheroidal modes ($n, l = 0, 2$) are allowed in the resonant Raman scattering involving DP exciton–phonon interaction. Scattering by quadrupolar ($n, l = 2$) modes is accompanied by hole spin-flip and gives rise to depolarization of the emitted light [167]. Whereas, for scattering by radial modes ($n, l = 0$) the incident and emitted polarizations are the same [167]. These predicted selection rules and polarization behavior are in agreement with experiments [122, 124, 139]. However, the relative intensities of the Raman scattering due to ($n, l = 2$) and ($n, l = 0$) modes and their dependence on the particle size and excitation energy still need to be investigated for a complete understanding of the experimental data [122–124, 139]. Moreover, matrix effects should be taken into account in the calculations of the exciton–phonon DP interaction and the Raman intensity, since the displacement field inside the particle strongly depends on the dot/matrix acoustic properties. As already suggested by *Murray* and *Saviot* [20] this can be achieved by coupling the CSM model with the Raman-efficiency calculations based on the approach of *Sirenko* et al. [25] and *Gupalov* and *Merkulov* [167].

4.3.2 Plasmon–Phonon Interaction in Metal Particles

In their earlier work, *Weitz* et al. [120] and *Gersten* et al. [150] studied experimentally the Raman scattering from rough metal surfaces produced by mild anodization of silver and copper electrodes. They observed an intense low-frequency peak that they ascribed to scattering by quadrupolar vibrations of size- and shape-distributed nanosized particles. For the interpretation of their data *Gersten* et al. [150] developed a model based on the resonant excitation of dipolar surface plasmons and on the plasmon-frequency modulation due to the change of the electronic density with the particle motion. Only quadrupolar ($n = 1, l = 2$) vibration modes were considered. Pure radial vibrations ($n, l = 0$) were not expected to produce a significant modulation of the plasmon frequency. The model of *Gersten* et al. [150] has been widely referred to because it gives good agreement between calculated and measured Raman frequencies and their dependence on excitation energy [120, 150]. In addition, it is consistent with the experimentally proven fact that quadrupolar vibrations dominate the Raman scattering in metal particles.

Nevertheless, coupling strengths of quadrupolar and radial vibrations with surface plasmons, and their relative contribution to the light-scattering efficiency, still need to be studied. The Raman experiments recently reported by *Courty* and *Pileni* [133] and *Portales* et al. [132], where both quadrupolar and pure radial vibrations are observed, open the way to interesting comparison of calculations with experiments. Such a comparison should allow the identification of the main coupling mechanisms between surface plasmons and confined phonons responsible for light scattering and for time modulation of the optical properties. Recently, *Bachelier* and *Mlayah* [170] reported calculations of the resonant Raman-scattering efficiency of silver particles. These calculations are based on the three-step light-scattering process discussed previously for semiconductor QD and QW. Various coupling mechanisms between confined vibrations and plasmons were discussed. Two of them, namely deformation-potential interaction and surface-orientation modulation were found to play important roles for Raman scattering as well as for the time dependence of the optical absorption.

Surface Plasmon–Polariton States

For silver particles excited in the visible range, the intermediate electronic states involved in the resonant light scattering are surface-plasmon–polariton states (SPP). They were calculated in [170] using the dielectric confinement model [171] that accounts well for the size dependence of the optical properties [136, 137, 172]. The metal dielectric response is given by

$$\chi(\omega, R_\mathrm{p}) = -\frac{\omega_\mathrm{p}^2}{\omega^2 + \mathrm{i}\omega\gamma(\omega, R_\mathrm{p})} + \chi^{\mathrm{ib}}(\omega)\,, \tag{21}$$

where ω_p is the bare plasmon frequency and $\gamma(\omega, R_\mathrm{p}) = \gamma_o + gs(\omega)v_\mathrm{F}/R_\mathrm{p}$ the effective Drude damping due to electron scattering by the nanoparticle surface [173]; v_F is the Fermi velocity and $gs(\omega)$ describes the frequency dependence of the electron-scattering rate by the particle surface. $\chi^{\mathrm{ib}}(\omega)$ is the interband dielectric response due to electronic transitons from the $4d$ valence band to the 5 s conduction band [174]. According to *Hovel* et al. [175]

$\chi^{\text{ib}}(\omega)$ is independent of the particle radius for $R_{\text{p}} > 1\,\text{nm}$, and can therefore be assumed to be equal to the interband susceptibility of bulk silver [176].

The electromagnetic field inside and outside the particle is computed from Maxwell's equations (i.e., including retardation effects). The usual boundary conditions, continuity of the normal and tangential components of the magnetic and electric fields, are, respectively, applied. The electromagnetic field $\boldsymbol{E}_{\omega,L,M}$ and polarization eigenvectors $\boldsymbol{P}_{\omega,L,M}$ form a continuum of states distributed over all frequencies ω; (L, M) are integer numbers associated with the spherical harmonics. The electric field inside the metal particle is maximum at the particle surface. The surface character of the SPP states increases with increasing L. Moreover, the electric field amplitude is frequency dependent. It reaches a maximum at discrete values of ω_L known as *Mie* resonances [172]. For spherical particles, ω_L is $(2L+1)$-fold degenerate. It is worthwhile to underline that estimation of the light-scattering efficiency requires a complete determination of the intermediate electronic states. In particular, the normalization of eigenstates is of prime importance for studying how the scattered intensity changes with particle size and excitation energy. As in the core–shell model developed by *Murray* and *Saviot* [20] for the particle vibrations, the SPP states in [170] were normalized by locating the particle at the center of a totally reflecting spherical vacuum cavity and applying the electromagnetic boundary conditions at both particle and cavity surfaces. The cavity radius was then made much larger than the particle radius. In this macroscopic limit, changing the cavity radius has no effect on the electromagnetic field-amplitudes.

Deformation-Potential Coupling

In the framework of the dielectric confinement model, the interaction between SPP states and confined phonons is treated in terms of modulation of the polarization vectors $\boldsymbol{P}_{\omega,L,M}$ by the particle motion. This interaction is responsible for the scattering of surface plasmons–polaritons from state (ω, L, M) to state (ω', L', M'). The matrix element for vibration-induced dipolar interaction between SPP states reads [170]

$$H_{\text{vib-SPP}} = -\int_{\text{particle}} \boldsymbol{E}_{\omega',L',M'} \delta_{n,l,m} \boldsymbol{P}_{\omega,L,M} \text{d}V\,, \qquad (22)$$

where $\delta_{n,l,m}\boldsymbol{P}_{\omega,L,M}$ is the polarization modulation of SPP state (ω, L, M) by the vibration eigenmode (n, l, m); $\boldsymbol{E}_{\omega',L',M'}$ is the electric field associated with the final SPP state (ω', L', M'). The integral in (22) runs over the particle volume only, if the surrounding medium is vacuum.

Equation (22) is the interaction step of the resonant Raman process involving the particle vibrations. $\delta_{n,l,m}\boldsymbol{P}_{\omega,L,M}$ could have different origins. For instance, *Del Fatti* et al. [158, 177] suggested that deformation-potential coupling [178] between lattice vibrations and individual electronic states can modulate the metal dielectric susceptibility and hence the SPP's polarization vectors. Electronic interband transitions lead indeed to strong screening of the bare-plasmon: the bare-plasmon energy in bulk silver is around 9 eV [179] whereas the plasmon absorption band occurs around 3 eV. Hence, any modulation of the interband dielectric susceptibility will reflect itself in the overall dielectric response, thus leading to a modulation of the polarization vectors $\delta_{n,l,m}\boldsymbol{P}_{\omega,L,M}$.

As mentioned above, size effects on the interband dielectric susceptibility ($\chi^{\mathrm{ib}}(\omega)$ in (21)) can be neglected for particles with $R_{\mathrm{p}} > 1\,\mathrm{nm}$. Hence, the modulation of $\chi^{\mathrm{ib}}(\omega)$ can be derived by assuming bulk $4d$ and 5 s electrons interacting with confined acoustic phonons via the DP mechanism. In this case, the modulation $\delta_{n,l,m}\boldsymbol{P}_{\omega,L,M}$ of the SPP polarization reads [170]

$$\delta_{n,l,m}\boldsymbol{P}_{\omega,L,M} = \varepsilon_0 \chi^{\mathrm{ib}}(\omega) \left(\frac{D^{\mathrm{e-ph}}}{\hbar\omega - \hbar\Omega^{\mathrm{ib}}} \boldsymbol{\nabla} \cdot \boldsymbol{D}_{n,l,m} \right) \boldsymbol{E}_{\omega,L,M}\,, \tag{23}$$

where $D^{\mathrm{e-ph}}$ (around $-1.55\,\mathrm{eV}$ for silver) is the DP energy averaged over the values calculated for zone center (Γ-point) and zone edge (L-point) electronic states [180]; $\hbar\Omega^{\mathrm{ib}}$ is the energy of the 4 s – $5d$ interband transitions. Equation (23) holds for bulk metals as well. The only difference with a QD is the displacement field $\boldsymbol{D}_{n,l,m}$ that accounts for the confinement of lattice vibrations. Replacing (23) into (22) allows us to estimate and compare the coupling strength of different vibration eigenmodes (n, l, m), provided the latter are properly normalized. For instance, $\delta_{1,2,m}\boldsymbol{P}_{\omega,L,M}$ is found to be one order of magnitude smaller than $\delta_{1,0,m}\boldsymbol{P}_{\omega,L,M}$ because of the weak displacement field gradient associated with quadrupolar vibrations. This will obviously reflect itself in the Raman activity of the vibration modes. Moreover, according to (23), the DP SPP-vibration coupling strength is inversely proportional to the energy difference $\hbar\omega - \hbar\Omega^{\mathrm{ib}}$. It is therefore expected to be much stronger when the interband transitions overlap with the surface plasmon excitations as in gold and copper. However, damping of the interband transitions, not taken into account in (23), should be an important limiting factor for the coupling strength.

Surface Coupling Mechanisms

An important characteristic of nanosized particles is the large surface/volume ratio. As discussed above, the surface atoms become very important for the vibrational properties of small clusters (Figs. 33, 34 and 35). The surface may also introduce new coupling mechanisms between vibrations and electronic excitations. This is well known for semiconductor quantum wells where the confined electronic states are modulated by the rippling motion of the QW/barrier interfaces [181]. In semiconductor quantum dots, the contribution of the ripple mechanism to the electron–acoustic-phonon interaction has been studied by *Alcalde* et al. [182]. In metal particles, the equivalent of the ripple mechanism would be the modulation of the SPP energies due to the change of the particle size while oscillating. This coupling mechanism is in fact negligible for nanosized particles because in the long-wavelength limit ($2\pi c/\omega \gg R_{\mathrm{p}}$), retardation effects are very weak and the SPP eigenfrequencies depend only little on the particle size.

Moreover, according to (21) the effective Drude damping term is determined by the collision rate of Fermi electrons on the particle surface. Therefore, when the particle surface oscillates one expects a modulation of the effective Drude damping term and hence of the intraband dielectric susceptibility. According to [173], for a silver particle with $R_{\mathrm{p}} = 3\,\mathrm{nm}$, $gs(\omega)v_{\mathrm{F}}/R_{\mathrm{p}} \approx 220\,\mathrm{meV}$ that is rather small compared to the first SPP Mie resonance (around $3\,\mathrm{eV}$). Hence, modulation of the intraband susceptibility via the Drude term should also be very weak.

As already pointed out by *Palpant* et al. [130], inhomogeneous broadening of the surface-plasmon absorption band is due to the particle-shape distribution rather than to size distribution [183]. Hence, shape modulation of the particle is expected to produce significant modulation of the surface plasmon. As recently shown in [170], changes in the particle shape can indeed modulate the SPP polarization vectors.

The polarization vectors $\boldsymbol{P}_{\omega,L,M}$ are associated with a distribution of polarization charges $\sigma_{\omega,L,M} = \boldsymbol{P}_{\omega,L,M}\boldsymbol{n}$ at the particle surface; ($\boldsymbol{n}$ being the normal to the surface). Hence, while the particle oscillates, in its (n,l,m) vibration mode, the surface orientation changes by $(\delta_{n,l,m}\boldsymbol{n})$; Therefore, the polarization charges are redistributed by $\delta_{n,l,m}\sigma_{\omega,L,M} = \boldsymbol{P}_{\omega,L,M}{\cdot}\delta_{n,l,m}\boldsymbol{n}$. With respect to the static situation (particle at rest) this can be interpreted in terms of a polarization modulation $\delta_{n,l,m}\sigma_{\omega,L,M} = \delta_{n,l,m}\boldsymbol{P}_{\omega,L,M}\boldsymbol{n}$.

To evaluate the polarization modulation $\delta_{n,l,m}\boldsymbol{P}_{\omega,L,M}$, $\delta_{n,l,m}\sigma_{\omega,L,M}$ is expanded in terms of the basis of the surface polarization charges $\sigma_{\omega,L',M'}$ of the motionless particle [170].

$$\delta_{n,l,m}\boldsymbol{P}_{\omega,L,M} = \varepsilon_0\chi(\omega)\sum_{L',M'}\frac{\int \sigma_{\omega,L',M'}\delta_{n,l,m}\sigma_{\omega,L,M}\cdot \mathrm{d}S}{\int \sigma^2_{\omega,L',M'}\cdot \mathrm{d}S}\boldsymbol{E}_{\omega,L',M'}\,. \tag{24}$$

The surface orientation (SO) mechanism [170] is very selective with respect to the vibration-mode symmetry. Pure rotational and pure radial vibrations ($n, l = 0$) do not change the shape of the particle surface. Therefore, they cannot interact with the SPP states via an SO mechanism. On the contrary, quadrupolar vibrations ($n, l = 2$) lead to a strong modulation of the polarization charge distribution at the particle surface.

In the SO mechanism, the particle vibrations modulate the polarization vector and not the dielectric susceptibility of the metal particle as in the DP mechanism. Moreover, the modulation is proportional to the total dielectric susceptibility $\chi(\omega)$ (24), whereas for the DP mechanism it is proportional to the interband dielectric susceptibility $\chi^{\mathrm{ib}}(\omega)$ (23). Hence, the relative importance of DP and SO mechanisms depends on the SPP state energy $\hbar\omega$.

Raman-Scattering Simulations

The Raman spectra of metal particles excited in resonance with SPP states are obtained by assuming the three-step light-scattering process [40] in which an SPP state (ω, L, M) is excited by the incoming photon, decays into another state (ω', L', M'), via emission or absorption of a confined (n, l, m) vibrational quantum, and a scattered photon is emitted. Since the spherical symmetry is preserved at the SPP-vibration interaction step, conservation of the angular momentum implies the selection rules

$$|L - L'| \leq l \leq L + L' , \tag{25a}$$

$$L + L' + l \quad \text{even} , \tag{25b}$$

$$M' - M = \pm m . \tag{25c}$$

On the other hand, there is no restriction on L and L' because the spherical symmetry is not preserved at the photon absorption and emission steps (interaction between plane waves and spherical harmonics).

According to Fermi's golden rule [184] the decay rate of surface plasmon–polaritons into electron–hole pairs is proportional to the electromagnetic field amplitude inside the metal particle. The latter increases with increasing L, i.e., with increasing confinement of the state. Hence, the SPP states with $L > 1$ are strongly damped [175, 184] and their contribution to the resonant Raman scattering can be neglected compared to that of the dipolar surface plasmon–polariton states $(\omega, L = 1, M)$. Thus, when considering only dipolar SPP states as the intermediate states in the resonant light-scattering process $(L = L' = 1)$ selection rules (25a) and (25b) indicate that only pure radial $(l = 0)$ and quadrupolar $(l = 2)$ vibrations are Raman active. These selection rules are the same as those deduced by *Duval* [166] from the symmetry of the polarizability tensor.

Figure 41 shows Raman spectra calculated for a silver particle with $R_{\mathrm{p}} = 2.5\,\mathrm{nm}$ (from [170]). For deformation potential SPP-vibration coupling, the Raman scattering (A) is dominated by the fundamental mode of pure radial vibrations $(n = 1, l = 0)$ and its first two harmonics. Quadrupolar vibrations $(n, l = 2)$ are allowed but their Raman intensity is very weak. *Del Fatti* et al. [177] suggested that modulation of the interband susceptibility by the particle vibrations via DP interaction is at the origin of the oscillations observed [158] in time-resolved absorption measurement (Fig. 39). These oscillations (Fig. 40) are caused by pure radial vibrations $(n = 1, l = 0)$. As a matter of fact, calculations of the DP interaction between the SPP states and the particle vibrations (23) indicate that pure radial modes couple more efficiently than quadrupolar modes. On the other hand, *Portales* et al. [132] proposed that the same modulation is responsible for the weak Raman bands attributed [132, 133, 159] to pure radial vibrations $(n = 1, l = 0)$. It is clear from Fig. 41 that DP coupling could be responsible for the weak Raman bands observed by *Portales* et al. [132] and *Courty* and *Pileni* [133]. However, it cannot explain the fact that Raman scattering in metal particles is dominated

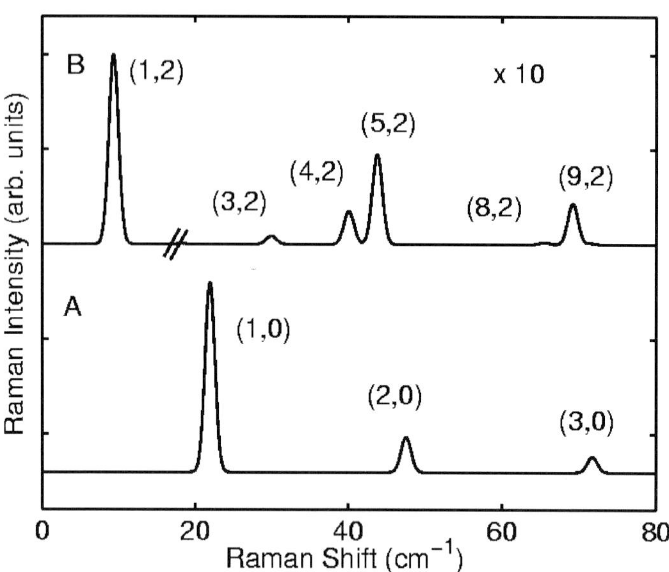

Fig. 41. Raman spectra calculated for a silver nanoparticles with $R_{\mathrm{p}} = 2.5\,\mathrm{nm}$ and for resonant excitation with the dipolar SPP state at 2.5 eV. The calculated spectra correspond to the coupling via (A) deformation potential and (B) surface orientation mechanisms. The frequency range 16–80 cm^{-1} is shown with $\times 10$ magnification for spectrum (B). The finite linewidths result from convolution with a Lorentzian function. From [170]

by quadrupolar vibrational modes [131–133, 150]. For SPP-vibration coupling via the SO mechanism, the calculated Raman spectrum (B in Fig. 41) exhibits peaks associated with the fundamental quadrupolar vibrational mode ($n = 1, l = 2$) and its harmonics. As mentioned above, pure radial vibrations are not active in the SO mechanism.

Since the acoustic displacement fields and plasmon–polariton electromagnetic fields were properly normalized, it is possible to compare the Raman efficiencies of DP and SO mechanisms. Figure 42 presents Raman spectra calculated assuming both DP and SO interactions between the SPP states and the particle vibrations. The particle radius is $R_{\mathrm{p}} = 2.5\,\mathrm{nm}$ and the spectra were generated for resonant excitation of the SPP states at 2.5 eV and 2 eV. Although matrix effects were not taken into account in the calculations of the vibration eigenfrequencies, a size-dependent damping [158, 185] was introduced. The intense low-frequency line in Fig. 42 is due to the fundamental quadrupolar vibrations ($n = 1, l = 2$) emitted by the resonant SPP states via SO coupling (see also spectrum B of Fig. 41). Scattering by the fundamental mode of pure radial vibrations ($n = 1, l = 0$) is responsible for the weaker band around 22 cm^{-1} in Fig. 42 (see also spectrum A of Fig. 41). The highest-frequency band has a mixed character: scattering by harmonic modes

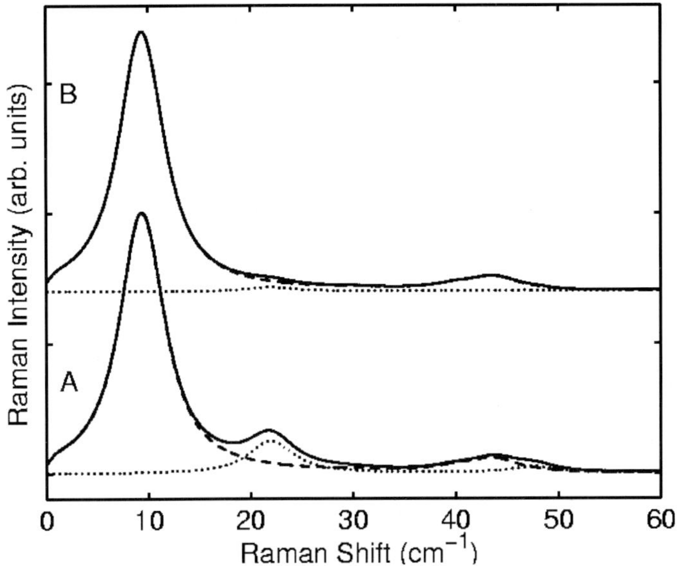

Fig. 42. Raman spectra calculated for a silver nanoparticle with $R_{\mathrm{p}} = 2.5\,\mathrm{nm}$ and for resonant excitation with the SPP states at (A) 2.5 eV and (B) 2 eV. The contribution of the deformation potentials and surface-orientation mechanisms are represented by *dotted* and *dashed lines*, respectively. The total scattered intensity is shown as *bold lines*. The linewidths result from the convolution with a Lorentzian function (finite experimental resolution) and from the size dependence of the vibrational lifetime. From [170]

of quadrupolar ($n = 4$ and 5, $l = 2$) and pure radial vibrations ($n = 2, l = 0$). Moreover, as mentioned previously, the DP/SO intensity ratio depends on $\chi^{\mathrm{ib}}(\omega)/\chi(\omega)$ (23 and 24). Hence, pure radial vibrations are more easily observed for resonant excitation of SPP states with energy close to the interband energy transition. The calculations of the resonant Raman spectra presented in [170] are in agreement with the spectral lineshapes and intensity ratios observed in silver particles [132, 133]. Moreover, the fact that scattering by pure radial modes is more pronounced for gold particles than for silver [132] is well explained by the smaller energy difference between surface-plasmon excitations and interband transitions in gold (larger $\chi^{\mathrm{ib}}(\omega)/\chi(\omega)$).

It is interesting to note that both SO and DP coupling mechanisms contribute to the Raman scattering. On the other hand, the dielectric susceptibility of the particle is not affected by the SO mechanism. Only the DP mechanism is expected to modulate the dielectric response of the particle. This could explain why, in metal particles, quadrupolar vibrations dominate the Raman scattering and pure radial vibrations the time evolution of the optical absorption. Another possible explanation is that only pure radial vibrations can be emitted by the uniform charge density excited by the pump laser beam [159]. The calculations in [170] were performed for the simple case of a spherical particle surrounded by vacuum. Nonsphericity and matrix effects should be taken into account for a detailed comparison with experiments. Such effects will affect both the vibrational and optical properties, as well as the coupling between surface-plasmon states and particle vibrations.

5 Conclusions

In this Chapter we have discussed selected recent theoretical and experimental works on Raman scattering in low-dimensional systems. The Chapter was limited to acoustic-phonon resonant Raman scattering in semiconductor and metal quantum dots. The central idea that guided us was the use of acoustic phonons as a means of studying electrons and plasmons. The latter are selected from the whole density of states using a resonant optical excitation. When light and sound propagate freely over distances comparable to typical lengths of localization and ordering of electronic states, diffraction and interference effects in the Raman-scattering process appear. Strong similarities between X-ray diffraction techniques and acoustic-phonon Raman scattering can be found. The spectral distribution of the low-frequency Raman intensity is the Fourier transform of the electronic-density distribution. It is, however, carried out by sound instead of light. Moreover, we showed that simulations of Raman spectra are an invaluable tool for extracting, from the experimental data, information on the spatial localization and spatial correlation of electronic states. Inverse Fourier transform analysis of the low-frequency Raman spectra is being developed in our laboratory for acoustic imaging of the probability density distribution of the excited electronic states. Throughout the Chapter we have underlined the spatially coherent nature of the scattering process. The transition from coherent to incoherent scattering is a very interesting aspect that still needs to be investigated in connection with phonon damping and localization.

The reflection of sound waves at a surface, and its effects on the electron–phonon interaction, open up interesting prospects for the control of exciton dephasing rates in quantum dots. For instance, surface patterning combined with self-ordering of quantum dots could be used to enhance or inhibit the electron–phonon interaction. Moreover, the wavelengths of acoustic phonons revealed by Raman-scattering range typically from 1 to 30 nm. Hence, they are specific probes of the surface roughness on the nanometer scale and the nature of the surface itself. In this case, the buried quantum dots can be viewed as built-in acoustic detectors. They only serve for scanning the reflection coefficient of the sound waves as a function of frequency.

For systems exhibiting strong acoustic-phonon confinement, the challenge lies in the description of the various coupling mechanisms between the vibrations and the electronic excitations. For semiconductor quantum dots, deformation-potential electron–phonon interaction has been already investigated by several groups; however, the relative Raman efficiencies due to the different confined vibration modes as well as matrix effects on the electron–phonon interaction still need to be investigated. For metal particles, in addition to deformation-potential electron–phonon coupling, the large dielectric mismatch between the particle and the matrix introduces a coupling mechanism related to the surface polarization charges. The dependence of the vibration–surface-plasmon interaction on particle shape, interparticle interaction and on the dielectric properties of the surrounding medium should be investigated both theoretically and experimentally.

For very small particles, containing less than a thousand atoms, the continuous elastic models and the concept of phonon confinement itself become questionable since the surface atoms predominate. As a result, not only microscopic approaches are required for the calculation of the vibration eigenmodes, but the electron–phonon coupling mechanisms themselves must be reconsidered.

Acknowledgements

We are greatly indebted to all the staff members of the Laboratoire de Physique des Solides at Université Paul Sabatier for their support. Special thanks to G. Bachelier, M. Cazayous and J.R. Huntzinger for their contribution during their PhD work. Thanks to N. Del Fatti, E. Duval, O. Kienzle, A. Milekhin, B. Palpant, L. Saviot, P. Yu and A. Zunger for useful discussions and for providing figures.

References

[1] C. Obermuller, A. Deisenrieder, G. Abstreiter, K. Karrai, S. Grosse, S. Manus, J. Feldmann, H. Lipsanen, M. Sopanen, J. Ahopelto: Appl. Phys. Lett. **74**, 3200 (1999)

[2] S. K. Zhang, H. J. Zhu, F. Lu, Z. M. Jiang, X. Wang: Phys. Rev. Lett. **80**, 3340 (1998)

[3] T. Inoshita, H. Sakaki: Physica B **227**, 373 (1996)

[4] R. Heitz, H. Born, F. Guffarth, O. Stier, A. Schliwa, A. Hoffmann, D. Bimberg: Phys. Rev. B **74**, 241305(R) (2001)

[5] S. Hameau, Y. Guldner, O. Verzelen, R. Ferreira, G. Bastard, J. Zeman, A. Lemaître, J. M. Gérard: Phys. Rev. Lett. **83**, 4152 (1999)

[6] S. Sauvage, P. Boucaud, R. P. S. M. Lobo, F. Bras, G. Fishman, R. Prazeres, F. Glotin, J. M. Ortega, J. M. Gérard: Phys. Rev. Lett. **88**, 177402–1 (2002)

[7] R. Ferreira, O. Verzelen, G. Bastard: Semicond. Sci. Technol. **19**, 85 (2004)

[8] D. S. Kop'ev, D. N. Mirlin, V. F. Sapega, A. A. Sirenko: JETP Lett. **51**, 708 (1990)

[9] V. F. Sapega, V. I. Belitsky, A. J. Shields, T. Ruf, M. Cardona, K. Ploog: Solid State Commun. **84**, 1039 (1992)

[10] D. N. Mirlin, I. A. Merkulov, V. I. Perel, I. I. Reshina, A. A. Sirenko, R. Planel: Solid State Commun. **84**, 1093 (1992)

[11] J. R. Huntzinger, J. Groenen, M. Cazayous, A. Mlayah, N. Bertru, C. Paranthoen, O. Dehaese, H. Carrère, E. Bedel, G. Armelles: Phys. Rev. B **61**, R10547 (2000)

[12] M. Cazayous, J. R. Huntzinger, J. Groenen, A. Mlayah, S. Christiansen, H. P. Strunk, O. G. Schmidt, K. Eberl: Phys. Rev. B **62**, 7243 (2000)

[13] M. Cazayous, J. Groenen, J. R. Huntzinger, A. Mlayah, O. G. Schmidt: Phys. Rev. B **64**, 33306 (2001)

[14] M. Cazayous, J. Groenen, A. Zwick, A. Mlayah, R. Carles, J. L. Bischoff, D. Dentel: Phys. Rev. B **66**, 195320 (2002)

[15] A. G. Milekhin, A. I. Nikiforov, O. G. Pchelyakov, S. Schulze, D. R. T. Zahn: JETP Lett. **73**, 461 (2001)

[16] A. G. Milekhin, A. I. Nikiforov, O. G. Pchelyakov, S. Schulze, D. R. T. Zahn: Nanotechnol. **13**, 55 (2002)

[17] A. G. Milekhin, A. I. Nikiforov, O. P. Pchelyakov, S. Schulze, D. R. T. Zahn: Physica E **13**, 982 (2002)

[18] H. Lamb: Proc. London Math. Soc. **13**, 189 (1882)

[19] A. E. H. Love: *A Treatrise on the Mathematical Theory of Elasticity* (Dover, New York 1944)

[20] D. B. Murray, L. Saviot: Phys. Rev. B **69**, 094305 (2004)

[21] W. Cheng, S. F. Ren, P. Y. Yu: Phys. Rev. B **68**, 193309 (2003)

[22] H. Fu, V. Ozoiins, A. Zunger: Phys. Rev. B **59**, 2881 (1999)

[23] C. Colvard, R. Merlin, M. V. Klein, A. C. Gossard: Phys. Rev. Lett. **45**, 298 (1980)

[24] T. Ruf: *Phonon Raman Scattering in Semiconductors Quantum Wells and Superlattices* (Springer, Berlin, Heidelberg 1998)

[25] A. A. Sirenko, V. I. Belitsky, T. Ruf, M. Cardona, A. I. Ekimov, C. Trallero-Giner: Phys. Rev. B **58**, 2077 (1998)

[26] V. F. Sapega, V. I. Belitsky, T. Ruf, H. D. Fuchs, M. Cardona, K. Ploog: Phys. Rev. B **46**, 16005 (1992)

[27] G. Goldoni, T. Ruf, V. F. Sapega, A. Fainstein, M. Cardona: Phys. Rev. B **51**, 14542 (1995)

[28] A. Fainstein, T. Ruf, M. Cardona, V. I. Belitsky, A. Cantarero: Phys. Rev. B **51**, 7064 (1995)

[29] A. Mlayah, A. Sayari, R. Grac, A. Zwick, R. Carles, M. Maaref, R. Planel: Phys. Rev. B **56**, 1486 (1997)

[30] T. Ruf, V. I. Belitsky, J. Spitzer, V. F. Sapega, M. Cardona, K. Ploog: Phys. Rev. Lett. **71**, 3035 (1993)

[31] T. Ruf, J. Spitzer, V. F. Sapega, V. I. Belitsky, M. Cardona, K. Ploog: Phys. Rev. B **50**, 1792 (1994)

[32] V. I. Belitsky, J. Spitzer, M. Cardona: Phys. Rev. B **49**, 8263 (1994)

[33] A. Mlayah, J. Groenen, G. Bachelier, F. Poinsotte, J. R. Huntzinger, M. Cazayous, E. Bedel-Pereira, A. Arnoult, O. G. Schmidt, N. Bertru, C. Paranthoen, O. Dehaese: in *SPIE Conf., Photonics Europe, Strasbourg* (2004)

[34] A. Mlayah, R. Grac, G. Armelles, R. Carles, A. Zwick, F. Briones: Phys. Rev. Lett. **78**, 4119 (1997)

[35] A. Mlayah, J. R. Huntzinger, G. Bachelier, R. Carles, A. Zwick: Physica E **17**, 204 (2003)

[36] J. A. Prieto, G. Armelles, M. E. Pistol, P. Castrillo, J. P. Silveira, F. Briones: Appl. Phys. Lett. **74**, 99 (1999)
[37] P. Castrillo, G. Armelles, J. P. Silveira, F. Briones, J. Balbolla: Appl. Phys. Lett. **71**, 1353 (1997)
[38] H. Bilz, W. Kreiss: *Phonon Dispersion Relations in Insulators* (Springer, Berlin 1979)
[39] Y. A. Goldbery: *Handbook Series on Semiconductor Parameters* (World Scientific, London 1996)
[40] P. Y. Yu, M. Cardona: *Fundamentals of Semiconductors* (Springer, Berlin 1999)
[41] T. Takagahara: Phys. Rev. B **60**, 2638 (1999)
[42] C. Kittel: *Quantum Theory of Solids* (Wiley, New York 1987)
[43] R. Zimmermann: Phys. Stat. Sol. B **173**, 129 (1992)
[44] R. Zimmermann: Nuovo Cimento D **17**, 1801 (1995)
[45] R. Zimmermann, E. Runge: in *Proc. 26th ICPS Edinburgh, Scotland, Institute of Physics* (2002)
[46] V. I. Belitsky, A. Cantarero, S. T. Pavlov, M. Gurioli, F. Bogani, A. Vinattieri, M. Colocci, M. Cardona: Phys. Rev. B **52**, 16665 (1995)
[47] S. Haacke: Rep. Prog. Phys. **64**, 737 (2001)
[48] B. R. Bennet, B. V. Shanabrook, R. Magno: Appl. Phys. Lett. **68**, 958 (1996)
[49] J. Groenen, A. Mlayah, R. Carles, A. Ponchet, A. L. Corre, S. Salaün: Appl. Phys. Lett. **69**, 954 (1996)
[50] P. D. Persans, P. W. Deelman, K. L. Stokes, L. J. Schowalter, A. Byrne, T. Thundat: Appl. Phys. Lett. **70**, 472 (1997)
[51] J. Groenen, R. Carles, S. Christiansen, M. Albrecht, W. Dorsch, H. P. Strunk, H. Wawra, G. Wagner: Appl. Phys. Lett. **71**, 3856 (1997)
[52] Y. A. Pusep, G. Zanelatto, S. W. da Silva, J. C. Galzerani, P. P. Galzerani, P. P. Gonzalez-Borrero, A. I. Toropov, P. Basmaji: Phys. Rev. B **58**, 1770 (1998)
[53] S. H. Kwok, P. Y. Yu, C. H. Tung, Y. H. Zang, M. F. Li, C. S. Peng, J. M. Zhou: Phys. Rev. B **59**, 4980 (1999)
[54] J. Groenen, C. Priester, R. Carles: Phys. Rev. B **60**, 16013 (1999)
[55] J. Gleize, J. Frandon, F. Demangeot, M. A. Renucci, C. Adelmann, B. Daudin, G. Feuillet, B. Damilano, N. Grandjean, J. Massies: Appl. Phys. Lett. **77**, 2174 (2000)
[56] J. Gleize, F. Demangeot, J. Frandon, M. A. Renucci, M. Kuball, B. Damilano, N. Grandjean, J. Massies: Appl. Phys. Lett. **79**, 686–688 (2001)
[57] A. G. Milekhin, N. P. Stepina, A. I. Yakimov, A. I. Nikiforov, S. Schulze, D. R. T. Zahn: Appl. Surf. Sci. **175–176**, 629 (2001)
[58] D. A. Tenne, V. A. Haisler, A. I. Toropov, A. K. Bakarov, A. K. Gutakovsky, D. R. T. Zahn, A. P. Shebanin: Phys. Rev. B **61**, 13785 (2000)
[59] A. G. Milekhin, A. I. Toropov, A. K. Bakarov, D. A. Tenne, G. Zanelatto, J. C. Galzerani, S. Schulze, D. R. T. Zahn: Phys. Rev. B **70**, 085314 (2004)
[60] J. L. Liu, G. Jin, Y. S. Tang, Y. H. Luo, K. L. Wang, D. P. Yu: Appl. Phys. Lett. **76**, 586 (2000)
[61] P. Y. Yu: Appl. Phys. Lett. **78**, 1160 (2001)
[62] J. L. Liu, G. Jin, Y. S. Tang, Y. H. Luo, K. L. Wang, D. P. Yu: Appl. Phys. Lett. **78**, 1162 (2001)
[63] M. Grundmann, O. Stier, D. Bimberg: Phys. Rev. B **52**, 11969 (1995)

[64] O. G. Schmidt, K. Eberl: Phys. Rev. B **13**, 721 (2000)
[65] O. Kienzle, F. Ernst, M. Rühle, O. G. Schmidt, K. Eberl: Appl. Phys. Lett. **74**, 269 (1999)
[66] Q. Xie, A. Madhukar, P. Chen, N. P. Kobayashi: Phys. Rev. Lett. **75**, 2542 (1995)
[67] G. S. Solomon, J. A. Trezza, A. F. Marshall, J. S. Harris: Phys. Rev. Lett. **76**, 952 (1996)
[68] J. Tersoff, C. Teichert, M. G. Lagally: Phys. Rev. Lett. **76**, 1675 (1996)
[69] C. Teichert, M. G. Lagally, L. J. Peticolas, J. C. Bean, J. Tersoff: Phys. Rev. B **53**, 16334 (1996)
[70] V. Le Thanh, V. Yam, P. Boucaud, F. Fortuna, C. Ulysse, D. Bouchier, L. Vervoort, J. M. Lourtioz: Phys. Rev. B **60**, 5851 (1999)
[71] V. Holy, G. Springholz, M. Pinczolits, G. Bauer: Phys. Rev. Lett. **83**, 356 (1999)
[72] G. Springholz, M. Pinczolits, P. Mayer, V. Holy, G. Bauer, H. H. Kang, L. Salamanca-Riba: Phys. Rev. Lett. **84**, 4669 (2000)
[73] J. Stangl, V. Holy, G. Bauer: Rev. Mod. Phys. **76**, 725 (2004)
[74] M. Colocci, A. Vinattieri, L. Lippi, F. Bogani, M. Rosa-clot, S. Taddei, A. Bosacchi, S. Franchi, K. Ploog: Appl. Phys. Lett. **74**, 564 (1999)
[75] M. Renucci, J. B. Renucci, M. Cardona: Solid State Commun. **9**, 1235 (1971)
[76] R. Carles, N. Saint-Cricq, J. B. Renucci, A. Zwick, M. A. Renucci: Phys. Rev. B **22**, 6120 (1980)
[77] T. P. Pearsall, F. H. Pollak, J. C. Bean, R. Hull: Phys. Rev. B **33**, 6821 (1986)
[78] M. Cazayous: Ph.D. thesis, University Paul Sabatier, Toulouse (2002)
[79] O. G. Schmidt, C. Lange, K. Eberl, O. Kienzle, F. Ernst: Appl. Phys. Lett. **71**, 2340 (1997)
[80] O. G. Schmidt, O. Kienzle, Y. Hao, K. Eberl, F. Ernst: Appl. Phys. Lett. **74**, 1272 (1999)
[81] M. Cazayous, J. Groenen, F. Demangeot, R. Sirvin, M. Caumont, T. Remmele, M. Albrecht, S. Christiansen, M. Becker, H. P. Strunk, H. Wawra: J. Appl. Phys. **91**, 6772 (2002)
[82] M. Born, E. Wolf: *Principles of Optics: Electromagnetic Theory of Propagation, Interference and Diffraction of Light* (Pergamon Press, London 1959)
[83] O. Pilla, V. Lemos, M. Montagna: Phys. Rev. B **50**, 11845 (1994)
[84] M. W. C. Dharma-wardana, P. X. Zhang, D. J. Lockwood: Phys. Rev. B **48**, 11960 (1993)
[85] S. M. Rytov: Akoust. Zh. **2**, 68 (1956)
[86] B. Jusserand, M. Cardona: Superlattices and other microstructures, in M. Cardona, G. Güntherodt (Eds.): *Light Scattering in Solids*, vol. V (Springer, Berlin 1989)
[87] C. Colvard, T. A. Gant, M. V. Klein, R. Merlin, R. Fischer, H. Morkoc, A. C. Gossard: Phys. Rev. B **31**, 2080 (1985)
[88] D. J. Lockwood, M. W. C. Dharma-wardana, J. M. Baribeau, D. C. Houghton: Phys. Rev. B **35**, 2243 (1987)
[89] J. Sapriel, J. He, B. Djafari-Rouhani, R. Azoulay, F. Mollot: Phys. Rev. B **37**, 4099 (1988)
[90] M. Giehler, T. Ruf, M. Cardona, K. Ploog: Phys. Rev. B **55**, 7124 (1997)
[91] M. W. C. Dharma-wardana, D. J. Lockwood, J. M. Baribeau, D. C. Houghton: Phys. Rev. B **34**, 3034 (1986)

[92] B. Jusserand, D. Paquet, F. Mollot, F. Alexandre, G. L. Roux: Phys. Rev. B **35**, 2808 (1987)
[93] P. Etchegoin, J. Kircher, M. Cardona: Phys. Rev. B **47**, 10292 (1993)
[94] Z. V. Popovic, J. Spitzer, T. Ruf, M. Cardona, R. Nötzel, K. Ploog: Phys. Rev. B **48**, 1659 (1993)
[95] J. He, B. Djafari-Rouhani, J. Sapriel: Phys. Rev. B **37**, 4086 (1988)
[96] J. Ménendez, M. Cardona: Phys. Rev. B **31**, 3696 (1985)
[97] A. Cantarero, C. Trallero-Giner, M. Cardona: Phys. Rev. B **40**, 12290 (1989)
[98] J. Stangl, T. Roch, G. Bauer, I. Kegel, T. H. Metzger, O. G. Schmidt, K. Eberl, O. Kienzle, F. Ernst: Appl. Phys. Lett. **77**, 3953 (2000)
[99] M. Cazayous, J. Groenen, J. Brault, M. Gendry, U. Denker, O. G. Schmidt: Physica E (2003)
[100] M. Cazayous, J. Groenen, J. R. Huntzinger, A. Mlayah, U. Denker, O. G. Schmidt: Mater. Sci. Eng. B **88**, 63 (2002)
[101] M. Gendry, J. Brault, C. Monat, J. Kapsa, M. P. Besland, G. Grenet, O. Marty, M. Pittaval, G. Hollinger: IPRM'01, Nara, in *Proc. of 13th Indium Phosphide and Related Materials Conference* (2001) p. 525
[102] B. Salem, G. Brémond, M. Hjiri, F. Hassen, H. Maaref, O. Marty, J. Brault, M. Gendry: Mater. Sci. Eng. B **101**, 259 (2003)
[103] M. Weyers, M. Sato, H. Ando: Jpn. J. Appl. Phys. **31**, L853 (1992)
[104] J. Salzman, H. Temkin: Mater. Sci. Eng. B **50**, 148 (1997)
[105] J. D. Perkins, A. Mascarenhas, Y. Zhang, J. F. Geisz, D. J. Friedman, J. M. Olson, S. R. Kurtz: Phys. Rev. Lett. **82**, 3312 (1999)
[106] Y. Zhang, A. Mascarenhas, H. P. Xin, C. W. Tu: Phys. Rev. B **61**, 7479 (2000)
[107] C. Skierbiszewski, P. Perlin, P. Wisniewski, W. Knap, T. Suski, W. Walukiewicz, W. Shan, K. M. Yu, J. W. Ager, E. E. Haller, F. Geisz, J. M. Olson: App. Phys. Lett. **76**, 2409 (2000)
[108] W. Shan, W. Walukiewicz, J. W. Ager III, E. E. Haller, J. F. Geisz, D. J. Friedman, J. M. Olson, S. R. Kurtz: Phys. Rev. Lett. **82**, 1221 (1999)
[109] E. D. Jones, N. A. Modine, A. A. Allerman, S. R. Kurtz, A. F. Wright, S. T. Tozer, X. Wei: Phys. Rev. B **60**, 4430 (1999)
[110] L. Bellaiche, S.-H. Wei, A. Zunger: Phys. Rev. B **54**, 17568 (1996)
[111] T. Mattila, Su-Huai Wei, A. Zunger: Phys. Rev. B **60**, R11245 (1999)
[112] P. R. C. Kent, A. Zunger: Phys. Rev. Lett. **86**, 2613 (2001)
[113] P. R. C. Kent, A. Zunger: Phys. Rev. B **64**, 115208 (2001)
[114] L. W. Wang: Appl. Phys. Lett. **78**, 1565 (2001)
[115] G. Bachelier, A. Mlayah, M. Cazayous, J. Groenen, A. Zwick, H. Carrere, E. Bedel-Pereira, A. Arnoult, A. Rocher, A. Ponchet: Phys. Rev. B **67**, 205325 (2003)
[116] W. Shan, W. Walukiewicz, K. M. Yu, J. Wu, J. W. Ager III, E. E. Haller, H. P. Xin, C. W. Tu: Appl. Phys. Lett. **76**, 3251 (2000)
[117] W. Walukiewicz, W. Shan, K. M. Yu, J. W. Ager III, E. E. Haller, I. Miotkowski, M. J. Seong, H. Alawadhi, A. K. Ramdas: Phys. Rev. Lett. **85**, 1552 (2000)
[118] M. Cazayous, J. Groenen, J. R. Huntzinger, G. Bachelier, A. Zwick, A. Mlayah, N. Bertru, C. Paranthoen, O. Dehaese, E. Bedel-Pereira, F. Négri, H. Carrère: Phys. Rev. B **69**, 125323 (2004)

[119] J. A. Prieto, G. Armelles, J. Groenen, R. Carles: Appl. Phys. Lett. **74**, 99 (1999)
[120] D. A. Weitz, T. J. Gramila, A. Z. Genack, J. I. Gersten: Phys. Rev. Lett. **45**, 355 (1980)
[121] E. Duval, A. Boukenter, B. Champagon: Phys. Rev. Lett. **56**, 2052 (1986)
[122] A. Tanaka, S. Onari, T. Arai: Phys. Rev. B **47**, 1237 (1993)
[123] A. Othmani, C. Bovier, J. C. Plenet, J. Dumas, B. Champagnon, C. Mai: Mater. Sci. Eng. A **168**, 263 (1993)
[124] L. Saviot, B. Champagnon, E. Duval, A. I. Ekimov: Phys. Rev. B **57**, 341 (1998)
[125] M. Fujii, Y. Kanzawa, S. Hayashi, K. Yamamoto: Phys. Rev. B **54**, 8373 (1996)
[126] M. Pauthe, E. Berstein, J. Dumas, L. Saviot, A. Pradel, M. Ribes: J. Mater. Chem. **9**, 187 (1999)
[127] N. N. Ovsyuk, E. B. Orokhov, V. V. Grishchenko, A. P. Shabanin: JETP Lett. **47**, 298 (1988)
[128] M. Fujii, S. Hayashi, K. Yamamoto: Phys. Rev. B **44**, 6243 (1991)
[129] G. Mariotto, M. Montagna, G. Viliani, E. Duval, S. Lefranc, E. Rzepka, C. Mai: Europhys. Lett. **6**, 239 (1988)
[130] B. Palpant, H. Portales, L. Saviot, J. Lerme, B. Prevel, M. Pellarin, E. Duval, A. Perez, M. Broyer: Phys. Rev. B **60**, 17107 (1999)
B. Palpant PhD thesis, Claude Bernard University.
[131] E. Duval, H. Portalès, L. Saviot, M. Fujii, K. Sumimoto, S. Hayashi: Phys. Rev. B **63**, 075405 (2001)
[132] H. Portalès, L. Saviot, E. Duval, M. Fujii, S. Hayashi, N. D. Fatti, F. Vallée: J. Chem. Phys. **115**, 3444 (2001)
[133] I. L. A. Courty, M. P. Pileni: J. Chem. Phys. **116**, 8074 (2002)
[134] A. Taleb, C. Petit, M. P. Pileni: Chem. Mater. **9**, 950 (1997)
[135] A. Courty, O. Araspin, C. Fermon, M. P. Pileni: Langmuir **17**, 1372 (2001)
[136] A. Taleb, C. Petit, M. P. Pileni: J. Phys. Chem. B **102**, 2214 (1998)
[137] A. Taleb, V. Russier, A. Courty, M. P. Pileni: Phys. Rev. B **59**, 13350 (1999)
[138] V. A. Dubrovski, V. Morochnik: Izv. Acad. Sci., USSR, Phys. Solid Earth **17**, 494 (1981)
[139] P. Verma, W. Cordts, G. Irmer, J. Monecke: Phys. Rev. B **60**, 5778 (1999)
[140] L. Saviot, D. B. Murray, M. del Carmen Marco de Lucas: Phys. Rev. B **69**, 113402 (2004)
[141] F. H. Stillinger, T. A. Weber: Phys. Rev. B **31**, 5262 (1985)
[142] P. N. Keating: Phys. Rev. **145**, 637 (1966)
[143] J. L. Martins, A. Zunger: Phys. Rev. B **30**, 6217 (1984)
[144] G. Qin, S. F. Ren: J. Phys.: Condens. Matter **14**, 8771 (2002)
[145] L. Saviot, B. Champagnon, E. Duval, I. A. Kurdiavtsev, A. I. Ekimov: J. Non-Cryst. Solids **197**, 238 (1996)
[146] D. Brill, G. Gaunaurd: J. Acoust. Soc. Am. **81**, 1 (1987)
[147] M. Montagna, R. Dusi: Phys. Rev. B **52**, 10080 (1995)
[148] H. Portales, L. Saviot, E. Duval, M. Gaudry, E. Cottancin, M. Pellarin, J. Lerme, M. Broyer: Phys. Rev. B **65**, 165422 (2002)
[149] L. Saviot, D. B. Murray: Phys. Rev. Lett. **93**, 055506 (2004)
[150] J. I. Gersten, D. A. Weitz, T. J. Gramila, A. Z. Genack: Phys. Rev. B **22**, 4562 (1980)

[151] A. Tamura, K. Higeta, T. Ichinokawa: J. Phys. C **15**, 4975 (1982)
[152] N. N. Ovsyuk, V. N. Novikov: Phys. Rev. B **53**, 3113 (1996)
[153] T. D. Krauss, F. W. Wise: Phys. Rev. Lett. **79**, 5102 (1997)
[154] E. R. Thoen, G. Steinmeyer, P. Langlois, E. P. Ippen, G. E. Tudury, C. H. Brito Cruz, L. C. Barbosa, C. L. Cesar: Appl. Phys. Lett. **73**, 2149 (1998)
[155] M. Nisoli, S. D. Silvestri, A. Cavalleri, A. M. Malvezzi, A. Stella, G. Lanzani, P. Cheyssac, R. Kofman: Phys. Rev. B **55**, R13424 (1997)
[156] J. H. Hodak, L. Martini, G. V. Hartland: J. Chem. Phys. **108**, 9210 (1998)
[157] B. Perrin, C. Rossignol, B. Bonello, J. C. Jeannet: Physica B **263–264**, 571 (1999)
[158] N. Del Fatti, C. Voisin, F. Chevy, F. Vallée, C. Flytzanis: J. Chem. Phys. **110**, 11484 (1999)
[159] A. Nelet, A. Crut, A. Arbouet, N. D. Fatti, F. Vallée, H. Portales, L. Saviot, E. Duval: Appl. Surf. Sci. **226**, 209 (2004)
[160] Y. Masumoto: J. Lumin. **70**, 386 (1996)
[161] L. Zimin, S. V. Nair, Y. Masumoto: Phys. Rev. Lett. **80**, 3105 (1998)
[162] J. Zhao, Y. Masumoto: Phys. Rev. B **60**, 4481 (1999)
[163] L. Besombes, K. Keng, L. Marsal, H. Mariette: Phys. Rev. B **63**, 155307 (2001)
[164] I. Favero, G. Cassabois, R. Ferreira, D. Darson, C. Voisin, J. Tignon, C. Delalande, G. Bastard, P. Roussignol, J. M. Gérard: Phys. Rev. B **68**, 233301 (2003)
[165] *Chronological Scientific Tables*, Technical report, National Astronomical Observatory Maruzen, Tokyo (1991)
[166] E. Duval: Phys. Rev. B **46**, 5795 (1992)
[167] S. V. Gupalov, I. A. Merkulov: Phys. Solid State **41**, 1349 (1999)
[168] M. Ikezawa, T. Okuno, Y. Yasumoto, A. A. Lipovskii: Phys. Rev. B **64**, 201315(R) (2001)
[169] M. Ivanda, K. Babocsi, C. Denn, M. Schmitt, M. Montagna, W. Kiefer: Phys. Rev. B **67**, 235329 (2003)
[170] G. Bachelier, A. Mlayah: Phys. Rev. B **69**, 205408 (2004)
[171] K. Arya, R. Zeyher: in M. Cardona, G. Guntherodt (Eds.): *Light Scattering in Solids*, vol. IV (Springer 1984)
[172] G. Mie: Ann. Phys. **25**, 377 (1908)
[173] F. Hache, D. Richard, C. Flytzanis: J. Opt. Soc. Am. B **3**, 1647 (1986)
[174] J. Lindhard, K. Dan: Vidensk. Selsk. Mat. Fys. Medd. **28**, 8 (1954)
[175] H. Hovel, S. Fritz, A. Hilger, U. Kreibig, M. Vollmer: Phys. Rev. B **48**, 18178 (1993)
[176] P. B. Johnson, R. W. Christy: Phys. Rev. B **6**, 4370 (1972)
[177] N. Del Fatti, C. Voisin, M. Achermann, S. Tzortsakis, D. Christofilos, F. Vallee: Phys. Rev. B **61**, 16956 (2000)
[178] J. Bardeen, W. Shockley: Phys. Rev. **80**, 72 (1950)
[179] U. Kriebig, M. Vollmer: *Optical Properties of Metal Clusters* (Berlin, Springer 1995)
[180] H. Tups, A. Otto, K. Syassen: Phys. Rev. B **29**, 5458 (1984)
[181] P. A. Knipp, T. L. Reinecke: Phys. Rev. B **52**, 5923 (1995)
[182] A. M. Alcalde, G. E. Marques, G. Weber, T. L. Reinecke: Brazilian J. Phys. **32**, 284 (2002)

[183] J. Lerme, B. Palpant, B. Prevel, M. Pellarin, M. Treilleux, J. L. Vialle, A. Perez, M. Broyer: Phys. Rev. Lett. **80**, 5105 (1998)
[184] R. A. Molina, D. Weinmann, R. A. Jalabert: Phys. Rev. B **65**, 155427 (2002)
[185] C. Voisin: Ph.D. thesis, Paris XI Orsay University (2001)

Index

acoustic
  bulk-like modes, 249, 250, 293, 298
  cavity effects, 277, 280, 282, 291
  damping, 239, 290, 294, 295, 304, 306
  displacement field, 238, 239, 291, 292, 298, 301, 304
  extended modes, 240
  image, 276, 277
  localization, 239, 287–289, 291, 293, 306
  mirror, 273
  phase, 256, 261, 267, 280–282
  strong confinement, 282
  surface-related modulation, 276, 277

complex frequency model (CFM), 290, 291
core shell model (CSM), 291–293, 296, 298
coupling
  deformation-potential, 300

electron–acoustic phonon
  interaction, 239, 246
electronic wavefunction
  distribution, 238, 240, 246, 247, 249, 255, 262, 264–266, 272, 273
  extension, 253, 272
  impurity-like, 272
  localization, 260
  ordering, 238, 306
  spatial correlation, 238, 240, 255, 262, 264
  spatial localization, 247
embedded particles, 293

in-plane ordering, 267

Lamb's model, 239, 283, 284, 286, 287, 296, 298

nanoparticle
  CdS, 297
  CdSe, 297

phonon
  density of states of GaP, 289
photoelastic
  constant, 261
plasmon-polariton states
  Drude damping, 299, 302
  interaction with acoustic vibrations
    deformation potential mechanism, 299, 300, 303, 305, 307
    surface orientation mechanism, 299, 302, 304, 305
  interband susceptibility, 300, 301, 303, 305
  intraband susceptibility, 293, 302
  Raman scattering mediated by, 303–306
  surface coupling, 301
  surface polarization charge, 302, 307

quantum dots
  Ge/Si, 256–259, 267, 274
  InAs/AlInAs, 268
  InAs/GaAs, 267
  InAs/InP, 252, 277
  ordering, 249
  self-assembled, 248
quantum well
  GaAs/GaP, 240
  multilayer, 265

Raman
  form factor, 247, 250, 252, 261, 262, 267
  interference contrast, 265, 266
  selection rule, 243, 295, 298

structure factor, 247, 255, 258, 262, 267, 273, 275, 277
Rytov's model, 259

scattering
efficiency, 245
selection
intermediate-state, 247
spatial localization, 250

transmission
time-resolved, 293, 295

valence force field model (VFF), 239, 284, 286, 287
vertical ordering, 262

Inelastic X-Ray Scattering from Phonons

Michael Krisch and Francesco Sette

European Synchrotron Radiation Facility, B.P. 220, 38043 Grenoble Cédex, France
{krisch,sette}@esrf.fr

Abstract. This Chapter presents the current status of phonon-dispersion studies using very high energy resolution inelastic X-ray scattering. The theoretical background and the instrumental principles are briefly summarized. This is followed by a representative selection of studies on single crystals and polycrystalline materials, including high-pressure work. The Chapter concludes with novel applications of the technique.

1 Introduction

The study of phonon dispersion in condensed matter at momentum transfers, $\boldsymbol{Q}$, and energies, E, characteristic of collective motions has been traditionally the domain of neutron spectroscopy. The experimental observable is the dynamic structure factor $S(\boldsymbol{Q}, E)$, which is the space and time Fourier transform of the density–density correlation function. Neutrons as the probing particle are particularly suitable, since 1. the neutron–nucleus scattering cross section is sufficiently weak to allow for a large penetration depth, 2. the energy of neutrons with wavelengths of the order of interparticle distances is about 100 meV, and therefore comparable to typical phonon energies, and 3. the momentum of the neutron allows the whole dispersion scheme to be probed out to several tens of nm^{-1}, in contrast to inelastic light-scattering techniques such as Brillouin and Raman scattering that can only determine acoustic and optical modes, respectively, at very small momentum transfers close to the Γ-point in reciprocal space. While it has been pointed out in several textbooks that X-rays can, in principle, also be utilized to determine the $S(\boldsymbol{Q}, E)$, it was stressed, however, that this would represent a formidable experimental challenge, mainly due to the fact that an X-ray instrument would have to provide an extremely high energy resolution. This is understood considering that photons with wavelength $\lambda = 0.1$ nm have an energy of about 12 keV. Therefore, the study of phonon excitations in condensed matter, which are in the meV region, requires a relative energy resolution of at least $\Delta E/E = 10^{-7}$.

On the other hand, there are situations where the use of photons has important advantages over neutrons. One specific case is based on the general

M. Cardona, R. Merlin (Eds.): Light Scattering in Solid IX,
Topics Appl. Physics **108**, 317–370 (2007)

consideration that it is not possible to study acoustic excitations propagating with a speed of sound v_s using a probe particle with a speed v smaller than v_s. This limitation is not particularly relevant for inelastic neutron scattering (INS) experiments of crystalline samples. Here, the translation invariance allows acoustic excitations in high-order Brillouin zones to be studied, thus overcoming the above-mentioned kinematics limit on phonon branches with steep dispersions. On the contrary, the situation is very different for topologically disordered systems such as liquids, glasses and gases. In these systems, in fact, the absence of periodicity imposes that the acoustic excitations must be measured at small momentum transfers. Thermal neutrons have a velocity in the range of 1000 m/s, and only in disordered materials with a speed of sound smaller than this value (mainly fluids and low-density gases) the acoustic dynamics can be effectively investigated. A second specific case, particularly relevant in the present context, concerns the necessary sample volume in order to perform an INS or IXS experiment. While INS experiments need sample volumes of typically several mm^3, IXS measurements can be performed utilizing volumes several orders of magnitude smaller (10^{-5} to 10^{-4} mm^3). This opens up possibilities to study materials only available in very small quantities and/or their investigation in extreme thermodynamic conditions, such as very high pressure.

The above-mentioned advantages of IXS motivated researchers in the 1980s to develop an X-ray instrument. Attempts utilizing an X-ray generator with a rotating anode yielded an instrumental energy resolution ΔE of 42 meV, but the photon flux was not sufficient to conduct an IXS experiment [1, 2]. Using synchrotron radiation from a bending magnet at HASYLAB provided the necessary high photon flux, and the first pioneering experiments were performed on graphite and beryllium with an energy resolution of $\Delta E = 55$ meV [3, 4]. This modest energy resolution and photon flux were significantly improved by the construction of a dedicated instrument located at a wiggler source at HASYLAB, where an instrumental resolution of 9 meV could be achieved [5]. With the advent of third-generation synchrotron radiation sources, namely the European Synchrotron Radiation Facility (ESRF) in Grenoble (France), the Advanced Photon Source (APS) at the Argonne National Laboratory (USA) and the Super Photon Ring (SPring-8) in Kansai (Japan) the IXS technique gained its full maturity within a few years. Thanks to the high brilliance of the undulator X-ray sources and important developments in X-ray optics, experiments are now routinely performed with an energy resolution of 1.5 meV. At present, there are worldwide four instruments dedicated to IXS from phonons: ID16 and ID28 at the ESRF [6–8], 3ID at the APS [9], and BL35XU at SPring-8 [10], and at least two further projects in an advanced design stage.

IXS is nowadays applied to a large variety of very different materials ranging from quantum liquids through high-T_c superconductors to biological aggregates. The present Chapter will focus on crystalline systems; for

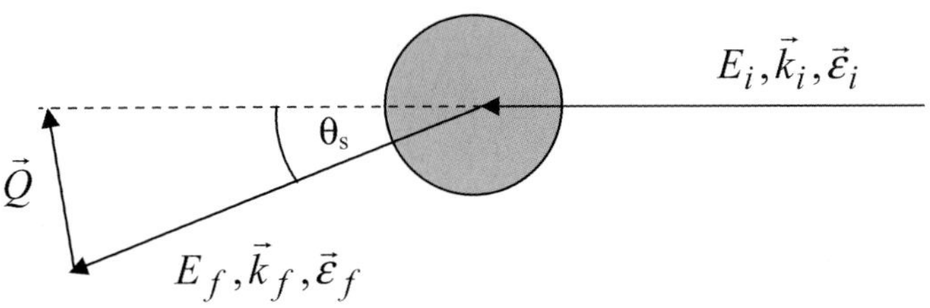

Fig. 1. The inelastic scattering process. $E_{\mathrm{i,f}}$, $\boldsymbol{k}_{\mathrm{i,f}}$ and $\varepsilon_{\mathrm{i,f}}$ denote the energy, the wavevector and the polarization vector of the incident (scattered) photon, and θ_{s} is the scattering angle. E and $\boldsymbol{Q}$ are the energy and momentum transfers, respectively ($\hbar = 1$)

complementary overviews the interested reader is referred to other review papers [5, 11–13].

2 Scattering Kinematics and Inelastic X-ray Cross Section

The inelastic scattering process is depicted schematically in Fig. 1. The momentum and energy conservation impose that:

$$\boldsymbol{Q} = k_{\mathrm{i}} - k_{\mathrm{f}} \tag{1a}$$

$$E = E_{\mathrm{i}} - E_{\mathrm{f}} \tag{1b}$$

$$\boldsymbol{Q}^2 = k_{\mathrm{i}}^2 + k_{\mathrm{f}}^2 - 2k_{\mathrm{i}}k_{\mathrm{f}} \cos(\theta_{\mathrm{s}}) , \tag{1c}$$

where θ_{s} is the scattering angle between the incident and scattered photons. The relation between momentum and energy in the case of photons is given by: $E(k) = \hbar ck$. Considering that the energy losses or gains associated with phonon-like excitations are always much smaller than the energy of the incident photon ($E \ll E_{\mathrm{i}}$), one obtains:

$$(Q/k_{\mathrm{i}}) = 2 \sin(\theta_{\mathrm{s}}/2) . \tag{2}$$

The ratio between the exchanged momentum and the incident photon momentum is completely determined by the scattering angle. Therefore, for IXS, there are no limitations in the energy transfer at a given momentum transfer for phonon-like excitations, in strong contrast to INS where a strong coupling between energy and momentum transfer exists.

The Hamiltonian, describing the electron–photon interaction in a scattering process, is composed, in the weak relativistic limit, of four terms [14]. Neglecting resonance phenomena close to X-ray absorption thresholds and the much weaker magnetic couplings, only the term arising from the Thomson interaction Hamiltonian has to be retained:

$$H_{\text{X-Th}} = \frac{1}{2} r_0 \sum_j \boldsymbol{A}^2(\boldsymbol{r}_j, t) , \tag{3}$$

where $r_0 = e^2/m_\mathrm{e}c^2$ is the classical electron radius, and $\boldsymbol{A}(\boldsymbol{r}_j, t)$ is the vector potential of the electromagnetic field in the $\boldsymbol{r}_j$, coordinate of the jth electron. The sum extends over all the electrons in the system.

The double differential cross section is proportional to the number of incident probe particles scattered within an energy range ΔE and momentum variation into a solid angle $\Delta\Omega$. In the process, where a photon of energy E_i, wavevector k_i, and polarization ε_i, is scattered into a final state of energy E_f, wavevector k_f, and polarization ε_f, and the electron system goes from the initial state $|I\rangle$ to the final state $|F\rangle$, we obtain:

$$\frac{\partial^2\sigma}{\partial\Omega\partial E} = r_0^2\left(\varepsilon_\mathrm{i}\cdot\varepsilon_\mathrm{f}\right)^2\frac{k_\mathrm{i}}{k_\mathrm{f}}\sum_{I,F}P_I\left|\langle F|\sum_j \mathrm{e}^{\mathrm{i}\boldsymbol{Q}\boldsymbol{r}_j}|I\rangle\right|^2\delta\left(E-E_\mathrm{f}-E_\mathrm{i}\right). \quad (4)$$

The sum over the initial and final states is the thermodynamic average, and P_I corresponds to the thermal population of the initial state. From this expression, which implicitly contains the correlation function of the electron density, one arrives at the correlation function of the atomic density, on the basis of the following considerations: 1. We assume the validity of the adiabatic approximation. This allows one to separate the system quantum state $|S\rangle$ into the product of an electronic part, $|S_\mathrm{e}\rangle$, which depends only parametrically from the nuclear coordinates, and a nuclear part, $|S_\mathrm{n}\rangle : |S\rangle = |S_\mathrm{e}\rangle|S_\mathrm{n}\rangle$. This approximation is particularly good for exchanged energies that are small with respect to the excitations energies of electrons in bound core states: this is indeed the case in basically any atomic species when considering phonon energies. In this approximation, one neglects the portion of the total electron density contributing to the delocalized bonding states in the valence-band region. 2. We limit ourselves to considering the case in which the electronic part of the total wavefunction is not changed by the scattering process, and therefore the difference between the initial state $|I\rangle = |I_\mathrm{e}\rangle|I_\mathrm{n}\rangle$ and the final state $|F\rangle = |I_\mathrm{e}\rangle|F_\mathrm{n}\rangle$ is due only to excitations associated with atomic density fluctuations. Using these two hypotheses we then obtain:

$$\frac{\partial^2\sigma}{\partial\Omega\partial E} = r_0^2\left(\varepsilon_\mathrm{i}\cdot\varepsilon_\mathrm{f}\right)^2\frac{k_\mathrm{i}}{k_\mathrm{f}} \times\left\{\sum_{I_\mathrm{n},F_\mathrm{n}}P_{I_n}\left|\langle F_\mathrm{n}|\sum_k f_k\left(Q\right)\mathrm{e}^{\mathrm{i}\boldsymbol{Q}\boldsymbol{R}_k}|I_\mathrm{n}\rangle\right|^2\delta\left(E-E_\mathrm{f}-E_\mathrm{i}\right)\right\}, \quad (5)$$

where $f_k(Q)$ is the atomic form factor of the atom k and $\boldsymbol{R}_k$ its position vector. The expression in the curly brackets contains the dynamical structure factor $S(\boldsymbol{Q}, E)$. Assuming that all the scattering units in the system are equal, this expression can be simplified by the factorization of the form factor of these scattering units:

$$\frac{\partial^2\sigma}{\partial\Omega\partial E} = r_0^2\left(\varepsilon_\mathrm{i}\cdot\varepsilon_\mathrm{f}\right)^2\frac{k_\mathrm{i}}{k_\mathrm{f}}\left|f(Q)\right|^2 S(\boldsymbol{Q}, E) = \left(\frac{\partial\sigma}{\partial\Omega}\right)_\mathrm{Th}\left|f(Q)\right|^2 S(\boldsymbol{Q}, E). \quad (6)$$

For this specific case, the coupling characteristics of the photons to the system, the Thomson-scattering cross section, is separated from the dynamical properties of the system, and the atomic form factor $f(Q)$ appears only as a multiplicative factor. The corresponding INS cross section is obtained by replacing the Thomson cross section and the atomic form factor by the coherent neutron scattering length b of the element under study. For this specific case, INS and IXS both probe the $S(\boldsymbol{Q}, E)$ in the same fashion.

In the case of single-crystal studies, and within the harmonic approximation, the general form of $S(\boldsymbol{Q}, E)$ for single-phonon scattering takes the following form:

$$S(E, \boldsymbol{Q}) = \sum_j \left\langle n(E) + \frac{1}{2} \pm \frac{1}{2} \right\rangle (E_j(\boldsymbol{q}))^{-1} F_{\text{in}}(\boldsymbol{Q}) \delta \left(E \pm E_j(\boldsymbol{q}) \right), \qquad (7)$$

where the sum extends over the $3k$ phonon modes of a crystal with k atoms per unit cell. The first term in angular brackets is the Bose factor, which describes the phonon population at temperature T, with the upper and the lower signs corresponding to phonon creation and annihilation, respectively. The inelastic structure factor $F_{\text{in}}(\boldsymbol{Q})$ involves a sum taken over all atoms in the unit cell:

$$F_{\text{in}}(\boldsymbol{Q}) = \left| \sum_k M_k^{-1/2} f_k(Q) \left[\boldsymbol{e}_k^j(\boldsymbol{q}) \cdot \boldsymbol{Q} \right] \exp(\mathrm{i}\boldsymbol{Q} \cdot \boldsymbol{r}_k) \exp(-w_k) \right|^2 , \qquad (8)$$

where M_k is the mass, $\exp(-w_k)$ is the Debye–Waller factor and $\boldsymbol{r}_k$ is the position of atom k in the unit cell. $\boldsymbol{e}_k^j(\boldsymbol{q})$ denotes the phonon eigenvector with wavevector $\boldsymbol{q}$ of atom k in mode j.

It is important to note that different atomic species contribute differently to the $S(\boldsymbol{Q}, E)$ for neutrons and X-rays. This arises from the fact that both the Q-dependence and the atomic number, Z, dependence of the X-ray form factor and of the neutron-scattering length are substantially different. The IXS cross section is proportional to $f_k(Q)^2$. In the limit $Q \to 0$, the form factor is equal to the number of electrons in the scattering atom, Z; for increasing values of Q, the form factor decays with a decay constant of the order of the inverse of the atomic wavefunction dimensions of the electrons in the atom. Furthermore, for X-rays that are linearly polarized in the horizontal scattering plane, $(\boldsymbol{\varepsilon}_{\mathrm{i}} \cdot \boldsymbol{\varepsilon}_{\mathrm{f}}) = \cos^2(\theta_{\mathrm{s}})$. Figure 2 shows the relative intensity as a function of Q for three different elements, namely aluminum, manganese and palladium [15]. The solid curve of each graph shows only the factor $f_k(Q)^2 \cdot Q^2$, whereas in the dashed curve the polarization factor is included. While for the relatively light element Al, the drop in the atomic form factor is not compensated by the Q^2-term, this is no longer true for the two heavier elements, where the Q^2-dependence becomes dominant. However, the polarization term ultimately leads to a functional dependence with a maximum that is displaced towards larger Q-values as a function of increasing atomic number Z.

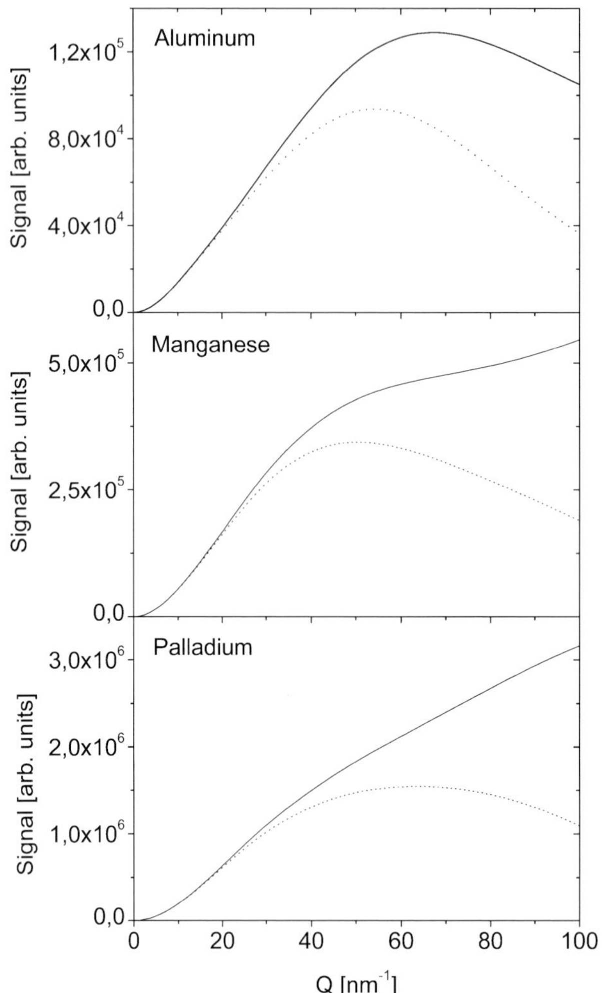

Fig. 2. The product $(f(Q) \cdot Q)^2$ (*solid line*) and $(f(Q) \cdot Q)^2 \cos^2(\theta_s)$ (*dashed line*) as a function of the momentum transfer Q at 22 keV (from [15])

To conclude this section, further similarities and differences between the IXS and INS cross section are summarized below:

- X-rays couple to the electrons of the system with a cross section proportional to the square of the classical electron radius, $r_0 = 2.82 \times 10^{-13}$ cm, i.e., with a strength comparable to the neutron–nucleus scattering cross section b.
- The total absorption cross section of X-rays above 10 keV energy is limited in almost all cases ($Z > 4$) by the photoelectric absorption process, and not by the Thomson-scattering process. The photoelectric absorp-

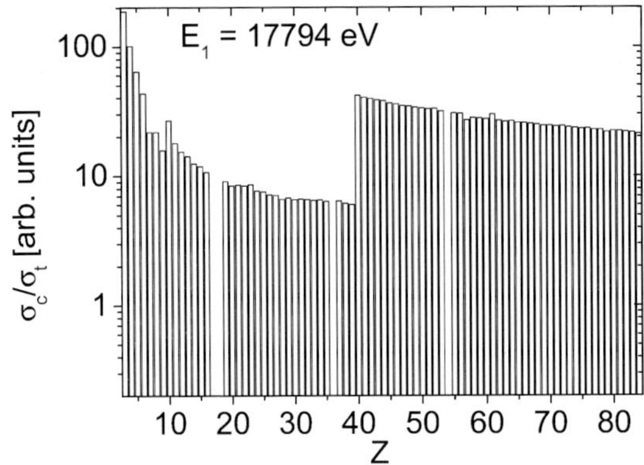

Fig. 3. Efficiency of the IXS technique. See text for details

tion, whose cross section, σ_t, is roughly proportional to Z^4, determines therefore the actual sample size along the scattering path. Consequently, the Thomson-scattering channel is not very efficient for systems with high Z in spite of the Z^2 dependence of its cross section, σ_c. In fact, the flux of scattered photons within an energy interval ΔE and within a solid angle $\Delta\Omega$ can be written as:

$$N = N_0 \, \frac{\partial\sigma}{\partial\Omega\partial E} \Delta\Omega\Delta E \rho L e^{-\mu L} \, , \tag{9}$$

where N_0 is the flux of the incident photons, ρ is the number of scattering units per unit volume, L is the sample length and μ is the total absorption coefficient. The quantity $\sigma_\mathrm{c}/\sigma_\mathrm{t}$ (with $\sigma_\mathrm{c} = (r_0 Z)^2$ and $\sigma_\mathrm{t} = \mu/\rho$) therefore provides a measure of the efficiency of the method for a given photon energy. This ratio is reported in Fig. 3 as a function of atomic number for monoatomic systems with an optimal sample length $L = 1/\mu$ and $f(Q) = Z$. The step between $Z = 39$ and $Z = 40$ signifies the energy of the K-shell absorption edge with respect to the incident photon energy (in this case 17.794 keV). If the photon energy is smaller than the K-shell binding energy, more sample can be probed, whereas in the other case an additional absorption channel opens up, and consequently the scattering volume is reduced. From the above, it is obvious that low momentum transfer studies, often performed on polycrystalline samples, become increasingly difficult for high-Z materials. For single-crystal materials this limitation can be overcome by working in higher Brillouin zones, where the Q^2-increase of the inelastic cross section partly compensates for the decrease of the $f(Q)$.

– The IXS cross section is highly coherent, contrary to neutrons where it is sometimes necessary to separate "a posteriori" the coherent, $S_\mathrm{c}(\boldsymbol{Q}, E)$, and incoherent, $S_\mathrm{s}(\boldsymbol{Q}, E)$, contributions.

– The magnetic cross section is negligible for IXS, whereas it is comparable to the nuclear cross section for neutrons. Therefore, X-rays are essentially insensitive to magnetic excitations, which can be helpful for separating magnetic and lattice contributions.

3 Instrumental Principles

All IXS instruments are based on the triple-axis spectrometer as developed by Brockhouse for INS. The three axes comprise the very high energy resolution monochromator (first axis), the sample goniometry (second axis) and the crystal analyzer (third axis). A detailed account of the instrumental developments and the two different types of very high energy resolution monochromators is given in [12]. Figure 4 shows a comparison of the schematic setup for the instruments at the ESRF, APS and SPring-8. The X-ray beam from the undulator source is premonochromatized by a silicon or a diamond (1,1,1) double-crystal monochromator to a relative bandwidth of $\Delta E/E \approx 2 \times 10^{-4}$. Due to the high heat load produced by the intense undulator beam, the silicon crystals have to be cooled down to cryogenic temperatures of about 125 K, where silicon displays a maximum in the thermal conductivity and a minimum in the linear thermal expansion coefficient. This allows the thermal deformation of the crystal to be kept below the limits where photon-flux losses occur. In the case of diamond conventional water cooling is sufficient thanks to its thermal and diffraction properties. The necessary very high energy resolution is then obtained by either a monochromator, operating very close to the backscattering geometry (ESRF and SPring-8) or by a multiple crystal inline monochromator (APS). The highly monochromatized X-rays are then focused onto the sample position by focusing optics. The scattered photons are energy analyzed by a spherical perfect silicon crystal analyzer, operated in the Rowland geometry, and at a Bragg angle of 89.98°. In the case of the most widely spread spectrometers, the momentum vector $\boldsymbol{Q}$ is selected by rotating the analyzer arm in the horizontal plane. Most of the IXS spectrometers are equipped with more than one analyzer so that spectra at several momentum transfers can be recorded simultaneously. The X-rays diffracted from the analyzer crystal are recorded by solid-state detectors, which provide a very low background.

The energy-resolution requirements of an IXS instrument are not only very demanding for the incident X-rays, but also for the energy analysis of the scattered photons. An effective method to obtain X-rays with high resolving power is based on Bragg reflection from perfect crystals. It can be shown that the resolving power ($E/\Delta E$) is given by [16]:

$$\left(\frac{\Delta E}{E}\right) = \frac{d_{hkl}}{\pi \Lambda_{\text{ext}}}, \qquad (10)$$

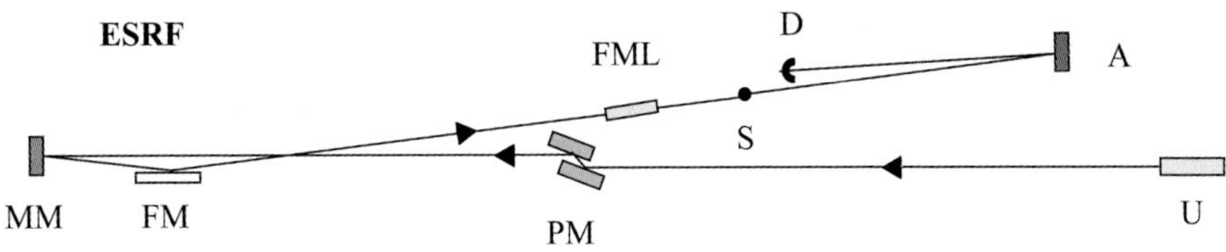

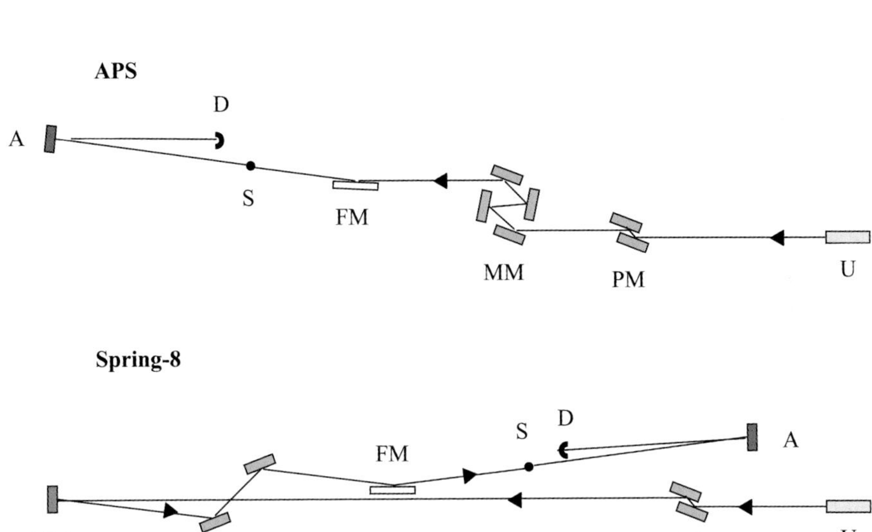

Fig. 4. Schematic optical setup of the IXS instruments at ESRF, APS and SPring-8. U: undulator, PM: high heat load premonochromator, MM: high-resolution main monochromator, LDPM: large vertical deflection Si (111) monochromator, FM: focusing mirror, FML: focusing multilayer, S: sample, A: analyzer crystal(s), D: detector. Sizes, lengths, and angles are not to scale

where d_{hkl} denotes the lattice spacing, associated with the (hkl) reflection order, and Λ_{ext} the primary extinction length, a quantity deduced within the framework of the *dynamical theory of X-ray diffraction* [16]. Λ_{ext} increases with increasing reflection order. In order to reach a high resolving power, it is therefore necessary to use high-order Bragg reflections, and to have highly perfect crystals. Such a perfect crystal can be defined as a periodic lattice without defects and/or distortions in the reflecting volume capable to induce relative variations of the distance between the diffracting planes, $\Delta d/d$, larger than the desired relative energy resolution: within this volume $\Delta d/d \ll (\Delta E/E)_{\text{h}} \approx 10^{-8}$. This stringent requirement practically limits the choice of the material to silicon.

Geometrical conditions, besides the energy-resolution issues, are another important aspect to use efficiently a high-order Bragg reflection. From the differentiation of Bragg's law, one obtains a contribution to the relative energy resolution due to the angular divergence $\Delta\theta$ of the beam impinging on the crystal: $\Delta E/E = \mathrm{ctg}(\theta_\mathrm{B}) \cdot \Delta\theta$. In order to reach the intrinsic energy resolution of the considered reflection, this angular contribution should be comparable to or smaller than the intrinsic energy resolution. In typical Bragg-reflection geometries $\mathrm{ctg}(\theta_\mathrm{B}) \approx 1$: consequently for high-order reflections with $(\Delta E/E)_\mathrm{h} \approx 10^{-8}$, the required angular divergence should be in the nRad range, i.e., values much smaller than the collimation of X-ray beams available even at third-generation synchrotron radiation sources. This geometrical configuration would induce a dramatic reduction of the number of photons Bragg reflected from the monochromator and analyzer crystals within the desired spectral bandwidth. An elegant solution to this problem is the *extreme backscattering* geometry, i.e., the use of Bragg angles very close to 90°. This provides very small values of $ctg(\theta_\mathrm{B})$ ($\theta_\mathrm{B} \approx 89.98°, ctg(\theta_\mathrm{B}) \approx 10^{-4}$). In such a way $\Delta\theta$ is increased to values well above $\sim$ 20 µRad, and therefore becomes comparable to the typical divergence of synchrotron radiation from an undulator source. Energy scans are conveniently done by changing the lattice constant via the temperature T: $\Delta E/E = \Delta d/d = \alpha\Delta T$, where $\alpha = 2.58 \times 10^{-6}\,\mathrm{K}^{-1}$ is the thermal expansion coefficient of silicon at room temperature [17]. In order to obtain an energy step of about one tenth of the energy resolution, i.e., $\Delta E/E \sim 10^{-9}$, it is necessary to control the monochromator crystal temperature with a precision of about 0.5 mK [18]. An alternative solution consists of using asymmetrically cut crystals, for which the extreme backscattering condition can be relaxed. This approach needs a multiple crystal setup where the "outer" crystal pair collimates the beam down to a typical divergence of 1 µRad, and the second "inner" pair of asymmetrically cut crystals, utilizing a high-reflection order, provides the very high energy resolution [19]. The energy tuning for such a four-crystal design is achieved by rotating the inner crystal pair with µRad precision.

Although the problems connected to the energy resolution are conceptually the same for the monochromator and for the analyzer, the required angular acceptances are very different. The monochromator can be realized with a flat perfect crystal. In the case of the analyzer crystal, however, the optimal angular acceptance is dictated by the desired momentum resolution. Considering typical values of ΔQ in the range of $0.2\,\mathrm{nm}^{-1}$ to $0.5\,\mathrm{nm}^{-1}$ the corresponding angular acceptance of the analyzer crystal must be $\sim$ 10 mRad or higher, which is again an angular range well above acceptable values, i.e., also larger than the deviation of the Bragg angle from 90°. The only way to obtain such large angular acceptance is by the use of a focusing system that, nevertheless, has to preserve the crystal-perfection properties, necessary to obtain the energy resolution. This automatically excludes considering elastically bent crystals. A solution consists in laying a large number of undistorted

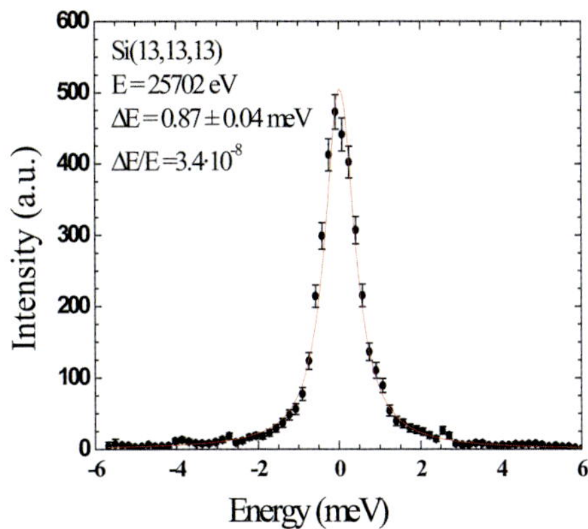

Fig. 5. Left panel: Picture of a 6.5 m crystal analyser. **Right panel**: IXS scan around zero energy transfer from a PMMA sample at $Q = 10\,\mathrm{nm}^{-1}$ and $T = 10\,\mathrm{K}$. The experimental data were fitted to a pseudo-Voigt function (*solid line*)

perfect flat crystals on a spherical surface, with the aim to use a 1:1 *pseudo*-Rowland circle geometry with aberrations kept such that the desired energy resolution is not degraded. Such a typical "*perfect* silicon crystal with spherical shape", consists of a spherical substrate of $R = 6500\,\mathrm{mm}$ and 100 mm diameter, on which approximately 12 000 perfect silicon crystals of surface size $0.6 \times 0.6\,\mathrm{mm}^2$ and thickness 3 mm are bonded [20, 21]. Figure 5 shows a picture of such an analyzer and the best energy resolution so far obtained (0.9 meV), utilizing the silicon (13,13,13) reflection order at 25 704 eV.

4 Phonons in Single Crystals

Inelastic neutron-scattering techniques are widely used to study lattice dynamics, and the use of IXS can only be justified in cases where INS techniques cannot be applied. This concerns, for example, experiments at very high pressure. These will be treated in a separate Chapter. Furthermore, this applies to samples that are only available in such small quantities that an INS experiment would not yield a sufficiently high signal, or samples that cannot be grown with the required single-crystal quality and/or homogeneous stoichiometry. Moreover, IXS is more suitable in cases where the constituent atom is a strong incoherent scatterer (for example, H and He^3), or where the natural relative isotope abundances conspire numerically with the isotope variation of the neutron-scattering length in such a fashion that the resulting total coherent scattering length is almost completely canceled out (vanadium).

As for INS studies, the interplay between experiment and theory is central, not only in the correct interpretation of the results, but also in the preparation phase, in order to determine the optimum scattering configuration.

Thanks to the rapid advances in computing power, ab-initio calculations can be performed for increasingly complex systems. While in quite some cases a remarkable agreement with theory is found, sometimes important discrepancies are found, stimulating further development of the numerical tools. In this context, the determination of phonon intensities (besides the phonon dispersion) is of particular interest, since it allows validation of phonon eigenvectors, thus providing a very stringent test of the calculations.

4.1 Large-Bandgap Materials

Large-bandgap materials such as silicon carbide (SiC) and III–V nitride semiconductors are very promising candidates for applications in high-temperature and high-frequency electronics, and for optical applications at short wavelengths. The behavior of carriers in such devices is affected by their interaction with phonons, and therefore the determination of the lattice-dynamical properties is of central importance in order to gain information on properties such as phonon-assisted photoemission, specific heat, thermal expansion and conductivity. While their electronic and optical properties are in general well studied and understood, the determination of the phonon spectrum was limited in most of the cases to the study of optical modes by Raman spectroscopy. From the theoretical side phenomenological and ab-initio methods are used to calculate the phonon-dispersion curves. They often yield very differing and controversial results, for example, concerning the Raman-active modes at the Γ-point.

4.1.1 SiC

SiC has generated interest not only because of its large bandgap, but also because of its large thermal conductivity, high breakdown voltage, and its outstanding mechanical and chemical stability. It exists in a large number of polytypes, which are difficult to grow as single-phase crystals. Despite the technical relevance of this material, little is known about its lattice-dynamical properties. INS experiments provided the complete set of phonon branches along the Γ–A direction and 14 out of 36 branches along the Γ–M direction in hexagonal (6H) SiC [22]. Further information along the Γ–L direction could be obtained by means of Raman spectroscopy on several polytypes (3C, 6H, and 15R), using the conjecture that the phonon dispersion along the Γ–A direction in the hexagonal modifications can be backfolded onto the Γ–L direction of the zincblende structure [23, 24].

IXS experiments were performed on zincblende 3C-SiC and hexagonal 4H-SiC single crystals of $5 \times 2 \times 1.5\,\mathrm{mm}^3$ and $4 \times 5 \times 0.8\,\mathrm{mm}^3$ size, respectively [25]. The instrumental energy and momentum resolution were set to 3 meV and $0.28\,\mathrm{nm}^{-1}$. Figure 6 shows the phonon frequencies measured for 3C-SiC (full circles) together with ab-initio calculations (solid lines). The open diamonds correspond to Raman data for 6H- and 15R-SiC [23]. There

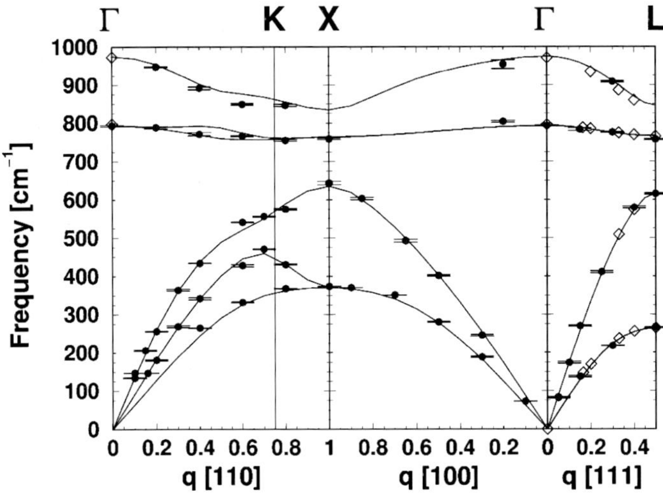

Fig. 6. Phonon dispersion of 3C–SiC. IXS data (*full circles*) are shown together with ab-initio calculations (*solid lines*) and Raman results (*open diamonds*) [24]. Figure taken from *Serrano* et al. [25]

is a very good agreement between the experimental data and the calculations with deviations less than 3 %. Figure 7 shows the IXS data for 3C-SiC along the Γ–L direction together with those obtained for the 4H-SiC, Raman data from other hexagonal and rhombohedral polytypes along the [001] direction, backfolded onto the 3C-SiC dispersion [23], and INS results on 6H-SiC [22]. The experimental results are reported together with ab-initio calculations for the 3C and 4H forms. In the latter case, the different stacking of the lattice planes along the c-axis gives rise to the observable discontinuity of the phonon branches at $q = 0.25\,[111]$.

The main observation is a substantial overlap of the phonon frequencies for hexagonal, rhombohedral and cubic polytypes. The direct comparison of the IXS results on 3C–SiC and 4H–SiC with previous Raman results confirms the validity of the backfolding technique. Agreement between experiment and theory is good. The strongest deviation is observed for the transverse acoustic branch in the 4H polytype, which might be related to the use of the local-coupling transfer approximation for the interatomic force constants.

4.1.2 GaN and AlN

The most common form of GaN is the hexagonal wurtzite structure, for which very good single crystals can be synthesized using a unique high-pressure growth technique. IXS experiments were performed on a small crystal of $2 \times 2 \times 0.2\,\mathrm{mm}^3$ size, using incident phonon energies of 17.79 keV and 15.82 keV with an instrumental resolution of 3.0 meV and 5.5 meV, respectively [26]. Figure 8 shows the experimentally determined phonon ener-

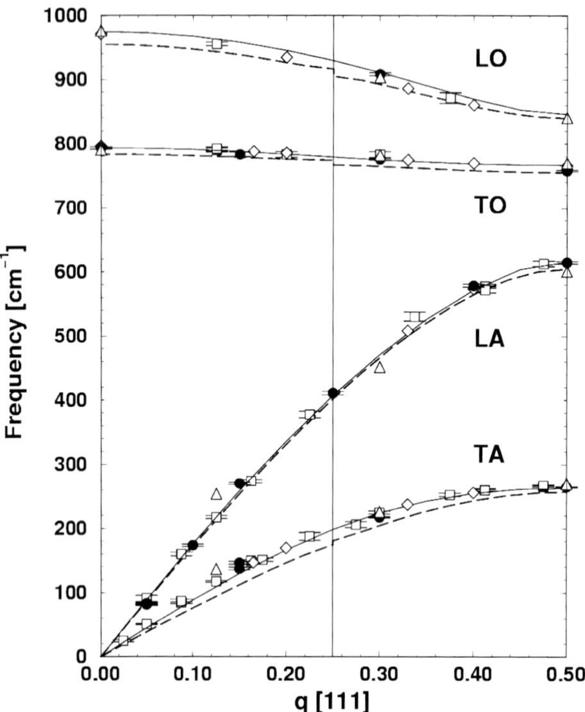

Fig. 7. Comparison between 3C- (*full circles*) and 4H-SiC (*open squares*) IXS data of the phonon frequencies along the c-axis and the [111] direction, respectively. *Open diamonds* correspond to Raman data [23], and *open triangles* represent INS data for 6H–SiC [22]. *Solid* and *dashed lines* show the calculated dispersion for 3C- and 4H-SiC. Figure taken from *Serrano* et al. [25]

gies together with results from ab-initio calculations and Raman results [27]. The theoretical calculations have been scaled by a factor of 0.97 in order to obtain optimum agreement with the experiment. The complete dispersion scheme along the Γ–A direction and several transverse branches along Γ–K–M and Γ–M could be determined. One main result of the study is the observation of the two silent B_1 modes at the Γ-point, which cannot be obtained by first-order Raman nor by infrared spectroscopy, and the determination of their dispersion throughout the Brillouin zone. These phonons are rather important input parameters for fits of phenomenological lattice-dynamical models. The IXS results allowed identification of some significant discrepancies in these models, as well as in previous ab-initio calculations, underlining that measurements are still indispensable.

Similar excellent agreement between experiment and theory could be obtained in the case of AlN in its wurtzite structure [28].

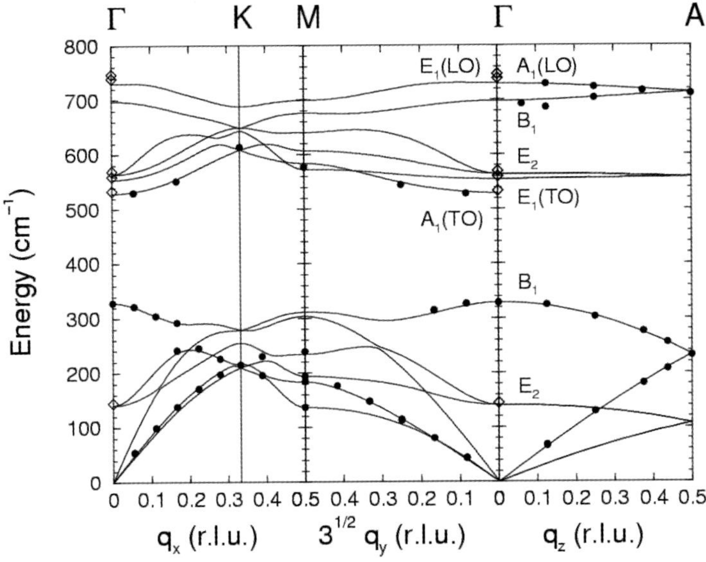

Fig. 8. Phonon dispersion of wurtzite-tupe GaN along several high-symmetry directions. IXS results (*full circles*), Raman results (*open diamonds*) [27], ab-initio calculations (*solid lines*). Figure taken from *Ruf* et al. [26]

4.2 Optical Phonons in Diamond

Despite the fact that diamond and silicon are neighboring elements in the group IV of the periodical table and possess the same crystal structure with tetrahedrally coordinated covalent bonds, their electronic band structure and lattice dynamics are profoundly different. For example, diamond is a perfect insulator of unparalleled hardness, while silicon is much softer and a semiconductor. One of the unusual properties of diamond, absent in silicon, which has intrigued scientists since its first observation in 1946, is the existence of a peak in the second-order Raman spectra above twice the one-phonon zone-center phonon energy [29, 30]. Various explanations were put forward, but only the advent of reliable ab-initio computational methods has allowed the reproduction of this unique feature. These calculations suggest that the peak is due to an "overbending" of the LO phonon branches with a minimum at the Γ-point, and maxima reached at some arbitrary points along all three high-symmetry directions [31]. This prediction stimulated more detailed experimental work on this phonon mode, and an inelastic neutron scattering (INS) study confirmed the LO overbending along the Γ-direction ([100]) [32]. A subsequent inelastic X-ray scattering (IXS) study claimed a quantitative disagreement with the INS results, and, furthermore, contested the presence of overbending in the other directions [33]. In order to resolve this controversy and apparent disagreement between the theoretical predictions and the IXS results, an accurate IXS study was performed at the ESRF [34].

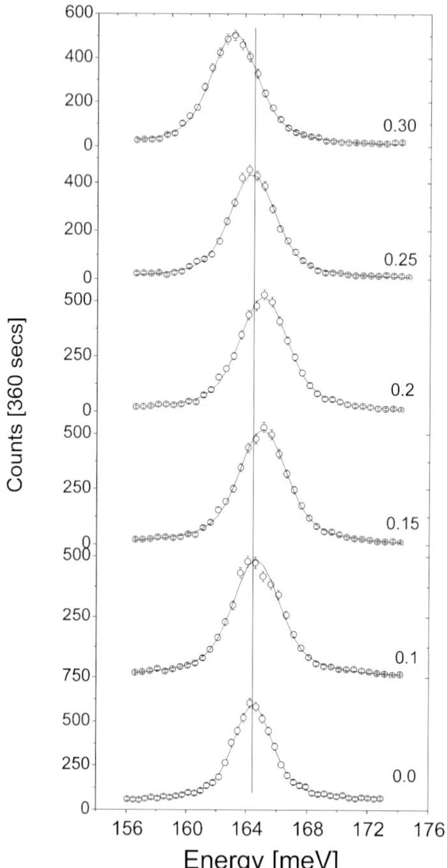

Fig. 9. IXS spectra (*circles*) at the indicated reduced q-values and their best fits (*solid lines*) for the longitudinal optical phonon of diamond along Λ. The *vertical line* indicates the position of the Γ-point phonon and serves to emphasize the overbending. Figure adapted from *Kulda* et al. [34]

The experiment was performed with a total energy resolution of 3 meV full-width half-maximum (at 17 794 eV incident energy). The diamond single crystal (4 mm diameter) was of excellent quality as testified by the narrow width of its rocking curve of only 0.004°. The stability and reproducibility of the instrument was checked by recording the Stokes–anti-Stokes pair of longitudinal acoustic phonons near the Brillouin-zone center, and by repeated scans of the Γ-point phonon. Figure 9 shows representative spectra at reduced momentum-transfer values, q, as indicated in the figure, obtained along the Λ direction ([111]), together with their best fits to a Lorentzian profile. The data clearly reveal an overbending for $0.15 < q < 0.25$. It can, furthermore, be noted that the measured Γ-point phonon energy of 164.75 meV is in excellent agreement with Raman data, which yield 165.18 meV.

Figure 10 shows the resulting dispersion relations for the Σ and Λ symmetry directions, together with the INS results and the ab-initio calculations. The overbending, as determined by IXS, amounts to 1.5 meV, 0.5 meV

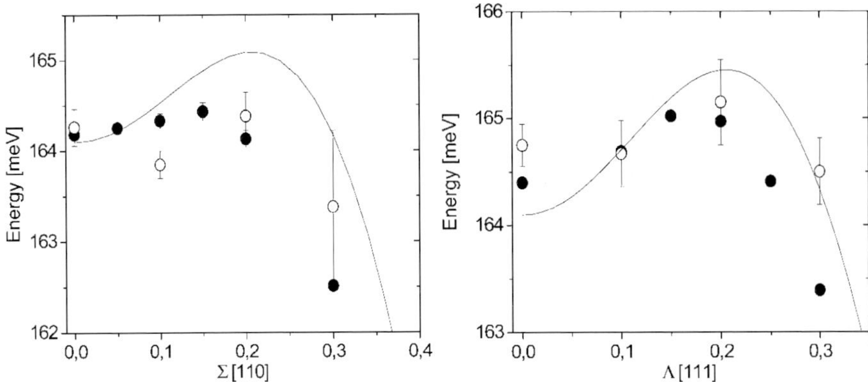

Fig. 10. LO-phonon dispersion of diamond along Σ and Λ. IXS (*full symbols*), INS (*open symbols*), and ab-initio calculations (*solid lines*). Figure adapted from *Kulda* et al. [34]

and 0.2 meV along the three high-symmetry directions (Δ, Σ and Λ). The IXS data are in very good agreement with both the INS results and the calculations. These results unambiguously reveal the LO overbending in all three high-symmetry directions, finally providing the physical origin for the anomalous peak in the two-phonon Raman spectrum.

4.3 Optical Phonon Dispersion in Graphite

Graphite is one of the very few elemental solids for which the complete phonon dispersion is not yet experimentally determined. While the phonon dispersion of the acoustic branches of graphite up to 400 cm^{-1} were determined by INS, data in the range of the optical branches (up to 1600 cm^{-1}) are scarce and widely scattered, and in the case of the K–M symmetry direction completely missing. Furthermore, existing calculations contradict each other qualitatively and quantitatively in frequencies, slopes and crossings of particular phonon branches. IXS data were recorded on a small microcrystal of $100 \times 200\,\mu\text{m}^2$ size for the longitudinal and transverse optical phonons along the in-plane Γ–M, Γ–K and K–M directions [35]. Figure 11 shows the IXS results together with ab-initio calculations that were performed in parallel. The theoretical frequencies were corrected by 1 % to obtain a better agreement with the experimental results. For the longitudinal branches the overbending, the small splitting of the LO and LA branch at the M point, and the local minimum of the lowest B_1 branch between K and M are very well described by the first-principles result. A good agreement for the TO branch is found along the Γ–M direction, while along the Γ–K–M direction the ab-initio dispersion is slightly less pronounced than the experimental result. The bandwidth of the TO branch amounts to 320 cm^{-1}, which is about 40 % larger than in a very recent ab-initio calculation [36]. The better agreement of these

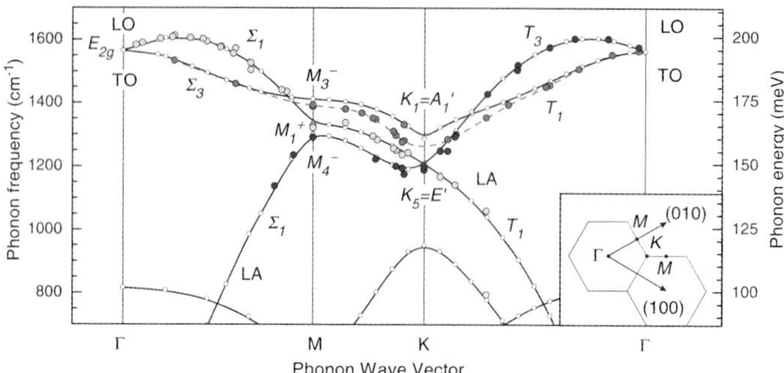

Fig. 11. Phonon dispersion of graphite. *Filled dots* are the experimental data for the LO, TO and the LA branches. The TO data are connected by a cubic spline extrapolation (*dashed line*). *Open dots* represent the calculated phonon frequencies. The *lines* are a spline extrapolation to the calculated points. The *inset* shows the graphite Brillouin zone in the basal plane. Figure taken from *Maultzsch* et al. [35]

calculations with respect to previous published work is thought to be only partly due to the generalized gradient approximation, but mainly due to the use of a larger supercell, which allows the long-range nature of the in-plane force constants of graphite to be properly taken into account.

4.4 Quantum Systems

Solid and liquid helium constitute the prototypical systems to study the influence of quantum effects in condensed matter thanks to the well-known interatomic potential and the availability of two different isotopes with different quantum statistics and a large relative mass difference. INS results are available only for ^{4}He, while studies on ^{3}He are hampered by the very high absorption cross section for thermal neutrons. This limitation does not apply to IXS, and motivated a comparative study of the lattice dynamics in ^{3}He and ^{4}He [37].

Single crystals were grown in situ in a polycrystalline Be cell of 1.2 mm inner diameter within a closed-cycle refrigerator. The Be cell was pressurized through a steel capillary using an external pressure-generating system. The cell was cooled below the freezing temperature of He at constant pressures of 70 MPa and 90 MPa. Longitudinal phonons were recorded at a density of $13.25\,\mathrm{cm}^3 \cdot \mathrm{mol}^{-1}$ for both isotopes along the [001] direction and along the [100] direction for ^{3}He. Typical IXS spectra of ^{3}He are shown in Fig. 12. These are shown together with their corresponding best fit, using a Lorentz function to account for the elastic line (due to the Be sample cell) and a damped harmonic oscillator model for the He LA phonons. The resulting two dispersion curves are shown in Fig. 13. They show a very similar shape, but the phonon energies in ^{3}He are in average about 11(5) % higher than in ^{4}He. This value is smaller than the ratio expected in a classical harmonic crystal,

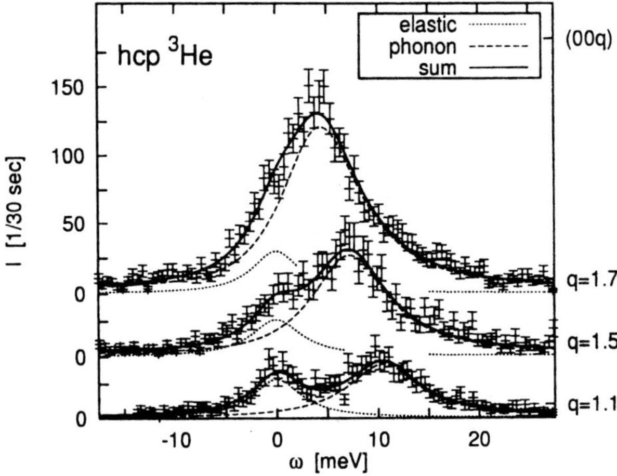

Fig. 12. IXS spectra of hcp ^{3}He for different momentum transfers as indicated on the *right side* of the figure along the [001] direction. The data were taken with an energy resolution of 5.9 meV. The experimental points and their corresponding error bars are shown together with the best-fit results as explained in the text. Figure taken from *Seyfert* et al. [37]

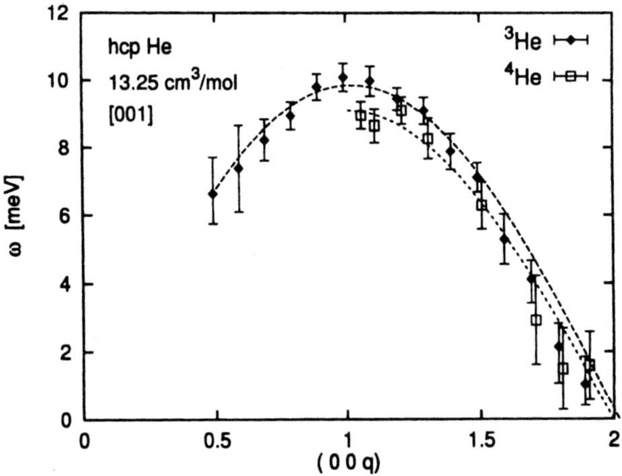

Fig. 13. Comparison of the LA-dispersion curves in hcp ^{3}He and ^{4}He along the [001] direction. The *dashed lines* result from a one-parameter fit to a monoatomic linear-chain model. Figure taken from *Seyfert* et al. [37]

i.e., the square root of the inverse masses, $\sqrt{4/3} = 1.155$. Thermodynamic measurements [38] and Raman spectroscopy [39] yield slightly higher values of 1.18 and 1.17, respectively, indicating that the frequency ratio might be mode dependent. Comparison of the experimental results with existing calculations clearly demonstrate that there is not yet a satisfactory agreement: The calculated phonon energies are about 20 – 30 % higher in the case of ^{3}He, and about 10 – 15 % for ^{4}He [40, 41].

4.5 Correlated Electron Systems

4.5.1 Superconductors

MgB_2

The discovery of superconductivity in MgB_2 has generated an immense amount of interest, since its T_c of $\sim$ 39 K is the highest for a simple metallic system and unusually high according to standard estimates. In the present understanding, MgB_2 is a phonon-mediated Eliashberg superconductor with multiple gaps. The presence of the large $\sim$ 7 meV isotropic gap on the sigma hole Fermi surface governs the superconducting properties and is to a large extent responsible for the unusually high T_c. The large gap is the result of an extremely strong electron–phonon coupling (EPC) of this Fermi surface to the basal-plane boron E_{2g} mode. Such a coupling to a single mode is quite unusual, and makes the investigation of this phonon mode particularly important. The simple crystallographic structure of MgB_2 with three atoms per unit cell and no magnetism allows reliable calculations to be performed, thus gaining important insight into the mechanism of superconductivity, when compared to experimental results.

The first IXS measurements using a sample of $400 \times 470 \times 40\,\mu m^3$ size determined several phonon modes along the Γ–M, Γ–A and A–L directions with particular emphasis on the determination of the energy dispersion and width of the crucial E_{2g} mode [42]. Figure 14 shows the resulting phonon dispersion together with a first-principles calculation. The top panel of the figure reports the phonon linewidths, which were obtained by deconvolving the experimental resolution function. Throughout the Γ–A direction the phonons are very broad with a width ranging between 20 meV and 28 meV. Calculations using density-functional theory in the generalized gradient approximation are in very good agreement with the data. The calculated linewidth due to electron–phonon interaction and the anharmonicity of the crystal potential reveal that the observed linewidth is almost exclusively due to EPC, and the anharmonic contributions are at most 1.2 meV at room temperature.

A subsequent experiment focused the attention on the E_{2g} mode along the Γ–M direction, and the temperature dependence of the phonon modes [43]. In contrast to the Γ–A direction, here the phonon width shows a pronounced q-dependence being at most 1 meV to 2 meV for $q > 0.6$ and then strongly

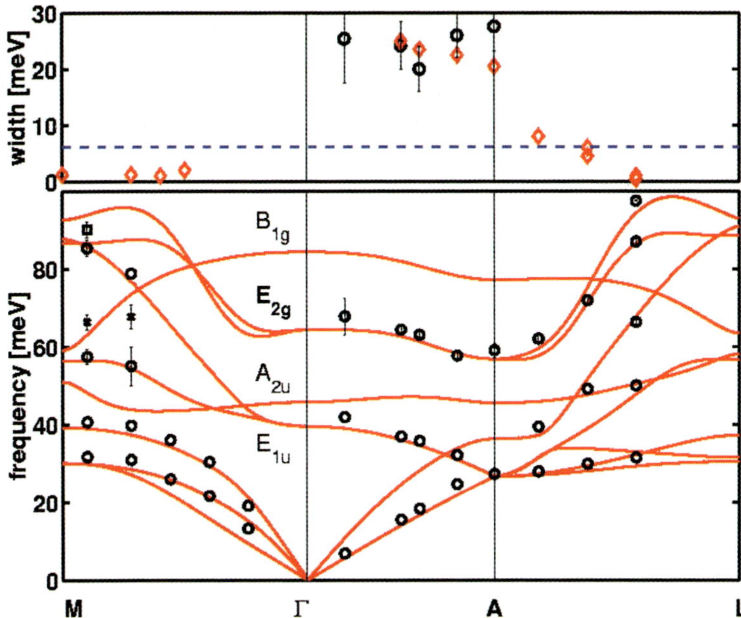

Fig. 14. Bottom panel: Experimental (*circles*) and theoretical phonon dispersion (*solid line*) in MgB_2 along Γ–A, Γ–M, and A–L. In the region near the M point, the probable detection of the E_{2g} mode is indicated with a square symbol. The *crosses* indicate a parasitic signal of unknown origin. **Top panel**: Intrinsic linewidth of the E_{2g} mode. The experimental widths (*circles*) are large along Γ–A and below the experimental resolution (*dashed line*) near L and M. The theoretical result (*diamonds*) for the electron–phonon coupling contribution to the linewidth is also shown. Along A–L and Γ–M, where the E_{2g} mode is nondegenerate, both theoretical values are shown, when different. Figure taken from *Shukla* et al. [42]

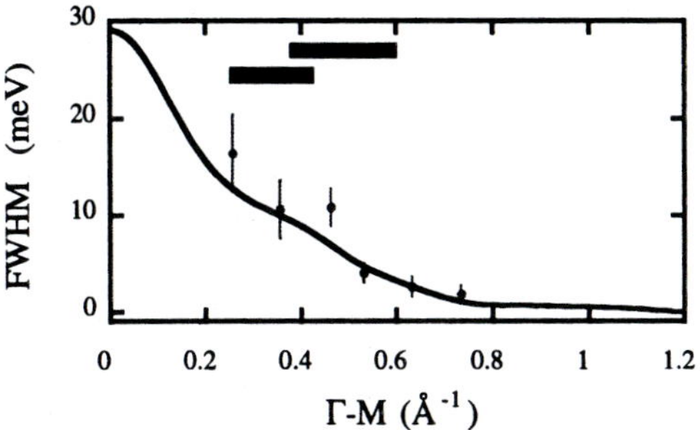

Fig. 15. Measured and calculated E_{2g} linewidth along Γ–M. The *horizontal bars* show the diameters of the sigma Fermi surfaces projected onto the plane perpendicular to the Γ–A axis. Figure taken from *Baron* et al. [43]

increasing on approaching the Γ-point (see Fig. 15). This strong dispersion is also confirmed by calculations, and can be understood by inspecting the topology of the Fermi surfaces. The two sigma surfaces are of approximately cylindrical symmetry with slightly different radii and their axes parallel to Γ–A. For $q < 2\boldsymbol{k}_\mathrm{F}$ ($\boldsymbol{k}_\mathrm{F}$ is the Fermi wavevector) electrons can be promoted from the filled states to the unfilled states across these surfaces, whereas for $q > 2\boldsymbol{k}_\mathrm{F}$ no electrons can be excited, and one observes consequently a narrowing of the phonon mode corresponding to an increased lifetime. Furthermore, the temperature dependence to below T_c of the E_{2g} mode at [2 1 0.5] was studied, and revealed that there is no observable change down to 16 K.

High-T_c Cuprates

Copper oxide superconductors exhibit the highest critical temperature found so far. Since their discovery in 1986, the microscopic mechanism at the origin of their superconductivity is still not fully understood [44]. While it is well established that in conventional superconductors the coupling between electrons and phonons (collective vibrations) leads to charge-carrier pairing and therefore superconductivity, the role of this coupling in copper-oxide superconducting compounds is still the subject of intense research efforts. Inelastic neutron-scattering studies on the hole-doped high-temperature copper oxide superconductors $YBa_2Cu_3O_{7-\delta}$ (YBCO) and $La_{2-x}Sr_xCuO_{4+\delta}$ (LSCO) systems revealed an anomalous behavior of the highest-energy optical phonon mode, related to the Cu–O in-plane bond-stretching vibrations [45, 46, and references therein]. This anomaly may be related to a strong coupling between the lattice vibrations and the charge carriers, as the conduction in the copper-oxide superconductors takes place by charge hopping along the Cu–O bond in the CuO_2 planes. Within this framework the optical-phonon anomaly is expected to be ubiquitous, and should therefore be observed in other high-T_c cuprates, and, furthermore, not only be limited to hole-doped compounds, but also be present in electron-doped compounds. A systematic study of the relevant phonon modes is therefore highly desirable in order to understand these anomalies and their role for the appearance of superconductivity.

For IXS, the study of the phonon modes related to the oxygen movements is particularly difficult. Besides the low intensity due to the high phonon energy ($1/\omega$ term in the dynamical structure factor) and the Bose factor, the scattering strength from the oxygen atoms is weak with respect to the other heavier atom species present in the sample (see (5)).

The first IXS study was conducted on the electron-doped compound $Nd_{1.86}Ce_{0.14}CuO_{4+\delta}$ (NCCO) and the parent compound Nd_2CuO_4 (NCO) [47, 48]. Figure 16 shows the phonon dispersion along the Δ [ξ 0 0] direction, together with results of the highest optical branch of NCO as determined by INS, and a lattice-dynamics calculation based on a shell model. While the calculations are in good agreement with experiment for the acoustic modes and the lower-energy optical modes, this is no longer true for the two highest optical branches. Along the Δ direction the highest branch, associated with the

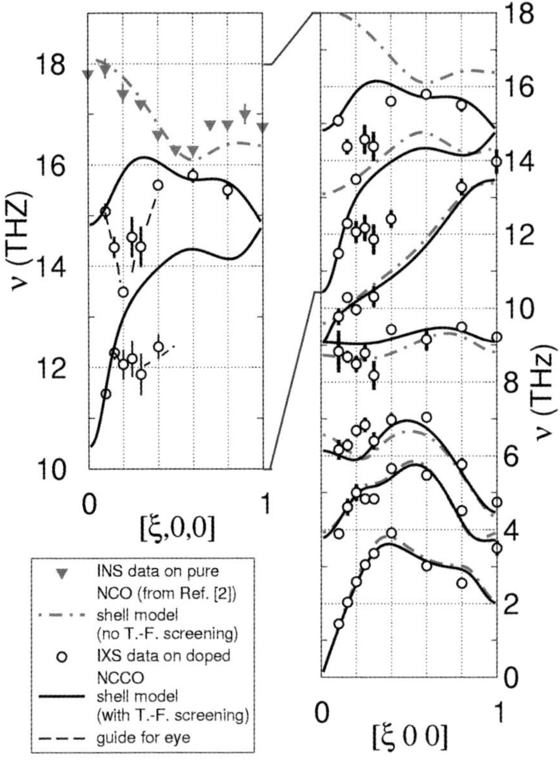

Fig. 16. Right panel: Dispersion of the longitudinal phonon modes in NCCO (*open circles*) from IXS spectra at $T = 15\,\mathrm{K}$ along the $[\xi\ 0\ 0]$ direction. *Solid* (*dot-dashed*) *lines* indicate the results of a lattice-dynamics calculation with a screened (unscreened) Coulomb interaction. **Left panel**: Magnification of the high-energy portion. For comparison, the dispersion of the highest LO mode in NCO by INS is shown as well (*full triangles*)

Cu–O bond stretching mode, is strongly renormalized with respect to the undoped parent compound, and, moreover, shows a softening of about 1.5 THz between $q = 0.1$ and 0.2. The observed shift is comparable to the anomalous shift observed in LSCO at $q = 0.25-0.3$ [49, 50], and might be linked to an interaction between the *in-plane* oxygen vibration and a charge modulation with a periodicity of about $(3-4) \cdot a$ in the CuO_2 planes. In the region between $q = 0.25$ and 0.3 the two modes are poorly defined in energy, which is consistent with the observations for LSCO [49, 50]. These results reveal that the anomalous softening previously observed in hole-doped compounds, is also present in the electron-doped cuprates, therefore suggesting that this is a generic feature of the high-temperature superconductors.

A further important step in establishing a potential systematic of electron–phonon coupling mechanisms was the study of phonon dispersion in a mercury-

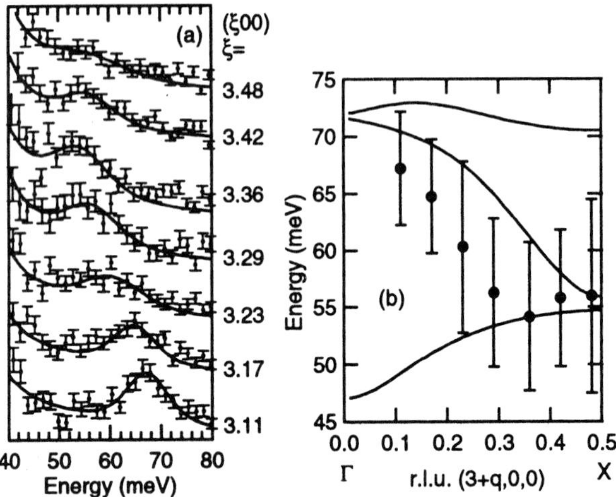

Fig. 17. Left panel (a): High-energy part of the IXS spectra around the bond-stretching mode in Hg-1201, recorded approximately along the [ξ 0 0] direction for the indicated values. The data and their error bars are shown together with the best-fit results (*solid line*). **Right panel (b)**: Experimentally determined phonon dispersion. The *vertical bars* indicate the FWHM of the phonon widths as determined from a fit to the IXS spectra. The results of a modified shell-model calculation (see text) are shown as *solid lines* and indicate (from *top* to *bottom*) the c-polarized apical oxygen mode, the a-polarized Cu–O bond stretching mode, and the a-polarized in-plane Cu–O bending mode, respectively. Figure taken from *Uchiyama* et al. [51]

based cuprate superconductor. The three-layer compound $HgBa_2Ca_2Cu_3O_{8+\delta}$ (Hg-1223) has the highest known T_c of about 136 K at ambient pressure, whereas the one-layer compound $HgBa_2CuO_{4+\delta}$ (Hg-1201) with its T_c of 97 K is quite outstanding, considering that other one-layer hole-doped cuprates have a T_c of 20 K to 40 K. At the same time it presents one of the simplest crystal structures, a tetragonal primitive P/4 mmm symmetry, with very few impurities, no distortions, and almost perfect square Cu–O_2 planes.

A first IXS study determined the phonon dispersion of some selected branches along the three main symmetry directions below 5 THz [52]. The highest energy Cu–O bond stretching mode could not be clearly identified, and was, moreover, only recorded for one q-value, so that no conclusion on its dispersion could be drawn. A subsequent study [51] overcame this limitation and provided the complete dispersion of the highest-energy mode along the Γ–X direction. Figure 17 shows an enlargement of the high-energy portion of the IXS spectra and the resulting dispersion, along with calculations, based on a shell model. A marked softening of the Cu–O bond stretching mode is observed, together with a substantial broadening of the phonon linewidth as shown by the vertical bars in the figure. A modified shell model, including next-neighbor oxygen interactions, which constitute a phenomenological way to describe a rhombic distortion of the CuO_2 squares, is capable of qualitatively describing this anomalous dispersion, though it fails to reproduce the shape.

4.5.2 Charge-Density Wave Systems

The physics of low-dimensional metals is the subject of active research due to their very peculiar electronic and elastic properties. The low dimensionality in the metal–metal interactions strongly enhances the electron–electron, electron–phonon and spin–phonon coupling strengths that, under suitable conditions result in the partial or complete condensation of the free carriers by formation of a modulated state. Several of these systems develop a charge-density wave (CDW), which is accompanied by a periodic lattice distortion. Above T_P – the transition temperature into such a state – the dynamics is generally described in the framework of weak electron–phonon coupling theory, which predicts a softening of the phonon mode coupled to the electrons at $2\boldsymbol{k}_F$ (Kohn anomaly). Below T_P, this soft mode then splits into two branches, corresponding to the CDW-amplitude (amplitudon, optical-like) and CDW-phase (phason, acoustic-like) excitations.

While prototypical CDW systems such as $K_{0.3}MoO_3$ (blue bronze), K_2–$Pt(CN)_4Br_{0.3}$, $3H_2O$ (KCP), and $(TaSe_4)_2I$ are available in good crystalline quality and sufficiently large sizes for INS measurements, high-quality single crystals of $NbSe_3$ can only be obtained in very small sizes. The only INS result was obtained utilizing a bunch-like assembly of several thousand $NbSe_3$ whiskers aligned parallel to each other along the b^* reciprocal lattice direction [54]. An IXS study on a single whisker of $0.3 \times 3 \times 0.003\,\mathrm{mm}^3$ size allowed access to the LA-phonon branch along $(0,k,0)$, studied by INS before, as well as the TA branch polarized along $(h,0,0)$, propagating along $(0,k,0)$ [53]. Here, the advantage over neutrons not only resides in the possibility to probe smaller volumes, but also in the significantly better Q-resolution, which is of importance for the study of sharp Kohn anomalies. These IXS experiments confirm the previous INS results and significantly extend the data further into the Brillouin zone. Moreover, for the first time, the phonon dispersion of the transverse acoustic branch could be obtained. This is of particular interest, since this TA branch is suspected to display an anomalous softening as the Peierls transition temperature, T_P, is approached, and should give rise to phason and amplitudon excitations below it, in analogy to observations in $K_{0.3}MoO_3$. As a matter of fact, a temperature-dependent IXS study in the vicinity of the X-ray-diffraction satellite position, located at a reduced lattice vector of $\boldsymbol{q} = [0; 0.241; 0]$, does not reveal any softening, but instead shows a significant broadening of the phonons (see Fig. 18). This result indicates differences in the formation and the nature of the CDW state of these two prototypical systems Nb_3Se and $K_{0.3}MoO_3$. These might be due to differences in the electronic properties ($NbSe_3$ remains partially

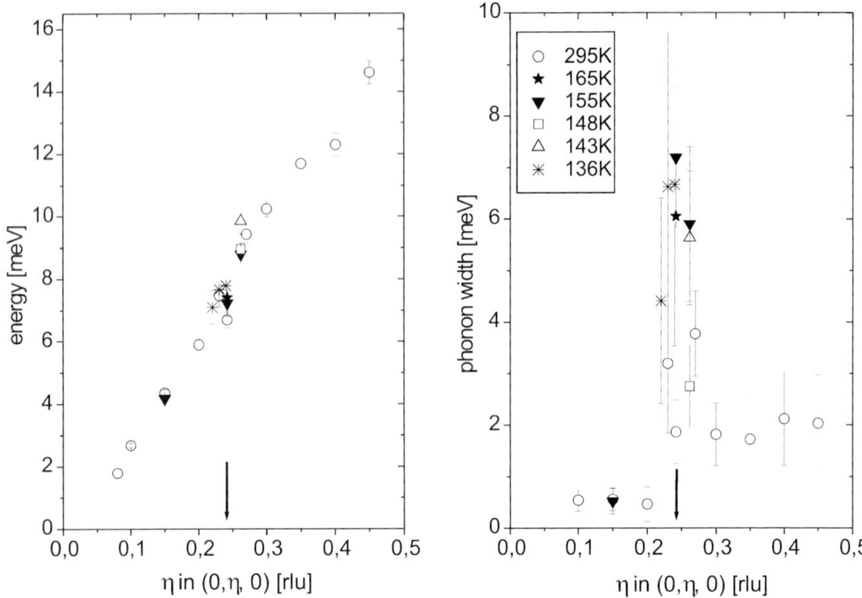

Fig. 18. Phonon dispersion (**left panel**) and linewidth (**right panel**) of the TA_1 branch in $NbSe_3$ around the CDW satellite position at different temperatures ranging between 295 K and 136 K ($= T_{P1} - 9$ K). The *black arrow* indicates the position of the q_1 CDW satellite. Figure adapted from *Requardt* et al. [53]

metallic, while $K_{0.3}MoO_3$ and KCP become semiconducting) or the strength of the electron–phonon interaction. As a matter of fact, a strong electron–phonon coupling theory predicts an order–disorder type of dynamics for the transition, due to the existence of bipolarons above T_P.

Another IXS study on $Rb_{0.3}MoO_3$ focused on the low-temperature behavior of the CDW-phase and -amplitude modes [55]. The experiment extended previous INS results, and the temperature-dependent study – at fixed momentum transfer Q – reveals that the phason mode is no longer visible at the lowest temperature of 40 K, indicating that it can not have the same dispersion as at higher temperatures. Possible explanations invoke a picture in which the phason mode becomes optical-like at low temperatures.

4.5.3 Actinides

Actinides often display peculiar physical and chemical properties, which are in one way or another associated with the unfilled 5f electronic shell. Experimental phonon-dispersion curves, however, exist only for a few elements and compounds. This has been largely due to the fact that many isotopes have a large neutron-absorption section, and even in cases where suitable isotopes exist, sufficiently large crystals for INS studies are not available. Further ob-

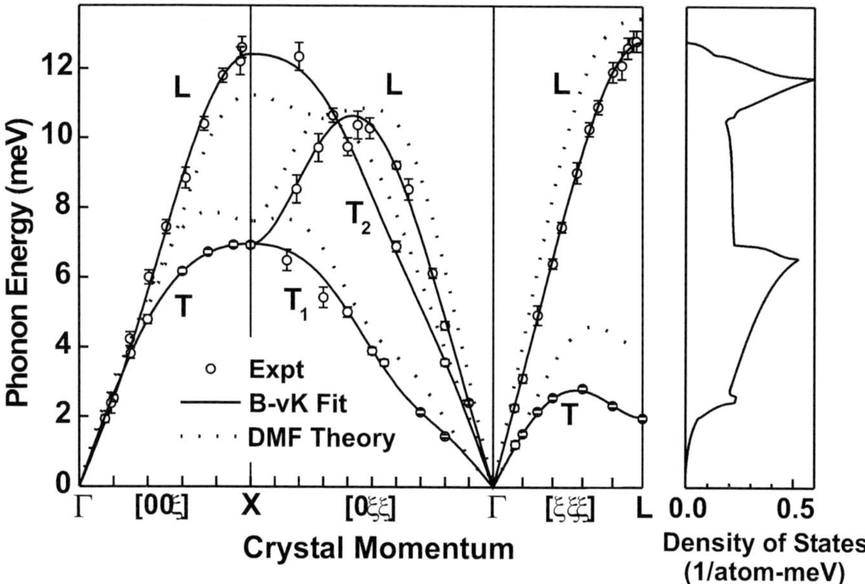

Fig. 19. Phonon dispersion along high-symmetry directions in a δ-Pu-0.6 wt% Ga alloy. The *solid curves* are the fourth-nearest-neighbor Born–van Kármán model fit. The derived phonon density of states, normalized to three states per atom, is plotted in the **right panel**. The *dashed curves* are DMFT calculations for pure δ-Pu [60]. Figure taken from *Wong* et al. [57]

stacles arise from the fact that actinide systems are often radioactive and highly toxic, thus requiring drastic safety and handling protocols, coupled to limitations in the largest amount of material to be studied from a legal point of view.

IXS studies have been performed so far on uranium [56] and plutonium [57, 58]. In particular, plutonium and its alloys are of central importance for many fundamental and applied issues, such as, for example, its safe handling and long-term storage. The many unusual properties of Pu, such as its numerous crystallographic phases, its low melting temperature, and the multitude of valence states in which Pu compounds can occur, are intimately linked to the 5f electrons, whose character is midway between itinerant (and therefore participating in the bonding) and localized. Small single crystals of Pu (alloyed with small amounts of either gallium or aluminum) have become recently available thanks to a strain-enhanced recrystallization technique [59]. The studied samples were large-grain polycrystalline specimens prepared from a fcc Pu–Ga alloy containing $\sim$ 0.6 wt% Ga. Typical single-grain sizes varied between 20 μm and 90 μm, with a sample thickness of about 10 μm.

The complete phonon-dispersion scheme is plotted in Fig. 19. The solid curves are calculated using a standard Born–von Kármán (B–vK) force-constant model. An adequate fit to the experimental data is obtained if interactions up to the fourth-nearest neighbors are included. The dashed curves are recent dynamical mean field theory (DMFT) results [60]. The elastic moduli calculated from the slopes of the experimental phonon-dispersion curves near the Γ point are: $C_{11} = 35.3 \pm 1.4$ GPa, $C_{12} = 25.5 \pm 1.5$ GPa and $C_{44} = 30.53 \pm 1.1$ GPa. These values are in excellent agreement with those of the only other measurement on a similar alloy (1 wt % Ga) using ultrasonic techniques as well as with those recently calculated from a combined DMFT and linear response theory for pure δ-Pu [60]. Several unusual features are observed, which can be related to the phase transitions of plutonium and to strong coupling between the lattice structure and the 5f valence instabilities. The shear moduli C_{44} and $C' = (C_{11}–C_{12})/2$ differ by a factor of 6, identifying Pu as the most elastically anisotropic fcc metal known. The T_1 branch along [011] exhibits a “kink”, suggesting a Kohn anomaly similar to that observed in the light actinide Th [61]. The most pronounced feature is a soft-mode behavior for the TA branch along [111]. This soft mode has been associated with the δ (fcc) to α' (monoclinic) structural transition. This α' phase, which is stable below $-110°$C, possesses a slightly distorted hcp structure, and can be reached by layer parallel shear perpendicular to the [111] fcc planes, in close analogy to the γ to β phase transition, observed in Ce and La, for which the same soft-mode behavior is observed. In the latter cases, the transition leads, instead, to a double-hcp structure, but the same general concept of layer parallel shear applies.

5 Phonons in Polycrystals

In polycrystalline samples, there are no distinct crystallographic directions as in single crystals, and therefore, the direction of the momentum transfer $\hbar \boldsymbol{Q}$ is not defined, but only its modulus. Keeping in mind that the dynamical structure factor is proportional to $(\boldsymbol{e_q} \cdot \boldsymbol{Q})^2$ (see (8)), it is obvious that in the general case, both longitudinal and transverse modes can be excited. The IXS spectra have a complex shape and resemble the phonon density of states (DOS). Formally, the DOS limit is only reached for very large momentum transfer and/or an appropriate sampling over a sufficiently large Q region at moderate Qs after proper subtraction of the multiphonon background [62]. Most studies on polycrystalline materials were performed in the first Brillouin zone, in which $\boldsymbol{Q} = \boldsymbol{q}$. Under these conditions, only phonon modes with an eigenvector component parallel to the $\boldsymbol{Q}$-vector acquire a finite intensity, and consequently, the IXS spectrum is dominated by longitudinal acoustic excitations. In an ideal, nontextured powder, the orientational averaged sound dispersion is measured.

Though the information content is much less than for single crystals, studies on polycrystals still provide important information in cases in which samples are only available in polycrystalline form, either because they can not be synthesized as a single crystal, or the single crystal is transformed into a polycrystal when driven through a structural phase transition (see section on high-pressure studies).

5.1 Ices

The water–ice system has attracted a lot of interest, not only due to its dominating abundance in nature, but also due to its many unusual and intriguing properties. Amongst these is amorphous polymorphism, i.e., the existence of two or more amorphous states in the phase diagram of a substance. In water, this phenomenon is of particular importance, since it is believed that the two amorphous ice phases can be identfied as the glassy counterpart of the postulated two distinct liquid phases in the deeply undercooled regime. IXS studies aimed at determining the dynamic signatures of the two amorphous phases and reveal possible relations to stable crystalline phases.

Experiments were performed on high-density amorphous (HDA, $\rho = 1.17\,\mathrm{g/cm^3}$), low-density amorphous (LDA, $\rho = 0.94\,\mathrm{g/cm^3}$) and cubic ice ($\mathrm{I_c}$, $\rho = 0.93\,\mathrm{g/cm^3}$) [63]. A selection of the spectra is shown in Fig. 20. In both LDA and HDA the resonances become broader with increasing momentum transfer, but the width of the resonances remain significantly smaller than the excitation energy over the whole spanned Q range, though qualitatively speaking, the HDA spectra look more disordered-like. This result is very different from the observation made so far in other glassy systems, where the linewidth displays an approximately quadratic dependence with Q up to a value for which $\Omega(Q) = Q$. At larger Q values, it is then no longer possible to observe well-defined exciations [64]. Comparison of the amorphous ice spectra with those for ice $\mathrm{I_c}$ reveals resemblances of LDA and $\mathrm{I_c}$. These concern the sharpness of the excitation and the existence of a second excitation at about 7 meV, which appears at higher Q. Though the similarity between the IXS spectra is less pronounced for HDA, these results reveal a surprisingly crystal-like dynamic response of the amorphous ices, indicating a high degree of local order, characterized by an intact hydrogen-bond network. This local order is seemingly more pronounced in LDA, maybe due to the fact that LDA is annealed, and not obtained via a fast quench from the liquid state.

The observed similarity in the dynamic response of LDA and $\mathrm{I_c}$ motivated a further study to identify an equivalent crystalline counterpart for HDA [65]. Due to their equivalent preparation procedure [66] and the close resemblance of their mass densities, ice IX ($\rho = 1.19\,\mathrm{g/cm^3}$) and ice XII ($\rho = 1.29\,\mathrm{g/cm^3}$) appear a priori as suitable reference systems. The IXS spectra of these two ice phases are shown together with the results for HDA in Fig. 21. They reveal that ice IX and ice XII possess a distinguishable dynamic response throughout the sampled Q space. Ice XII has a single, well-defined excitation, identified as the LA mode up to $5\,\mathrm{nm^{-1}}$, and shows above $Q = 6.5\,\mathrm{nm^{-1}}$, a second, nondispersive feature, associated with transverse optical modes. In stark contrast, the inelastic response of ice IX has a sophisticated spectral

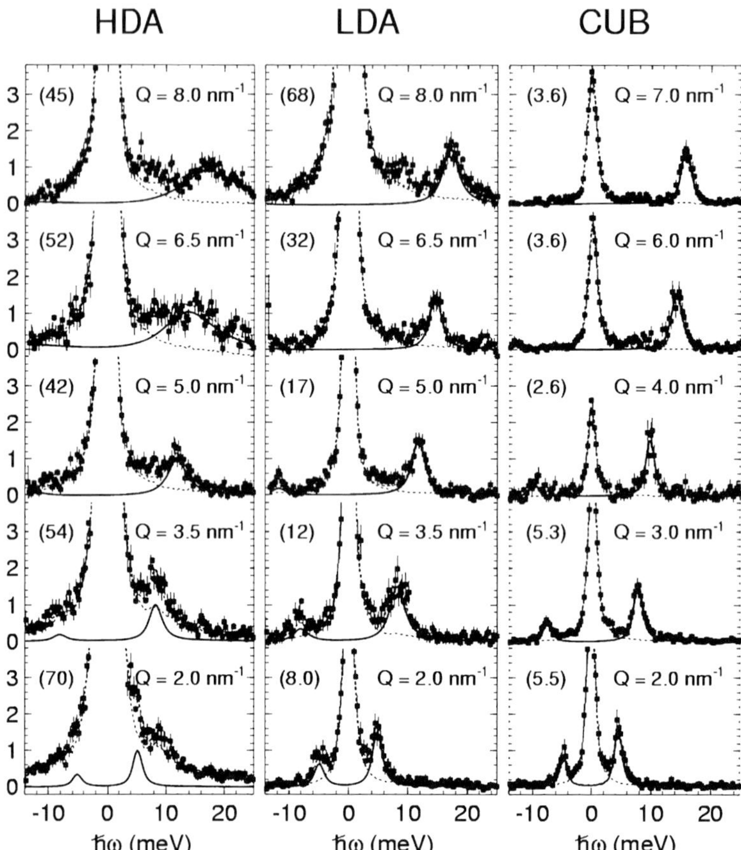

Fig. 20. IXS spectra of HDA, LDA and ice I_c at the indicated Q values. The *dashed lines* are fits to the signal using Lorentzian lineshapes convoluted with the experimental resolution function (*dotted line*). The *solid lines* represent the inelastic contribution to the total fits. The *numbers in brackets* on the left of the elastic line give the elastic intensities in arbitrary units. Figure taken from *Schober* et al. [63]

shape already at the lowest Q values. The essential ice IX spectral features are also found in HDA, and, furthermore, quantities such as the sound velocity and the energy of the nondispersive, optical modes are identical in ice IX and HDA. These properties are clearly distinguishable from the dynamics of the ice XII, ice I_c and LDA phases.

5.2 Ice Clathrates

Clathrate hydrates have attracted a large amount of interest in recent years since they are considered a model system for the study of hydrophobic interactions and as a potential future energy resource, as well as one of the possible

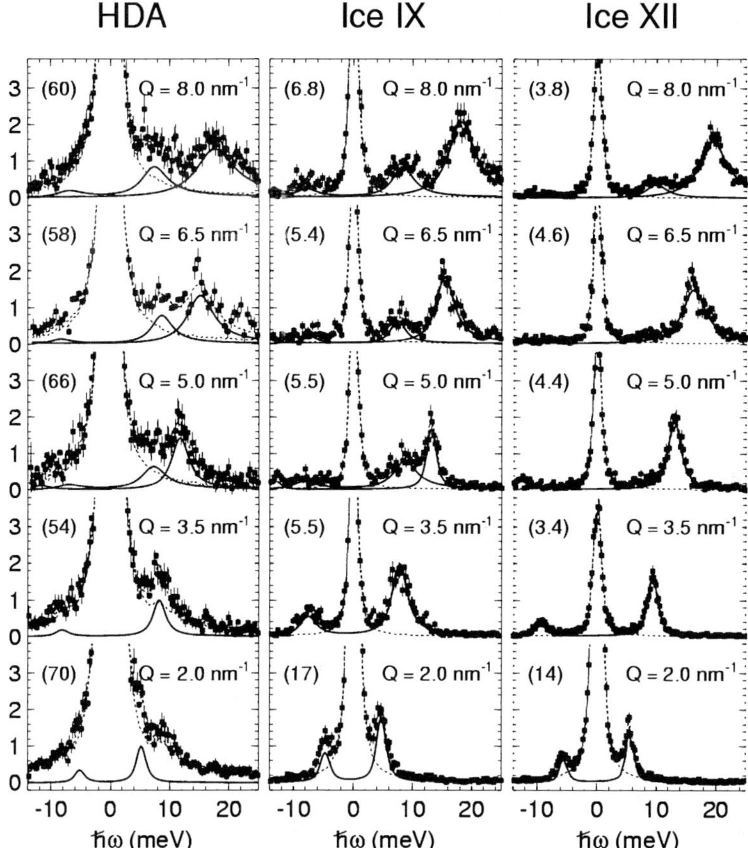

Fig. 21. IXS spectra of HDA, ice IX and ice XII at the indicated Q values. The *dashed lines* are fits to the signal using Lorentzian lineshapes convoluted with the experimental resolution function (*dotted line*). The *solid lines* represent the inelastic contribution to the total fits. The *numbers in brackets* on the left of the elastic line give the elastic intensities in arbitrary units. Figure taken from *Koza* et al. [65]

contributors to the greenhouse effect. In methane and xenon hydrate, the gas molecules are situated in cages formed by an ice-like network of H-bonded water molecules. They possess a cubic structure, forming six large ellipsoidal and two small spherical cages in the unit cell. In the case of methane hydrate each case is occupied by one gas guest molecule. Despite the fact that gas hydrates consist of more than 80 % of hydrogen-bonded water molecules, the temperature dependence of the thermal conductivity is very different from that of ordinary ice, and resembles that of glasses. This behavior was attributed to a resonant scattering of the acoustic phonons by the localized guest vibrations in the cage [67].

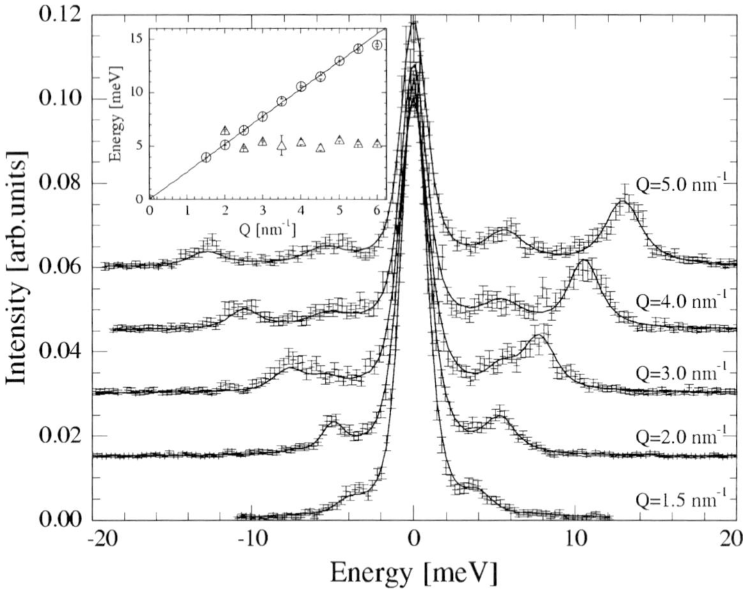

Fig. 22. IXS spectra of methane hydrate at the indicated Q values, recorded at 100 K. The *lines* are fits to the experimental data using Lorentzian functions convoluted with the experimental resolution. The spectra are normalized to their integrated intensity. The *inset* shows the respective dispersion relation of the two distinct inelastic features. Figure taken from *Baumert* et al. [68]

Figure 22 shows a selection of the IXS spectra for methane hydrate, recorded with 1.6 meV overall energy resolution [68]. The spectra display a well-defined dispersive mode, and, for $Q > 3\,\mathrm{nm}^{-1}$, a second nondispersive feature. The dispersive excitation can be unambiguously identified with the LA host lattice branch, while the correct assignment of the second excitation, centred around 5 meV, could be done thanks to previous IXS results on polycrystalline ice I_h [69, 70], and density-of-states measurements on CH_4–D_2O by INS [71]. It is attributed to the guest vibrations inside the large cage. These vibrations become visible in the spectrum after their intersection with the LA lattice mode at about $Q = 2.5\,\mathrm{nm}^{-1}$. This behavior is providing strong evidence for the validity of the resonant scattering model, which predicts an avoided crossing between acoustic (host)-lattice phonons and localized guest modes of the same symmetry leading to a mixing of guest and host modes with a subsequent energy exchange. At this avoided crossing, the guest molecules receive a strong longitudinal polarization through a mixing of the eigenvectors of the guest and host modes. At these Q values the guest vibrations are no longer independent rattlers in the cages, but their vibrations are modulated by the host lattice, leading to a transfer of scattered intensity from the acoustic lattice mode to the guest vibrations, and thus to the observed optical modes. The resulting intensity decrease of the LA lattice mode, which is closely related to the thermal conductivity, can therefore account for its unusual temperature dependence. This interpretation of the experimental results is further validated by lattice-dynamics calculations and a comparative study on xenon clathrate [68], which reproduce very well the dynamical properties of the hydrate clathrates. Further studies, focusing on the dynamics of the high-pressure methane clathrate phases II and III unveiled the evolution of the effective elastic and shear modulus. The elastic properties were found to differ substantially from the ones of high pressure ice structures, while strongly reflecting the transition from a cage clathrate (phase II) to a filled ice structure (phase III) [72].

6 Phonon Studies at High Pressure

The investigation of materials under high-pressure conditions is one of the challenging topics in condensed-matter research, both for understanding basic physical phenomena and in applied sciences with an important interdisciplinary impact in chemistry, biology, earth science, materials science, physics and engineering. INS studies are, despite important progress in the development of high-pressure devices, limited typically to pressures around 10 GPa due to the relatively large sample size of at least $10\,\mathrm{mm}^3$, necessary to have sufficient signal to conduct the experiment [73]. This limitation does not exist for IXS. Thanks to very efficient optics X-rays can nowadays be focused down to micrometer sizes, and even submicrometer sizes have been obtained. These dimensions are well compatible with the constraints of diamond anvil cells (DAC) [74], which allow extremely high pressures above 100 GPa to be reached. Figure 23 shows a schematic view of such a device. It consists of two opposed diamonds of about 0.2−0.5 carats, which compress a thin metal gasket. The sample is placed into a small hole of the metal gasket, together with the pressure-transmitting medium. Thanks to the large surface ratio of the upper to the lower flat part of the diamonds (up to 10^3) and its hardness, a relatively small pressure exerted on the diamonds leads to very high pressures at the sample location. The acceptable sample size and thickness depends on the pressure range needed, and varies typically between 100 µm and 20 µm in diameter, and 40 µm to 10 µm in thickness, corresponding to a volume between $3 \times 10^{-4}\,\mathrm{mm}^3$ and $3 \times 10^{-6}\,\mathrm{mm}^3$. The optimum IXS signal is obtained, if the X-ray beam fully intercepts the sample, and the sample thickness $t_{\mathrm{opt}} = 1/\mu$ (see Fig. 3). Consequently, high-pressure studies become increasingly more difficult for light elements, for which t_{opt} can be as much as 1 mm, or even more. This is illustrated in Fig. 24, where the signal ratio for a sample of 20 µm thickness versus a sample of optimum thickness is displayed. While indeed, the inefficiency for low-Z materials is evident, for $Z > 20$, the size constraints imposed by the DAC, lead only to moder-

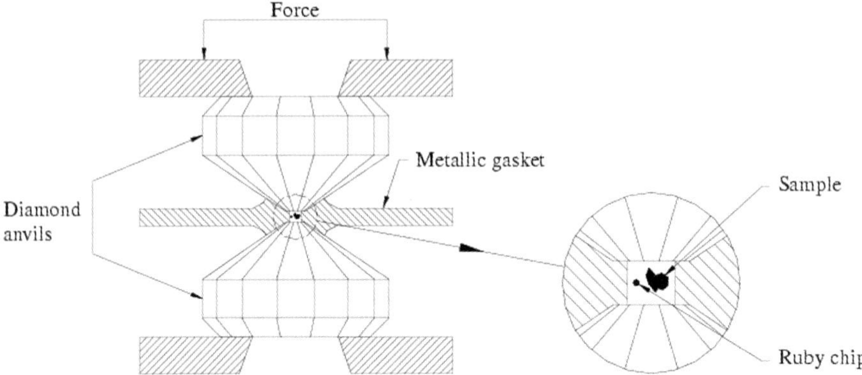

Fig. 23. Schematic view of a diamond anvil cell and an enlarged view of the sample chamber. Typical sizes are a diameter of 50 mm with a height of 30 mm

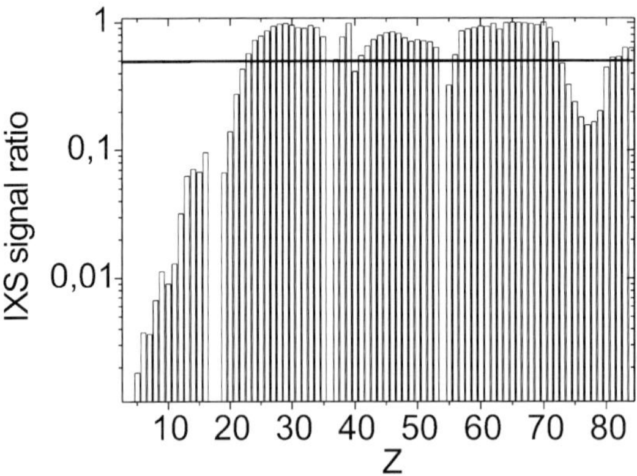

Fig. 24. Ratio of IXS signal at 17 794 eV and small scattering angles ($f(Q) = Z$) for a sample thickness of 20 µm over optimum thickness $t_{\mathrm{opt}} = 1/\mu$ as a function of Z. The *horizontal line* indicates a loss of a factor two

ate losses. Despite this, IXS experiments could be successfully conducted on low-Z materials such as H_2O [75] and MgO [76].

Other high-pressure devices such as the Paris–Edinburgh cell [77] can accommodate significantly larger volumes, and can therefore overcome the above-mentioned limitations, at least to pressures up to 10 GPa. The first test experiments were performed with such a device and an IXS spectrum from a single crystal of silicon (2.5 mm^3 volume) at 4 GPa could be recorded [78]. The first experiments in a DAC were conducted on polycrystalline CdTe [79] and on a single crystal of Ar [80]. In the following, several representative examples of more recent studies will be presented.

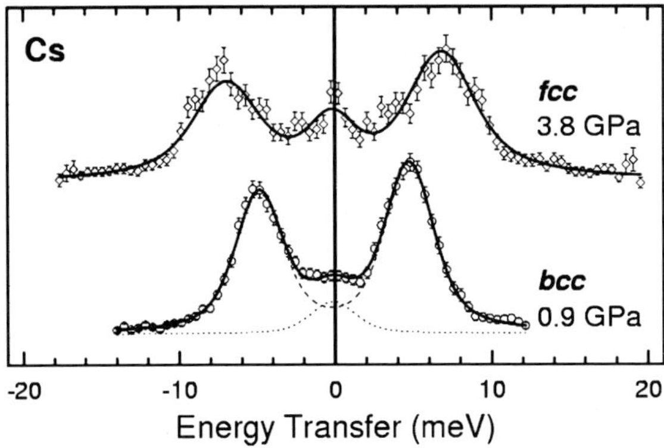

Fig. 25. IXS spectra of Cs at $Q = (100)$ in the bcc phase at 0.9 GPa and in the fcc phase at 3.8 GPa ($T = 300$ K). The *open symbols* represent the experimental data, the *solid lines* the fitted spectra, and the *dashed* and *dotted lines* the phonon and elastic contributions, respectively. Figure taken from *Loa* et al. [81]

6.1 Zone-Boundary Modes in Cesium

Pressure-induced structural phase transitions in alkali metals have been studied in much detail, and theoretical treatments have been quite successful in reproducing the observed structural sequences as a function of pressure, and in many cases they also gave transition pressures in good agreement with the experimental data. Most of these theoretical studies considered the static lattice. Structural transitions, may, however, also be driven by dynamical instabilities due to phonon softening. In the case of bcc and fcc cesium recent theoretical work predicted a pronounced phonon softening near the upper limit of the fcc-Cs stability range that could lead to a dynamical instability of the fcc lattice with increasing pressure. This phonon softening is predicted to be accompanied by a decrease of the shear elastic constant $C' = (C_{11}–C_{12})/2$, associated with a tetragonal deformation, that would lead to an elastic instability when C' becomes negative.

An IXS study of the LA phonon at $Q = (100)$ in the bcc and fcc phases of Cs was performed up to 4 GPa [81]. The crystals were grown in a DAC by heating the sample close to its melting line and then cooling it slowly to ambient temperature. The experiments were complemented by first-principles calculations. Figure 25 shows representative IXS spectra of Cs in the bcc and fcc phases. The pressure dependences of the zone-boundary phonon are reported in Fig. 26. In the bcc phase, the phonon frequency initially increases as expected with pressure, but with a slope approaching zero near the bcc–fcc transition. In the fcc phase the LA (100) phonon frequency is pressure independent within the experimental uncertainty.

The phonon-frequency calculations also indicate an anomaly near the bcc-fcc transition, but not quite as pronounced as in the experiment. This deviation may arise from differences in the thermal population of the electronic

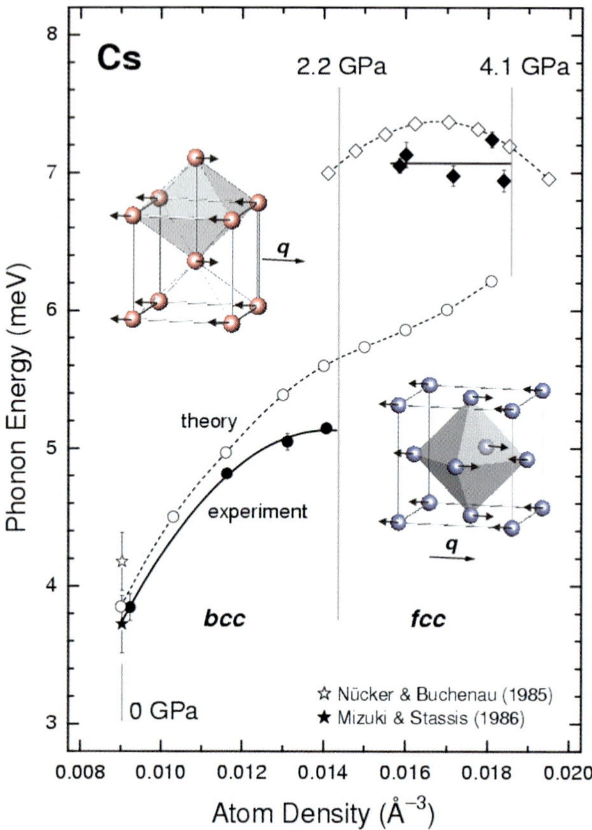

Fig. 26. Experimental and calculated phonon frequencies of Cs in the bcc and fcc phases. *Solid symbols* and *lines* refer to the experimental data, *open symbols* and *dashed lines* to the theoretical data. The experiment was performed at $T = 300\,\mathrm{K}$; finite-temperature effects were not included in the theory. Also shown (by *stars*) are the results of two ambient-pressure INS experiments [82, 83]. Figure taken from *Loa* et al. [81]

levels, as the experiment was performed at $T = 300\,\mathrm{K}$, while the theory corresponds to $T = 0$. For the fcc phase, the calculation shows a small overall variation with pressure, with a maximum near the center of the fcc stability range. The overall mode Grüneisen parameter for the fcc phase is almost zero, while its value amounts to 1.3 in the bcc phase. The general picture that emerges from the experimental and theoretical results is that of 1. a gradual decrease of the pressure dependence of the bcc LA (100) phonon frequency when the bcc–fcc transition is approached, and 2. an anomalously small pressure dependence in the fcc phase, with a small softening near the upper fcc stability limit in the calculation.

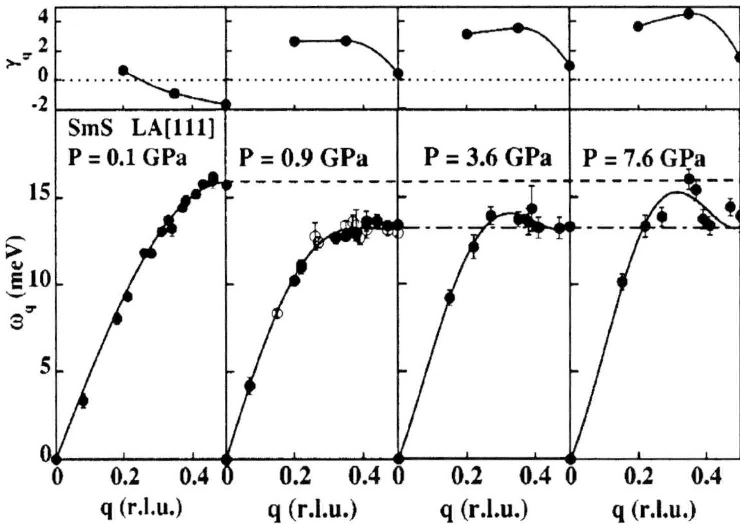

Fig. 27. Dispersion relation for the LA [111] branch of SmS at $T = 300\,\mathrm{K}$ for several pressures. *Lines* through the points are guides for the eye. The *open circles* in the second panel correspond to data obtained at 1.2 GPa. The *dashed line* and the *dot-dashed line* show the ZB value at ambient and high pressures, respectively. The *upper part* shows the variation with q of the mode Grüneisen parameter γ_q. The *dotted line* indicates $\gamma_q = 0$. Figure taken from *Raymond* et al. [85]

6.2 Acoustic Branch Softening in SmS

SmS belongs to the class of strongly correlated electron systems and possesses a complex temperature–pressure phase diagram. It undergoes a first-order isostructural (NaCl-type) pressure-induced semiconductor (black phase) to metal (gold phase) transition at a pressure of only 0.65 GPa. In this phase, the Sm ion has an intermediate valence achieved by promoting a 4f electron into the conduction band: $\mathrm{Sm}^{2+} \leftrightarrow \mathrm{Sm}^{3+} + 5d$. This leads to a volume reduction of 15 % at the transition, and to strong anomalies in the phonon spectrum. These anomalies were investigated by inelastic neutron scattering to 0.7 GPa [84], and recently extended to 7.6 GPa by IXS [85].

The IXS measurements were carried out on a single crystal of SmS (size: $150 \times 100 \times 40\,\mu\mathrm{m}^3$) placed in a DAC with an overall energy resolution of 3 meV. The dispersion of longitudinal acoustic (LA) phonons along the [111] direction were recorded for pressures between 0.1 GPa and 7.6 GPa, and they are displayed in Fig. 27. A softening of the mode from halfway and up to the zone boundary (ZB) is observed when entering the metallic phase. The anomalous phonon-softening effects observed are attributed to the electron–phonon interaction occurring at the valence transition of SmS where the delocalization of a 4f electron into the conduction band induces a breathing of the Sm atom that couples resonantly to lattice vibrations.

Furthermore, the ZB LA [111] phonon energy does not evolve further up to 7.6 GPa while a gradual hardening of the low and intermediate q modes occur in parallel. The first effect is due to the unusually strong pressure dependence of the bulk modulus in the metallic phase, whereas the latter is probably linked to an increasing density of states (DOS) at the Fermi level with pressure parallel to (or in cooperation with) the change of valence.

6.3 Sound Velocities in hcp Iron

The knowledge of the elastic properties of iron is of central importance to further constrain existing models of the Earth's core, which is thought to be predominantly composed of hcp-iron alloyed with some light elements. Theoretical attempts to calculate the elastic moduli of iron with the aim to reproduce the seismic observations yield very different results, and, furthermore, are in disagreement with the scarce experimental results [86, 87]. Possible causes for this are: 1. the difficulty to perform reliable calculations at high pressures and temperatures, and 2. the impossibility to carry out experiments on single-crystal hcp-iron, because of the bcc-to-hcp transition at about 13 GPa, which does not allow preservation of the single-crystalline state. As already stressed previously, experiments on polycrystalline systems allow only extraction of averaged properties. If performed within the 1. Brillouin zone these measurements yield the orientation-averaged LA dispersion.

IXS experiments were carried out at high pressure on iron with an overall energy resolution of 5.5 meV [87, 88]. The dispersion of longitudinal acoustic phonons has been measured from 0.2 GPa to 110 GPa at momentum transfers Q varying from $4\,\mathrm{nm}^{-1}$ to $12\,\mathrm{nm}^{-1}$ on polycrystalline iron compressed in a DAC. Typical IXS spectra and their corresponding fits are shown in Fig. 28. The strong inelastic signal at high energy transfers at the lowest recorded Q-value corresponds to the transverse acoustic (TA) phonon of diamond, whereas the remaining peak can be attributed to the longitudinal acoustic (LA) phonon of iron. These inelastic contributions shift towards higher energies with increasing Q-values, so that the diamond phonon is rapidly out of the energy-transfer window at Q-values larger than $4\,\mathrm{nm}^{-1}$ due to its higher sound speed with respect to iron. A sine function was fitted to the experimental data points, and the sound velocity V_{P} was determined from the initial slope of the so-fitted dispersion curve. The Q-value (Q_{max}) at which the sine function attains its maximum, was kept either free in the fit, or fixed to the value determined from an orientation average of the Wigner–Seitz–Brillouin zone [87].

Figure 29 reports the density evolution of V_{P}, together with the results obtained by other experimental techniques and calculations. Here, the density–velocity representation has been chosen to display the data. Within the quasiharmonic approximation, this allows comparison of results, which were obtained at different pressure or temperature conditions. The IXS values for V_{P} compare well with the ultrasonic measurements reported at low

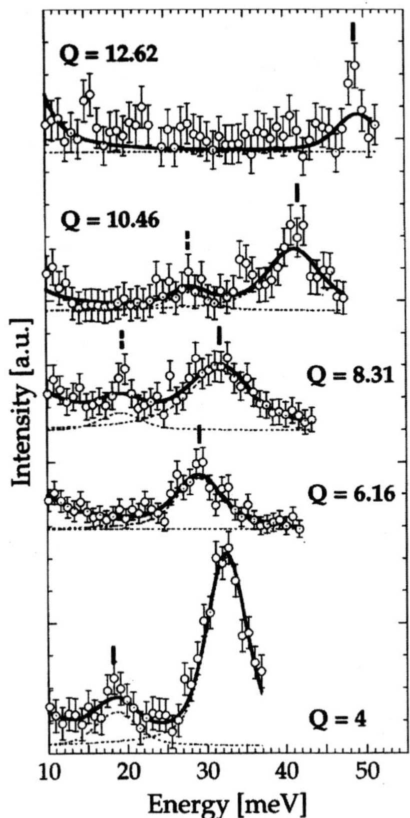

Fig. 28. Representative IXS spectra of polycrystalline hcp-iron at 28 GPa and at the indicated Q-values in nm^{-1}. The data are shown together with the best-fit results (*solid line*) and their individual components (*dashed line*). The LA phonons of iron are indicated by *ticks*, while a possible additional excitation is marked by *broken ticks*. Figure taken from *Fiquet* et al. [88]

pressure [91, 92], and follow a nearly linear evolution with density. A linear extrapolation to higher densities is close to the V_P values given by the PREM seismic model for the Earth's inner core [99]. As a matter of fact, the linear extrapolations of the two sets of IXS data (Q_{max} free and fixed) straddle the seismic data. The IXS results are in reasonably good agreement with results obtained by impulsive stimulated light scattering (ISLS) [100], as well as with nuclear resonant inelastic X-ray scattering (NRIXS) [89] and X-ray radial diffraction (XRD) [90–92], though disagreements can be observed at low densities. The longitudinal sound velocities derived from shock-wave measurements [93] are lower by $3-4\,\%$ than all other experimental data determined at room temperature, indicating that anharmonic effects gain an increasing importance at high temperature. In contrast, velocities determined

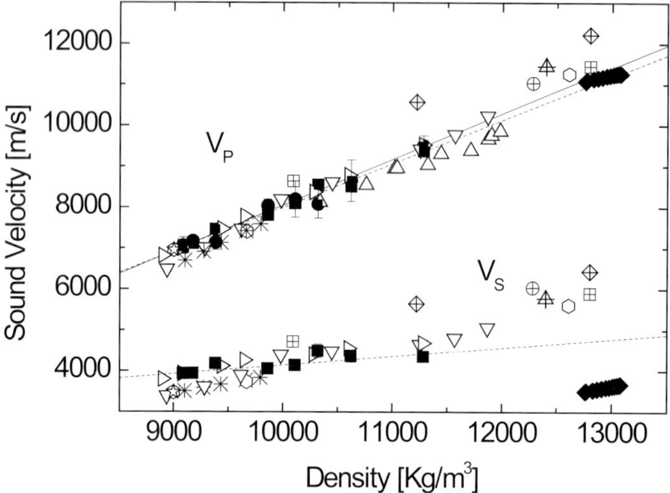

Fig. 29. Aggregate compressional (V_P) and shear (V_S) velocities as a function of density. No texture and temperature effects have been considered. *Solid circles* (*squares*): IXS results with the parameter Q_{max} left free (Q_{max} fixed); the *lines* represent linear fits to the density evolution of the IXS data: *solid line* (Q_{max} free), *dashed line* (Q_{max} fixed). Measurements: *inverted triangles*, NRIXS data [89]; *hexagons* and *asterisks*, XRD data [90–92]; *triangles*, shock-wave Hugoniot measurements [93]; *star*, ultrasonic measurements [91,92]. Calculations: *crossed squares* [94, 95]; *crossed circles* [96] and *crossed diamonds* [97, 98]. Preliminary reference Earth model seismic data [99] are reported for comparison as *full diamonds*. Figure taken from *Antonangeli* et al. [87]

through theoretical calculations are systematically higher than experimental data [87].

Combining the results for V_P with the measured density ρ and the bulk modulus K obtained from the ε-iron equation of state, the shear velocity V_S can be derived:

$$V_S^2 = \frac{3}{4}\left(V_P^2 - \frac{K}{\rho}\right) . \tag{11}$$

The resulting density evolution of V_S is also reported in Fig. 29. For clarity, only the IXS data set for which V_P was determined keeping Q_{max} fixed is reported (employing the other method gives results that differ by a few per cent). In the case of V_S, a substantial agreement with other experimental results is also observed. In the density region between 9000 kg/m^3 and 10 000 kg/m^3 the IXS data lie slightly above the NRIXS and XRD results, as a direct consequence of the higher V_P values in this density region. The overall agreement is, however, still satisfactory, although a linear fit to the IXS data shows a reduced slope as a function of density with respect the trend of NRIXS and XRD. Independently of the particular experimental data set used, an extrapolation to core densities leads to V_S values that are significantly higher than what is shown by the PREM model. This observation is in agreement with theoretical results [101], which showed that the shear modulus for ε-iron at core pressures is reduced by 70 % going from room temperature to 5400 K. It has also been speculated that partial melting could occur in the inner core, dramatically reducing the composite shear velocities (with respect to pure crystalline hcp iron) in that region.

Further studies on hcp-iron aimed at determining differences in the sound velocity, exploiting the strong texture development of iron, and different orientations of the momentum transfer with respect to the loading axis of the DAC, revealed a 4−5 % difference [87]. This difference is of the same order as that of compressional seismic waves that travel along the Earth's rotation axis and in the equatorial plane, respectively. These results therefore suggest that already a very moderate alignment of hcp-iron is sufficient to explain the seismic observations, in stark contrast to previously proposed models of an almost perfectly aligned single crystal. In addition, a series of measurements that show the influence of light elements such as sulfur, silicon, or oxygen on acoustic sound velocities of iron alloys at high-pressure was conducted [86]. Combining these results with X-ray diffraction measurements and seismic observations should provide constraints on the amount of light elements in the Earth's inner core.

6.4 Elasticity of cobalt to 39 GPa

In contrast to *hcp* iron the *hcp* phase of cobalt is stable at ambient temperature to about 100 GPa [102]. The mechanical and thermal properties of these two elements are close, and most importantly, *ab initio* calculations [95] show that they display a very similar pressure evolution of the elastic moduli. Cobalt can therefore serve as an analogue to hcp iron, and the availability of high-quality single crystals allows the determination of the five independent elastic moduli via the linear part of the acoustic phonon branches and the Christoffel equation [103].

The single crystals (45 to 85 μm diameter, 20 μm thickness) were prepared at the Lawrence Livermore National Laboratory, using femtosecond laser cutting and state-of-the-art polishing techniques [106]. The samples were loaded in DACs, using helium as pressure transmitting medium in order to ensure hydrostatic pressure conditions. Typical mosaic spreads amounted to 0.1−0.2°, and no degradation was observed up to the highest pressure point of 39 GPa. All crystals were cut with the surface normal parallel to the [110] direction. By simple rotation of the cell around the incident beam direction the longitudinal acoustic (LA) phonon branches along the [100] and [001] direction, the two transverse acoustic (TA) branches along the [110] direction, and the quasi-longitudinal branch along the [101] direction could be determined [105]. For the first four directions, the sound velocity is directly related to a specific elastic modulus, while the longitudinal sound velocity

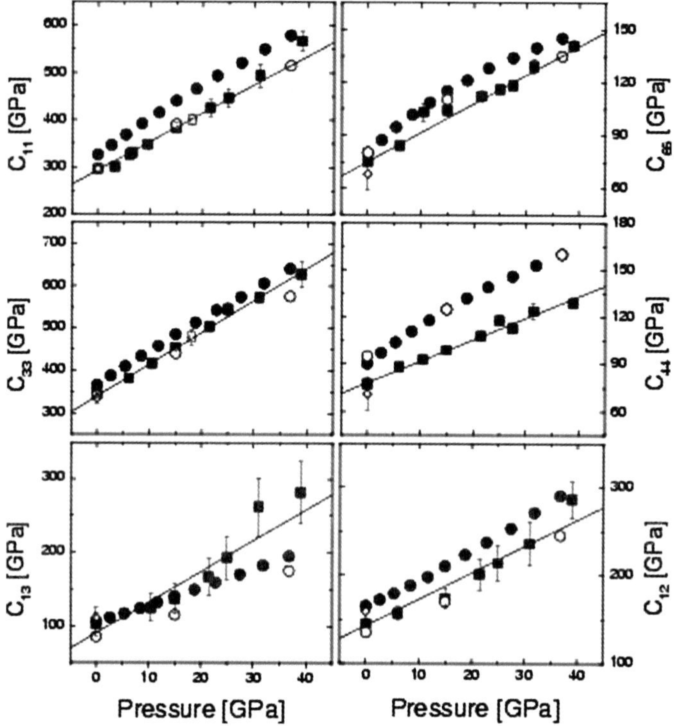

Fig. 30. Pressure evolution of the *hcp* Co elastic moduli. IXS results, obtained using helium (*full squares*) and neon (*open squares*) as pressure transmitting medium; ambient pressure ultrasonic measurements (*open diamonds*) [104]; *ab initio* calculations within the GGA (*full circles*) and LDA approximation (*open circles*) [95]. The *solid line* is a linear fit to the data. Figure taken from *Antonangeli* et al. [105]

along the [101] direction is a function of all five elastic moduli, so that the knowledge of the first four allows determining C_{13}. Figure 30 reports the pressure evolution of the five independent elastic moduli C_{11}, C_{33}, C_{44}, C_{12}, and C_{13} (as well as $C_{66} = 1/2(C_{11} - C_{12})$, together with ambient pressure ultrasonic measurements [104] and ab initio calculations [95]. At ambient conditions there is a good agreement between IXS and ultrasonic results, except for C_{12}, where the difference amounts to 11 %. Results from *ab initio* calculations yield mostly higher values, with the exception of C_{13}, while the respective pressure derivatives are close to the experimental ones for C_{11}, C_{33}, C_{66}, and C_{12}, but $\partial C_{44}/\partial P$ and $\partial C_{13}/\partial P$ are, respectively, too high and too low. A better quantitative agreement is observed with calculations performed within the local density approximation (LDA), especially for C_{11} and C_{66}, rather than within the generalised gradient approximation (GGA), while the IXS values for C_{33} and C_{12} lie in between the two approximations.

These results illustrate the power of IXS as a tool to study elasticity under extreme conditions with an accuracy of a few percent. The technique can be applied to a large variety of samples (opaque or transparent, insulating or metallic), therefore complementing and extending other experimental methods such as Brillouin light scattering and ultrasonics.

6.5 Lattice Dynamics of Molybdenum

This study was mainly motivated by the open questions, related to the H-point anomaly of the LA branch, as revealed by INS measurements at 1 atm and room temperature [107], and as a function of temperature to 1203 K [108]. With increasing temperature, the H-point phonon displays anomalous stiffening that has been proposed to arise from either intrinsic anharmonicity of the inter-atomic potential or electron phonon coupling. To address the nature of these phonon anomalies, a high-pressure experimental study was undertaken, since lattice compression provides a very convenient way to probe the inter-atomic potential as well as electron-phonon interactions.

The sample was a high-quality molybdenum single-crystal, cut with the surface normal parallel to the [110] direction, $\sim 60\,\mu$m diameter and $20\,\mu$m thickness. It was loaded in a DAC, using helium as pressure transmitting medium. Phonon dispersions of all the branches were recorded at a pressure of 17 GPa, and a partial set of dispersions (lacking only the transverse modes along [001] and [111]) at 37 GPa with an overall instrumental resolution of 3 meV [109]. The experimental results were confronted with *ab initio* calculations within density functional theory (DFT) and linear response theory, using the ABINIT code [110, 111].

Figure 31 shows the pressure evolution of the LA[00ξ] phonon branch, which displays, in analogy to the high-temperature INS results, a significant decrease in the relative magnitude of the H-point phonon anomaly upon compression. This is further highlighted in the inset, where the differing density evolutions of the Grüneisen parameters at the maximum of the dispersion ($q \sim 0.65$) and at the H-point ($q = 1.0$) are reported. The calculations capture quite well, both qualitatively and quantitatively, the experimental phonon dispersion branch.

The joint study provides several arguments that the explanation for the disappearance of the H-point anomaly with pressure is a decrease in the magnitude of the electron-phonon coupling. Responsible are the nested p-like bands, which are located near the Fermi level, ε_F, at one atmosphere. Firstly, the calculated H-point anomaly showed to be very sensitive to the Gaussian broadening of the bands. Secondly, on compression band broadening also occurs, but now solely due to the perturbation of the electronic structure in the compressed lattice. Furthermore, since the Fermi level increases under compression, at high pressure, the energy of the nested bands will be lower relative to ε_F (effectively moving outside of k_BT from ε_F). Therefore, both the pressure induced band broadening, and the lowering of nested bands relative

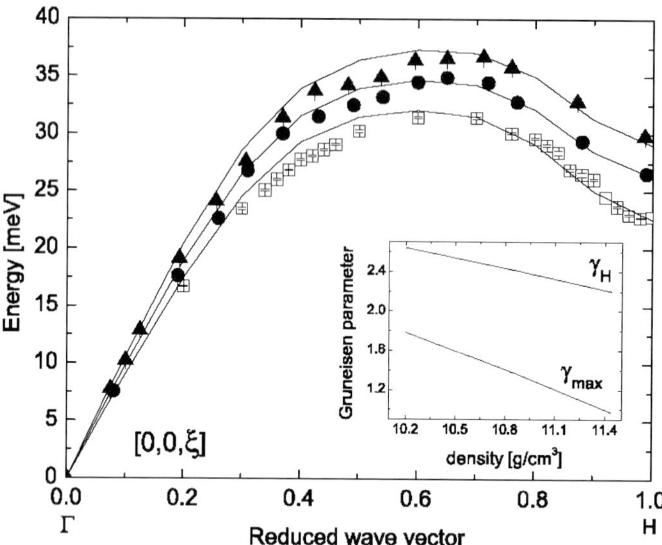

Fig. 31. Phonon dispersions along [001] as a function of pressure. Filled symbols are IXS data (*circles*, 17 GPa; *triangles*, 37 GPa), the *unfilled squares* are inelastic neutron scattering data (one atmosphere). Calculations are shown as *solid lines*. The inset shows the Grüneisen parameter as a function of density for a $q \sim 0.65$, corresponding to the maximum in the dispersion, and at the H-point ($q = 1.0$). Figure taken from *Farber* et al. [109]

to ε_F, should favor a decrease in the H-point phonon anomaly, yielding a more 'normal' bcc phonon spectrum, as experimentally observed. Finally, important anharmonic effects can be ruled out, since the calculations, performed within the quasi-harmonic approximation, reproduce the experimental results quite accurately.

7 Further Applications

7.1 Determination of the Phonon Density-of-States

The experimental determination of the energy distribution function $g(\mathrm{E})$, or vibrational density-of-states (VDOS), gives important insight into the physical properties of materials, since it allows the derivation of many thermodynamic and elastic properties. If single crystals are available, the VDOS can in principle be derived from a Born-von Kármán (BvK) fit to the experimentally determined phonon dispersion curves. In cases, where sufficiently perfect single crystals are not available, or for non-crystalline systems such as liquids and glasses, $g(\mathrm{E})$ has to be determined directly. This is commonly done by inelastic neutron scattering (INS), or more recently as well by IXS [62]. As

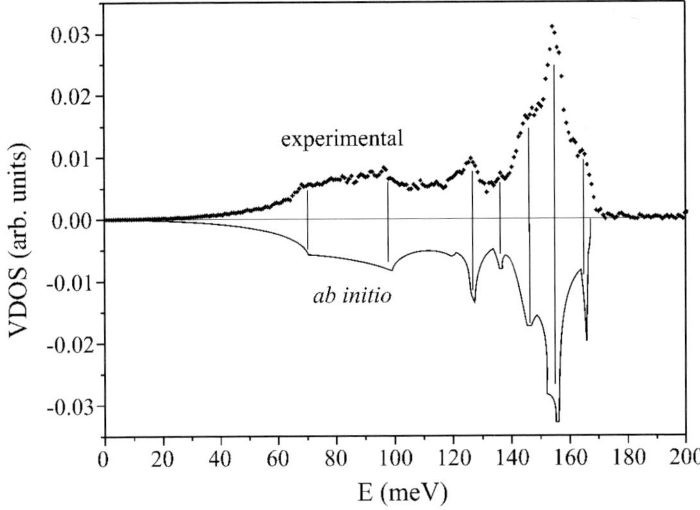

Fig. 32. Reconstructed vibrational density of states of diamond versus calculated *ab initio* results [115]. The data are normalised to equal surfaces. Figure taken from *Bossak* et al. [62]

inelastic X-ray scattering (IXS) from phonons is essentially a coherent scattering process, the same incoherent approximation as for coherent INS has to be applied. This necessitates a correct 1. directional averaging in polycrystalline samples and 2. an appropriate reciprocal space sampling [112–114]. The sampling procedure of this so-called "incoherent approximation" is usually obtained empirically. For systems with different atomic species, only the generalized density of states can be obtained, for which the individual contributions of the different constituent atoms are weighted by their corresponding scattering strength.

Several aspects have to be considered in order to ensure a correct VDOS approximation. The momentum transfer range corresponding to the first Brillouin zone has to be excluded, since in this case the total momentum transfer $\boldsymbol{Q}$ is equal to the phonon wave vector $\boldsymbol{q}$, and only phonon modes with a eigenvector component parallel to $\boldsymbol{Q}$ acquire finite intensity. Another aspect concerns the thickness of the integration shell, since for X-rays the atomic form factor f(Q) displays a pronounced Q-dependence with an approximately exponential decay. This leads to a distortion of the VDOS, if the integration is performed over a large Q-range. In general, no recipe exists for the choice of the shell sampling, but it was be shown that quantitative criteria for a uniform sampling can be established, which are independent of any specific lattice dynamics model, and only result from simple symmetry considerations [62].

The validity of the above approach was checked by comparison of results for polycrystalline diamond with *ab initio* lattice dynamics calculations and macroscopic thermodynamic and elastic measurements [62]. Ten IXS spectra were recorded, covering a Q-range of about $15\,\mathrm{nm}^{-1}$ (approximately from 60 to $75\,\mathrm{nm}^{-1}$) with a momentum resolution of $0.7\,\mathrm{nm}^{-1}$ both in horizontal and vertical directions. Prior to summing, the individual IXS spectra were corrected for the individual crystal analyser efficiency, the polarisation factor of the X-rays (see equation (4)) and the diamond form factor. The further data treatment followed the same approach as for nuclear inelastic scattering, where the multiphonon term is eliminated simultaneously with the deconvolution of the data with the instrumental function [116]. Figure 32 shows the thus obtained VDOS, together with the result of an *ab initio* calculation [115]. The agreement is quite remarkable: the position of special points is nearly identical, and the high-energy peak, due to the overbending of the optical phonon branch, is clearly seen. Several macroscopic parameters such as, for example, the specific heat at constant volume, the low- and high–temperature limit of the Debye temperature θ_D, and the average sound velocity v_D were derived from the VDOS, and yielded good agreement with available thermodynamic and elastic results [62].

7.2 Phonons in Surface-Sensitive Geometry

Inelastic X-ray scattering offers the unique possibility to study surface and bulk dynamics in a single experiment. This can be achieved by setting the sample in grazing incidence condition thus holding the incoming X-rays below the critical angle of total external reflection. In this case, the incident electromagnetic field displays an exponential decay in the sample bulk with a typical probing depth of only a few nm. By increasing the incidence angle beyond the critical angle one can penetrate deep into the sample bulk. Such a surface sensitive set-up was utilised to study the lattice dynamics of $NbSe_2$, a quasi-two-dimensional van der Waals layered material [117]. A metal at room temperature, $NbSe_2$ undergoes a charge density wave (CDW) transition at about 33 K and a superconducting transition at about 5 K. This CDW transition is interesting on many fronts, not least as it is an incommensurate, continuous phase transition and the CDW phase remains incommensurate right down to the formation of the superconducting phase [118, 119].

The IXS experiment was performed with an energy resolution of 3 meV, and the depth sensitivity was varied from 4 nm – the topmost four atomic layers contribute 50 % to the scattered signal – to 100 nm. IXS data were recorded along the $[\xi 00]$ direction for longitudinal acoustic and optical branches around the (200) surface reflection. Figure 33 shows an overview of the determined phonon dispersions and a comparison with previous INS results [118, 119] as well as calculations [120, 121]. The overall form of the experimental dispersion relation – with exception of the surface anomaly – agrees well with the bulk calculations, which included an effective ion-ion interaction. In both configurations the longitudinal $\Sigma_1\ \omega_2$ mode displays significant softening in energy at 2/3 of the Brillouin zone. This points to a

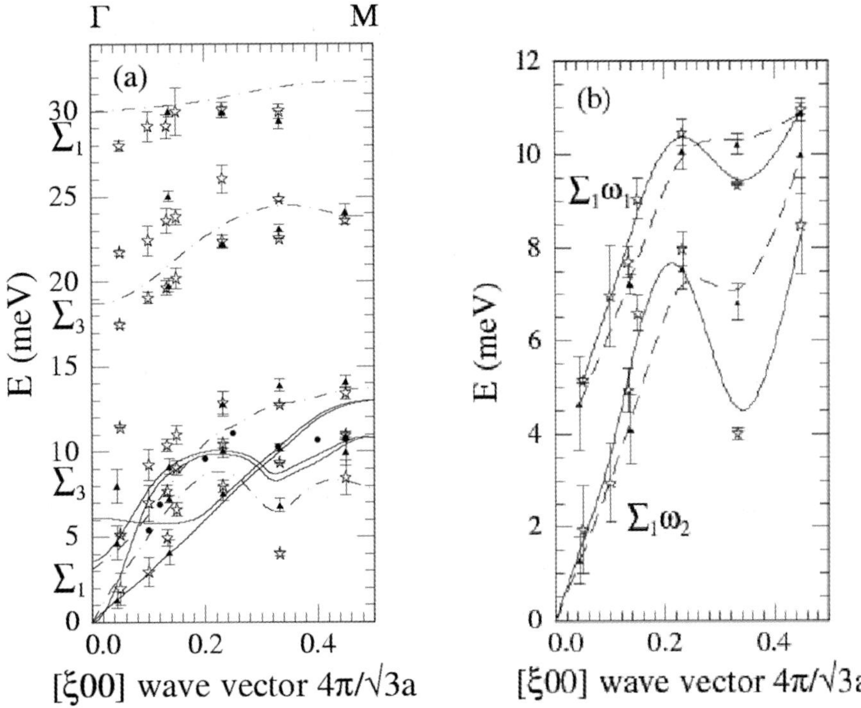

Fig. 33. **(a)** Room temperature surface (*open stars*) and bulk (*filled triangles*) phonon dispersion curve of 2H$-$NbSe$_2$, obtained by grazing incidence IXS. Bulk sensitive INS data [118, 119] (*solid circles*) are shown for comparison. The calculations by *Motizuki* et al. [120] and *Feldman* [121] are displayed by *solid* and *dashed lines*, respectively. **(b)** Enlarged view of the low energy part of the phonon dispersion (lines are guide to the eye). Figure taken from *Murphy* et al. [117]

strong interaction between the phonons and the conduction electrons. The increased anomaly at the surface is evidence that some change is occurring in the topmost layers. One possibility is that the surface electronic state cuts the Fermi surface at a different position than the bulk Fermi wave vector, as deduced by *Benedek* et al. from Helium scattering results [122]. Another possibility would involve an increased electron-phonon interaction, which is consistent with the change of symmetry and coordination of the upper atomic layers with respect to that of the bulk.

8 Concluding Remarks

It was the aim of this Chapter to give a representative overview on the various applications of phonon studies in crystalline solids using inelastic X-ray scattering with very high energy resolution. The Chapter was not meant to be a complete compendium, and therefore several studies were not discussed. These comprise work, for example, on quasicrystals [123–126], SiO_2 [127] and biological systems [128–130], as well as the large number of studies on liquids and amorphous materials that are outside the scope of the present Chapter.

Within the last ten years the IXS technique has gained a high level of maturity and has made important contributions to various fields of research. At present, the number of instruments around the world is limited to only four, but several projects are in an advanced design stage or are planned, reflecting the interest of the scientific community. These new instruments will benefit from the latest advances in storage-ring design, X-ray optics, spectrometers and detectors. In turn, the existing instruments are being continuously improved. Further gains in photon flux by the increase of the storage-ring current and undulators with a shorter magnetic period, as well as the implementation of advanced focusing schemes, aiming at a further reduction of the X-ray beam size on the sample, will enhance the current capabilities of IXS. Even smaller samples will be studied, utilizing an increasingly complex sample environment. One example concerns high-pressure studies at very high temperatures, using laser heating techniques. Finally, the advent of X-ray lasers will also pave new avenues for IXS, making studies possible that are today only visions.

References

[1] W. G. Hofmann: Ph.D. thesis, University of Bayreuth (1989)
[2] H. Egger, W. Hofmann, J. Kalus: Appl. Phys. A **35**, 41 (1984)
[3] E. Burkel, J. Peisl, B. Dorner: Europhys. Lett. **3 (8)**, 957 (1987)
[4] B. Dorner, E. Burkel, T. Illini, J. Peisl: Z. Phys. B: Condens. Matter **69**, 179 (1987)
[5] E. Burkel: in *Inelastic Scattering of X-rays with Very High Energy Resolution*, vol. 125, Springer Tracts in Modern Physics (Springer Verlag, Berlin, Heidelberg 1991)
[6] F. Sette, M. Krisch: ESRF Newsletter **22**, 15 (1994)
[7] F. Sette, G. Ruocco, M. Krisch, C. Masciovecchio, R. Verbeni: Phys. Scr. **T66**, 48 (1996)
[8] M. Krisch, M. Lorenzen: (1999), unpublished
[9] M. Schwoerer-Böhning, A. T. Macrander: Rev. Sci. Instrum. **69**, 3109 (1998)
[10] A. Q. R. Baron, Y. Tanaka, S. Goto, K. Takeshita, T. Matsushita, T. Ishikawa: J. Phys. Chem. Sol. **61**, 461 (2000)
[11] G. Ruocco, F. Sette: J. Phys.: Condens. Matter **11**, R259 (1999)
[12] E. Burkel: Rep. Prog. Phys. **63**, 171 (2000)
[13] G. Ruocco, F. Sette: J. Phys.: Condens. Matter **13**, 9141 (2001)
[14] W. Schülke: in G. Brown, D. E. Moncton (Eds.): *Handbook on Synchrotron Radiation*, vol. 3 (Elsevier Science Publ., Amsterdam 1991) Chap. 15
[15] M. Krisch, R. A. Brand, M. Chernikov, H. R. Ott: Phys. Rev. B **65**, 134201 (2002)

[16] H. Zachariasen: *Theory of X-ray Diffraction in Crystals* (Dover, New York 1944)
[17] A. Bergamin, G. Cavagnero, G. Mana, G. Zosi: Eur. Phys. J. B **9**, 225 (1999)
[18] R. Verbeni, F. Sette, M. H. Krisch, U. Bergmann, B. Gorges, C. Halcoussis, K. Martel, C. Masciovecchio, J. F. Ribois, G. Ruocco, H. Sinn: J. Synchr. Radiat. **3**, 62 (1996)
[19] T. Toellner: Hyperfine Interact. **125**, 3 (2000)
[20] C. Masciovecchio, U. Bergmann, M. Krisch, G. Ruocco, F. Sette, R. Verbeni: Nucl. Instrum. Methods B **117**, 339 (1996)
[21] C. Masciovecchio, U. Bergmann, M. Krisch, G. Ruocco, F. Sette, R. Verbeni: Nucl. Instrum. Methods B **111**, 181 (1996)
[22] B. Dorner, H. Schober, A. Wonhas, M. Schmitt, D. Strauch: Eur. Phys. B **5**, 839 (1998)
[23] F. Widulle, T. Ruf, O. Buresch, A. Debernardi, M. Cardona: Phys. Rev. Lett. **82**, 3089 (1999)
[24] D. W. Feldman, J. H. J. Parker, W. J. Choyke, L. Patrick: Phys. Rev. **173**, 787 (1968)
[25] J. Serrano, J. Strempfer, M. Cardona, M. Schwoerer-Böhning, H. Requardt, M. Lorenzen, B. Stojetz, P. Pavone, W. J. Choyke: Appl. Phys. Lett. **80**, 4360 (2002)
[26] T. Ruf, J. Serrano, M. Cardona, P. Pavone, M. Pabst, M. Krisch, M. D'Astuto, T. Suski, I. Grzegory, M. Leszczynski: Phys. Rev. Lett. **86**, 906 (2001)
[27] J. M. Zhang, T. Ruf, M. Cardona, O. Ambacher, M. Stutzmann, J.-M. Wagner, F. Bechstedt: Phys. Rev. B **56**, 14399 (1997)
[28] M. Schwoerer-Böhning, A. T. Macrander, M. Pabst, P. Pavone: phys. stat. sol. (b) **215**, 177 (1999)
[29] R. S. Krishnan: Proc. Indian Acad. Sci., Sect. A **24A**, 45 (1946)
[30] R. S. Krishnan: Proc. Indian Acad. Sci., Sect. A **26A**, 399 (1947)
[31] W. Windl, P. Pavone, K. Karch, O. Schütt, D. Strauch, P. Gianozzi, S. Baroni: Phys. Rev. B **48**, 3164 (1993)
[32] J. Kulda, B. Dorner, B. Roessli, H. Sterner, R. Bauer, T. May, K. Karch, P. Pavone, D. Strauch: Solid State Commun. **99**, 799 (1996)
[33] M. Schwoerer-Böhning, A. T. Macrander, D. A. Arms: Phys. Rev. Lett. **80**, 5572 (1998)
[34] J. Kulda, H. Kainzmaier, D. Strauch, B. Dorner, M. Lorenzen, M. Krisch: Phys. Rev. B **66**, 241202 (2002)
[35] J. Maultzsch, S. Reich, C. Thomsen, H. Requardt, P. Ordejón: Phys. Rev. Lett. **92**, 075501 (2004)
[36] O. Dubay, G. Kresse: Phys. Rev. B **67**, 035401 (2003)
[37] C. Seyfert, R. O. Simmons, H. Sinn, D. A. Arms, E. Burkel: J. Phys.: Condens. Matter **11**, 3501 (1999)
[38] H. H. Sample, C. A. Swenson: Phys. Rev. **158**, 188 (1967)
[39] R. E. Slusher, C. M. Surko: Phys. Rev. B **13**, 1086 (1976)
[40] G. L. Morley, K. L. Kliewer: Phys. Rev. **180**, 245 (1969)
[41] N. S. Gillis, T. R. Koehler, N. R. Werthamer: Phys. Rev. **175**, 1110 (1968)
[42] A. Shukla, M. Calandra, M. D'Astuto, M. Lazzeri, F. Mauri, C. Bellin, M. Krisch, J. Karpinski, S. M. Kazakov, J. Jun, D. Daghero, K. Parlinski: Phys. Rev. Lett. **90**, 095506 (2003)

[43] A. Q. R. Baron, H. Uchiyama, Y. Tanaka, S. Tsutsui, D. Ishikawa, S. Lee, R. Heid, K.-P. Bohnen, S. Tajima, T. Ishikawa: Phys. Rev. Lett. **92**, 197004 (2004)

[44] J. Orenstein, A. J. Millis: Science **288**, 468 (2000)

[45] J.-H. Chung, T. Egami, R. J. McQueeney, M. Yethiraj, M. Arai, T. Yokoo, Y. Petrov, H. A. Mook, Y. Endoh, S. Tajima, C. Frost, F. Dogan: Phys. Rev. B **67**, 014517 (2003)

[46] L. Pintschovius, D. Reznik, W. Reichardt, Y. Endoh, H. Hiraka, J. M. Tranquada, H. Ushiyama, T. Masui, S. Tajima: Phys. Rev. B **69**, 214506 (2004)

[47] M. D'Astuto, P. K. Mang, P. Giura, A. Shukla, P. Ghigna, A. Mirone, M. Braden, M. Greven, M. Krisch, F. Sette: Phys. Rev. Lett. **88**, 167002 (2002)

[48] M. D'Astuto, P. K. Mang, P. Giura, A. Shukla, A. Mirone, M. Krisch, F. Sette, P. Ghigna, M. Braden, M. Greven: Int. J. Mod. Phys. B **17**, 484 (2003)

[49] R. McQueeney, Y. Petrov, T. Egami, M. Yethiraj, G. Shirane, Y. Endoh: Phys. Rev. Lett. **82**, 628 (1999)

[50] L. Pintschovius, M. Braden: Phys. Rev. B **60**, R15039 (1999)

[51] H. Uchiyama, A. Q. R. Baron, S. Tsutsui, Y. Tanaka, W.-Z. Hu, A. Yamamoto, S. Tajima, Y. Endoh: Phys. Rev. Lett. **92**, 197005 (2004)

[52] M. D'Astuto, A. Mirone, P. Giura, D. Colson, A. Forget, M. Krisch: J. Phys.: Condens. Matter **15**, 8827 (2003)

[53] H. Requardt, J. E. Lorenzo, P. Monceau, R. Currat, M. Krisch: Phys. Rev. B **66**, 214303 (2002)

[54] P. Monceau, L. Bernard, R. Currat, F. Levy, J. Rouxel: Synth. Met. **19**, 819 (1987)

[55] S. Ravy, H. Requardt, D. Le Bolloc'h, P. Foury-Leylekian, J.-P. Pouget, R. Currat, P. Monceau, M. Krisch: Phys. Rev. B **69**, 115113 (2004)

[56] M. E. Manley, G. H. Lander, H. Sinn, A. Alatas, W. L. Hults, R. J. McQueeney, J. L. Smith, J. Willit: Phys. Rev. B **67**, 052302 (2003)

[57] J. Wong, M. Krisch, D. L. Farber, F. Occelli, A. J. Schwartz, T.-C. Chiang, M. Wall, C. Boro, R. Xu: Science **301**, 1078 (2003)

[58] J. Wong, M. Krisch, D. L. Farber, F. Occelli, R. Xu, T.-C. Chiang, D. Clatterbuck, A. J. Schwartz, M. Wall, C. Boro: Phys. Rev. **72**, 064115 (2005)

[59] J. C. Lashley, M. Stout, R. A. Pereyra, M. S. Blau, J. D. Embury: Scr. Mater. **44**, 2815 (2001)

[60] X. Dai, S. Y. Savrasov, G. Kotliar, A. Migliori, H. Ledbetter, E. Abrahams: Science **300**, 953 (2003)

[61] R. A. Reese, S. K. Sinha, D. T. Peterson: Phys. Rev. B **8**, 1332 (1973)

[62] A. Bossak, M. Krisch: Phys. Rev. B **72**, 224305 (2004)

[63] H. Schober, M. M. Koza, M. M. Tölle, C. Masciovecchio, F. Sette, F. Fujara: Phys. Rev. Lett. **85**, 4100 (2000)

[64] F. Sette, M. H. Krisch, C. Masciovecchio, G. Ruocco, G. Monaco: Science **280**, 1550 (1998)

[65] M. M. Koza, H. Schober, B. Geil, M. Lorenzen, H. Requardt: Phys. Rev. B **69**, 024204 (2004)

[66] M. M. Koza, H. Schober, T. Hansen, A. Tölle, F. Fujara: Phys. Rev. Lett. **84**, 4112 (2000)

[67] J. S. Tse, M. A. White: J. Phys. Chem. **92**, 5006 (1988)

[68] J. Baumert, C. Gutt, V. P. Shpakov, J. S. Tse, M. Krisch, M. Müller, H. Requardt, D. D. Klug, S. Janssen, W. Press: Phys. Rev. B **68**, 174301 (2003)
[69] F. Sette, G. Ruocco, M. Krisch, C. Masciovecchio, R. Verbeni, U. Bergmann: Phys. Rev. Lett. **77**, 83 (1996)
[70] G. Ruocco, F. Sette, M. Krisch, U. Bergmann, C. Masciovecchio, R. Verbeni: Phys. Rev. B **54**, 14892 (1996)
[71] J. Baumert, C. Gutt, M. R. Johnson, W. Press, J. S. Tse: *Annual Report*, Institute Laue-Langevin (2002)
[72] J. Baumert, C. Gutt, M. Krisch, H. Requardt, M. Müller, J. S. Tse, D. D. Klug, W. Press: Phys. Rev. B **72**, 054302 (2005)
[73] S. Klotz, M. Braden: Phys. Rev. Lett. **85**, 3209 (2000)
[74] A. Jayaraman: Rev. Mod. Phys. **55**, 65 (1983)
[75] M. Krisch, P. Loubeyre, G. Ruocco, F. Sette, A. Cunsolo, M. D'Astuto, R. Le Toullec, M. Lorenzen, A. Mermet, G. Monaco, R. Verbeni: Phys. Rev. Lett. **89**, 125502 (2002)
[76] S. Ghose, M. Krisch, A. Oganov, A. Beraud, A. Bossak, R. Gulve, R. Seelaboyina, H. Yang, S. K. Saxena: Phys. Rev. Lett. **96**, 035507 (2006)
[77] J. M. Besson, R. J. Nelmes, G. Hamel, J. S. Loveday, G. Weill, S. Hull: Physica B **180&181**, 907 (1992)
[78] E. Burkel, M. Schwoerer-Böhning, C. Seyfert, R. O. Simmons: Physica B **219&220**, 98 (1996)
[79] M. H. Krisch, A. Mermet, A. San Miguel, F. Sette, C. Masciovecchio, G. Ruocco, R. Verbeni: Phys. Rev. B **56**, 8691 (1997)
[80] F. Occelli, M. Krisch, P. Loubeyre, F. Sette, R. Le Toullec, C. Masciovecchio, J. P. Rueff: Phys. Rev. B **63**, 224306 (2001)
[81] I. Loa, K. Kunc, K. Syassen, M. Krisch, A. Mermet, M. Hanfland: High Press. Res. **23**, 1 (2003)
[82] N. Nücker, U. Buchenau: Phys. Rev. B **31**, 5479 (1985)
[83] J. Mizuki, C. Stassis: Phys. Rev. B **34**, 5890 (1986)
[84] H. Mook, D. B. McWhan, F. Holtzberg: Phys. Rev. B **25**, 4321 (1982)
[85] S. Raymond, J.-P. Rueff, M. D'Astuto, D. Braithwaite, M. Krisch, J. Flouquet: Phys. Rev. B **66**, 220301 (R) (2002)
[86] G. Fiquet, J. Badro, F. Guyot, C. Bellin, M. Krisch, D. Antonangeli, A. Mermet, H. Requardt, D. Farber, J. Zhang: Phys. Earth Planet. Int. **143–144**, 5 (2004)
[87] D. Antonangeli, F. Occelli, H. Requardt, J. Badro, G. Fiquet, M. Krisch: Earth Planet. Sci. Lett. (2004)
[88] G. Fiquet, J. Badro, F. Guyot, H. Requardt, M. Krisch: Science **291**, 468 (2001)
[89] H. K. Mao, J. Xu, V. V. Struzhkin, J. Shu, R. J. Hemley, W. Sturhahn, M. Y. Hu, E. E. Alp, L. Vocadlo, D. Alfè, G. D. Price, M. J. Gillan, M. Schwoerer–Böhning, D. Häusermann, P. Eng, G. Shen, H. Giefers, R. Lübbers, G. Wortmann: Science **292**, 914 (2001)
[90] L. S. Dubrovinsky, N. A. Dubrovinskaia, S. K. Saxena, S. Rehki, T. LeBihan: J. Alloys Compd. **297**, 156 (2000)
[91] H. K. Mao, J. Shu, G. Shen, R. J. Hemley, B. Li, A. K. Singh: Nature **396**, 741 (1998)
[92] H. K. Mao, J. Shu, G. Shen, R. J. Hemley, B. Li, A. K. Singh: Nature **399**, 280 (1999) correction of [91]

[93] J. M. Brown, R. G. McQueen: J. Geophys. Res. **91**, 7485 (1986)
[94] L. Stixrude, R. E. Cohen: Science **267**, 1972 (1995)
[95] G. Steinle-Neumann, L. Stixrude, R. E. Cohen: Phys. Rev. B **60**, 791 (1999)
[96] P. Söderlind, J. A. Moriarty, J. M. Wills: Phys. Rev. B **53**, 14063 (1996)
[97] R. E. Cohen, L. Stixrude, E. Wasserman: Phys. Rev. B **56**, 8575 (1997)
[98] R. E. Cohen, L. Stixrude, E. Wasserman: Phys. Rev. B **58**, 5873 (1997) errata [97]
[99] A. M. Dziewonsky, D. L. Anderson: Phys. Earth Planet. Inter. **25**, 297 (1981)
[100] J. C. Crowhurst, A. F. Goncharov, J. M. Zaug: J. Phys.: Condens. Matter **16**, S1137 (2004)
[101] A. Laio, S. Bernard, G. L. Chiarotti, S. Scandolo, E. Tosatti: Science **287**, 1027 (2000)
[102] C. S. Yoo, H. Cynn, P. Söderlind, V. Iota: Phys. Rev. Lett. **84**, 4132 (2000)
[103] B. A. Auld: *Acoustic Fields and Waves in Solids*, vol. 1 (J. Wiley and Sons, New York 1973)
[104] H. R. Schober, H. Dederichs: in *Elastic Piezoelectric Pyroelectric Piezooptic Electrooptic Constants and Nonlinear Dielectric Susceptibilities of Crystals*, vol. 11a, Landolt-Börnstedt New Series III (Springer, Berlin 1979)
[105] D. Antonangeli, M. Krisch, G. Fiquet, D. L. Farber, C. M. Aracne, J. Badro, F. Occelli, H. Requardt: Phys. Rev. Lett. **93**, 215505 (2004)
[106] D. L. Farber, D. Antonangeli, C. M. Aracne, J. Benterau: High Pressure Research **26**, 1 (2006)
[107] C. M. Varma, W. Weber: Phys. Rev. Lett. **39**, 1094 (1977)
[108] J. Zarestky, C. Stassis, B. N. Harmon, B. N. Harmon, K.-M. Ho, C. L. Fu: Phys. Rev. B **28**, 697 (1983)
[109] D. L. Farber, M. Krisch, D. Antonangeli, A. Beraud, J. Badro, F. Occelli, D. Orlikowski: Phys. Rev. Lett. **96**, 115502 (2006)
[110] X. Gonze, S.-M. Beuken, R. Caracas, F. Detraux, M. Fuchs, G.-M. Rignanese, L. Sindic, M. Verstraete, G. Zerah, F. Jollet, M. Torrent, A. Roy, M. Mikami, P. Ghosez, S.-Y. Raty, D. C. Allan: Comp. Mat. Sci. **25**, 478 (2002)
[111] S. Goedecker: SIAM J. Sci. Comp. **18**, 1605 (1997)
[112] L. M. Needham, M. Cutroni, A. J. Dianoux, H. M. Rosenberg: J. Phys. Condens. Matter **5**, 637 (1993)
[113] F. W. de Wette, A. Rahman: Phys. Rev. **176**, 784 (1968)
[114] G. L. Squires: *Introduction to the Theory of Thermal Neutron Scattering* (Cambridge University Press 1978)
[115] P. Pavone, K. Karch, O. Schütt, W. Windl, D. Strauch, P. Gianozzi, S. Baroni: Phys. Rev. B **48**, 3156 (1993)
[116] V. Kohn, A. Chumakov: Hyperfine Interactions **125**, 205 (2000)
[117] B. M. Murphy, H. Requardt, J. Stettner, J. Serrano, M. Krisch, M. Müller, W. Press: Phys. Rev. Lett. **95**, 256104 (2005)
[118] D. E. Moncton, J. D. Axe, F. J. DiSalvo: Phys. Rev. Lett. **34**, 734 (1975)
[119] D. E. Moncton, J. D. Axe, F. J. DiSalvo: Phys. Rev. B **16**, 801 (1977)
[120] K. Motizuki, K. Kimura, E. Ando, N. Suzuki: J. Phys. Soc. Jpn. **53**, 1078 (1984)
[121] J. L. Feldman: Phys. Rev. B **25**, 7132 (1982)
[122] G. Benedek, L. Miglio, G. Seriani: in E. Hulpke (Ed.): *Helium Atom Scattering from Surfaces*, vol. 27, Springer Series in Surface Science (Springer 1992) p. 208

[123] R. A. Brand, M. Krisch, M. Chernikov, H. R. Ott: Ferroelectrics **250**, 233 (2001)
[124] M. Krisch, R. A. Brand, M. Chernikov, H. R. Ott: Phys. Rev. B **65**, 134201 (2002)
[125] E. Burkel, R. Nicula, U. Ponkratz, H. Sinn, A. Alatas, E. E. Alp: Nucl. Instrum. Methods B **199**, 151 (2003)
[126] R. A. Brand, J. Voss, F. Hippert, M. Krisch, R. Sterzel, I. R. Fischer: J. Non-Cryst. Solids **334&335**, 207 (2004)
[127] E. Burkel, C. Seyfert, C. Halcoussis, H. Sinn, R. O. Simmons: Physica B **263–264**, 412 (1999)
[128] S. H. Chen, C. Y. Liao, H. W. Huang, T. M. Weiss, M. C. Bellisent-Funel, F. Sette: Phys. Rev. Lett. **86**, 740 (2001)
[129] T. M. Weiss, P.-J. Chen, H. Sinn, E. E. Alp, S.-H. Chen, H. W. Huang: Biophys. J. **84**, 3767 (2003)
[130] Y. Liu, D. Berti, A. Faraone, W.-R. Chen, A. Alatas, H. Sinn, E. E. Alp, A. Said, P. Baglioni, S.-H. Chen: Phys. Chem. Chem. Phys. **6**, 1499 (2004)

Index

acoustic
  modes, 317
actinides, 342, 343
amorphous ice, 345
amplitudon, 341
Ar, 350

Brillouin
  scattering, 317

CdTe, 350
cesium, 351
charge-density wave systems, 341, 342
clathrate hydrates, 346

density–density correlation function, 317
diamond, 331
double differential cross section, 317, 320, 321

elastic moduli, 354

graphite, 333

hcp-cobalt, 357
hcp-iron, 354
helium, 334
high-pressure, 317
high-pressure
  studies, 350

impulsive stimulated light scattering (ISLS), 355
inelastic X-ray scattering, 317
IXS instruments, 324

large-bandgap materials, 328, 329
lattice dynamics, 317, 327, 328, 336, 338, 343, 351, 354
  in-plane, 334

MgB_2, 336

neutron spectroscopy, 317

optical modes, 317

phason, 341
Plutonium, 343
polycrystals, 344

Raman
  scattering, 317, 331

SiC, 328
superconductivity, 336, 338
synchrotron radiation, 324

uranium, 343

Ultrafast X-Ray Scattering in Solids

David A. Reis[1] and Aaron M. Lindenberg[2]

[1] FOCUS Center and Department of Physics, University of Michigan, 450 Church Street, Ann Arbor, MI 48109-1040, USA
dreis@umich.edu

[2] Stanford Synchrotron Radiation Laboratory/Stanford Linear Accelerator Center (SLAC), Menlo Park, CA 94025, USA
aaronl@slac.stanford.edu

Abstract. X-rays are a valuable probe for studying structural dynamics in solids because of their short wavelength, long penetration depth and relatively strong interaction with core electrons. Recent advances in accelerator- and laser-based pulsed X-ray sources have opened up the possibility of probing nonequilibrium dynamics in real time with atomic-scale spatial resolution. The timescale of interest is a single vibrational period, which can be as fast as a few femtoseconds. To date, almost all such experiments on this timescale have been carried out optically, which only indirectly measure atomic motion through changes in the dielectric function. X-rays have the advantage that they are a direct probe of the atomic positions.

1 Introduction

The field of ultrafast X-ray physics is still in its infancy. Ultrafast sources of X-ray radiation are significantly less developed than lasers. The advancement of ultrafast X-ray science shows many parallels to ultrafast optical science. As was the case with the development of femtosecond lasers, a strong synergy exists between improvements in sources and experimental techniques that increases our ability to answer ever more sophisticated scientific questions. There are a number of examples of important experiments that have helped propel ultrafast X-ray science. For example, researchers have used time-resolved X-ray scattering from solids to study transient dynamics in metals [1], organic solids and biomolecules [2–4], laser-generated shock waves [5–7], impulsive strain generation and melting in semiconductors [8–21] and nanoparticles [22, 23], electron-hole plasma diffusion [24], solid-solid phase transitions [25–27], folded acoustic phonons [28, 29] and coherent optical phonons [30]. A number of early reviews and surveys exist in the literature, for example [31–33].

We have not attempted here to give a complete account of the progress in the field. A number of important experiments are notably absent. For example, we do not discuss molecular dynamics [34–39] or spectroscopic experiments in the solid state [27, 40, 41]. Instead, we highlight a number of seminal experiments in ultrafast X-ray scattering from solids with the goal of giving a reader who may not be familiar with the field a flavor of its progress

M. Cardona, R. Merlin (Eds.): Light Scattering in Solid IX,
Topics Appl. Physics **108**, 371–422 (2007)

and potential. The Chapter is organized as follows: first, in Sect. 2, we present a brief survey of ultrafast X-ray sources. These sources fall under two broad categories, namely, laser based and accelerator based, and each has its own advantages and disadvantages. While this work emphasizes past experiments, it would not be complete without a reference to the X-ray free-electron laser (FEL), the first of which is currently under construction at the Stanford Linear Accelerator Center (SLAC). The X-ray FEL is expected to operate with seven orders of magnitude enhancement in peak brightness compared to the brightest existing source, the Subpicosecond Pulse Source (SPPS) also at SLAC. The hard X-ray radiation will have full transverse coherence – and potentially has the capability of attosecond pulse generation with full longitudinal coherence. After the discussion of sources, in Sect. 3, we discuss time-resolved diffraction experiments on coherent phonons. Here, the vast majority of experiments have looked at the generation and propagation of coherent acoustic phonons (i.e., ultrasound) in semiconductors, in more and more detail. Recent experiments have also looked at phonon propagation across single heterostructure boundaries and in superlattices as well as at coherent optical-phonon excitations in semimetals. Ultrafast diffraction can be an important tool for studying phase transitions, and in Sect. 4, we present experiments on the so-called nonthermal melting transition including a number of recent experiments at SPPS that are shedding new light on the solid-liquid transition. We then discuss two experiments on solid-solid transitions, one in a correlated electron system (VO_2) and the other a paraelectric to ferroelectric transition in a charge-transfer molecular crystal (TTF-CA). We conclude, in Sect. 5, with a few brief remarks.

2 Ultrafast X-Ray Sources

Advances in the understanding of ultrafast phenomena have depended crucially on technological advances in both femtosecond optical and X-ray sources. There now exist a wide variety of sources of picosecond and subpicosecond X-ray pulses that have been used to probe structural dynamics in solids. These vary in size from table-top to many kilometers in length. In this section, we briefly review some of the methods used for producing short bursts of X-rays, focusing on recent developments in the field.

Laser-produced plasmas are table-top sources in which an intense laser pulse incident on a solid creates a plasma of high-energy electrons that, through collisions with the solid, produces bremsstrahlung and characteristic X-ray line radiation. The duration of the X-ray pulses can be comparable to the excitation pulse [42–45] although, at the present time, no direct measure of the pulse duration of these sources has been made. The fluorescent lines from a plasma source are emitted isotropically, such that many experiments require collection and focusing optics, although a large divergence angle can be used to access a wide range of reciprocal space. State-of-the-art laser-plasma sources produce on the order of 10^6 photons/s in a focal spot on a sample to be probed [46,47]. We refer the reader to the review by *Sokolowski-Tinten* et al. [46] for more detail on these sources.

Third-generation X-ray sources are orders of magnitude brighter than laser-based sources and offer greater tunability. However, their ability to resolve ultrafast dynamics is limited by their pulse duration, on the order of 100 ps (FWHM), set by the equilibrium electron-bunch length in the storage ring [48, 49]. Reviews describing these sources can be found in [50–54]. X-ray streak cameras are able to extend the temporal resolution achievable at synchrotron sources to a single picosecond [55–58]. Although the quantum efficiency of these detectors is low, work is ongoing on improving this [59]. A variety of other techniques have been proposed to produce femtosecond X-ray pulses at synchrotron sources including electron [60–62] and X-ray [63] beam-slicing techniques. Electron beam slicing is now being developed at a number of synchrotrons worldwide. In this technique, a short laser pulse co-propagates with the electron beam through a wiggler to produce a significant energy modulation (on the order of 10 MeV under resonant conditions [60]) of those electrons that temporally overlap the laser pulse. The energy-modulated electrons are then spatially separated from the main bunch by a dispersive section and used to produce femtosecond X-ray pulses in a bending magnet or undulator, leaving a femtosecond hole behind. A source under development at the Advanced Light Source at Lawrence Berkeley National Laboratory is expected to produce on the order of 10^6 photons/s in a narrow bandwidth, albeit with significant background scattering from the spatial wings of the long X-ray pulse. The first experiments using a lower-flux version of this source have recently been carried out by *Cavalleri* et al. [27].

In recent years, sources of ultrafast X-rays based on linear-accelerator technology have been developed that in many ways compete favorably with the sources described above. Research and development on X-ray free-electron lasers is underway at several places around the world. These sources are expected to produce intense, coherent, ultrafast X-ray pulses, based on self-amplified spontaneous emission in a long undulator [64–66]. The basic principle makes use of a highly relativistic, compressed electron bunch propagating through an undulator (a periodic magnetic field). Under conditions similar to those found at third-generation synchrotrons worldwide, the electrons within a bunch radiate independently. But under the right conditions (sufficiently optimized emittance, energy spread, current) the emitted radiation acts back on the electron bunch (in a process similar to the electron beam slicing described above) leading to longitudinal density modulations with period corresponding to the wavelength of the undulator fundamental. The constructive interference of the radiation from the bunched electrons leads to further bunching and exponential gain. Calculations indicate one may expect gains on the order of 10^8 in a 100 m undulator, corresponding to $\sim 10^{12}$ photons/pulse with full transverse coherence. Construction of the first X-ray FEL, the Linac Coherent Light Source (LCLS) project at the

Stanford Linear Accelerator Center (SLAC), is currently underway and hard X-rays should be available by 2009. A soft-X-ray free-electron laser at the Deutsches Elektronen-Synchrotron (DESY) in Hamburg Germany is now operational [67, 68], also with a planned extension to the hard-X-ray regime. A review by *Feldhaus* et al. [65] provides a good introduction to the basic principles of hard-X-ray free-electron lasers.

While many of the sources mentioned above are still years from completion, R&D efforts for the LCLS X-ray FEL have led to the development of the Sub-Picosecond Pulse Source (SPPS) at SLAC [19, 69–71]. SPPS uses the same linac-based acceleration and electron-bunch-compression schemes as those of future X-ray FELs, but the X-rays are produced through spontaneous emission in a short undulator. The electron bunches are compressed in two stages by first chirping them (i.e., creating a correlation between energy and length along the bunch) and then passing them through a dispersive section consisting of a series of bend magnets such that the back of the bunch catches up with the front. In the first compression stage, the chirp is applied by an RF field with optimized phase with respect to the electron bunch. In the second stage, the chirp on the electron bunch is self-induced by the wakefields it produces acting back on itself. Calculations predict an electron bunch of FWHM 80 fs with a peak current of 30 kA. After the undulator, we measure approximately 10^8 photons/s in a 1.5 % bandwidth at 10 Hz, tunable between 8–10 keV. Recent experiments conducted with this source are described in this Chapter.

One of the key issues with accelerator-based sources is the inability to precisely time an external laser pulse to the X-rays. This is typically not an issue with laser-based sources because the same laser provides both the pump and probe pulses. It is also not typically an issue at third-generation synchrotrons, where, with appropriate care, external lasers can be synchronized to the source to much better than the pulse duration. However, the pulse duration at the SPPS and LCLS is considerably shorter than the level to which an external laser can be synchronized. *Cavalieri* et al. [71] demonstrated that the arrival time of the X-rays could be measured on a shot-by-shot basis using single-shot electro-optic sampling of the ultrarelativistically enhanced field of the SPPS electron bunch.

Figure 1 compares various accelerated sources discussed above in terms of peak brightness and pulse duration. Laser plasma-based sources are considerably less bright than the sources shown here, in part due to their large divergence; however, we note that peak brightness may not always be the most appropriate figure of merit. The oval marked ERL refers to proposed energy-recovery linac schemes [64]. The line represents a line of constant single-pulse time-integrated brightness passing through current third-generation sources, for comparison. One observes that FELs represent many orders of magnitude increase in brightness compared to third-generation sources, even accounting for the shorter pulse duration, and should make possible significant new advances in the field of ultrafast X-ray scattering.

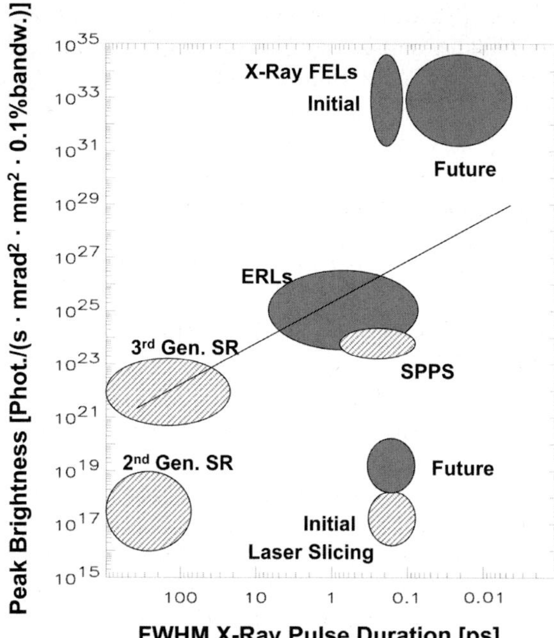

Fig. 1. Peak brightness for various accelerator X-ray sources as a function of pulse duration. Current sources are *hashed* and future sources are *shaded*. Also shown is a line of constant single-pulse time-integrated brightness. Modified from a graph courtesy of H. Winick and H. Nuhn

3 X-Ray Diffraction from Coherent Phonons

In the following sections we will present a number of experiments on picosecond time-resolved X-ray diffraction from crystalline solids. In these experiments an ultrafast laser pulse creates a transient distortion in the crystal. In order to understand what information can be obtained about the ionic motion in these experiments, we first consider the effect of lattice vibrations on ordinary X-ray diffraction.

Recall that X-rays interact with the electronic density of a solid primarily through Thomson scattering and are thus very sensitive to the atomic positions (see, for example, [72, 73]). The scattered electric field, E_{s}, at a distance R from the solid can be considered as a sum of partial waves due to the scattering from the electron densities of the individual atoms (ions) j, that for an incident plane wave, of amplitude, E_0, and angular frequency, ω_0, is

$$E_{\mathrm{s}} = \frac{e^2}{mc^2 R} E_0 \mathrm{e}^{\mathrm{i}\omega_0 t} \sum_j f_j \mathrm{e}^{\mathrm{i}((\boldsymbol{k}_0 - \boldsymbol{k}') \cdot \boldsymbol{r}_j)} \,. \tag{1}$$

Here, $\boldsymbol{k}_0, \boldsymbol{k}'$ are the incident and scattered wavevectors, $\boldsymbol{r}_j$ and f_j are the position and atomic form factor (which depends on energy, momentum transfer and the X-ray polarization) for the jth atom, respectively, and e^2/mc^2 is the classical electron radius. The scattered intensity is $I = (c/8\pi)E_\mathrm{s}E_\mathrm{s}^*$ and involves a double sum over all the ions. For an infinite crystal (ignoring zero-point motion) this intensity is zero except for Bragg diffraction, i.e., in directions where the momentum transfer equals a reciprocal lattice vector,

$$\boldsymbol{k}' - \boldsymbol{k}_0 = \boldsymbol{G} . \tag{2}$$

In the kinematic limit of weak scattering, the scattered field is thus the three-dimensional Fourier transform of the electron density. In the presence of lattice vibrations, the positions of the ions differ from their equilibrium position $\boldsymbol{R}_j$ such that $\boldsymbol{r}_j = \boldsymbol{R}_j + \boldsymbol{u}_j$. The primary effect of thermal vibrations is a reduction in the Bragg-scattered intensity, the Debye–Waller effect. Inelastic scattering from one or more phonons contributes to scattering outside the Bragg peak in what is known as thermal diffuse scattering.

Unlike neutron scattering, inelastic X-ray scattering involves little change in X-ray energy and extremely high resolution spectrometers are required to observe the small energy shifts associated with the absorption or emission of a phonon. In X-ray diffuse scattering, it is not the energy shifts, but the momentum transfer that is measured. Alternatively, in the time domain, the energy shifts from coherent lattice motion can be observed through the heterodyne detection of inelastically and elastically scattered photons at a given momentum transfer. In this case, the difference frequency between X-rays scattered through stimulated Brillouin or Raman scattering and those that are elastically scattered is the phonon frequency. The inelastically scattered X-rays satisfy energy and crystal-momentum conservation. For a given momentum transfer, nonzero scattering occurs when

$$\boldsymbol{k}' - \boldsymbol{k}_0 = \boldsymbol{G} \pm \boldsymbol{q} , \tag{3}$$

where $\boldsymbol{q}$ is the phonon wavevector and the $+(-)$ corresponds to phonon absorption (emission). This leads to sidebands of the Bragg peak, as shown schematically in Fig. 2. For the case where $\boldsymbol{q}$ is parallel to $\boldsymbol{G}$, the angular offset, $\Delta\theta$, of the sideband (relative to the Bragg angle, θ_B) is [74]

$$\Delta\theta = \pm \frac{q}{G} \tan\theta_\mathrm{B} , \tag{4}$$

while the frequency of the inelastically scattered photons is

$$\omega' = \omega_0 \pm \Omega(\boldsymbol{q}) , \tag{5}$$

where $\Omega(\boldsymbol{q})$ is the angular frequency corresponding to a phonon of wavevector $\boldsymbol{q}$.

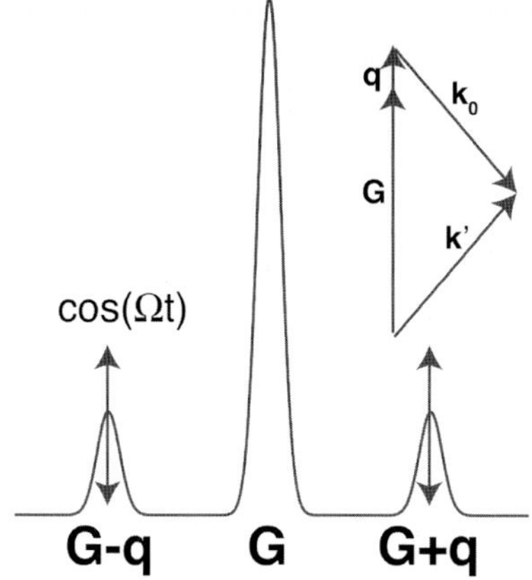

Fig. 2. Schematic of sidebands on a Bragg peak due to a single-phonon mode of wavevector $\boldsymbol{q}$. The central peak is the zero-phonon contribution corresponding to $\boldsymbol{k}_0 - \boldsymbol{k}' = \boldsymbol{G}$. The two subpeaks correspond to scattering that involves the emission or absorption of a single phonon of wavevector $\boldsymbol{q}$. The diagram in the *upper right* shows the phase-matching condition for absorption of a phonon (in this case with $\boldsymbol{q}$ parallel to $\boldsymbol{G}$). The peak height of the sidebands oscillates at the frequency of the phonon

At the detector, we expect to see a signal that is the combination of elastically and inelastically scattered photons. The contribution from a single-phonon mode with angular frequency Ω is

$$\begin{aligned} I(\Omega, t) &= \frac{c}{8\pi} |E_\mathrm{i} + E_\mathrm{e}|^2 \\ &= \frac{c}{8\pi} \left(|E_\mathrm{i}|^2 + |E_\mathrm{e}|^2 + 2|E_\mathrm{i}||E_\mathrm{e}| \cos(\Omega t + \phi) \right) . \end{aligned} \tag{6}$$

Here, $E_{(\mathrm{i})\mathrm{e}}$ is the (in)elastic scattered amplitude, which may itself be time dependent, and ϕ is an overall phase difference. We therefore see that the scattered signal has a term that oscillates with the frequency of the phonon. In practice, we do not pick out a single mode, but a collection, each of which may have a random phase and a different scattering amplitude. Thus, time-domain inelastic scattering, as in the optical case [75], is sensitive to the coherent phonon signal due to a large number of modes with the same initial phase. Note that it is possible for either E_i or E_e to be comparable to E_0, when scattering occurs close to the Bragg peak.

In the case of coherent acoustic phonons and perfect crystals, kinematic diffraction theory is often not sufficient. In this case, the dynamical theory of X-ray diffraction [73, 76] is modified to incorporate the phonons in the form of strain. In the case where the X-rays are nearly phase matched for diffraction from a single set of planes, $H = (hkl)$, *Takagi* [77] and, independently, *Taupin* [78], developed an eikonal approximation for the field amplitudes in the presence of static strain. The electric displacement amplitudes are given by two coupled differential equations,

$$i\frac{\lambda}{\pi}\boldsymbol{\beta}_0 \cdot \nabla D_0(r) = \psi_0 D_0(r) + \psi_H D_H(r)\,, and \tag{7}$$

$$i\frac{\lambda}{\pi}\boldsymbol{\beta}_H \cdot \nabla D_H(r) = \psi_0 D_H(r) + \psi_H D_0(r) - \alpha_H D_H(r)\,, \tag{8}$$

where λ is the X-ray wavelength in vacuum, $D_{0,H}$ are the electric displacement fields inside the crystal, $\alpha_H = -2(\theta - \theta_\mathrm{B}) \sin 2\theta_\mathrm{B}$ is the offset from the Bragg condition, $\psi_{0,H}$ are the (000) and $H = (hkl)$ Fourier coefficients of the periodic electric susceptibility (directly related to the structure factor $F_{0,H}$), and $\beta_{0,H}$ are the wavevectors inside the crystal. For simplicity, we have assumed that the crystal is centrosymmetric. Strain is taken into account by modifying α_H to represent a crystal with a slightly modified lattice spacing. *Wie* et al. [79] give a single differential equation for the ratio of the field amplitudes $x = D_H/D_0$ in the Bragg geometry. Since the index of refraction in the X-ray region of the spectrum is very nearly unity, $D_H \approx E_H$, such that $|x|^2$ is, to a good approximation, the X-ray reflectivity (I/I_0).

Because the difference in the X-ray and phonon energies is so great, it is often an excellent approximation to treat the scattering process as elastic, placing all time dependence ad hoc into the elastically scattered amplitudes. In particular, the Takagi–Taupin approximation is valid when the spatial strain gradient varies slowly compared with the spacing of the diffracting planes (i.e., $|q| \ll |G|$) a condition that is well satisfied for laser-generated strains. In addition, we require that the time derivatives of the field amplitudes be negligible, corresponding to the limits that the phonon frequency is small compared to the X-ray frequency and that the X-ray propagation time through the crystal is small compared to the X-ray pulse duration [80, 81]. The time dependence is reintroduced by solving these equations at different time delays between the X-ray probe and the laser pump with the associated strain propagation.

Things are very different in the case of optical-phonon distortions. For small crystals, or other cases where the kinematic limit to diffraction is valid, the measured intensities will depend on the square of the time-dependent structure factor, $|F_H(t)|^2$. However, for strong scattering from perfect crystals, the usual Takagi–Taupin equations are not valid because the distortions occur within a unit cell. In this case, the time dependence is in the Fourier components of the electronic susceptibilities. Recently, *Sondhauss* et al. [82] and *Adams* [83] have extended the Takagi–Taupin formalism to include optical-phonon distortions.

3.1 Coherent Acoustic Phonons

Time-resolved X-ray diffraction is well suited for investigation of the dynamics of high-frequency coherent acoustic phonons. In general, provided the momentum transfer is large compared to the Darwin width [72, 73, 76], the coherent lattice motion adds sidebands to ordinary Bragg reflection peaks due to X-ray Brillouin scattering. For strong reflections from perfect crystals this corresponds to phonons of GHz frequency. Such sidebands were observed many years ago using acoustoelectrically amplified phonons and a conventional X-ray tube [84, 85]; however, in these experiments there was not sufficient temporal resolution to observe the expected oscillations. An alternate method of producing high-frequency coherent acoustic phonons is through the absorption of femtosecond laser pulses. When a femtosecond laser excites electrons in an opaque material, acoustic phonons are generated in the form of a wavepacket with a spatial extent on the order of the absorption depth of the light. Consequently, the wavepacket frequency components often extend to a fraction of a THz. In this case, we expect a spectrum of sidebands each corresponding to a particular phonon mode. Under appropriate conditions, oscillations of the sidebands can be observed, and details of the phonon generation and propagation can be obtained from the frequency, phase and amplitude of the sidebands.

The physics of photoexcited carrier and lattice dynamics following intense ultrafast laser excitation are complex and intricately coupled. The dynamics occur over a wide range of timescales from subpicosecond to microsecond and longer (see, for example, [86, 87]). Large uniaxial stresses are produced that can easily exceed tens of MPa under modest optical excitation densities. Typically, the stress develops on a timescale much shorter than the acoustic propagation time across the laser absorption depth. As a result, the strain comprises a broad spectrum of coherent acoustic phonons with frequencies extending upward of a THz, limited by electron–phonon coupling. Such ultrafast acoustic pulses have applications in the study of thin films and interfaces as well as bulk materials and have been studied extensively in optical experiments [88–98]. Phonons produced by femtosecond lasers have found use in all-optical measurements of the quality of bonded interfaces [92], thermal conductivity [96, 97] and thermal boundary resistance of solid/solid [91] and solid/liquid interfaces [90]. In each case the laser plays a dual role in both creating the phonons (coherent or incoherent) and detecting them indirectly through changes in the optical properties.

In semiconductors, the phonons are generated by a variety of mechanisms including the volume deformation potential interaction, optical-phonon emission and subsequent decay and Auger heating [99]. Both the electron-hole plasma (through the deformation potential) and the heated lattice (through thermal expansion) contribute to the stress. Consider a cubic crystal. In the limit when the laser-spot dimensions are large compared to the penetration depth, the stress will be approximately uniaxial. Accordingly, the strain will be uniaxial for a (100), (111) or other high-symmetry surface orientation. Let-ting the z-direction be along the inward pointing surface normal, the stress is

$$\sigma_{zz}(z,t) = -3B\beta\Delta T(z,t) - B\frac{\partial E_{\mathrm{g}}}{\partial P}\Delta n(z,t) + \rho v^2\eta_{zz}(z,t)\,, \tag{9}$$

where B is the bulk modulus, β the linear expansion coefficient, T is the lattice temperature, n is the photoexcited carrier density, ρ is the mass density, v is the longitudinal speed of sound and η_{zz} is the resultant strain [89]. The last term is the elastic response of the crystal. The stress and strain are related to the atomic displacements (u_z) by

$$\frac{\partial\sigma_{zz}}{\partial z} = \rho\frac{\partial^2 u_z}{\partial t^2} \quad \text{and} \tag{10}$$

$$\eta_{zz} = \frac{\partial u_z}{\partial z}\,, \tag{11}$$

subject to the boundary conditions that the strain is initally zero everywhere, $\eta(z,0) = 0$, and the stress is always zero at the (free) surface, $\sigma(0,t) = 0$. The stress is time dependent because of thermal and carrier relaxation. Assuming that the time evolution is governed by diffusion, and that Auger recombination instantaneously heats the lattice,

$$\frac{\mathrm{d}n}{\mathrm{d}t} = D_{\mathrm{p}}\frac{\mathrm{d}^2 n}{\mathrm{d}z^2} - An^3 \quad \text{and} \tag{12}$$

$$\frac{\mathrm{d}T}{\mathrm{d}t} = D_T\frac{\mathrm{d}^2 T}{\mathrm{d}z^2} + An^3\frac{E_{\mathrm{g}}}{C_v}\,, \tag{13}$$

where $D_{\mathrm{p}}(D_T)$ is the ambipolar carrier (thermal) diffusivity, A is the Auger recombination rate, E_{g} is the bandgap and C_v is the specific heat. Here, we assume that the carriers thermalize quickly giving their excess energy to the lattice such that, for laser photon energy E_{l},

$$T(z,t=0) = n(z,t=0)\frac{E_{\mathrm{l}} - E_{\mathrm{g}}}{C_v}\,. \tag{14}$$

In the simple case of linear absorption the electron-hole plasma density follows an exponential decay from the surface,

$$n(z,t=0) = \frac{(1-R)F}{E_{\mathrm{l}}\zeta}\mathrm{e}^{-z/\zeta}\,, \tag{15}$$

where R is the reflectivity of the surface, F is the incident fluence, and ζ is the absorption length of the light. If the laser pulse duration and the electron–phonon relaxation time are both short compared to the acoustic transit time across the penetration depth, then a nearly instantaneous stress is set up that leads to impulsive strain generation (a coherent acoustic pulse). In this case,

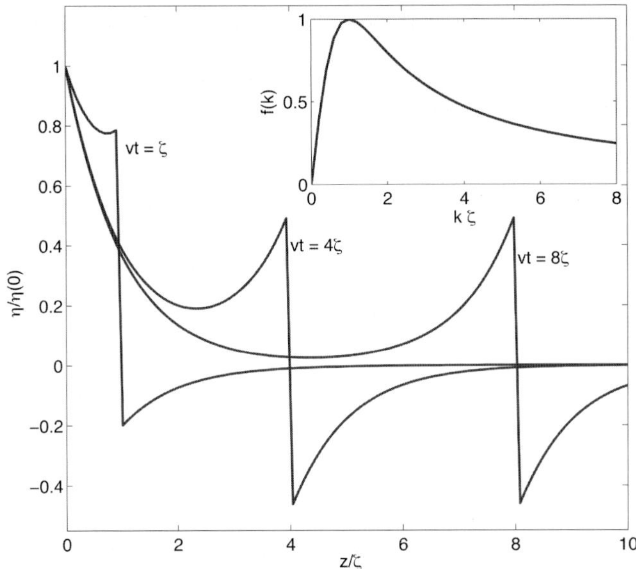

Fig. 3. Normalized strain due to an instantaneous and uniaxial stress that follows an exponential spatial profile. A few representative times are shown. The spectrum of the acoustic pulse, peaked at the inverse of the absorption length, is shown in the inset

the work of *Thomsen* et al. [89] provides a simple analytical solution to the resultant strain for a purely thermal stress, and we have

$$\eta_{33}(z,t) = (1-R)\frac{F\beta}{\zeta C_v}\frac{v^2\rho}{3B}\left[\mathrm{e}^{-z/\zeta} - \frac{1}{2}\left(\mathrm{e}^{-(z+vt)/\zeta} + \mathrm{e}^{-|z-vt|/\zeta}\,\mathrm{sgn}(z-vt)\right)\right]. \quad (16)$$

This equation contains two separate components: a static thermal layer proportional to the instantaneous stress and a bipolar coherent acoustic pulse that travels away from the surface at the speed of sound: see Fig. 3. Note that the spectrum of the acoustic pulse is broad and that it is peaked at the inverse of the optical penetration depth. Deviations from an instantaneous stress lead to a time-dependent surface strain and an asymmetric acoustic pulse with different spectral content. For example, diffusion can significantly alter the shape of the acoustic pulse when $D/v\zeta \gtrsim 1$, i.e., when the diffusion length is comparable to the penetration length in a time given by the propagation of sound across the same distance.

The first time-domain experiment on picosecond coherent acoustic phonons was carried out by *Rose-Petruck* et al. [11] using a single femtosecond laser both to generate a large acoustic strain in GaAs and produce the short-pulse X rays [11]. The strain is produced by the absorption of the laser in a

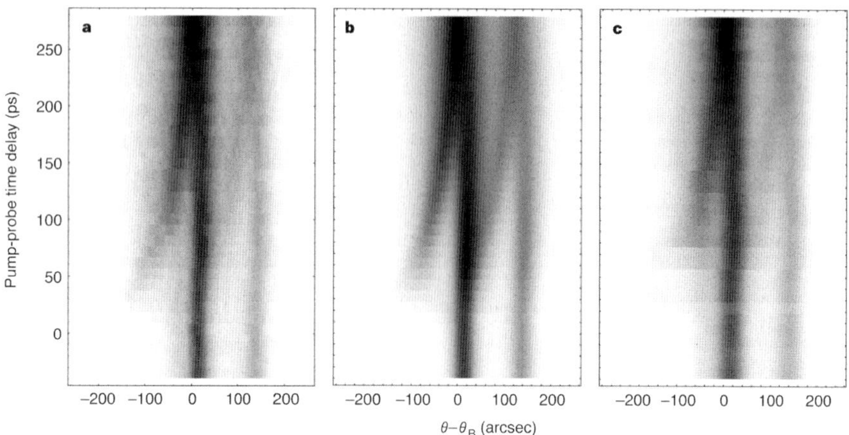

Fig. 4. Experimentally measured (**a**), theoretically calculated (**b**) and genetic algorithm inverted (**c**) time- and angle-resolved diffraction curves from optically excited GaAs [11]

thin region near the surface of the single crystal, as described above, while the X-rays are K-shell flourescence produced by focusing a portion of the laser on a moving copper wire. These X-rays are then diffracted from the strained crystal at various time delays and detected by an X-ray CCD camera. Figure 4a shows the results with 2400 shots accumulated for each time delay. In this figure, one can clearly see diffraction of both the K_α and K_β fluorescence lines from the (111) lattice planes modified by the large acoustic response of the material. Figure 4b shows theoretical calculations of the diffraction profile using the Takagi–Taupin equations and the phenomenological model for the strain generation and propagation put forward by Thomsen et al. Qualitatively, there is excellent agreement with the thermoelastic model. The authors also compared other strain profiles, using an iterative genetic-algorithm to best fit the data. The calculated diffraction pattern for the best results is shown in Fig. 4c. While the resultant strain produced a calculated diffraction profile that was similar to both the data and the theoretical calculations, the strain profile was significantly different from that predicted by Thomsen et al. in that it comprises a static surface expansion and a unipolar tensile pulse without the associated compression.

The previous results, while consistent with the production of a coherent acoustic pulse did not show temporal oscillations in the X-ray reflectivity that one could have expected from the arguments we previously presented. This is most likely due to the mismatch between the large penetration depth of the near-infrared laser in GaAs and the finite penetration depth of the X-rays. In a similar experiment by *Lindenberg* et al. [13] using synchrotron X-rays, and laser-excited InSb, temporal oscillations were observed. These experiments utilized a picosecond X-ray streak camera triggered by a photoconductive switch [56] in order to resolve temporal features that occurred within the diffraction of a single $\sim$ 80 ps pulse. The Ti:sapphire laser used to trigger the streak camera and excite the crystal was synchronized to $\sim$ 5 ps with an individual electron bunch in the Advanced Light Source storage ring. Because of the small bandgap, the penetration depth of InSb is $\sim$ 100 nm at the excitation of 1.55 eV used in the experiment. In an attempt to better match the X-ray penetration depth and that of the laser, an asymmetric geometry was used in which the X-rays exit the crystal at 3° [13]. The results show striking oscillations corresponding to coherent acoustic-phonon sidebands, as can be seen in Fig. 5. The data are taken at different angles relative to the Bragg peak, corresponding to a compressed lattice spacing. From the measured oscillation frequencies as a function of angle, the phonon dispersion can be determined. Within experimental error, *Lindenberg* et al. found a linear dispersion with a slope corresponding to the longitudinal speed of sound in the [111] direction (3900 m/s), as expected for a coherent acoustic-phonon pulse.

The results were compared with calculations based on dynamical diffraction theory, as presented earlier. The authors found that their results could be accounted for by a two-component stress, an instantaneous component attributed to photoexcited carriers and a delayed component attributed to thermal stress (due to the finite electron–phonon coupling time). As can be seen in Fig. 5, the calculations are a good match to the data. The measured strain corresponds to a 0.17 % thermal and 0.08 % nonthermal component within a 100 nm attenuation length, and with an electron–phonon coupling time of 12 ps. It was not necessary to include the effects of carrier or thermal diffusion to match the experiment.

Reis et al. [16] also looked at impulsive-strain generation in InSb using hard X-rays from the Advanced Photon Source synchrotron. As with other synchrotron experiments, the excitation laser is synchronized to a single X-ray pulse in the ring. Unlike in *Lindenberg* et al. [13], where an X-ray streak camera was used, Reis et al. made use of an avalanche photodiode for pump-probe experiments. In this experiment, a symmetric X-ray diffraction geometry was chosen where the lattice planes are parallel to the (111) surface. Because the X-ray penetration depth (which is limited by the shorter of extinction or absorption) is much longer than the optical penetration depth, the strain can be followed as it propagates into the bulk of the crystal. Although the temporal resolution is worse than in previous experiments, this allows for a more complete view of the strain dynamics and for the surface and bulk contributions of the strain to be separately resolved. Because of the symmetric geometry, the data revealed both the compression and expansion associated with the strain wave. Figures 7a and d show the results for two diffraction peaks: the strong (111) and nearly forbidden (222). A two-dimensional fast Fourier transform of the data shows clear signs of acoustic-phonon dispersion.

Dynamical-diffraction calculations were performed assuming that the strain results from thermoelastic effects. The data are best fit with a peak

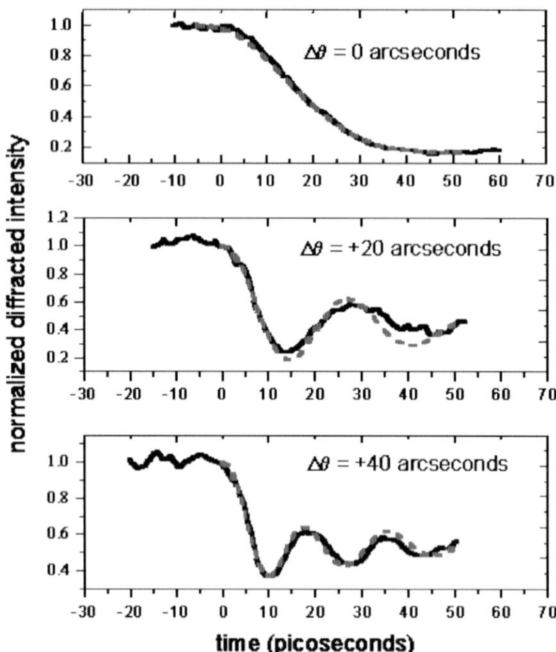

Fig. 5. Experimentally measured (*solid line*) and simulated (*dashed line*) time-resolved diffracted intensity at crystal angles of 0, 120, and 140 arcsecond from the InSb (111) Bragg peak [13]

strain of 0.51 % across a 100 nm absorption depth and a 0.51 % peak-to-peak acoustic pulse. The effects of a delayed stress, as seen by *Lindenberg* et al. and others [10–12], could not be resolved with the 100 ps resolution of these experiments. The results are presented in Fig. 7. Even though the theoretical expressions qualitatively reproduce many of the features of the data, Fig. 8 shows that they significantly underestimate the fraction of the strain in the coherent acoustic pulse. In fact, a least squares fit to the data that allowed for separate partitioning into the surface and bulk contributions to the strain gives a 0.39 % surface strain with a nearly doubled 0.76 % peak-to-peak acoustic pulse. It is likely that this discrepancy is due to photoexcited carrier-induced stress.

For the most part, time-resolved diffraction experiments concentrate on features in the vicinity of strong Bragg reflections and, as such, they are limited to the study of strain within the near-surface region defined by the X-ray extinction depth. This is typically $\sim 1\,\mu$m but can be considerably smaller for asymmetric geometries. A quasiforbidden reflection such as (222) in InSb (and other zincblende materials) increases the probe length by an order of magnitude or more to be limited by X-ray absorption. However, the diffraction efficiency of these reflections is small, as is the spectral acceptance,

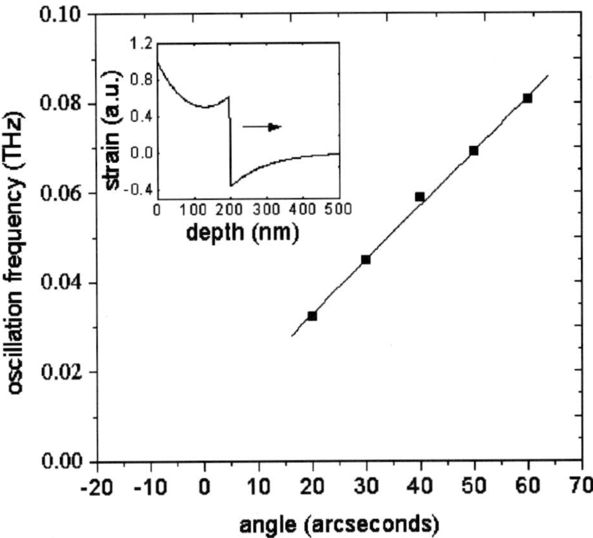

Fig. 6. From [13] Experimentally observed oscillation frequency of the diffracted intensity as a function of crystal angle for laser-excited InSb. The *solid line* is a fit to the data. *Inset*: Calculated laser-induced strain profile, which propagates at the sound velocity into the material

such that it is difficult to make precise measurements of the strain with X-ray fluorescence lines or conventional double-crystal monochromators.

A major advantage of synchrotron-based sources is that arbitrary time delays can be achieved by purely electronic means because the laser pump and the X-ray probe are independent (the mechanical delay stage required to optically separate two pulses to introduce a variable delay from picoseconds to microseconds or longer would be too long to implement). At the same time, this can also be a major disadvantage of accelerator-based sources, for there is a lack of intrinsic synchronization between the pump and probe that can lead to reduced temporal resolution due to timing jitter [71]. Figure 9 from *DeCamp* et al. [33] shows the X-ray diffraction intensity as a function of time following an impulsive laser excitation of the surface of a 280 μm thick single-crystal Ge (001) sample. The maximum time delay exceeds a microsecond while maintaining the $\sim$ 100 ps resolution limit of the X-ray pulse duration. The sharp temporal peaks correspond to diffraction when the acoustic pulse is traveling within the probe region of the X-rays diffracting from the (004) lattice planes. The timed delay between subsequent peaks corresponds to the roundtrip propagation through the crystal (reflecting off the highly polished rear surface and back), $\Delta t = 2L/v = 110\,\mathrm{ns}$, where L is the crystal thickness and $v \approx 5000\,\mathrm{m/s}$ is the longitudinal speed of sound. One can easily make out the contribution to the time-resolved diffraction from the surface component of the strain as the slowly varying background that decays over hundreds

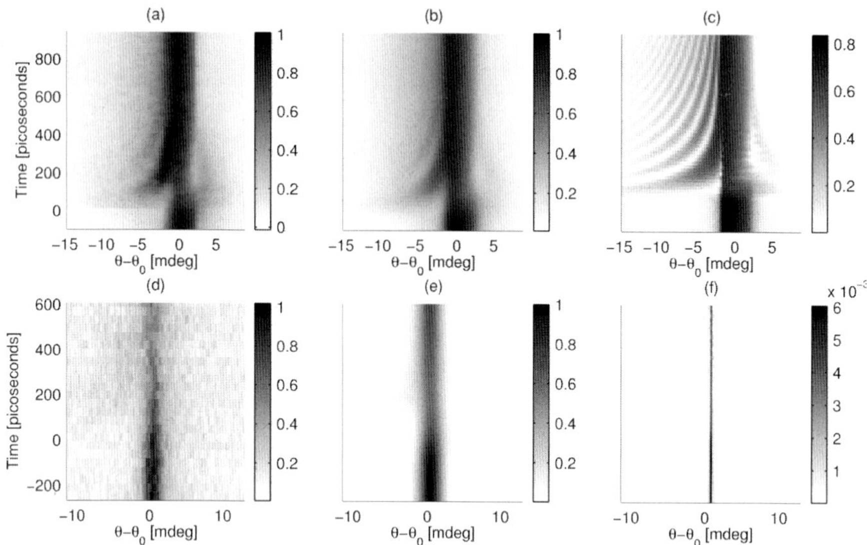

Fig. 7. Time-resolved rocking curves of impulsively strained InSb. (**a**),(**d**) Data and (**b**),(**e**) dynamical diffraction simulation for a 0.51 % peak strain with a 100 nm laser absorption depth and a 100 ps, 1.25 mdeg (2.5 mdeg for the (222)) FWHM Gaussian resolution functions (normalized to the peak unstrained signal); and (**c**),(**f**) dynamical diffraction simulation for the reflectivity of a monochromatic X-ray pulse with ideal temporal resolution for the (111) and (222) reflections, respectively [16]

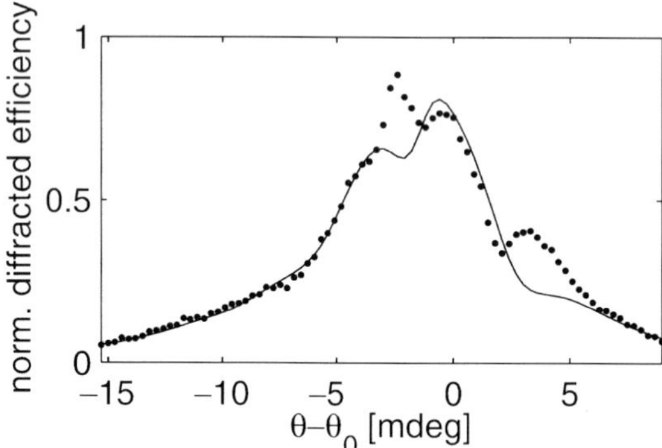

Fig. 8. Rocking curve of the (111) peak at 150 ps time delay (normalized to the peak unstrained value). The *solid circles* are the data and the *line* is the simulation. The discrepancy is ascribed to an excess population of coherent acoustic modes [16]

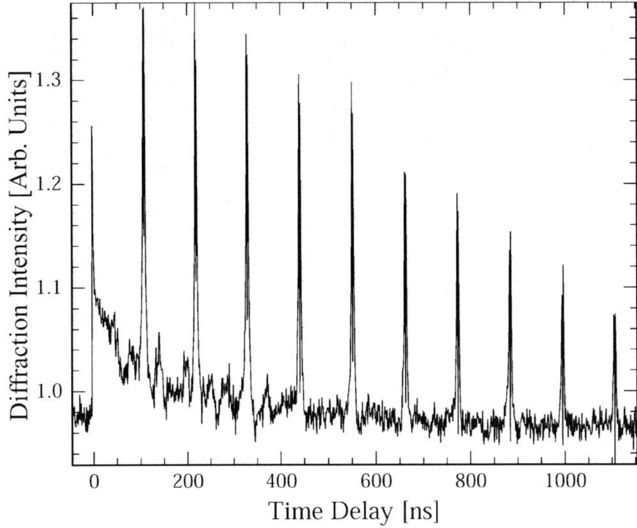

Fig. 9. Relative reflected intensity as a function of time delay, for the symmetric (004) Bragg peak in Ge following an impulsive excitation [33]

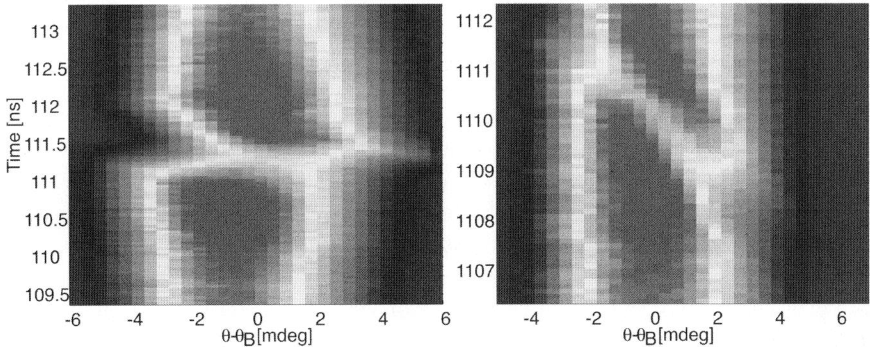

Fig. 10. Diffraction image of an acoustic pulse in Ge after propagating one pass through the 280 μm crystal (*left*) and 10 passes through the crystal (*right*) [33]

of nanoseconds due to thermal diffusion. The contribution of the surface strain can be removed by measuring the time-resolved diffraction pattern due to the acoustic pulse off the opposite face of the crystal. In this geometry, *DeCamp* et al. obtained detailed time- and angle-resolved diffraction patterns that show the evolution of different frequency components of the strain pulse leading to significant pulse broadening [33], as seen in Fig. 10.

All of the experiments listed above were performed in the Bragg or reflection geometry for which the diffracted X-rays are directed away from the crystal. However, it is also possible to have the diffracted beam directed such that it propagates through the bulk crystal. This is known as the Laue geometry [72, 73] (not to be confused with the Laue or white-beam diffraction technique in which a broad spectrum of X-rays is used to simultaneously meet the Bragg condition for numerous sets of diffracting planes). In general, photoelectric absorption will limit the transmission to relatively thin crystals. High-energy X-rays can mitigate these effects because the absorption cross section decreases with energy (far from an absorption edge). In this geometry, *Liss* et al. observed X-ray diffraction from MHz frequency acoustic waves using 30 keV X-rays [100].

In the case of strong diffraction, the X-rays can still propagate through many ordinary (incoherent) absorption lengths in what is known as the Borrmann effect, or X-ray anomalous transmission [73, 76]. Here, two diffracted beams are produced, the forward diffracted, or 0-beam, in the direction of the incident X-rays and the deflected diffracted, or H-beam, in the direction given by Bragg's law (2), as shown schematically in Fig. 11. The amplitude of the forward and diffracted beams do not represent the eigensolutions for the electric field inside the medium. When the boundary conditions are such that a single diffraction condition is met, the solutions to Maxwell's equations in a periodic media are two waves that propagate along the lattice planes with a transverse profile corresponding to mutually orthogonal standing waves. This is the essence of dynamical-diffraction theory. The only difference between this and the Bragg case are the particular solutions that are excited, as determined by the boundary conditions. It is also formally equivalent to electron band theory except that the electric and magnetic fields are vector quantities [101]. The solutions are analogous to the symmetric and asymmetric wavefunctions and the opening of a bandgap. In fact, the width of the Bragg peaks is determined by the Fourier coefficients of the electron density in much the same way that the bandgaps are determined by the Fourier coefficients of the periodic crystal potential in which the electrons move.

In the Borrmann effect, the wavelength of the transverse standing waves is twice the planar spacing. In the case of strong diffraction, one of the solutions, α, has nodes in regions of high electron density and the other, β, in the regions of low electron density. Since the dominant absorption process is photoelectric, the α solution has low absorption (the anomalous transmission solution) whereas the β solution has enhanced absorption. The two modes α and β oscillate in and out of phase as they propagate through the crystal. The wavelength of the interference, $\Lambda = |\boldsymbol{k}_\alpha - \boldsymbol{k}_\beta|^{-1}$, is known as the Pendellösung length that is typically a few micrometers and is often shorter than the absorption length. The external, forward and diffracted, amplitudes are linear combinations of these two solutions. In the case of a thick crystal (compared to the β absorption length), the output beams are both derived from the α solution and, except for possible geometrical differences, they are identical. In the case of a thin crystal, the relative diffraction efficiency of the two beams depends on the phase difference between the two modes at the exit face.

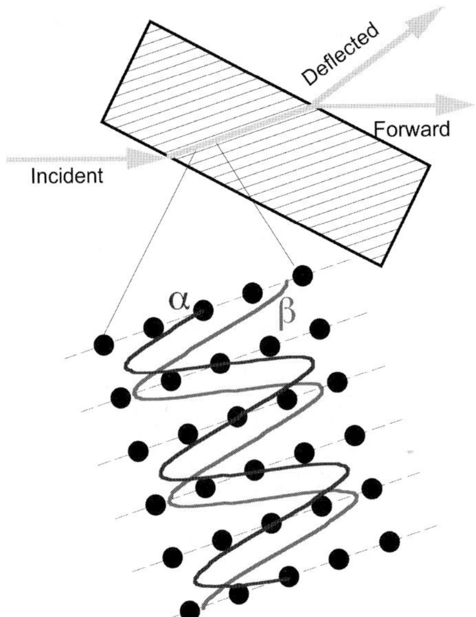

Fig. 11. Schematic of Borrmann Effect, adapted [17]

Anomalous transmission is extremely sensitive to the crystalline quality. Borrmann's original experiments demonstrated the loss of X-ray anomalous transmission due to a small thermal gradient [102]. In these experiments, a thick calcite crystal was exposed to a temperature gradient of 0.2 °C/cm that was sufficient to remove the anomaly. The thermal strain was enough to disrupt the X-ray wave-guiding effect causing complete absorption. More recent experiments show that acoustic waves can spatially modulate [103] and strongly attenuate [85, 104, 105] the anomalous transmission. This also applies to lattice dislocations and other deffects [106, 107]. These distortions mix the α and β solutions, an effect that has been used as a means to control the relative diffraction efficiency into the forward and diffracted beams [17].

In these experiments by *DeCamp* et al. [17], a laser-induced acoustic pulse produces a localized disturbance in a thick Ge crystal, as was discussed earlier in the Bragg experiments on multiple acoustic reflections. In this case, the acoustic pulse modulates the ratio of the two anomalous transmission beams for the asymmetric $(20\bar{2})$ diffraction condition by coupling the α and β solutions. The relative time delay between the X-rays and the acoustic pulse is used as a means to control the exit beams to preferentially switch energy from one beam to another, to modulate the two beams together, or even to enhance the anomalous transmission, as shown in Fig. 12. Before $t = 0$, the energy in the two beams is mainly due to the α solution. At $t = 0$ approximately 75 % of the energy in the forward beam is switched to the deflected beam. The boundary conditions determine whether the switching

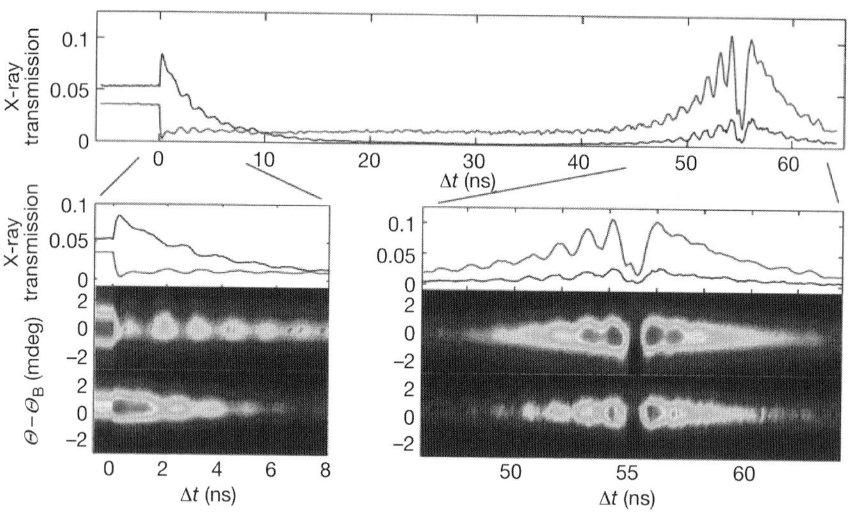

Fig. 12. Time-resolved diffraction measurement of the $(20\bar{2})$ asymmetric Laue diffraction peak following laser-impulsive excitation of an acoustic wave on the exit face of a thick Ge (001) crystal. *Top*: Forward and deflected X-ray intensity as a function of time following excitation. *Bottom*: expanded scale showing the regions where the acoustic pulse is near the exit or entrance crystal surface as well as time- and angle-resolved diffraction images [17]

occurs between the forward and deflected, or vice versa, as would be the case if the $(\bar{2}02)$ diffraction conditions were met. Following the initial switch, Pendellösung oscillations are apparent in both the forward and deflected beams. The oscillations appear to be 180° out of phase with each other and occur with a frequency of ~ 1 GHz. This frequency is consistent with the passage of an acoustic impulse that redistributes energy from the α wave to a linear combination of α and β as it moves along the perpendicular to the (001) surface, at the speed of sound. These oscillations decay with a time constant ~ 5 ns, determined by the absorption length of the β solution and the speed of sound. At approximately 55 ns the acoustic pulse approaches the X-ray input face of the crystal. At this point, the diffraction efficiency increases because the initial β wave has not yet decayed. As a result, more than 50 % of the input intensity couples to the α wave. What is left of the β wave will be nearly completely attenuated upon propagation through the bulk of the crystal. Thus, the Pendellösung oscillations are in phase because the exit beams are now determined solely by the α solution, as modified at the interface.

The initial transient shown in Fig. 12 occurs too fast to be accounted for by an acoustic wave produced only by instantaneous stress. The authors suggest that the strain may be modified significantly by the ambipolar diffusion of the dense electron-hole plasma. In a follow-up experiment, *DeCamp* et al. [24] used the time-resolved Borrmann and Pendellösung effects to study the details of the strain-generation process. Deviations from purely impulsive-strain generation have been seen in a number of X-ray diffraction experiments including [12,13,16,25]. In particular, *Cavalleri* et al. [12,25] studied coherent strain near the melting threshold in Ge utilizing X-ray Bragg diffraction. They concluded that the strain is produced over a region that is thick compared to the optical penetration depth, likely due to ambipolar diffusion. However, their experiment was only sensitive to structural changes in the near-surface region due to the short X-ray extinction depth. In the Laue geometry, deviations from impulsive-strain generation are evident in the relative phase of the Pendellösung oscillations and the amplitude of the transient, providing information about the strain-generation process at times shorter than the X-ray probe duration. By studying these effects at different excitation densities and comparing the results to dynamical diffraction simulations, DeCamp et al. were able to demonstrate that the development of the coherent strain pulse is dominated by rapid ambipolar diffusion and coupling to the lattice by the deformation potential. Because the electron-hole plasma diffusion is not limited by the sound speed of the material, the strain *front* initially advances at a speed greater than that of sound. The period of the Pendellösung oscillations is unchanged as the resulting acoustic pulse travels into the crystal bulk at the longitudinal sound speed. While electron-hole plasma dynamics have been studied extensively in ultrafast optical experiments such as in [108–116], this experiment represents a true bulk measurement of the diffusion.

In addition to studying bulk properties, ultrafast X-ray diffraction is also a useful tool to probe dynamics on the nanoscale [12, 22, 23, 29, 117, 118]. The physics of nanoscale thermal transport is of fundamental and practical interest where interfaces play an especially important role [98]. Ultimately, phonons transfer heat between materials and their scattering is particularly sensitive to the details of interfaces [119, 120]. Early studies of phonon propagation across a superlattice used superconducting tunnel junctions as the phonon source and detector [121]. Recently, femtosecond lasers have found use in nanoscale thermal transport measurements as the source and detection of phonons. Examples include the all-optical measurements of thermal conductivity [96, 97] and thermal boundary resistance of solid/solid [91] and solid/liquid interfaces [90]. In contrast to all optical techniques, X-ray methods yield quantitative information about atomic positions. In this approach, the laser is used as the source of the phonons and picosecond X-ray diffraction is the probe. In the case of thin films and interfaces, picosecond diffraction has been used to study phonon damping [12], the generation of folded acoustic modes in superlattices [29, 117] and in the reflection and transmission of phonons across a single heterostructure interface [118].

In the novel experiments of *Lee* et al. [118], picosecond X-ray diffraction was used to study the propagation of acoustic phonons across the nearly ideal $Al_{1-x}Ga_xAs$/GaAs interface. As with previously discussed experiments, a femtosecond laser creates a coherent acoustic wavepacket. However, because the particular alloy of the $Al_{1-x}Ga_xAs$ film ($x = 0.3$) was chosen to be trans-

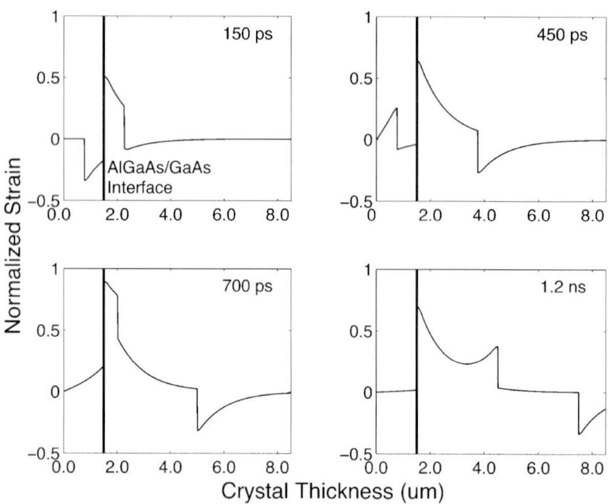

Fig. 13. Simulations of a the coherent acoustic wavepacket produced by impulsive stress localized on the bulk side of a heterostructure interface [118]

parent to the laser, the strain generation is localized at the GaAs substrate. In this case, the boundary conditions that give rise to the generation of a bipolar acoustic pulse (Fig. 3) no longer hold; instead the solutions to the elastic equations give the pulse shown in Fig. 13.

Even though the lattice constants of the $Al_{1-x}Ga_xAs$ and GaAs are very close, high-resolution X-ray diffraction can distinguish between the bulk and the substrate, as can be seen by the two peaks of the static rocking curve on the left of Fig. 14. Note that the fringes, due to a finite thickness and interference between the wings of the $Al_{1-x}Ga_xAs$ and GaAs peaks, are predicted by the dynamical theory of X-ray diffraction. Because the two peaks are well separated, it is possible to use X-ray diffraction as a means to isolate and measure the strain in either the film or the bulk. On the right-hand side of Fig. 14, we show slices of the time-resolved X-ray diffraction from the high- and low-angle side of the $Al_{1-x}Ga_xAs$ peak and the low-angle side of the GaAs peak. The data are consistent with dynamical-diffraction calculations for two counterpropagating unipolar compression pulses, as shown in Fig. 13, and with an acoustic-mismatch model of the heterostructure and film/air interfaces [119, 121]. These interfaces represent the extreme limits of a near-perfect acoustic match between the film and substrate (near unity transmission) and a near-infinite acoustic mismatch between the film and air (near unity reflection, with a π phase shift). It is also important to note that the data are consistent with a strain profile produced predominantly by an impulsive stress. This is to be expected because diffusion is slow on the timescale of the experiments and the photoexcited carriers were confined to the substrate.

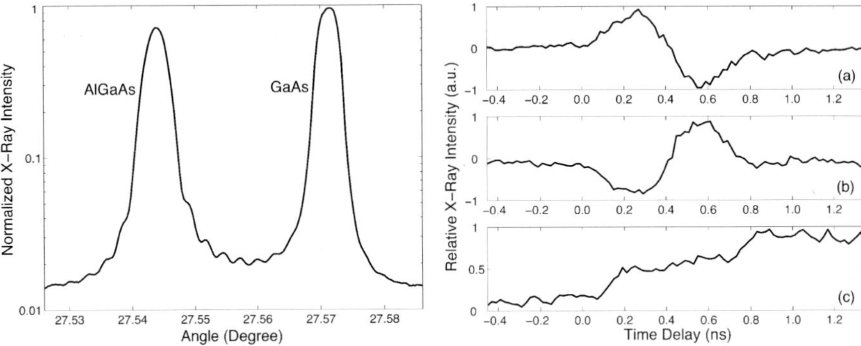

Fig. 14. (*Left*) High-resolution diffraction image from the (004) Bragg peak of a single AlGaAs/GaAs heterostructure, showing the individual peaks of the AlGaAs film and the GaAs substrate. (*Right*) Relative time-resolved X-ray diffraction intensity for three different angles: (**a**) 27.546 0, on the high-angle side of the AlGaAs peak; (**b**) 27.541 5, on the low-angle side of the AlGaAs peak; and (**c**) 27.569 0, on the low-angle side of the GaAs peak [118]

Experiments using shorter-wavelength, higher-frequency coherent phonons, could provide frequency-dependent reflection and transmission coefficients. Here, multiple quantum wells or metallic films can be used as the source. In addition to providing further insight into the strain-generation process, the added specificity of X-ray diffraction combined with the ability to measure phonons at a given wavevector, as well as measure the local temperature, should prove a useful tool in measuring frequency-dependent thermal and mechanical properties of thin films and interfaces. For complete information of the phonon dynamics, time-resolved diffuse scattering holds tremendous promise [122].

3.2 Coherent Optical Phonons

We have seen that time-resolved X-ray diffraction is a particularly sensitive probe of coherent motion. Most of the ultrafast experiments to date have centered around coherent acoustic modes in bulk materials, although the technique has recently been used to study folded acoustic-phonon modes in an artificial superlattice [29, 117]. Conspicuously, the literature is almost devoid of studies of coherent optical modes in bulk materials. Optical modes are, of course, fundamental in carrier-relaxation processes and are known to play an important role in various phase transitions [123]. There are two main difficulties in the direct observations of the atomic motion associated with coherent optical phonons. One is the fact that the typical phonon period is considerably shorter than even the shortest X-ray sources. The other is that typically the phonon amplitudes are very small, making it difficult to separate the effect of near-zone-center phonons from the usual diffraction.

This is not to say that there hasn't been significant interest in X-ray scattering from optical phonons. There have even been proposals to use optical phonons as the basis of an ultrafast X-ray switch, in which a Bragg mirror has its reflectivity switched on and off on an ultrafast timescale. One advantage of such a device would be that it could be used as an *insertion device* to an existing X-ray beamline at a synchrotron to significantly shorten the X-ray pulse duration for time-resolved experiments. This idea was proposed by *Bucksbaum* and *Merlin* [63]. Their phonon Bragg switch would work by setting up a relatively large wavevector phonon field using a pair of ultrafast laser pulses. The phonons are generated by impulsive stimulated Raman scattering with a wavevector sufficiently large to put sidebands on the Bragg peak for a perfect crystal. The periodicity of the phonon field produces a superlattice on top of a crystal lattice, producing sidebands on the ordinary X-ray diffraction similar to that seen in the case of the acoustic-phonon scattering. By tuning the crystal to a sideband, a high contrast can be achieved between the elastically scattered X-rays in the tail of the Bragg peak and the phase-matched inelastically scattered X-rays. In order to make a switch, a second set of laser pulses is timed such that the coherent-phonon amplitude is completely cancelled. In this manner, the reflectivity of the Bragg mirror can be switched on and off in roughly half a phonon period. In the case of GaAs and for tranverse optical phonons, this can be as short as 50 fs (in principle, shorter pulses can be made, using laser pulses of unequal amplitude; albeit at a reduced efficiency) and as long as the phonon lifetime. It has been noted that the anomalous transmission experiments of *DeCamp* et al. [17] (using coherent acoustic phonons) can also be scaled to optical-phonon excitations and may offer a number of advantages over the Bragg geometry. Recently, *Nazarkin* et al. [124] proposed that high-order Raman scattering in this geometry would lead to even faster switching. As noted previously, in calculating the effect of an optical phonon field on dynamical X-ray diffraction, the usual Takagi–Taupin equations do not hold and must be modified [82, 83].

The first experiments on time-resolved X-ray diffraction from coherent optical phonons were performed by *Sokolowski-Tinten* et al. on photoexcited bismuth [30]. This system is well known in that it exhibits very large amplitude and relatively low-frequency oscillations. Bismuth is not well suited for the Bragg switch described above because it is difficult to grow high-quality single crystals and the photoexcited phonons are typically produced in a very small penetration depth. However, bismuth is an almost ideal system for studying a number of high-density excitation effects. Despite the fact that high-amplitude coherent-phonon generation in bismuth has been well studied using ultrafast laser pump-probe techniques techniques [125–133], there is much controversy over the very strong softening observed in the experiments [134, 135]. In a number of experiments, the authors have claimed to probe anharmonicity in the form of an amplitude-dependent phonon frequency [130], quantum effects such as collapse and revival [131] and

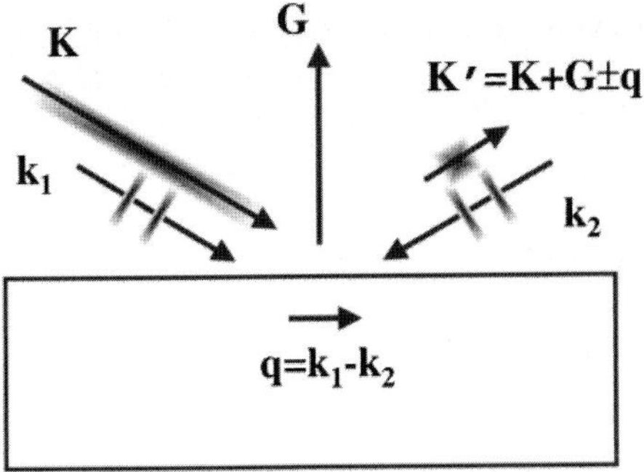

Fig. 15. Geometry for the Bragg switch. Optical pulse trains with wavevectors $\boldsymbol{k}_1$ and $\boldsymbol{k}_2$ turn on and off a transient standing-wave phonon lattice with periodicity $2\pi/|q|$. An ultrafast slice of the X-rays with incident wavevector $\boldsymbol{K}$ can Bragg scatter from the superlattice if it satisfies the Bragg condition $\Delta\boldsymbol{K} = G \pm q$ [63]

even Bose–Einstein condensation [132]. *Murray* et al. [133] appear to have resolved this controversy showing both theoretically and experimentally (using two-pulse coherent control) that the observed effects are due primarily to electronic softening of the interatomic bonds and the complex dynamics of the photoexcited electron-hole plasma. While certainly present, anharmonicity plays a minor role in the motion of the ions for the optically generated amplitudes. X-ray diffraction could play an important role in studying the complex dynamics because of its sensitivity to the atomic positions.

The structure of bismuth is rhombohedral (A7), which is a Peierls distortion of the simple cubic structure, with two atoms per unit cell [136]. The three zone-center modes transform like the A_{1g} mode and the doubly degenerate E_g mode. Optical excitation of electron-hole pairs by an ultrafast laser pulse generate large-amplitude coherent A_{1g} phonons by displacive excitation [137], which has recently been shown to be a special case of stimulated Raman scattering involving the imaginary component of the susceptibility [138]. The A_{1g} mode corresponds to the beating of the two atoms in the unit cell against each other, resulting in a modulation of the X-ray structure factor. In this case, the hklth Fourier component is:

$$F_{hkl}(t) = 2 f_{\mathrm{Bi}} \cos\left(2\pi\left(h+k+l\right)\left(\tfrac{1}{4} - \delta(t)\right)\right), \tag{17}$$

where f_{Bi} is the scattering factor of a bismuth ion and δ is the fractional displacement of each ion along the trigonal axis with respect to the simple cubic case. The equilibrium position at room temperature has $\delta = 0.0159$, and we see that the structure is very slightly distorted.

In the experiments by *Sokolowski-Tinten* et al. [30], Ti K_α radiation at $\lambda = 2.74$ Å is produced at 10 Hz using a laser-produced plasma source. These X-rays were focused onto a thin bismuth film using a toroidally bent silicon

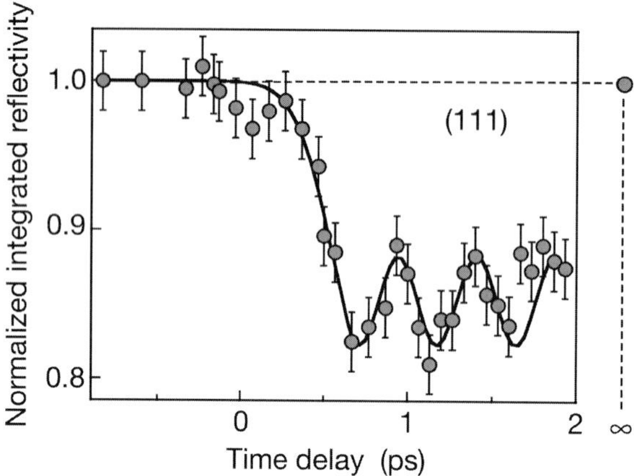

Fig. 16. Time-resolved X-ray diffraction from coherent A_{1g} phonons in bismuth as seen by the (111) Bragg peak, from [30]

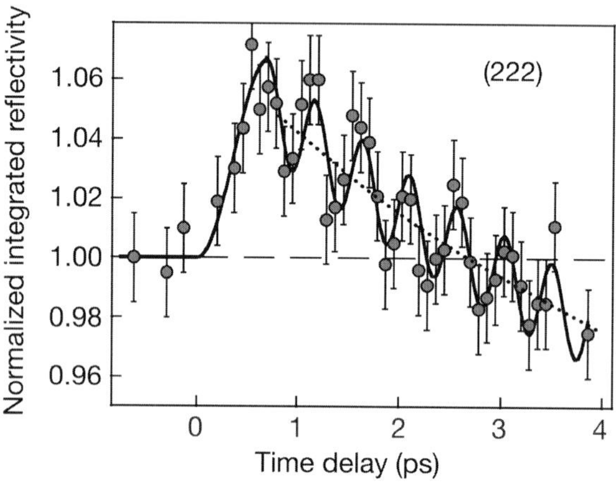

Fig. 17. Time-resolved X-ray diffraction from coherent A_{1g} phonons in bismuth as seen by the (222) Bragg peak. The slow decay was attributed to the Debye–Waller effect, i.e., the reduction of the X-ray intensity due to an increase in the root-mean squared displacement of the ions [30]

crystal. Because the X-ray penetration depth is much longer than the optical penetration depth ($\sim$ 10–20 nm), these authors used a thin film to restrict the X-ray probe region. The bismuth film used in these experiments was grown epitaxially on (111) silicon and had its c-axis perpendicular to the surface. The film thickness was 50 nm, such that the X-rays still see a nonuniform excitation density. Time-resolved X-ray diffraction following laser-excitation (800 nm, 100 fs, 6 mJ/cm^2) are shown in Fig. 16 and Fig. 17 for the (111) and (222) Bragg peaks. As can be seen from (17), the (111) peak becomes

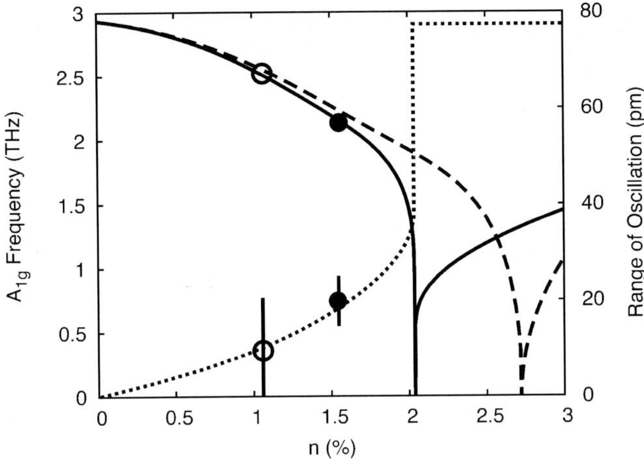

Fig. 18. Comparison of experimental results with calculations, modified from [133]. The *solid line* shows the phonon frequency including the effects of anharmonicity as a function of per cent valence exciation, while the *dashed line* shows only the harmonic frequency. The *dotted line* shows the full range of motion. The *filled circle* is the results of [30] for a softening to 2.12 THz. The *open circle* is a crude estimate from typical optical data showing a peak softening of $\sim$ 2.5 THz

completely forbidden in the case of a simple cubic crystal ($\delta = 0$), whereas the (222) attains its maximally allowed value. Because the films are thin, the diffraction occurs in the kinematic limit where the diffracted intensity is proportional to the square of the structure factor.

The data are consistent with the scenario whereby the optical excitation suddenly drives the bismuth towards a higher-symmetry structure by removing the Peierls distortion. The ions oscillate coherently about their new equilibrium positions at a significantly softened frequency (2.12 THz, compared to the 2.92 THz for the low-amplitude mode). The phonon amplitude was estimated to be 5–8 % of the nearest-neighbor separation, with the uncertainty limited by the depth-dependent excitation and the finite duration of the X-ray probe (estimated at 300 fs). At higher fluences, corresponding to motion of more than 10 % of the nearest-neighbor separation, no oscillations were observed. The strong reduction of the X-ray intensity was attributed to a loss of long-range order and most probably melting.

The softening seen in both optical experiments [125–133] and the X-ray experiments of *Sokolowski-Tinten* et al. [30] were compared with density-functional theory calculations by *Murray* et al. [133]. The full results of these calculations are shown in Fig. 18. In this figure, we indicate the results of the X-ray and typical optical experiments. For optical experiments, which cannot measure the atomic displacement directly, one can make a crude estimate based on the changes in the reflectivity [129]: a 1 % relative change corresponds approximately to a displacement of 10 pm. For the X-ray experiment, the observed softening of 2.12 THz coincides with a carrier density corresponding to excitation of 1.5 % of the valence electrons. At this density, the calculations indicate that the amplitude of motion would correspond to 18 pm, which is consistent with the measured amplitude of 20 ± 5 pm. At the excitation densities of the optical experiments the effect of anharmonicity on the phonon period is expected to be negligible. Even for the strong softening seen in the X-ray experiments, Murray et al. expect that anharmonicity will manifest itself in the phonon frequency at the few per cent level, which is much smaller than the nearly 30 % effect on the frequency due to electronic softening.

4 X-Ray Studies of Ultrafast Phase Transitions

4.1 Nonthermal Melting

The development of X-ray probes with femtosecond time resolution have led to the capability of capturing the first dynamical steps during structural phase transitions. Here, one is interested in not just the initial and final states, but in the pathway the atoms follow, the transition states, and associated timescales. In this context, a new class of "nonthermal" processes governing the ultrafast solid-liquid melting transition have recently emerged, as supported by time-resolved optical [111, 139–141] and X-ray [15, 18, 21] experiments, and with technological applications ranging from micromachining to eye surgery [142]. Femtosecond excitation of direct-bandgap semiconductor materials, in particular III-V polar semiconductors, is usually understood in terms of a phenomological model involving the initial excitation from valence to conduction band followed by carrier relaxation involving many optical-phonon emission processes (see [141, 143] for good reviews on this topic). However, theoretical work [144–150] also indicates that excitation of a dense electron-hole plasma leads to accompanying dramatic changes in the interatomic potential, i.e., softening of the potential-energy surface and resulting large-amplitude atomic motion on timescales shorter than the electron–phonon coupling times [10, 151–153].

Advances in the understanding of these phenomena have followed the development of new sources and new ways of probing. The first optical experiments, carried out by *Shank* and coworkers [109, 139] measured the time-resolved reflectivity or second-harmonic generation efficiency following intense femtosecond excitation. Changes in the optical reflectivity and reflected second harmonic (expected to vanish during the transition from crystalline to centrosymmetric liquid state) provided the first indications, albeit indirect, of a transition to the liquid state on timescales faster than electron–phonon coupling timescales. It was later observed [111, 140, 154] that the changes in the second harmonic (i.e., changes in the second-order susceptibility, $\chi^{(2)}$)

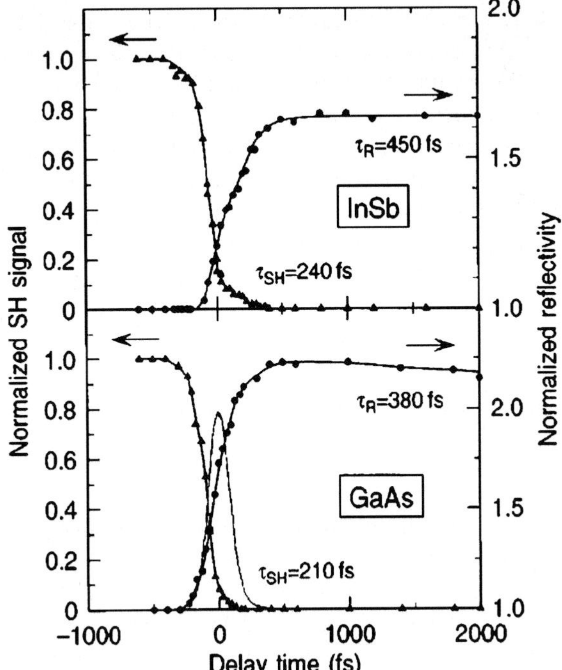

Fig. 19. Time-resolved changes in reflectivity and second-harmonic generation in GaAs and InSb [140]

occurred on faster timescales than the changes in the optical reflectivity, highlighting the uncertainty of interpreting the optical measurements in terms of structural changes. Later work by *Mazur* and coworkers [155, 156] probed reflectivity changes over a range of wavelengths. They were able to extract changes in the real and imaginary parts of the dielectric function, providing evidence for a semiconductor-to-metal transition at high fluences. Nevertheless, the pathways the atoms followed during the disordering process remained unclear.

In the past decade, short-pulse hard-X-ray probes were applied to this problem, and began to provide the first direct evidence for ultrafast structural changes. Femtosecond X-ray pulses produced by laser-generated plasmas were used to measure changes in the intensity of a single Bragg peak following excitation, confirming the interpretation that loss of order was occurring on subpicosecond timescales [4, 15, 18, 21, 157] (as we show below, one must be careful in how one interprets the disappearence of a single Bragg peak). Experiments by *Siders* et al. [21] and *Sokolowski-Tinten* et al. [15] made use of semiconductor heterostructures in which a thin epitaxial layer of germanium on silicon was excited at 800 nm and probed by a short X-ray pulse at variable delays, measuring only the (111) reflection. Figure 20 shows

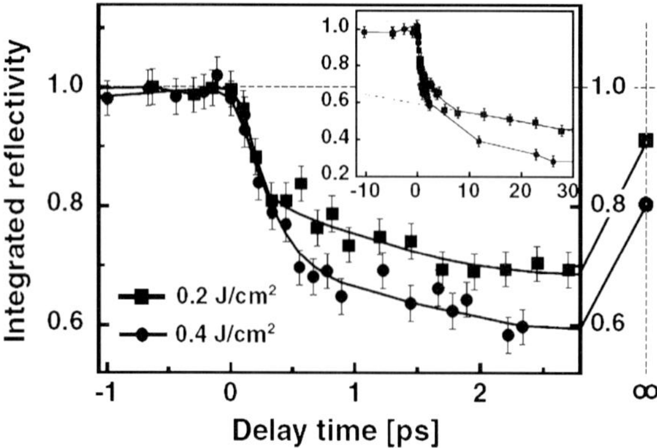

Fig. 20. Integrated (111) reflectivity on germanium as a function of time for two difference fluences following femtosecond excitation. (*inset*) Extended timescale [15]

measurements by *Sokolowski-Tinten* et al. of the (111) X-ray diffraction efficiency for two fluences. A decrease on a timescale of order 300 fs is observed. Similar measurements on InSb were carried out by *Rousse* et al. [18] using bulk crystals and asymmetric reflections in order to match pump and probe penetration depths. Changes in the (111) diffraction peak were observed to occur on the scale of 400 fs. From this, the dependence of the thickness of the molten film as a function of laser fluence could be extracted. However, the question of how the disordering occurred or what the extracted timescales correspond to remained unanswered.

In parallel with the development of laser-plasma X-ray sources, sources of ultrafast X-rays based on accelerator technology have been developed that in many ways compete favorably with laser-plasma sources. Linac-based sources, in particular, in which electron bunches are compressed to 100 fs duration and then used to produce ultrashort X-ray pulses have recently experienced impressive technological advances. In the form of X-ray free-electron lasers (FEL), research and development is underway at several places around the world to produce intense, coherent, ultrafast X-ray pulses using self-amplified spontaneous emission in a long undulator [64]. While these sources are still years from completion, research efforts directed towards the Linac Coherent Light Source (LCLS) X-ray FEL have led to the development of the Sub-Picosecond Pulse Source (SPPS) [19, 69, 71], using the same linac-based acceleration and electron-bunch compression schemes as future X-ray FELs, but spontaneous production of X-rays in a short undulator. In the following, we describe the first application of this new source, in which the ultrafast structural changes associated with high carrier density excitation were revisited [19].

Experiments focused on InSb, and measurements of the time-resolved diffracted intensity were made at two different X-ray reflections. An extreme asymmetric geometry [158] was used in order to match X-ray and laser pen-

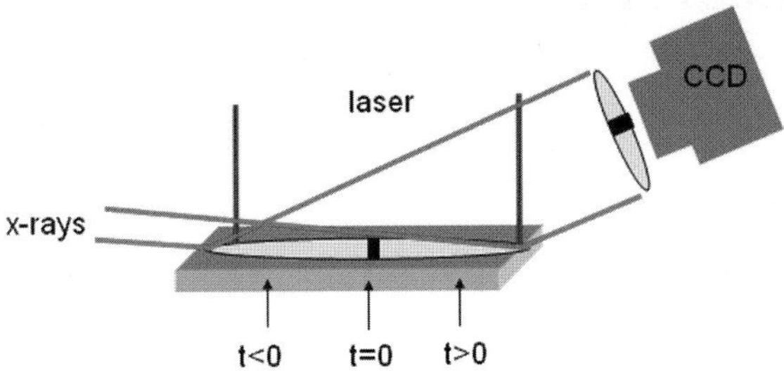

Fig. 21. Experimental setup showing a technique in which a range of times can be measured, in a single shot, by use of a crossbeam topography

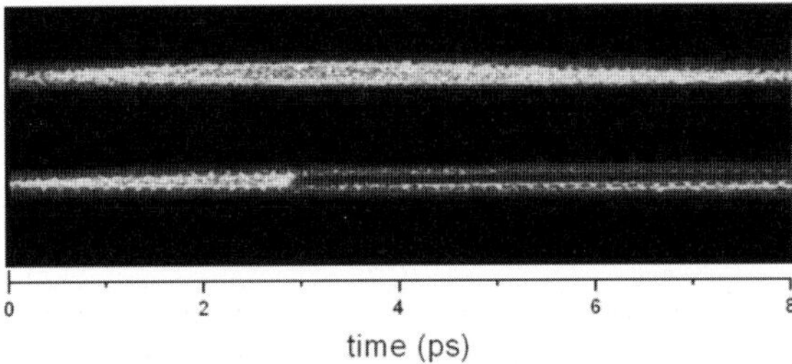

Fig. 22. Single-shot images of scattered X-rays from an InSb sample. The *top image* is measured from the unperturbed sample. The *bottom image* captures the transient changes in the diffracted intensity over a time window of $\sim 8\,\mathrm{ps}$. Time runs from left to right and the *dashed curves* show the region excited by the laser pulse

etration depths. By adjusting the azimuthal angle of an asymmetrically cut crystal, the asymmetry angle for different reflections can be continuously tuned in such a way that different reflections can be studied on the same crystal under identical excitation conditions (with laser/X-ray incidence angle unchanged). In this way, the X-ray incidence angle was fixed at 0.4 degrees, near the critical angle for total external reflection. Experiments were conducted in a crossbeam geometry in order to overcome the intrinsic jitter between the independent pump and probe sources [71, 159, 160]. With the optical pump pulse incident at an angle with respect to the X-ray probe pulse, as shown in Fig. 21, a temporal sweep is created along the crystal surface. In this way, temporal information is transformed into spatial information as a result of the difference in propagation times across the sample surface. By imaging the diffracted X-ray spot with a CCD camera, the complete time history around $t = 0$ is obtained in a single shot.

Figure 22 shows typical single-shot images. The top image displays X-rays diffracted from the unperturbed sample. The bottom image is obtained when the pump and probe pulses overlap in space and time on the sample.

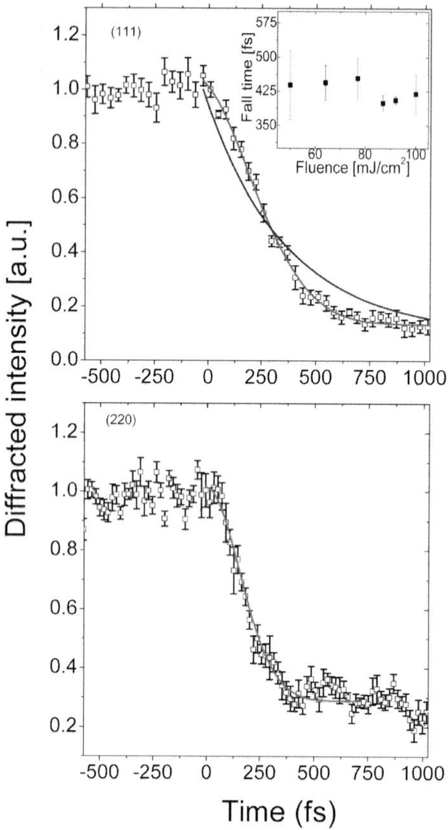

Fig. 23. Time-dependent diffracted intensity for the (111) (*top*) and (220) (*bottom*) reflections of InSb, measured under identical excitation conditions. Gaussian fits to the data are displayed, corresponding to 10–90 fall times of 430 fs and 280 fs, respectively. For comparison, an exponential fit to the (111) reflection is also shown. *Inset*: Fluence dependence of the time constant for (111) data

The sharp edge in the bottom image indicates time zero. To the left of this edge, X-rays scatter from the unperturbed sample before the optical excitation pulse arrives. To the right, X-rays probe the optically induced liquid state, resulting in a strong decrease in the diffracted intensity. Note that the temporal resolution of the measurement is set not only by the pulse duration of the pump and probe, but also the accuracy with which the surface topography can be imaged; an estimate obtained by imaging the edge of a razor blade placed on the sample surface gives ~ 130 fs for the (111) reflection and ~ 80 fs for the (220) reflection by imaging the sharp edge of a razor blade placed on the sample surface. The finite bandwidth of the X-ray source leads to a divergence in the diffracted beams, which in turn, requires the placement of the detector as close as possible to the sample.

Figure 23 shows the time-dependent intensity $I(Q,t)$ for both the (111) and (220) reflections, averaged over 10 single-shot images, measured on the same crystal under identical fluence conditions. The data are shown fit to both exponential and Gaussian decays. One observes, first, that for short times after excitation, the diffracted intensity is nonexponential and well fitted by a Gaussian with peak centered at the excitation time. Moreover, the (220) reflection decays with a time constant qualitatively faster than the (111). The time for the intensity to fall from 90 % to 10 % of its initial value for the (111) and (220) reflections is 430 and 280 fs, respectively, with ratio $\tau_{111}/\tau_{220} = 1.5 \pm 0.2$. This value is equal (within experimental error) to the ratio of the magnitude of the reciprocal lattice vectors for the two reflections ($Q_{220}/Q_{111} = \sqrt{8/3}$).

This inverse-Q-dependent scaling and Gaussian time dependence strongly implies uncorrelated atomic motion and suggests that the data can be described using a time-dependent Debye–Waller (DW) model, which relates the time-dependent decrease in scattered intensity to a time-dependent rms displacement:

$$I(Q,t) \sim \mathrm{e}^{-Q^2\langle u^2(t)\rangle/3} \,. \tag{18}$$

Here, Q is the reciprocal lattice vector for the reflection probed, and $\langle u^2(t)\rangle$ is the time-dependent mean-square displacement of the photoexcited atoms, averaged spatially over the sample [72, 161]. This is the same factor normally used to describe the effect of thermal fluctuations of atoms about their equilibrium lattice sites on the intensity of a Bragg peak. Although the DW factor is usually thought of as a time-averaged quantity, it may be shown [72, 162] that (18) is exact for the case of a Gaussian distribution of atomic displacements. Deviations from a Gaussian distribution due, for example, to anharmonicity in the potentials, lead to Q-dependent correction factors, as described in [163–165]. Thus, time-dependent changes of, for example, the width of the Gaussian distribution of atomic displacements may be modeled exactly. As a model for the time-dependent phenomena described above, this makes physical sense if the first step in the disordering process corresponds to an effective amplification of the rms displacements characteristic of a room-temperature thermal distribution, preserving $\langle u(t)\rangle = 0$. For the case of a sudden, impulsive softening of the interatomic potential, a time-dependent increase in the rms displacement is indeed expected [145, 150]. In contrast to displacive excitation, in which the atomic motion is induced by impulsively shifting a potential, this effect corresponds to a sudden change in the frequency of the potential. Moreover, although the exact change in the atomic-distribution function depends sensitively on the exact time-dependent shape of the interatomic potential, to first order in time, the atoms will continue to move with velocities set by their initial conditions. This is just a restatement of Newton's first law saying that inertial dynamics dominate for small times. In this approximation, with $\langle u^2(t)\rangle^{1/2} = v_{\rm rms}t$, the time-resolved diffracted intensity $I(Q,t)$ is then expected to decay as

$$I(Q,t) \sim \mathrm{e}^{\frac{-Q^2 v_{\rm rms}^2 t^2}{3}} , \tag{19}$$

i.e., Gaussian in both Q and t, as observed. Taking $v_{\rm rms} = (3k_{\rm B}T/M)^{1/2}$ in accordance with equipartition, we extract a (1/e) Gaussian time constant from (19) of the form

$$\tau = \left(\frac{M}{Q^2 k_{\rm B}T}\right)^{1/2} . \tag{20}$$

This corresponds to values of 410 fs and 250 fs for the (111) and (220) reflections, respectively, in good agreement with the experimental data. We note that a similar time-dependent DW model has been applied to describe visible-light Bragg scattering off atoms trapped in optical lattices [166, 167]. On turning off the light-induced potential, the rms displacement of the atoms again increases inertially, with the velocity set by the initial temperature of the lattice. This leads to a Gaussian-like decay in the diffracted light, with a time constant that depends on the initial lattice temperature.

The validity of the model sketched out above was checked in three different ways: 1. Comparison of the extracted rms displacements for the (111) and (220) reflections; 2. Comparison of the measured velocity of the atoms to room-temperature excitations; 3. Fluence dependence of the time constants of the decay of the diffracted intensity. Extracted rms displacements for the (111) and (220) reflections are shown in Fig. 24, obtained by using (18) to invert the raw $I(Q,t)$ data. For short times, the two curves overlap within error, indicating an initial, isotropic disordering process, independent of reciprocal lattice vector. Second, for short times, we observe that the rms displacements are linear in t, with slope (corresponding to a velocity) 2.3 Å/ps. This value is in good agreement with room-temperature rms velocities in InSb – $(3k_{\rm B}T/M)^{1/2} = 2.5$ Å/ps where M is the average mass of InSb and $T = 300$ K [168]. Finally, the inset to Fig. 23 shows the fluence dependence of the time constants for the (111) reflection. There is little variation with fluence (a similar result is obtained for the 220 reflection). Thus, as might be expected, for short times, it is difficult to tell the difference between a flat potential (corresponding to a free particle) and a motion driven by a softened phonon with an imaginary frequency (corresponding to a saddle point on the potential energy surface). From the data, we can place both an upper and lower bound on the extent of the softening of the potential surface. In effect, the amplitude of the inertial response determines the extent of the softening. At very large softening, we expect to observe acceleration of the atoms as they fall down the potential-energy surface. Because we observe that the dynamics remain inertial for at least 250 fs, it can be shown, neglecting anharmonicities in the potential, that this requires a negative curvature $\leq 1\,\mathrm{eV/Å^2}$.

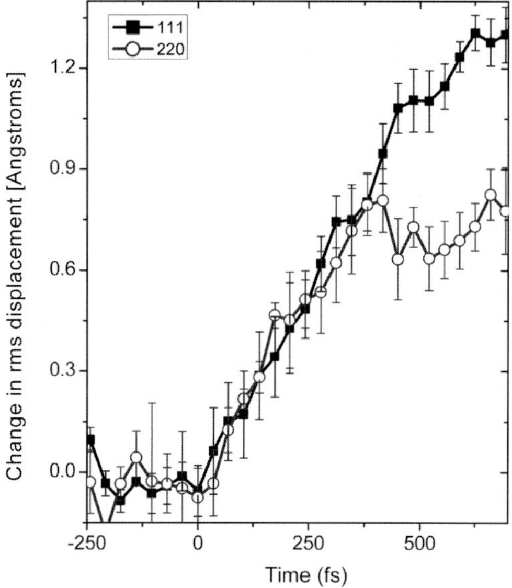

Fig. 24. Change in the root-mean-square displacement of InSb, extracted from the time-dependent diffracted intensity. For short times, the measured displacements agree well for both the (111) and (220) X-ray reflections. The slope corresponds to a velocity in good agreement with the room-temperature rms atomic velocity

Similarly, the fact that we observe displacements of order 1 Å from the equilibrium position implies a maximum positive curvature (corresponding to a real frequency) of $\sim 0.1\,\text{eV}/\text{Å}^2$, which represents a dramatic change in the interatomic potential. Figure 25 illustrates the changes in the interatomic potential and atomic-scale displacements, based on a simple one-dimensional model. We emphasize that the timescale of the change of the diffracted intensity is set by the initial equilibrium thermal fluctuations in the lattice, despite the highly nonequilibrium nature of the transition. Prior to these experiments, there was considerable confusion as to how large-amplitude atomic motion could occur on a timescale faster than that of the electron–phonon coupling time.

An elegant means of describing the changes in the atomic-distribution function is to depict the dynamics in a velocity vs. position phase space. The insets to Fig. 26 show the distribution of atomic displacements and velocities for various times following impulsive softening of the potential, in a simplified one-dimensional model. Initally, the distribution is set by the unperturbed harmonic-oscillator potential. The average displacement and velocity of each atom is zero but there exists a Gaussian distribution of displacements and velocities about their equilibrium values, indicated by the horizontal and vertical extent of the phase-space ellipse. Following the softening of the potential, in the limit where the frequency goes to zero, the distribution of

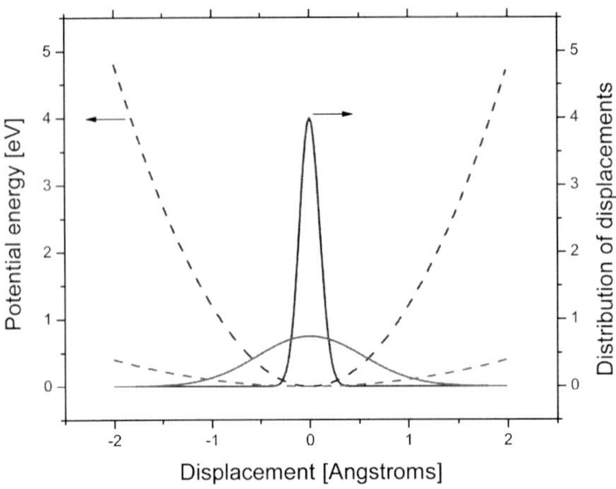

Fig. 25. One-dimensional illustration of changes in the interatomic potential (*dashed*) and distribution of atomic displacements (*solid*) from the equilibrium state in InSb. At $t = 200\,\mathrm{fs}$, we measure rms displacements $\sim 0.5\,\text{Å}$ corresponding to a substantial fraction of the interatomic spacing. The softened potential shown corresponds to the mininum softening required ($0.1\,\mathrm{eV}/\text{Å}^2$) to be consistent with the experimental results

atomic displacements broadens due to the ballistic motion of the atoms, but the distribution is unchanged in velocity space as there are no forces acting on the atoms. As the phase-space density remains consant, this gives rise to a shear deformation of the distribution as indicated in the successive insets in Fig. 26. The rms displacement as a function of time is given by

$$\sigma(t) = \sqrt{\sigma_0^2 + v^2 t^2}\,, \tag{21}$$

where σ_0 is the initial rms displacement before excitation. This curve is displayed in Fig. 26. One observes that the curve rapidly approaches the ballistic limit, with the rms width proportional to time. However, note that for $t \leq \sigma_0/v$ (of order 50 fs for InSb), $\sigma(t)$ is actually quadratic in t. In these experiments, the temporal resolution to observe this effect is insufficient.

For the case in which the frequency does not go all the way to zero, one would expect impulsive softening to induce oscillations in the rms displacement about the new equilibrium value. Such oscillations have been observed for atoms trapped in optical lattices [167, 169, 170]. The oscillations probed in the Bragg-scattered light correspond to breathing-mode excitations in atomic wavepackets. Such impulsive changes in frequency are a means of generating classical squeezed states, in which the rms distribution may be focused or compressed for a fraction of a vibrational period.

The interpretation of the data becomes more difficult for long times. As can be observed in Fig. 24, after $\sim$ 350 fs, the extracted rms displacements for the (111) and (220) reflections no longer agree. This may be a result of

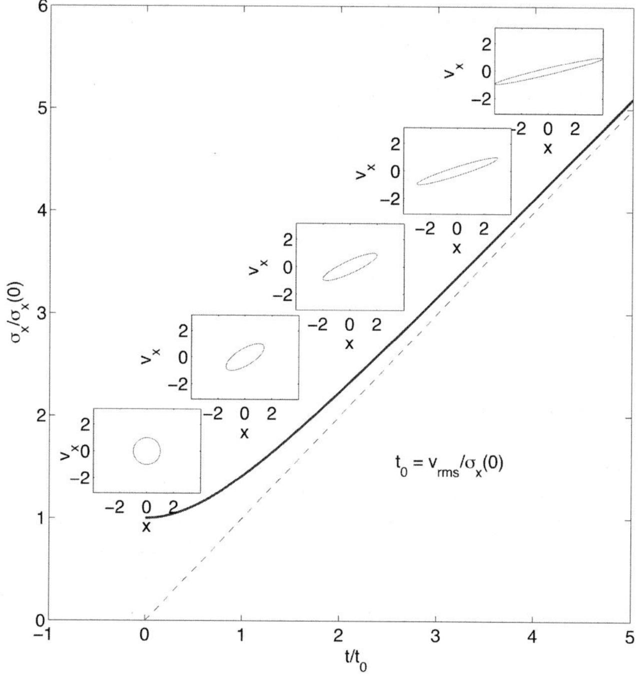

Fig. 26. Calculated change in rms displacements in the limit of perfect softening. The curve rapidly approaches the ballistic limit. *Insets* show phase-space plots of the atomic distributions as a function of time

non-Gaussian corrections arising as the atoms sense anharmonicity in the potential at large displacements. Alternatively, this may be interpreted as indicative of anisotropic softening of the potential. The force an atom feels depends on its nearest neighbors, and, at large displacements, the surrounding cage of atoms may influence the dynamics, such that the independent atom approximation breaks down. One may gain some insight into the behavior at longer times by comparing the X-ray results to frequency-domain inelastic neutron-scattering measurements or molecular-dynamics simulations [171–174] in studies of liquid-state dynamics. Here, the width of the self-correlation function $G_s(r,t)$ is found to be linear in time at small t (representing inertial dynamics) but crosses over to square-root behavior for times significantly larger than a typical collision time. This represents a transition to diffusive dynamics as a result of multiple collisions with surrounding atoms. Indeed, one may show that the mean-square displacement in a liquid [175] can be expanded as

$$\langle u^2(t)\rangle = \frac{3k_\mathrm{B}T}{M}t^2 - \frac{k_\mathrm{B}T}{4M}\Omega^2 t^4 + \cdots , \tag{22}$$

where Ω is the collision frequency, and the second term is the first correction factor to the ballistic motion of the atoms. Perfect diffusive motion, with $\langle u^2(t)\rangle \sim t$ would manifest itself in the tail of the decay in the diffracted intensity as an exponential rather than Gaussian time response [20]. It is only in this limit that the memory of the initial crystalline atomic positions is lost. Thus, after a few hundred femtoseconds, although we observe an inertially driven divergence of the mean-square displacements [176], the transition to the liquid state is not yet complete.

4.2 Solid-Solid Phase Transitions

In this section, we briefly review some of the recent ultrafast X-ray scattering investigations probing solid-solid phase transitions. These studies are motivated by the goal of creating novel transient structures with light. In these cases, one does not simply measure the disappearance of a Bragg peak but also the appearance of new ones due to the different symmetry of the transient state. The early experiments have involved materials that exhibit strong coupling between lattice and electronic degrees of freedom, as well as cooperative structural effects.

Cavalleri et al. [25] investigated subpicosecond structural changes in VO_2, a material with strongly correlated electrons, which undergoes a thermodynamic insulator-to-metal phase transition. This transition can also be driven by optical excitation. In both case, changes in the electronic band structure are associated with changes in the atomic structure. Time-resolved techniques may be able to decouple the electronic from the structural degrees of freedom. Measurements on the (110) diffraction peak were made on a thin 200 nm film with initial state in the low-temperature phase. Following optical excitation at 800 nm, a shoulder was observed on the low-angle side of the diffraction peak, corresponding to growth of the high-temperature phase, appearing on a timescale of a few hundred femtoseconds. The evidence thus points to the appearance of the new optically induced structural phase on subpicosecond timescales. Further experiments with higher time resolution are necessary to elucidate the correlation between the electronic and structural degrees of freedom. We note that very recently, measurements under similar optical excitation conditions were made using femtosecond X-ray absorption spectroscopy [27].

Collet et al. [26, 177] studied optically induced solid-solid phase transitions in the charge-transfer molecular material tetrathiafulvalene-p-chloranil (TTF-CA, $C_6H_4S_4C_6Cl_4O_2$) using 100 ps synchrotron-based X-ray pulses. In this system, optical excitation leads to a transition from a paraelectric stuctural phase to a macroscopically ordered ferroelectric phase on a 500 ps timescale. Absorption by this molecule is followed by an intermolecular charge-transfer process that, through an unusual cooperative effect, leads to a three-dimensional ordered ionic state. Interestingly [177], the transient phase is the lower-symmetry phase. Thus, this experiment illustrates the power of

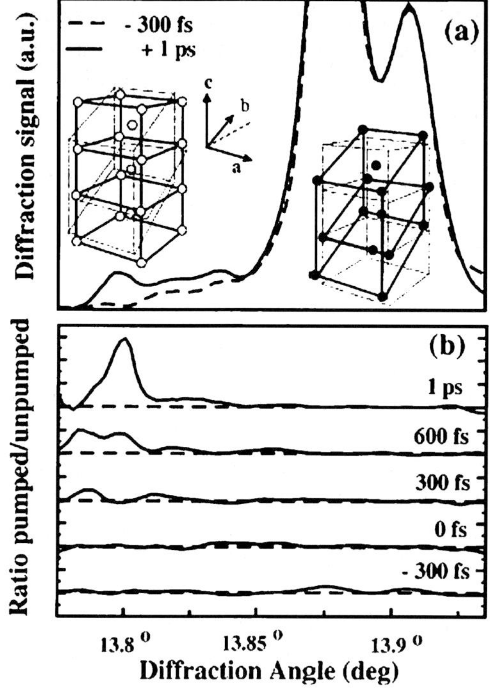

Fig. 27. (*top*) X-ray diffracted signal from VO_2 at time delays of −300 fs and 1 ps. The low-temperature and high-temperature structural phases are also shown. (*bottom*) X-ray reflectivity ratio between excited and unexcited states shown at various pump-probe delay times [25]

using atomic-scale probes in conjuction with optical excitation to control the evolution of complex systems. Figure 28 shows the raw data for this experiment, showing changes in a variety of reflections on 100 ps timescales.

As in the previous example, more work is required to fully explore the fundamental dynamics associated with this phase transition. In particular, higher time-resolution measurements are needed to capture the first steps in the transition, and their relationship to the large-scale cooperative effects observed.

5 Concluding Remarks

We hope the reader finds this a valuable overview of ultrafast X-ray scattering in solids. A number of seminal experiments were discussed including the diffraction of X-rays from coherent acoustic and optical phonons and the study of ultrafast solid-liquid and solid-solid phase transitions. The experiments presented here have been made possible by advances in both ultrafast

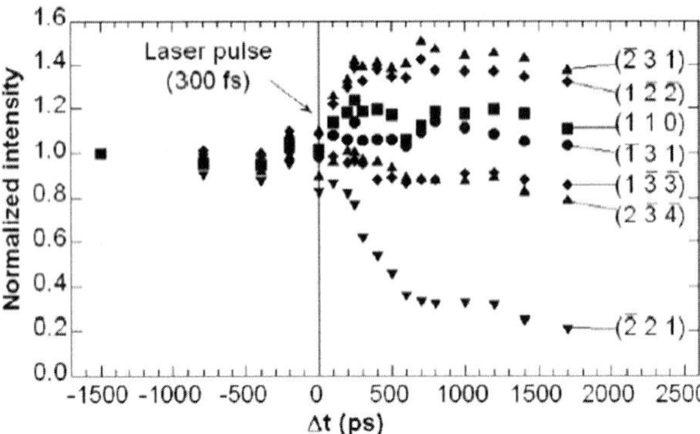

Fig. 28. Intensity of various Bragg reflections in TTF-CA as a function of time following excitation. Changes are associated with a symmetry-lowering neutral-to-ionic transformation [26]

laser technology and X-ray sources. In particular, fourth-generation sources such as the LCLS will provide femtosecond pulses that are orders of magnitude brighter than existing sources [64–66]. These pulses will be of sufficient intensity to drive nonlinear processes in atoms and of sufficient flux to image nonperiodic structures in a single shot. In addition, plans for future upgrades to third-generation synchrotrons promise to provide few-picosecond X-rays with high-average brightness [178, 179]. This is an exciting and pivotal time for ultrafast X-ray science. As we refine our techniques and new high-brightness short-pulse sources become available, we expect that ultrafast X-ray scattering will provide useful insights into the motion of atoms on the fastest timescales.

Acknowledgements

We thank the entire Subpicosecond Pulse Source Collaboration and our collaborators at the Advanced Light Source and Advanced Photon Source. Special thanks to Phil Bucksbaum, Matt DeCamp, Jerry Hastings and Roger Falcone. Partial support was provided by the United States Department of Energy, Basic Energy Sciences and the National Science Foundation. Finally, we thank Roberto Merlin and Manuel Cardona for, among other things, their patience and careful reading of our manuscript.

References

[1] P. Chen, I. V. Tomov, P. M. Rentzepis: Time resolved heat propagation in a gold crystal by means of picosecond X-ray diffraction, J. Chem. Phys. **24**, 10001–10007 (1996)

[2] V. Srajer, T.-Y. Teng, T. Ursby, C. Pradervand, Z. Ren, S.-I. Adachi, W. Schildkamp, D. Bourgeois, M. Wulff, K. Moffat: Photolysis of the carbon monoxide complex of myoglobin: Nanosecond time-resolved crystallography, Science **274**, 1726–1729 (1996)

[3] S. Techert, F. Schotte, M. Wulff: Picosecond X-ray diffraction probed transient structural changes in organic solids, Phys. Rev. Lett. **86**, 2030–2033 (2001)

[4] C. Rischel, A. Rousse, I. Uschmann, P.-A. Albouy, J.-P. Geindre, P. Audebert, J.-C. Gauthier, E. Froster, J.-L. Martin, A. Antonetti: Femtosecond time-resolved X-ray diffraction from laser-heated organic films, Nature **390**, 490–492 (1997)

[5] J. S. Wark, N. C. Woolsey, R. R. Whitlock: Novel measurements of high-dynamic crystal strength by picosecond X-ray diffraction, Appl. Phys. Lett. **61**, 651–653 (1992)

[6] J. S. Wark, R. R. Whitlock, A. A. Hauer, J. E. Swain, P. J. Solone: Subnanosecond X-ray diffraction from laser-shocked crystals, Phys. Rev. B **40**, 5705–5714 (1989)

[7] A. Loveridge-Smith, A. Allen, J. Belak, T. Boehly, A. Hauer, B. Holian, D. Kalantar, G. Kryala, R. W. Lee, P. Lomdahl, M. A. Meyers, D. Paisley, S. Pollarin, B. Remington, D. C. Swift, S. Weber, J. S. Wark: Anomalous elastic response of silicon to uniaxial shock compression on nanosecond time scales, Phys. Rev. Lett. **86**, 2349–2352 (2001)

[8] B. C. Larson, C. W. White, T. S. Noggle, D. Mills: Synchrotron X-ray diffraction study of silicon during pulsed-laser annealing, Phys. Rev. Lett. **48**, 337–340 (1982)

[9] B. C. Larson, C. W. White, T. S. Noggle, J. F. Barhorst, D. M. Mills: Time-resolved X-ray diffraction measurement of the temperature and temperature gradients in silicon during pulsed laser annealing, Appl. Phys. Lett. **42**, 282–284 (1983)

[10] A. H. Chin, R. W. Schoenlein, T. E. Glover, P. Balling, W. P. Leemans, C. V. Shank: Ultrafast structural dynamics in InSb probed by time-resolved X-ray diffraction, Phys. Rev. Lett. **83**, 336–339 (1999)

[11] C. Rose-Petruck, R. Jimenez, T. Guo, A. Cavalleri, C. W. Siders, F. Raksi, J. A. Squier, B. C. Walker, K. R. Wilson, C. P. J. Barty: Picosecond-milliangstrom lattice dynamics measured by ultrafast X-ray diffraction, Nature **398**, 310–313 (1999)

[12] A. Cavalleri, C. W. Siders, F. L. H. Brown, D. M. Leitner, C. Toth, J. A. Squier, C. P. J. Barty, K. R. Wilson, K. Sokolowski-Tinten, M. Horn von Hoegen, D. von der Linde, M. Kammler: Anharmonic lattice dynamics in germanium measured with ultrafast X-ray diffraction, Phys. Rev. Lett. **85**, 586–589 (2000)

[13] A. M. Lindenberg, I. Kang, S. L. Johnson, T. Missalla, P. A. Heimann, Z. Chang, J. Larsson, P. H. Bucksbaum, H. C. Kapteyn, H. A. Padmore, R. W. Lee, J. S. Wark, R. W. Falcone: Time-resolved X-ray diffraction from coherent phonons during a laser-induced phase transition, Phys. Rev. Lett. **84**, 111–114 (2000)

[14] A. Cavalleri, C. W. Siders, C. Rose-Petruck, R. Jimenez, C. Toth, J. Squier, C. P. J. Barty, K. R. Wilson, K. Sokolowski-Tinten, M. Horn von Hoegen, D. von der Linde: Ultrafast X-ray measurement of laser heating in semiconductors: Parameters determining the melting threshold, Phys. Rev. B **63**, 193306 (2001)

[15] K. Sokolowski-Tinten, C. Blome, C. Dietrich, A. Tarasevitch, M. Horn von Hoegen, D. von der Linde, A. Cavalleri, J. Squier, M. Kammler: Femtosecond X-ray measurement of ultrafast melting and large acoustic transients, Phys. Rev. Lett. **87**, 225701 (2001)

[16] D. A. Reis, M. F. DeCamp, P. H. Bucksbaum, R. Clarke, E. Dufresne, M. Hertlein, R. Merlin, R. Falcone, H. Kapteyn, M. M. Murnane, J. Larsson, T. Missalla, J. S. Wark: Probing impulsive strain propagation with X-ray pulses, Phys. Rev. Lett. **86**, 3072–3075 (2001)

[17] M. F. DeCamp, D. A. Reis, P. H. Bucksbaum, B. Adams, J. M. Caraher, R. Clarke, C. W. S. Conover, E. M. Dufresne, R. Merlin, V. Stoica, J. K. Wahlstrand: Coherent control of pulsed X-ray beams, Nature **413**, 825–828 (2001)

[18] A. Rousse, C. Rischel, S. Fourmaux, I. Uschmann, S. Sebban, G. Grillon, P. Balcou, E. Förster, J. P. Geindres, P. Audebert, J. C. Gauthier, D. Hulin: Non-thermal melting in semiconductors measured at femtosecond resolution, Nature **410**, 65–68 (2001)

[19] A. M. Lindenberg, J. Larsson, K. Sokolowski-Tinten, K. J. Gaffney, C. Blome, O. Synnergren, J. Sheppard, C. Caleman, A. G. MacPhee, D. Weinstein, D. P. Lowney, T. K. Allison, T. Matthews, R. W. Falcone, A. L. Cavalieri, D. M. Fritz, S. H. Lee, P. H. Bucksbaum, D. A. Reis, J. Rudati, P. H. Fuoss, C. C. Kao, D. P. Siddons, R. Pahl, J. Als-Nielson, S. Duerster, R. Ischebeck, H. Schlarb, H. Schulte-Schrepping, T. Tschentscher, J. Schneider, D. von der Linde, O. Hignette, F. Sette, H. N. Chapman, R. W. Lee, T. N. Hansen, S. Techert, J. S. Wark, M. Bergh, G. Huldt, D. van der Spoel, N. Timneanu, J. Hajdu, R. A. Akre, E. Bong, P. Krejcik, J. Arthur, S. Brennan, K. Luening, J. B. Hastings: Atomic-scale visualization of inertial dynamics, Science **308**, 392–395 (2005)

[20] K. J. Gaffney, A. M. Lindenberg, J. Larsson, K. Sokolowski-Tinten, C. Blome, O. Synnergren, J. Sheppard, C. Caleman, A. G. MacPhee, D. Weinstein, D. P. Lowney, T. Allison, T. Matthews, R. W. Falcone, A. L. Cavalieri, D. M. Fritz, S. H. Lee, P. H. Bucksbaum, D. A. Reis, J. Rudati, A. T. Macrander, P. H. Fuoss, C. C. Kao, D. P. Siddons, R. Pahl, K. Moffat, J. Als-Nielsen, S. Duesterer, R. Ischebeck, H. Schlarb, H. Schulte-Schrepping, J. Schneider, D. von der Linde, O. Hignette, F. Sette, H. N. Chapman, R. W. Lee, T. N. Hansen, J. S. Wark, M. Bergh, G. Huldt, D. van der Spoel, N. Timneanu, J. Hajdu, R. A. Akre, E. Bong, P. Krejcik, J. Arthur, S. Brennan, K. Luening, J. B. Hastings: Observation of structural anisotropy and the onset of liquidlike motion during the nonthermal melting of InSb, Phys. Rev. Lett. **95**, 125701–125704 (2005)

[21] C. W. Siders, A. Cavalleri, K. Sokolowski-Tinten, C. Toth, T. Guo, M. Kammler, M. Horn von Hoegen, K. R. Wilson, D. von der Linde, C. P. J. Barty: Detection of nonthermal melting by ultrafast X-ray diffraction, Science **286**, 1340–1342 (1999)
[22] A. Plech, S. Kürbitz, K. J. Berg, H. Graener, G. Berg, S. Gresillon, M. Kaempfe, J. Feldmann, M. Wulff, G. von Plessen: Time-resolved X-ray diffraction on laser-excited metal nanoparticles, Europhys. Lett. **61**, 762–768 (2003)
[23] A. Plech, V. Kotaidis, S. Gresillon, C. Dahmen, G. von Plessen: Laser-induced heating and melting of gold nanoparticles studied by time-resolved X-ray scattering, Phys. Rev. B **70**, 195423 (2004)
[24] M. F. DeCamp, D. A. Reis, A. Cavalieri, P. H. Bucksbaum, R. Clarke, R. Merlin, E. M. Dufresne, D. A. Arms, A. M. Lindenberg, A. G. MacPhee, Z. Chang, B. Lings, J. S. Wark, S. Fahy: Transient strain driven by a dense electron–hole plasma, Phys. Rev. Lett. **91**, 165502 (2003)
[25] A. Cavalleri, C. Toth, C. W. Siders, J. A. Squier, F. Raksi, P. Forget, J. C. Kieffer: Femtosecond structural dynamics in VO_2 during an ultrafast solid-solid phase transition, Phys. Rev. Lett. **87**, 237401 (2001)
[26] E. Collet, M.-H. Lemee-Cailleau, M. Buron-Le Cointe, H. Cailleau, M. Wulff, T. Luty, S.-Y. Koshihara, M. Meyer, L. Toupet, P. Rabiller, S. Techert: Laser-induced ferroelectric structural order in an organic charge-transfer crystal, Science **300**, 612–615 (2003)
[27] A. Cavalleri, M. Rini, H. H. W. Chong, S. Fourmaux, T. E. Glover, P. A. Heimann, J. C. Kieffer, R. W. Schoenlein: Band-selective measurements of electron dynamics in VO_2 using femtosecond near-edge X-ray absorption, Phys. Rev. Lett. **95**, 067405 (2005)
[28] M. Bargheer, N. Zhavoronkov, Y. Gristal, J. C. Woo, D. S. Kim, M. Woerner, T. Elsaesser: Coherent atomic motions in a nanostructure studied by femtosecond X-ray diffraction, Science **306**, 1771–1773 (2004)
[29] P. Sondhauss, J. Larsson, M. Harbst, G. A. Naylor, A. Plech, K. Scheidt, O. Synnergren, M. Wulff, J. S. Wark: Picosecond X-ray studies of coherent folded acoustic phonons in a multiple quantum well, Phys. Rev. Lett. **94**, 125509 (2005)
[30] K. Sokolowski-Tinten, C. Blome, J. Blums, A. Cavalleri, C. Dietrich, A. Tarasevitch, I. Uschmann, E. Förster, M. Kamller, M. Horn von Hoegen, D. von der Linde: Femtosecond X-ray measurement of coherent lattice vibrations near the Lindenmann stability limit, Nature **422**, 287–289 (2003)
[31] A. Rousse, C. Rischel, J. C. Gauthier: Colloquium: Femtosecond X-ray crystallography, Rev. Mod. Phys. **73**, 17–31 (2001)
[32] J. S. Wark, A. M. Allen, P. C. Ansbro, P. H. Bucksbaum, Z. Chang, M. DeCamp, R. W. Falcone, P. A. Heimann, S. L. Johnson, I. Kang, H. C. Kapteyn, J. Larsson, R. W. Lee, A. M. Lindenberg, R. Merlin, T. Missalla, G. Naylor, H. A. Padmore, D. A. Reis, K. Scheidt, A. Sjoegren, P. C. Sondhauss, M. Wulff: Femtosecond X-ray diffraction: experiments and limits, in D. M. Mills, H. Schulte-Schrepping, J. R. Arthur (Eds.): *X-ray FEL Optics and Instrumentation, SPIE International Symposium on Optical Science and Technology, San Diego, CA, 2000*, vol. 4143 (Proceedings of the SPIE 2001) pp. 26–37

[33] M. F. DeCamp, D. A. Reis, D. M. Fritz, P. H. Bucksbaum, E. M. Dufresne, R. Clarke: X-ray synchrotron studies of ultrafast crystalline dynamics, J. Synch. Radiat. **12**, 172–192 (2005)

[34] H. Ihee, M. Lorenc, T. K. Kim, Q. Y. Kong, M. Cammarata, J. H. Lee, S. Bratos, M. Wulff: Ultrafast X-ray diffraction of transient molecular structures in solution, Science **309**, 1223–1227 (2005)

[35] A. M. Lindenberg, Y. Acremann, D. P. Lowney, P. A. Heimann, T. K. Allison, T. Matthews, R. W. Falcone: Time-resolved measurements of the structure of water at constant density, J. Chem. Phys. **122**, 204507 (2005)

[36] T. Lee, F. Benesch, Y. Jiang, C. Rose-Petruck: Structure of solvated Fe(Co)(5): XANES and EXAFS measurements using ultrafast laser plasma and conventional X-ray sources, Chem. Phys. **299**, 233–245 (2004)

[37] M. Saes, C. Bressler, R. Abela, D. Grolimund, S. L. Johnson, P. A. Heimann, M. Chergui: Observing photochemical transients by ultrafast X-ray absorption spectroscopy, Phys. Rev. Lett. **90**, 047403 (2003)

[38] A. Plech, M. Wulff, S. Bratos, F. Mirloup, R. Vuilleumier, F. Schotte, P. A. Anfinrud: Visualizing chemical reactions in solution by picosecond X-ray diffraction, Phys. Rev. Lett. **92**, 125505 (2004)

[39] L. X. Chen, W. J. H. Jager, G. Jennings, D. J. Gosztola, A. Munkholm, J. Hessler: Capturing a photoexcited molecular structure through time-domain X-ray absorption fine structure, Science **292**, 262–264 (2001)

[40] S. L. Johnson, P. A. Heimann, A. G. MacPhee, O. R. Monteiro, Z. Chang, R. W. Lee, R. W. Falcone: Bonding in liquid carbon studied by time-resolved X-ray absorption spectroscopy, Phys. Rev. Lett. **94**, 057407 (2005)

[41] S. L. Johnson, P. A. Heimann, A. M. Lindenberg, H. O. Jeschke, M. E. Garcia, Z. Chang, R. W. Lee, J. J. Rehr, R. W. Falcone: Properties of liquid silicon observed by time-resolved X-ray absorption spectroscopy, Phys. Rev. Lett. **91**, 157403 (2003)

[42] M. M. Murnane, H. C. Kapteyn, M. D. Rosen, R. W. Falcone: Ultrafast X-ray pulses from laser-produced plasmas, Science **251**, 531–536 (1991)

[43] A. Rousse, P. Audebert, J. P. Geindre, F. Fallies, J. C. Gauthier, A. Mysyrowicz, G. Grillon, A. Antonetti: Efficient Kα X-ray source from femtosecond laser-produced plasmas, Phys. Rev. E **50**, 2200–2207 (1994)

[44] J. C. Kieffer, M. Chaker, J. P. Matte, H. Pepin, C. Y. Cote, Y. Beaudoin, T. W. Johnston, C. Y. Chien, S. Coe, G. Mourou, O. Peyrusse: Ultrafast X-ray sources, Phys. Fluids B **5**, 2676–2681 (1993)

[45] C. Reich, P. Gibbon, I. Uschmann, E. Forster: Yield optimization and time structure of femtosecond laser plasma K-alpha sources, Phys. Rev. Lett. **84**, 4846–4849 (2000)

[46] K. Sokolowski-Tinten, D. von der Linde: Ultrafast phase transitions and lattice dynamics probed using laser-produced X-ray pulses, J. Phys. Condens. Matter **16**, R1517–R1536 (2004)

[47] A. Rousse, C. Rischel, J. Gauthier: Femtosecond X-ray crystallography, Rev. Mod. Phys. **73**, 17–31 (2001)

[48] H. Wiedemann: *Particle Accelerator Physics I: Basic Principles and Linear Beam Dynamics* (Springer, Berlin, Heidelberg 1999)

[49] H. Wiedemann: *Particle Accelerator Physics II: Nonlinear and Higher-Order Beam Dynamics* (Springer, Berlin, Heidelberg 1999)

[50] W. Barletta, H. Winick: Introduction to special section on future light sources, Nucl. Instrum. Meth. A **500**, 1–10 (2003)
[51] J. Corbett, T. Rabedeau: Intermediate-energy light sources, Nucl. Instrum. Meth. A **500**, 11–17 (2003)
[52] C. Pellegrini, J. Stohr: X-ray free-electron lasers – principles, properties and applications, Nucl. Instrum. Methods A **500**, 33–40 (2003)
[53] D. Attwood: *Soft X-Rays and Extreme Ultraviolet Radiation: Principles and Applications* (Cambridge University Press, Cambridge 1999)
[54] J. Als-Nielsen, D. McMorrow: *Elements of Modern X-Ray Physics* (John Wiley and Sons, Ltd., New York 2001)
[55] A. M. Lindenberg, I. Kang, S. L. Johnson, T. Missalla, P. A. Heimann, Z. Chang, J. Larsson, P. H. Bucksbaum, H. C. Kapteyn, H. A. Padmore, R. W. Lee, J. S. Wark, R. W. Falcone: Time-resolved X-ray diffraction from coherent phonons during a laser-induced phase transition, Phys. Rev. Lett. **84**, 111–114 (2000)
[56] Z. Chang, A. Rundquist, J. Zhou, M. M. Murnane, H. C. Kapteyn, X. Liu, B. Shan, J. Liu, L. Niu, M. Gong, X. Zhang: Demonstration of a sub-picosecond X-ray streak camera, Appl. Phys. Lett. **69**, 133–135 (1996)
[57] M. M. Shakya, Z. Chang: Achieving 280 fs resolution with a streak camera by reducing the deflection dispersion, Appl. Phys. Lett. **87**, 041103 (2005)
[58] J. Larsson, Z. Chang, E. Judd, P. J. Schuck, R. W. Falcone, P. A. Heimann, H. A. Padmore, H. C. Kapteyn, P. H. Bucksbaum, M. M. Murnane, R. W. Lee, A. Machacek, J. S. Wark, X. Liu, B. Shan: Ultrafast X-ray diffraction using a streak-camera detector in averaging mode, Opt. Lett. **22**, 1012–1014 (1997)
[59] D. P. Lowney, P. A. Heimann, H. A. Padmore, E. M. Gullikson, A. G. MacPhee, R. W. Falcone: Characterization of cSi photocathodes at grazing incidence for use in a unit quantum efficiency X-ray streak camera, Rev. Sci. Instrum. **75**, 3131 (2005)
[60] R. W. Schoenlein, S. Chattopadhyay, H. H. W. Chong, T. E. Glover, P. A. Heimann, C. V. Shank, A. A. Zholents, M. S. Zolotorev: Generation of femtosecond pulses of synchrotron radiation, Science **287**, 2237–2240 (2000)
[61] A. Cavalleri, R. Schoenlein: Femtosecond X-rays and structural dynamics in condensed matter, in K.-T. Tsen (Ed.): *Ultrafast Dynamical Processes in Semiconductors*, vol. 92, Topics in Applied Physics (Springer, Berlin, Heidelberg 2004) pp. 309–338
[62] A. A. Zholents, M. S. Zolotorev: Femtosecond X-ray pulses of synchrotron radiation, Phys. Rev. Lett. **76**, 912–915 (1996)
[63] P. H. Bucksbaum, R. Merlin: The phonon Bragg switch: A proposal to generate sub-picosecond X-ray pulses, Solid State Commun. **111**, 535–539 (1999)
[64] R. F. Service: Battle to become the next-generation X-ray source, Science **298**, 1356–1358 (2002)
[65] J. A. J. Feldhaus, J. B. Hastings: X-ray free electron lasers, J. Phys. B: At. Mol. Opt. Phys. **38**, S799–S819 (2004)
[66] M. Cornacchia, J. Arthur, K. Bane, P. Bolton, R. Carr, F. J. Decker, P. Emma, J. Galayda, J. Hastings, K. Hodgson, Z. Huang, I. Lindau, H. D. Nuhn, J. Paterson, C. Pellegrini, S. Reiche, H. Schlarb, J. Stohr, G. Stupakov, D. Walz, H. Winick: Future possibilities of the linac coherent light source, J. Synch. Radiat. **11**, 227–238 (2004)
[67] H. C. Kapteyn, T. Ditmire: Ultraviolet upset, Nature **420**, 467–468 (2002)

[68] H. Wabnitz, L. Bittner, A. R. B. de Castro, R. Dohrmann, P. Gurtler, T. Laarmann, W. Laasch, J. Schulz, A. Swiderski, K. von Haeften, T. Moller, B. Faatz, A. Fateev, J. Feldhaus, C. Gerth, U. Hahn, E. Saldin, E. Schneidmiller, K. Sytchev, K. Tiedtke, R. Treusch, M. Yurkov: Multiple ionization of atom clusters by intense soft X-rays from a free-electron laser, Nature **420**, 482–485 (2002)

[69] P. Emma, R. Iverson, P. Krejcik, J. Raimondi, J. Safranek: Femtosecond electron bunch lengths in the SLAC FFTB beamline, Proceedings of the 2001 Particle Accelerator Conference (PAC2001) pp. 4038–4040 (2001)

[70] P. Krejcik, F.-J. Decker, P. Emma, K. Hacker, L. Hendrickson, C. L. O'Connel, H. Schlarb, H. Smith, M. Stanek: Commissioning of the SPPS linac bunch compressor, Proceedings of the 2003 Particle Accelerator Conference (PAC2003) pp. 423–425 (2003)

[71] A. L. Cavalieri, D. M. Fritz, S. H. Lee, P. H. Bucksbaum, D. A. Reis, J. Rudati, D. M. Mills, P. H. Fuoss, G. B. Stephenson, C. C. Kao, D. P. Siddons, D. P. Lowney, A. G. MacPhee, D. Weinstein, R. W. Falcone, R. Pahl, J. Als-Nielsen, C. Blome, S. Dürsterer, R. Ischebeck, H. Schlarb, H. Schulte-Schrepping, T. Tschentscher, J. Schneider, O. Hignette, F. Sette, K. Sokolowski-Tinten, H. N. Chapman, R. W. Lee, T. N. Hansen, O. Synnergren, J. Larsson, S. Techert, J. Sheppard, J. S. Wark, M. Bergh, C. Caleman, G. Huldt, D. van der Spoel, N. Timneanu, J. Hajdu, R. A. Akre, E. Bong, P. Emma, P. Krejcik, J. Arthur, S. Brennan, K. J. Gaffney, A. M. Lindenberg, K. Luening, J. B. Hastings: Clocking femtosecond X-rays, Phys. Rev. Lett. **94**, 114801 (2005)

[72] B. Warren: *X-Ray Diffraction* (Addison Wesley Pub. Co., Reading, MA 1969)

[73] B. W. Batterman, H. Cole: Dynamical diffraction of X rays by perfect crystals, Rev. Mod. Phys. **36**, 681–716 (1964)

[74] J. Larsson, A. Allen, P. H. Bucksbaum, R. W. Falcone, A. Lindenberg, G. Naylor, T. Missalla, D. A. Reis, K. Scheidt, A. Sjögren, P. Sondhauss, M. Wulff, J. S. Wark: Picosecond X-ray diffraction studies of laser-excited acoustic phonons in InSb, Appl. Phys. A. **75**, 467–478 (2002)

[75] R. Merlin: Generating coherent THz phonons with light pulses, Solid State Commun. **102**, 207–220 (1997)

[76] W. H. Zachariasen: *Theory of X-Ray Diffraction in Crystals* (John Wiley and Sons, Inc., New York 1945)

[77] S. Takagi: Dynamical theory of diffraction applicable to crystals with any kind of small distortion, Acta Cryst. **15**, 1311 (1962)

[78] D. Taupin: Théorie dynamique de la diffraction des rayons X par les cristaux déformé, Bull. Soc. Franç. Minér. Crist. **87**, 469–511 (1964)

[79] C. R. Wie, T. A. Tombrello, T. Vreeland, Jr.: Dynamical X-ray diffraction from nonuniform crystalline films: Application to X-ray rocking curve analysis, J. Appl. Phys. **59**, 3743–3746 (1985)

[80] J. S. Wark, H. He: Subpicosecond X-ray-diffraction, Laser and Particle Beams **12**, 507–513 (1994)

[81] F. N. Chukhovskii, E. Förster: Time-dependent X-ray Bragg-diffraction, Acta Cryst. A **51**, 668–672 (1995)

[82] P. Sondhauss, J. S. Wark: Extension of the time-dependent dynamical diffraction theory to 'optical phonon'-type distortions: Application to diffraction from coherent acoustic and optical phonons, Acta. Cryst. **A59**, 7–13 (2002)

[83] B. W. Adams: Time-dependent Takagi–Taupin eikonal theory of X-ray diffraction in rapidly changing crystal structures, Acta Cryst. A **60**, 120–133 (2004)
[84] S. D. LeRoux, R. Colella, R. Bray: X-ray-diffraction studies of acoustoelectrically amplified phonon beams, Phys. Rev. Lett. **35**, 230–234 (1975)
[85] L. D. Chapman, R. Colella, R. Bray: X-ray-diffraction studies of acoustoelectrically amplified phonons, Phys. Rev. B **27**, 2264–2277 (1983)
[86] E. J. Yoffa: Dynamics of dense laser-induced plasmas, Phys. Rev. B **21**, 2415–2425 (1980)
[87] J. F. Young, H. M. van Driel: Ambipolar diffusion of high-density electrons and holes in Ge, Si, and GaAs: Many-body effects, Phys. Rev. B **26**, 2147–2158 (1982)
[88] C. Thomsen, J. Strait, Z. Vardeny, H. J. Maris, J. Tauc, J. J. Hauser: Coherent phonon generation and detection by picosecond light pulses, Phys. Rev. Lett. **53**, 989–992 (1984)
[89] C. Thomsen, H. T. Grahn, H. J. Maris, J. Tauc: Surface generation and detection of phonons by picosecond light pulses, Phys. Rev. B **34**, 4129–4138 (1986)
[90] G. Tas, H. J. Maris: Picosecond ultrasonic study of phonon reflection from solid–liquid interfaces, Phys. Rev. B **55**, 1852–1857 (1997)
[91] R. J. Stoner, H. J. Maris: Kapitza conductance and heat flow between solids at temperatures from 50 to 300 K, Phys. Rev. B **48**, 16373–16387 (1993)
[92] H. Y. Hao, H. J. Maris, D. K. Sadana: Nondestructive evaluation of interfaces in bonded silicon-on-insulator structures using picosecond ultrasonics technique, Electrochem. Solid-State Lett. **1**, 54–55 (1998)
[93] H. Y. Hao, H. J. Maris: Study of phonon dispersion in silicon and germanium at long wavelengths using picosecond ultrasonics, Phys. Rev. Lett. **84**, 5556–5559 (2000)
[94] H. Y. Hao, H. J. Maris: Dispersion of the long-wavelength phonons in Ge, Si, GaAs, quartz, and sapphire, Phys. Rev. B **63**, 224301 (2001)
[95] H. Y. Hao, H. J. Maris: Experiments with acoustic solitons in crystalline solids, Phys. Rev. B **64**, 064302 (2001)
[96] B. C. Daly, H. J. Maris, W. K. Ford, G. A. Antonelli, L. Wong, E. Andideh: Optical pump and probe measurement of thermal conductivity of low-k dielectric thin films, J. Appl. Phys. **92**, 6005–6009 (2002)
[97] R. M. Costescu, M. A. Wall, D. G. Cahill: Thermal conductance of epitaxial interfaces, Phys. Rev. B **67**, 054302 (2003)
[98] D. G. Cahill, W. K. Ford, K. E. Goodson, G. D. Mahan, A. Majumdar, H. J. Maris, R. Merlin, S. R. Phillpot: Nanoscale thermal transport, J. Appl. Phys. **93**, 793–818 (2003)
[99] P. Y. Yu, M. Cardona: *Fundamentals of Semiconductors: Physics and Materials Properties*, 3rd ed. (Springer, Berlin, Heidelberg 2004)
[100] K. D. Liss, A. Magerl, R. Hock, A. Remhof, B. Waibel: Towards a new (Q, t) regime by time-resolved X-ray diffraction: Ultra-sound excited crystals as an example, Europhys. Lett. **40**, 369–374 (1997)
[101] J. C. Slater: Interaction of waves in crystals, Rev. Mod. Phys. **30**, 197–222 (1958)

[102] G. Borrmann, G. Hildebrandt: Röntgen-Wellenfelder in grossen Kalkspatkristallen und die Wirkung einer Deformation, Z. Naturforsch. A **11**, 585–587 (1956)

[103] A. Hauer, S. J. Burns: Observation of an X-ray shuttering mechanism utilizing acoustic interruption of the Borrmann effect, Appl. Phys. Lett. **27**, 524–526 (1975)

[104] I. R. Entin, E. V. Suvorov, N. P. Kobelev, Y. M. Soifer: X-ray acoustic resonance in a perfect silicon crystal, Fizikia Tverdogo Tela **20**, 1311–1315 (1978)

[105] S. D. LeRoux, R. Colella, R. Bray: Effect of acoustoelectric phonons on anomalous transmission of X-rays, Phys. Rev. Lett. **37**, 1056–1059 (1976)

[106] A. Authier, S. Lagomarsino, B. K. Tanner (Eds.): *X-ray and Neutron Dynamical Diffraction: Theory and Applications*, NATO ASI Series (Plenum, New York 1996)

[107] H. Klapper, I. L. Smolsky: Borrmann-effect topography of thich potassium dihydrogen phosphate (kdp) crystals, Cryst. Res. Technol. **33**, 605–611 (1998)

[108] C. V. Shank, D. H. Auston: Picosecond ellipsometry of transient electron–hole plamsas in germanium, Phys. Rev. Lett. **32**, 1120–1123 (1974)

[109] C. V. Shank, R. Yen, C. Hirlimann: Femtosecond-time-resolved surface structural dynamics of optically excited silicon, Phys. Rev. Lett. **51**, 900–903 (1983)

[110] H. W. K. Tom, G. D. Aumiller, C. H. Brito-Cruz: Time-resolved study of laser-induced disorder of Si surfaces, Phys. Rev. Lett. **60**, 1438–1441 (1988)

[111] K. Sokolowski-Tinten, J. Bialkowski, D. von der Linde: Ultrafast laser-induced order-disorder transitions in semiconductors, Phys. Rev. B **51**, 14186–14198 (1995)

[112] K. Sokolowski-Tinten, J. Bialkowski, M. Boing, A. Cavalleri, D. von der Linde: Thermal and nonthermal melting of gallium arsenide after femtosecond laser excitation, Phys. Rev. B **58**, R11805–R11808 (1998)

[113] K. Solokolowski-Tinten, A. Cavalleri, D. von der Linde: Single-pulse time- and fluence-resolved optical measurements at femtosecond excited surfaces, Appl. Phys. A **69**, 577–579 (1999)

[114] K. Sokolowski-Tinten, D. von der Linde: Generation of dense electron–hole plasmas in silicon, Phys. Rev. B **61**, 2643–2650 (2000)

[115] N. V. Chigarev, D. Yu, Y. Paraschuk, X. Y. Pan, V. E. Gusev: Coherent phonon emission in the supersonic expansion of photoexcited electron–hole plasma in Ge, Phys. Rev. B **61**, 15837–15840 (2000)

[116] O. B. Wright, B. Perrin, O. Matsuda, V. E. Gusev: Ultrafast carrier diffusion in gallium arsenide probed with picosecond acoustic pulses, Phys. Rev. B **64**, 081202 (2001)

[117] M. Bargheer, N. Zhavoronkov, Y. Gritsai, J. C. Woo, D. S. Kim, M. Woerner, T. Elsaesser: Coherent atomic motions in a nanostructure studied by femtosecond X-ray diffraction, Science **306**, 1771–1773 (2004)

[118] S. H. Lee, A. L. Cavalieri, D. M. Fritz, M. C. Swan, R. S. Hegde, M. Reason, R. S. Goldman, D. A. Reis: Generation and propagation of a picosecond acoustic pulse at a buried interface: Time-resolved X-ray diffraction measurements, Phys. Rev. Lett. **95**, 246104 (2005)

[119] E. T. Swartz, R. O. Pohl: Thermal boundary resistance, Rev. Mod. Phys. **61**, 605–668 (1989)

[120] J. P. Wolfe: *Imaging Phonons: Acoustic Wave Propagation in Solids* (Cambridge University Press, Cambridge 1998)
[121] V. Narayanamurti, H. L. Störmer, M. A. Chin, A. C. Gossard, W. Wiegmann: Selective transmission of high-frequency phonons by a superlattice: The "dielectric" phonon filter, Phys. Rev. Lett. **43**, 2012–2016 (1979)
[122] D. B. McWhan, P. Hu, M. A. Chin, V. Narayanamurti: Observation of optically excited near-zone-edge phonons in GaAs by diffuse X-ray scattering, Phys. Rev. B **26**, 4774–4776 (1982)
[123] J. F. Scott: Soft-mode spectroscopy: Experimental studies of structural phase transitions, Rev. Mod. Phys. **46**, 83–128 (1974)
[124] A. Nazarkin, I. Uschmann, E. Förster, R. Sauerbrey: High-order Raman scattering of X-rays by optical phonons and generation of ultrafast X-ray transients, Phys. Rev. Lett. **93**, 207401 (2004)
[125] T. K. Cheng, S. D. Brorson, A. S. Kazeroonian, J. S. Moodera, G. Dresselhaus, M. S. Dresselhaus, E. P. Ippen: Impulsive excitation of coherent phonons observed in reflection in bismuth and antimony, Appl. Phys. Lett. **57**, 1004 (1990)
[126] T. K. Cheng, J. Vidal, H. J. Zeiger, G. Dresselhaus, M. S. Dresselhaus, E. P. Ippen: Mechanism for displacive excitation of coherent phonons in Sb, Bi, Te and Ti_2O_3, Appl. Phys. Lett. **59**, 1923 (1991)
[127] M. Hase, K. Mizoguchi, H. Harima, S. Nakashima, M. Tani, K. Sakai, M. Hangyo: Optical control of coherent optical phonons in bismuth films, Appl. Phys. Lett. **69**, 2474–2476 (1996)
[128] M. Hase, K. Mizoguchi, H. Harima, S. Nakashima, K. Sakai: Dynamics of coherent phonons in bismuth generated by ultrashort laser pulses, Phys. Rev. B **58**, 5448 (1998)
[129] M. F. DeCamp, D. A. Reis, P. H. Bucksbaum, R. Merlin: Dynamics and coherent control of high-amplitude optical phonons in bismuth, Phys. Rev. B **64**, 92301 (2001)
[130] M. Hase, M. Kitajima, S. Nakashima, K. Mizoguchi: Dynamics of coherent anharmonic phonons in bismuth using high density photoexcitation, Phys. Rev. Lett. **88**, 67401 (2002)
[131] O. V. Misochko, M. Hase, K. Ishioka, M. Kitajima: Observation of an amplitude collapse and revival of chirped coherent phonons in bismuth, Phys. Rev. Lett. **92**, 197401 (2004)
[132] O. V. Misochko, M. Hase, K. Ishioka, M. Kitajima: Transient Bose–Einstein condensation of phonons, Phys. Lett. A **321**, 381–387 (2004)
[133] E. D. Murray, D. M. Fritz, J. K. Wahlstrand, S. Fahy, D. A. Reis: Effect of lattice anharmonicity on high-amplitude phonon dynamics in photoexcited bismuth, Phys. Rev. B **72**, 060301 (2005)
[134] S. Fahy, D. A. Reis: Coherent phonons: Electronic softening or anharmonicity?, Phys. Rev. Lett. **93**, 109701 (2004)
[135] M. Hase, M. Kitajima, S. Nakashima, K. Mizoguchi: Coherent phonons: Electronic softening or anharmonicity? Hase et al. reply, Phys. Rev. Lett. **93**, 109702 (2004)
[136] X. Gonze, J.-P. Michenaud, J.-P. Vigneron: First-principles study of As, Sb, and Bi electronic properties, Phys. Rev. B **41**, 11827 (1990)

[137] H. J. Zeiger, J. Vidal, T. K. Cheng, E. P. Ippen, G. Dresselhaus, M. S. Dresselhaus: Theory for displacive excitation of coherent phonons, Phys. Rev. B **45**, 768 (1992)
[138] T. E. Stevens, J. Kuhl, R. Merlin: Coherent phonon generation and the two stimulated Raman tensors, Phys. Rev. B **65**, 144304 (2002)
[139] C. V. Shank, R. Yen, C. Hirlimann: Femtosecond-time-resolved surface structural dynamics of optically excited silicon, Phys. Rev. Lett. **51**, 900–902 (1983)
[140] I. L. Shumay, U. Hofer: Phase transformation of an InSb surface induced by strong femtosecond laser pulses, Phys. Rev. B **53**, 15878–15884 (1996)
[141] S. K. Sundaram, E. Mazur: Inducing and probing non-thermal transitions in semiconductors, Nature Mater. **1**, 217–224 (2002)
[142] X. Lui, D. Du, G. Mourou: Laser ablation and micromachining with ultrashort laser pulses, IEEE J. Quantum Electron. **33**, 1706 (1997)
[143] F. Rossi, T. Kuhn: Theory of ultrafast phenomena in photoexcited semiconductors, Rev. Mod. Phys. **74**, 895 (2002)
[144] J. A. V. Vechten, R. Tsu: Nonthermal pulsed laser annealing of Si; plasma annealing, Phys. Lett. A **74**, 422–426 (1979)
[145] P. Stampfli, K. H. Bennemann: Time dependence of the laser-induced femtosecond lattice instability of Si and GaAs: Role of longitudinal optical distortions, Phys. Rev. B **49**, 7299–7305 (1994)
[146] P. Stampfli, K. H. Bennemann: Dynamical theory of the laser-induced lattice instability of silicon, Phys. Rev. B **46**, 10686–10692 (1992)
[147] P. Stampfli, K. H. Bennemann: Theory for the instability of the diamond structure of Si, Ge, and C induced by a dense electron–hole plasma, Phys. Rev. B **42**, 7163–7173 (1990)
[148] J. S. Graves, R. E. Allen: Response of GaAs to fast intense laser pulses, Phys. Rev. B **58**, 13627 (1998)
[149] P. L. Silvestrelli, A. Alavi, M. Parrinello, D. Frenkel: Ab initio molecular dynamics simulations of laser melting of silicon, Phys. Rev. Lett. **77**, 3149–3152 (1996)
[150] T. Dumitrica, A. Burzo, Y. Dou, R. E. Allen: Response of Si and InSb to ultrafast laser pulses, Phys. Stat. Sol. (b) **241**, 2331–2342 (2004)
[151] D. von der Linde, J. Kuhl, H. Klingenberg: Raman scattering from nonequilibrium LO phonons with picosecond resolution, Phys. Rev. Lett. **44**, 1505 (1980)
[152] S. S. Prabhu, A. S. Vengurlekar, S. K. Roy, J. Shah: Nonequilibrium dynamics of hot carriers and hot phonons in CdSe and GaAs, Phys. Rev. B **51**, 14233–14246 (1995)
[153] J. A. Kash, J. C. Tsang, J. M. Hvam: Subpicosecond time-resolved Raman spectroscopy of LO phonons in GaAs, Phys. Rev. Lett. **54**, 2151 (1985)
[154] P. Saeta, J.-K. Wang, Y. Siegal, N. Bloembergen, E. Mazur: Ultrafast electronic disordering during femtosecond laser melting of GaAs, Phys. Rev. Lett. **67**, 1023–1026 (1991)
[155] L. Huang, J. P. Callan, E. N. Glezer, E. Mazur: GaAs under intense ultrafast excitation: Response of the dielectric function, Phys. Rev. Lett. **80**, 185–188 (1998)

[156] J. P. Callan, A. M.-T. Kim, C. A. D. Roeser, E. Mazur: Universal dynamics during and after ultrafast laser-induced semicondctor-to-metal transitions, Phys. Rev. B **64**, 073201 (2001)
[157] K. Sokolowski-Tinten, D. von der Linde: Ultrafast phase transitions and lattice dynamics probed using laser-produced X-ray pulses, J. Phys. Condens. Matter **16**, R1517–R1536 (2004)
[158] J. Larsson, P. A. Heimann, A. M. Lindenberg, P. J. Schuck, P. H. Bucksbaum, R. W. Lee, H. A. Padmore, J. S. Wark, R. W. Falcone: Ultrafast structural changes measured by time-resolved X-ray diffraction, Appl. Phys. A **66**, 587–591 (1998)
[159] O. Synnergren, M. Harbst, T. Missalla, J. Larsson, G. Katona, R. Neutze, R. Wouts: Projecting picosecond lattice dynamics through X-ray topography, Appl. Phys. Lett. **80**, 3727–3730 (2002)
[160] R. Neutze, J. Hajdu: Femtosecond time resolution in X-ray diffraction experiments, Proceedings of the National Academy of Science **94**, 5651–5655 (1997)
[161] P. C. Liu, P. R. Okamoto, N. J. Zaluzec, M. Meshii: Boron-implantation-induced crystalline-to-amorphous transition in nickel: An experimental assessment of the generalized Lindemann melting criterion, Phys. Rev. B **60**, 800–814 (1999)
[162] E. E. Castellana, P. Main: On the classical interpretation of thermal probability ellipsoids and the Debye–Waller factor, Acta Cryst. A **41**, 156–157 (1985)
[163] D. A. Arms, R. S. Shah, R. O. Simmons: X-ray Debye–Waller factor measurements of solid ^{3}He and ^{4}He, Phys. Rev. B **67**, 094303 (2003)
[164] E. W. Draeger, D. M. Ceperley: Debye–Waller factor in solid ^{3}He and ^{4}He, Phys. Rev. B **61**, 12094 (2000)
[165] W. F. Kuhs: Generalized atomic displacements in crystallographic structure analysis, Acta Cryst. A **48**, 80–98 (1992)
[166] G. Birkl, M. Gatzke, I. H. Deutsch, S. L. Rolston, W. D. Phillips: Bragg scattering from atoms in optical lattices, Phys. Rev. Lett. **75**, 2823–2826 (1995)
[167] J. W. Gadzuk: Breathing mode excitation in near-harmonic systems: Resonant mass capture, desorption and atoms in optical lattices, J. Phys. B: At. Mol. Opt. Phys. **31**, 4061–4084 (1998)
[168] G. L. Squires: *Introduction to the Theory of Thermal Neutron Scattering* (Cambridge Univ. Press, Cambridge 1978)
[169] G. Raithel, G. Birkl, W. D. Phillips, S. L. Rolston: Compression and parametric driving of atoms in optical lattices, Phys. Rev. Lett. **78**, 2928–2931 (1997)
[170] M. Leibscher, I. S. Averbukh: Squeezing of atoms in a pulsed optical lattice, Phys. Rev. A **65**, 053816 (2002)
[171] B. A. Dasannacharya, K. R. Rao: Neutron scattering from liquid argon, Phys. Rev. **137**, A417–427 (1965)
[172] A. Paskin, A. Rahman: Effects of a long-range oscillatory potential on the radial distribution function and the constant of self-diffusion in liquid Na, Phys. Rev. Lett. **16**, 300–303 (1966)
[173] K. Skold, J. M. Rowe, G. Ostrowski, P. D. Randolph: Coherent and incoherent scattering laws of liquid argon, Phys. Rev. A **6**, 1107–1131 (1972)

[174] G. H. Vineyard: Scattering of slow neutrons by a liquid, Phys. Rev. **110**, 999–1010 (1958)
[175] U. Balucani, M. Zoppi: *Dynamics of the Liquid State* (Clarendon Press, Oxford 1994)
[176] F. A. Lindemann: Über die Berechnung molekularer Eigenfrequenzen, Phys. Z. **11**, 609–612 (1910)
[177] C. W. Siders, A. Cavalleri: Creating transient crystal structures with light, Science **300**, 591–592 (2003)
[178] A. Zholentz, P. Heimann, M. Zolotorev, J. Byrd: Generation of subpicosecond X-ray pulses using rf orbit deflection, NIM A **425**, 385–389 (1999)
[179] M. Borland: Simulation and analysis of using deflecting cavities to produce short X-ray pulses with the advanced photon source, Phys. Rev. S.T. **8**, 074001 (2005)

Index

Auger recombination, 379–380

beam-slicing
  electron, 373
  X-ray, 373, 394
bismuth, 394–398
Borrmann effect, 388–390
breathing-mode excitation, 406

coherent phonon, 375
  acoustic, 377–393
  folded, 391–393
  optical, 378, 393–398

Debye–Waller, 376, 403
deformation potential, 379, 383, 391
diffuse scattering, 376, 393
  thermal, 376
dynamical diffraction theory, 377–378

electron beam compression, 374

inelastic scattering, 376, 394

Laue geometry, 388, 390–391

melting, nonthermal, 398–408

Peierls distortion, 395, 397

Pendellösung length, 388, 390–391
phonon
  sideband, 376, 379, 383, 394
  softening, 394–398, 403–406

self-correlation function, 407
solid-solid phase transition, 408–409
stimulated
  Brillouin scattering, 376, 379
  Raman scattering, 376, 394–395
streak camera, 373, 382
stress/strain
  ultrafast laser generated, 379–381

Takagi–Taupin equations, 378
timing jitter, linac-based source, 374, 401
TTF-CA, 408

VO_2, 408

wakefield, 374

X-ray sources
  energy-recovery linac, 374
  free-electron laser, 372–374
  laser-produced plasma, 372
  sub-picosecond pulse source, 374
  synchrotron, 373

Index

2D hydrogen model, 158

absorption, 20
acoustic
(ZA) branch, 141
of graphene, 141
bulk-like modes, 249, 250, 293, 298
cavities
raman spectra, 78
cavity effects, 277, 280, 282, 291
damping, 239, 290, 294, 295, 304, 306
displacement field, 238, 239, 291, 292, 298, 301, 304
extended modes, 240
image, 276, 277
impedance
mismatch, 75
localization, 239, 287–289, 291, 293, 306
mirror, 273
modes, 317
phase, 256, 261, 267, 280–282
phonons
finite-size effects, 68
strong confinement, 282
surface-related modulation, 276, 277
actinides, 342, 343
additional boundary conditions (ABCs), 84, 90
adiabatic approximation, 35
Airy function, 28
AlAs, 29, 31, 36, 60
AlAs/$Al_{0.33}Ga_{0.67}As$, 52
AlGaAs/AlAs, 45
amorphous ice, 345
amplification
light–sound interaction, 76
optical resonant, 43
amplitudon, 341
angle tuning, 36, 37
one-mode cavities, 39
angular selectivity, 50
anisotropic polarizability, 153, 174
antenna effect, 44
anti-Stokes process, 33
anticrossings, 67
multibranch, 98
Ar, 350
arc-discharge, 122
armchair
nanotubes, 140
tubes, 133
atomic-microscope image
carbon nanotube, 209
Auger recombination, 379–380

backscattering, 79
band-to-band transition, 151, 167
Bangalore, 8
$BaTiO_3$, 76
beam-slicing
electron, 373
X-ray, 373, 394
benzene, 8, 21
Bethe–Salpeter equation, 159
bismuth, 394–398
Bloch oscillation
electron, 33
photons, 33
Borrmann effect, 388–390
boundary conditions
electrostatic, 58
mechanical, 58
Bragg
microcavity
reflectivity, 30

- mirror, 29, 88
  - angular dependence of the reflectivity, 26–27
- reflector, 26
  - distibuted (DBR), 19, 21, 23

breathing-mode excitation, 406
Brillouin
- scattering, 317
- spectroscopy, 1
- zone
  - center, 58
  - edge, 58
  - extended scheme, 25
  - schemes, 68

broadening
- homogeneous, 90
- inhomogeneous, 38

buckypaper, 194

C_{60} molecules, 125
C_{60} peapod, 125
Carbon Nanotubes, 115
carbon nanotubes, 140
- metallic, 168
- optical selection rules, 153
- single-wall, 140

cavity
- acoustic, 17
- cartography, 38
- double, 36
- double structure, 32
- GaAs, 38
- high-Q, 52, 57
- high-finesse, 39
- II–VI, 98
- low-Q, 52
- multimode, 31
- multiple coupled, 32
- piezoelectrically controlled, 39
- polariton-mediated scattering, 91
- polaritons, 22
- Q-factors, 21
- Raman spectra, 53
- resonance condition, 37
- two-mode, 40

$Cd_{1-x}Mg_xTe$, 36
$Cd_{1-x}Mn_xTe$, 36
CdTe, 36, 350
- QW, 97, 99
- QW mode, 97

cesium, 351
character table, 129
- D_{qh}, 128

charge-density wave systems, 341, 342
chemical vapor deposition, 122, 123
chirality assignment, 177
clathrate hydrates, 346
coherent phonon, 375
- acoustic, 377–393
- folded, 391–393
- optical, 378, 393–398

collection cone, 90
colloids, 19
Combination Scattering, 8
complex frequency model (CFM), 290, 291
confinement
- photon, 17

conservation
- energy, 35
- wavevector, 35

core shell model (CSM), 291–293, 296, 298
coulomb polarization fields, 62
coupled microcavities, 32
coupled oscillators, 87
coupling
- between cavity-photons and excitons, 37
- deformation-potential, 300
- strong, 17
- weak, 17

crystal-field transitions, 18
crystal-momentum conservation, 68
curvature, 185
CVD, 122
cyclic voltammetry, 194

D mode, 136, 163, 207, 208, 213
D^* mode, 213
damping
- polariton, 102

DBR, 23
- acoustic, 23
- GaAs/AlAs, 60
- penetration depth, 30
- reflectivity, 26
- structure, 26

transmission, 26
Debye–Waller, 376, 403
defects, 214
deformation potential, 379, 383, 391
density–density correlation function, 317
dephasing time, 93
depolarization
effect, 174
ratio, 175
diamond, 331
dielectric
constant, 158
continuum model, 58, 59
function, 152
resonant critical point, 18
model, 67
diffuse scattering, 376, 393
thermal, 376
dipole
approximation, 152
damping constant, 44
dispersion relations
phonon, 59
distributed Bragg reflectors
optical phonons, 58
double differential cross section, 317, 320, 321
double optical resonance (DOR), 65, 76
double-resonance, 35
conditions, 171
process, 192
double-resonant
angle-tuning, 93
Raman scattering, 169
scattering, 215
double-resonators
light–sound, 76
double-walled nanotubes, 134
dynamical diffraction theory, 377–378

eigenmodes
of nanotubes, 134
eigenvector
of nanotube vibrations, 136
Raman-active, 173
Einstein, 5, 12
elastic moduli, 354
electrochemical doping, 194
electron beam compression, 374
electron–acoustic phonon interaction, 239, 246
electron–phonon coupling, 191, 196, 201
electronic band structure, 148
electronic excitations
inter- and intraband, 18
electronic transitions
dipolar allowed, 18
electronic wavefunction
distribution, 238, 240, 246, 247, 249, 255, 262, 264–266, 272, 273
extension, 253, 272
impurity-like, 272
localization, 260
ordering, 238, 306
spatial correlation, 238, 240, 255, 262, 264
spatial localization, 247
embedded particles, 293
emission
boser, 21
dipole, 20
single-photon, 21
spontaneous, 20, 45
stimulated, 20
energy conservation
in the scattering process, 89
excitations
plasmon, 18
exciton, 151, 158
binding energy, 158–160
exciton–phonon excitations, 161
of the QWs, 38
phonon side bands, 161
excitonic transitions, 168
extended tight-binding model, 148

Fabry–Perot
confocal resonator, 27
performance, 28
Q, 28
resonators, 44
standing wave, 27
Fano lineshape, 190
far field, 45
Fermi level, 195
Fermi's golden rule, 20

final states
  accessible, 90
finesse, 49, 75
  effective, 30
finite-size effects, 81
folded
  acoustic phonons, 52
forward-scattered modes, 64
forward-scattering, 76, 79
frequencies
  LO, 59
  TO, 59
Fresnel reflection coefficients, 45
frozen-phonon approximation, 197, 199

G line, 136
G mode, 209, 213
$Ga_{0.8}Al_{0.2}As$, 41
$Ga_{0.8}Al_{0.2}As/AlAs$, 38, 41
$Ga_{1-x}Al_xAs$, 36
GaAs, 24, 29, 31, 36, 47, 50, 61
  substrate, 56
GaAs-like
  optical phonon, 61
  vibrations, 64
GaAs/AlAs, 24, 26, 41, 45, 75, 91
  MQW, 92, 94
  parameters, 27
  quarter-wave layers, 49
  Stokes and anti-Stokes spectra, 70
Ge, 207
Ginzburg, 6
graphene, 117, 118, 126, 199
  cylinder limit, 206
  electronic band structure, 146
  empirical formula, 186
  Kohn anomalies, 143
  semimetal, 145
  trigonal distortion, 155
graphene LO mode
  overbending, 140
graphite, 116, 137, 140, 168, 333
  D mode, 116, 172
  optical modes, 139
  phonon dispersion, 137
  Raman spectrum, 212
  sound velocities, 131
  tight-binding parameters, 116
Gross, 6

H_2
  antibonding states, 42
  bonding states, 42
  molecule, 42
half-cavity
  low-Q, 21
half-width at half maximum (HWHM), 51
Hamiltonian
  exciton plus cavity photon, 88
  of light–matter interaction, 34
hcp-cobalt, 357
hcp-iron, 354
helical quantum numbers, 143
helium, 334
high-energy mode (HEM), 188, 191
  single-walled tubes, 163
high-energy phonon, 130
high-pressure, 317
high-pressure
  studies, 350
hypersound, 17

ICORS, 1, 3
Iijima, 115
impulsive stimulated light scattering (ISLS), 355
$In_{0.14}Ga_{0.86}As$, 91
in-plane ordering, 267
inelastic light scattering
  polariton-mediated, 22
inelastic scattering, 376, 394
inelastic X-ray data
  on graphite, 215
inelastic X-ray scattering, 317
inorganic molecules, 125
interband transitions, 18
interface modes, 59
interferences, 168
interlayer coupling, 134
intermediate states, 165
isogonal point groups, 127
isotope shift, 164
IXS instruments, 324

joint density of electronic states, 152

Kataura plot, 154, 155, 160, 178, 180, 187, 204

Kohn anomaly, 139, 140, 190, 196, 204, 206
  graphene, 143
Krishnan, 11

Lamb's model, 239, 283, 284, 286, 287, 296, 298
Lamb-shift
  atom-vacuum-field, 20
large-bandgap materials, 328, 329
laser, 74
  ablation, 122
  excitation
    self-interference, 46
  HeNe, 38
  Ti-sapphire, 38
  tunable, 38
lateral nanostructuring, 104
lattice dynamics, 317, 327, 328, 336, 338, 343, 351, 354
  in-plane, 334
Laue geometry, 388, 390–391
lifetime
  detuning dependent, 90
light–matter interaction, 17, 74
  absorption, 20
  emission, 20
line groups, 126
linewidths
  cavity-photon, 99
  exciton, 99
LO GaAs-like phonons
  GaAlAs, 65
LO phonon, 51
  in bulk GaAs, 99
  Stokes frequency, 50
Lorentzian probability function, 90
luminescence, 20, 42

magnons, 18
Mandelstam–Brillouin scattering, 6
Mandelstam–Landsberg, 6
matrix element, 203
  electron–phonon, 197, 207
  squared, 34
melting, nonthermal, 398–408
metallic tips, 20
metallic tubes, 155
  condition for, 150
MgB_2, 336
microcavity
  amplification power, 36
  confinement, 35
  confocal, 22
  for Raman enhancement, 17
  high-finesse, 23
  II–VI, 96
  III–V, 93
  optical, 17
  piezoelectrically controlled, 21
  planar, 17, 22, 39, 90
  Polaritons, 86
  semiconductor, 29
  strongly coupled, 19, 83, 86
microcavity modes
  parity, 63
minigaps, 58
mirror
  dispersive, 23
  losses, 93
  low-reflection, 32
  superlattices, 81
mode anticrossing, 86
molecular beam epitaxy (MBE), 19, 36
multiple quantum well (MQW), 37

nanocavities
  acoustic, 19, 74
nanoparticle
  CdS, 297
  CdSe, 297
  self-assembled, 19
nanostructures
  metallic (SERS), 18
nanotube, 118
  achiral, 121
  armchair, 121, 130
  axial vibrations, 130
  branches and families, 182
  bundled, 191
  bundles, 123
  charge transfer, 193
  chiral, 121
    angle, 118
    index, 118
  chiralities, 123
  circumferential vibrations, 130
  classifications, 157

curvature, 132, 150
diameter, 118
doping, 193
double-walled, 135
families, 157
and branches, 155
forest, 124
interference effects, 202
light-absorption, 174
luminescence, 177
metallic, 121, 149, 187, 193
multiwalled, 174
neighbor rule, 157
phonon branch, 192
plasmons, 190
point group, 129
Raman active, 130
semiconducting, 121, 149
single-walled, 127
structural parameters, 119
symmetry, 126
operations, 127
zigzag, 121, 130
near-field optical microscopy, 20
neutron spectroscopy, 317
Nobel Prize, 10

optical absorption, 152
optical matrix elements, 152
optical modes, 317
optical phonon
double-cavity dispersion, 67
in finite superlattices, 64
optical reflectivity
for different angles, 40
optical resonance
double (DOR), 35
optical resonant, 22
optical selection rules
carbon nanotubes, 153
overbending, 138
graphene LO branches, 138
graphene LO mode, 140
overtone, 172, 173, 212
Raman spectrum, 163

peapods, 125
Peierls distortion, 204, 395, 397
Peierls-like
mechanism, 190
softening, 196
Pendellösung length, 388, 390–391
perturbation theory
second-order, 18
third-order, 18
phason, 341
phonon, 18
cavity, 76
cavity design, 75
confined, 59
density of states of GaP, 289
dispersion
bulk-like, 73
graphite, 215
inplane wavevector, 62, 63
dispersion curves, 142, 171
armachair tube, 142
eigenvectors, 129
engineering, 74
forbidden and allowed, 82
interface, 50, 59
lasing, 74
longitudinal optical, 50
phonon
m confined, 66
mirror reflectivity, 75
odd and even, 81
optical, 47
plasmon coupled modes, 188, 190, 194
sideband, 376, 379, 383, 394
softening, 394–398, 403–406
symmetries, 129
carbon nanotubes, 129
photoelastic
constant, 69, 82, 261
mechanism, 69
model, 72
photoluminescence, 38
photon
confined, 48
confinement, 17
photonic
bandgap materials, 21
molecule, 42
structures, 21
π wavefunctions, 145
picotubes, 211

pillars, 104
plasmon, 18
plasmon-polariton states
  Drude damping, 299, 302
  interaction with acoustic vibrations
    deformation potential mechanism, 299, 300, 303, 305, 307
    surface orientation mechanism, 299, 302, 304, 305
  interband susceptibility, 300, 301, 303, 305
  intraband susceptibility, 293, 302
  Raman scattering mediated by, 303–306
  surface coupling, 301
  surface polarization charge, 302, 307
Plutonium, 343
polar dielectric materials, 58
polariton
  cavity, 17, 19
  damping, 89
  damping factors, 102
  dephasing, 89
  dynamics, 85
  exciton, 83
  gap, 84
  inplane dispersion, 37
  low (LP), 88, 101
  medium (MP), 101
  radiatively stable, 86
  upper, 88
polariton-mediated scattering
  in bulk materials, 90
polycrystals, 344
porous silicon, 21
process
  Stokes, 33
pyramidalization angle, 132

Q-factor, 21, 35
quantization condition
  electronic band structure, 144
quantum dots, 83
  Ge/Si, 256–259, 267, 274
  InAs/AlInAs, 268
  InAs/GaAs, 267
  InAs/InP, 252, 277
  ordering, 249
  self-assembled, 248
quantum well
  GaAs/GaP, 240
  multilayer, 265
quantum wells, 83
quantum wires, 83
quartz, 8
quasimetallic tubes, 140

Rabi gap, 91, 96
  large, 94
  small, 96
Rabi oscillations
  coherent, 86
Rabi splitting, 87, 89, 91, 93, 99
radial-breathing mode (RBM), 130, 162, 177, 196
radial-breathing vibration, 130
radiation pattern
  azimuthal angle, 46
Raman
  amplification, 17, 21, 48
    by optical confinement, 33
  cross section, 18, 20, 165
    total, 48
  double resonance, 211
    in Ge, 172
  double-resonance, 55
  effect, 11
    carbon nanotubes, 162
  efficiency, 103, 167
    amplifications, 21
    is enhanced, 103
    maximum for zero detuning, 98
    photoelastic model, 68, 78
    Stokes scattering, 165
    theoretical, 57
  enhancement, 20
    by optical confinement, 19
  form factor, 247, 250, 252, 261, 262, 267
  intensity
    single-resonance scans, 55
  interference contrast, 265, 266
  luminescence, 77
  matrix element, 165, 166, 170, 188
  optical resonances, 23
  photons and luminescence, 93
  scattering, 22, 317, 331
    cavity-polariton-mediated, 17, 104

confined photons, 80
DOR, 41
double optical resonance, 91
double optical resonant, 21
double-resonant, 163
double-resonance, 54
droplets, 21
efficiency, 96
efficiency enhancement, 30
forward, 74
in cavities, 17
in microcavities, 43
interference, 99
interference-enhanced (IERS), 21
lower polariton, 93
on droplets, 39
optical resonant, 17, 19
optically confined structures, 22
organic molecules, 21
planar cavities, 88
polariton-mediated, 85, 88
resonant (RRS), 17, 18, 22, 35
stimulated, 39, 103
surface-enhanced (SERS), 18
tensor, 35
under optical confinement, 17
upper polariton, 93
scattering amplitude
in microcavities, 35
selection rule, 243, 295, 298
spectra
single-walled nanotubes, 162
spectroscopy
of single low-dimensional nanostructures, 103
optically confined, 57
structure factor, 247, 255, 258, 262, 267, 273, 275, 277
Raman-active vibrations, 136
ratio problem, 179
Rayleigh scattering, 6, 8
reflection
cavity at boundaries, 29
refractive index
n_{eff}, 29
contrast, 23
modulation, 25
resonance
incoming, 165
morphology-dependent, 21
outgoing, 165
profiles
of nanotubes, 180
Raman scattering, 162
resonant
Raman
experiments outgoing, 96
Raman effect, 150
Raman scattering
ingoing, 84
optical, 35
outgoing, 84
resonators
Fabry–Perot (FP), 27
Rytov's model, 70, 259

scattering
deformation-potential, 63
efficiency, 245
probabilities, 18
volumes
small, 18
scissors operator, 182
second-harmonic generation, 31
secondary gap, 140
selection
intermediate-state, 247
rule, 148, 152
$\mathcal{R} = 3$, 210
self-correlation function, 407
semiconducting tubes, 155
condition for, 150
semiconductor
II–VI, 36, 85
III–V, 36, 85
semimetal
graphene, 145
SERS
chemical effect, 19
electromagnetic effect, 19
hot-spots, 19
shape anisotropy, 174
shape depolarization, 212
Si/Ge, 75
SiC, 328
single resonance, 164
Smekal, 7
sodium dodecyl sulfate, 177

solid-solid phase transition, 408–409
space-resolved imaging, 209
spacer, 45
spatial localization, 250
spectral density, 90
spectroelectrochemistry, 193
$SrTiO_3$, 76
standing optical vibrations, 64
standing waves
  optical, 22
states
  antibonding, 42
  bonding, 42
stimulated
  Brillouin scattering, 376, 379
  emission
    of coherent phonons, 104
  Raman scattering, 376, 394–395
Stokes and anti-Stokes frequencies, 172
Stokes energy, 50, 52
Stokes peaks, 21
Stokes-shifted photon, 34
stop band, 23, 31, 42
  edge, 79
  reflectivity, 23
  width, 23
stratified media, 45
streak camera, 373, 382
stress/strain
  ultrafast laser generated, 379–381
strong coupling, 30
  polaritonic, 34
superconductivity, 336, 338
superlattices
  mirror-like, 68
surface
  chemically roughened, 19
  modes, 58
  vibrations, 59
synchrotron radiation, 13, 324

Takagi–Taupin equations, 378
tangential modes, 136
Third Reich, 10
third-neighbor interaction, 145
three-oscillator model, 101
tight-binding
  approximation, 197
  energies, 185
  model, 145, 159
timing jitter, linac-based source, 374, 401
Toronto arc, 2
transfer-matrix formalism, 44
transmission
  factor
    bulk, 89
  time-resolved, 293, 295
transport properties, 189
triple spectrometer
  additive mode, 79
  subtractive mode, 80
TTF-CA, 408
tunability
  of energies, 37
tunnel barriers, 32
two-photon absorption, 160

ultrasound
  THz, 23
uranium, 343

vacuum-field fluctuations, 20
valence force field model (VFF), 134, 239, 284, 286, 287
van-der-Waals
  forces, 123
  interaction, 185
van-Hove singularity, 154
vertical ordering, 262
virtual process, 35
VO_2, 408

wakefield, 374
waves
  evanescent, 45
  leaky, 45
wavevector
  partial conservation, 64
  scan, 171
Web of Science, 5
whispering-gallery modes, 21, 104
WS_2, 176

X-ray sources
  energy-recovery linac, 374
  free-electron laser, 372–374
  laser-produced plasma, 372

sub-picosecond pulse source, 374

synchrotron, 373

zeolite, 174
zigzag tubes, 133
zone folding, 137, 138, 144, 155, 185

Topics in Applied Physics

91 **Vortex Electronis and SQUIDs**
By T. Kobayashi, H. Hayakawa, M. Tonouchi (Eds.) 2003, 259 Figs. XII, 302 pages

92 **Ultrafast Dynamical Processes in Semiconductors**
By K.-T. Tsen (Ed.) 2004, 190 Figs. XI, 400 pages

93 **Ferroelectric Random Access Memories**
Fundamentals and Applications
By H. Ishiwara, M. Okuyama, Y. Arimoto (Eds.) 2004, Approx. 200 Figs. XIV, 288 pages

94 **Silicon Photonics**
By L. Pavesi, D.J. Lockwood (Eds.) 2004, 262 Figs. XVI, 397 pages

95 **Few-Cycle Laser Pulse Generation and Its Applications**
By Franz X. Kärtner (Ed.) 2004, 209 Figs. XIV, 448 pages

96 **Femtsosecond Technology for Technical and Medical Applications**
By F. Dausinger, F. Lichtner, H. Lubatschowski (Eds.) 2004, 224 Figs. XIII 326 pages

97 **Terahertz Optoelectronics**
By K. Sakai (Ed.) 2005, 270 Figs. XIII, 387 pages

98 **Ferroelectric Thin Films**
Basic Properies and Device Physics for Memory Applications
By M. Okuyama, Y. Ishibashi (Eds.) 2005, 172 Figs. XIII, 244 pages

99 **Cryogenic Particle Detection**
By Ch. Enss (Ed.) 2005, 238 Figs. XVI, 509 pages

100 **Carbon**
The Future Material for Advanced Technology Applications
By G. Messina, S. Santangelo (Eds.) 2006, 245 Figs. XXII, 529 pages

101 **Spin Dynamics in Confined Magnetic Structures III**
By B. Hillebrands, A. Thiaville (Eds.) 2006, 164 Figs. XIV, 345 pages

102 **Quantum Computation and Information**
From Theory to Experiment
By H. Imai, M. Hayashi (Eds.) 2006, 49 Figs. XV, 281 pages

103 **Surface-Enhanced Raman Scattering**
Physics and Applications
By K. Kneipp, M. Moskovits, H. Kneipp (Eds.) 2006, 221 Figs. XVIII, 464 pages

104 **Theory of Defects in Semiconductors**
By D. A. Drabold, S. K. Estreicher (Eds.) 2007, 60 Figs. XIII, 297 pages

105 **Modern Ferroelectrics**
By K. Rabe, Ch. H. Ahn, J.-M. Triscone (Eds.) 2007, 65 Figs. XII, 250 pages

106 **Rare Earth Oxide Thin Films**
Growth, Characterization, and Applications
By M. Fanciulli, G. Scarel (Eds.) 2007, 209 Figs. XVI, 426 pages

107 **Microscale and Nanoscale Heat Transfer**
By S. Volz (Ed.) 2007, 144 Figs. XIV, 370 pages

108 **Light Scattering in Solids IX**
Novel Materials and Techniques
By M. Cardona, R. Merlin (Eds.) 2007, 215 Figs. XIV, 404 pages

Printing: Krips bv, Meppel
Binding: Stürtz, Würzburg